DEVELOPMENTS IN INTEGRATED ENVIRONMENTAL ASSESSMENT

VOLUME 1B

INTEGRATED AND PARTICIPATORY WATER RESOURCES MANAGEMENT: PRACTICE

Contents

Elsevier
Radarweg 29, PO Box 211, 1000 AE Amsterdam, The Netherlands
The Boulevard, Langford Lane, Kidlington, Oxford OX5 1GB, UK

First edition 2007

Library of Congress Cataloging-in-Publication Data
A catalog record for this book is available from the Library of Congress

British Library Cataloguing in Publication Data
A catalogue record for this book is available from the British Library

ISBN-13: 978-0-444-53012-7

For information on all Elsevier publications
visit our website at books.elsevier.com

Printed and bound in The Netherlands

07 08 09 10 11   10 9 8 7 6 5 4 3 2 1

DEVELOPMENTS IN INTEGRATED ENVIRONMENTAL ASSESSMENT

VOLUME 1B

INTEGRATED AND PARTICIPATORY WATER RESOURCES MANAGEMENT: PRACTICE

by

Rodolfo Soncini-Sessa

with

Francesca Cellina
Francesca Pianosi
Enrico Weber

Dipartimento di Elettronica e Informazione
Politecnico di Milano
Milano, Italy

ELSEVIER

2007

Amsterdam • Boston • Heidelberg • London • New York • Oxford
Paris • San Diego • San Francisco • Singapore • Sydney • Tokyo

Introduction

A few years ago at a conference, a Swiss insurance man displayed the chart in Figure 1. It shows the damages paid, year by year from 1966 to 2000, by the Swiss insurance companies for the damage produced by the forces of nature, 60% of which were from floods. The data refers to the whole Swiss Confederation but those produced in the Ticino Canton by Lake Maggiore floods are 2.5 times higher than the national average. Flooding in this lake is a recurring phenomenon. The last, in October 2000, brought the lake level up to 4 m above the average, causing damages that in Ticino Canton alone amounted to more than 100 million euro. At the peak of the flood the flow into the lake exceeded 380 million cubic meters per day, compared to an 'operational' storage of 420 million cubic meters: as if almost a whole lake was entering into the lake. But the 2000 flood, even if significant, is not an isolated case.

Let us look again at the chart in Figure 1: the most impressive thing is not the spikes of the most recent years but the behavior of the phenomenon as a whole, which seems to proceed with exponential growth. A small part of this growth is maybe due to inflation (they are raw data), but the most part? There are two possible causes. First, urbanization has spread to areas that were once exclusively the domain of the lake. Compare the urban plans for a lake-side city in different times, for example the plans for Locarno in 1897 and 2000

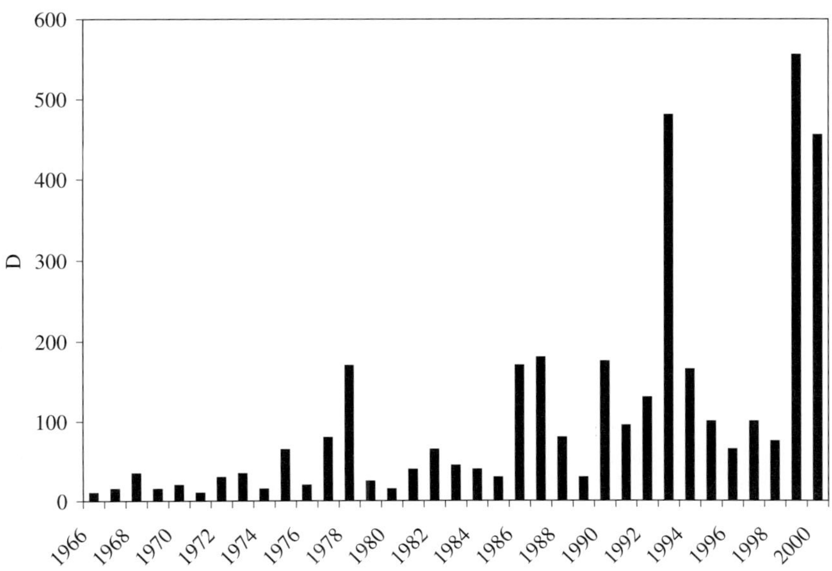

Figure 1: Annual damages paid by Swiss insurance companies for the damage caused by the forces of nature.

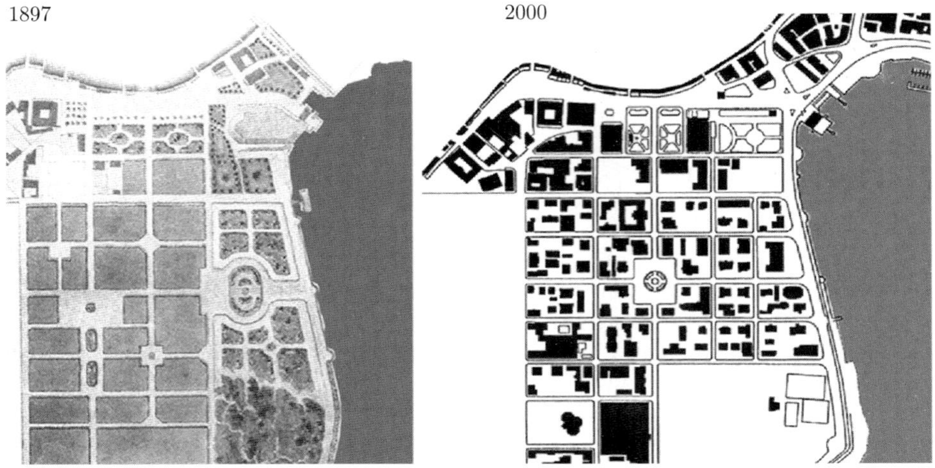

Figure 2: Comparison of two urban plans of Locarno City (by kind permission of the Archivio Comunale di Locarno).

(Figure 2). The urbanization of the areas overlooking the lake is evident: this lake-side city, once able to live in harmony with Verbano,[1] was transformed into a city looking *onto* a lake, as a view just outside the window, hoping that it will remain outside. Second, the apparent worsening of extreme meteorological phenomena, that may be due to climate change.

Facing the problem by acting on these causes would possibly give results only in the long-term, while public opinion asks for quick response. A solution might be to excavate the lake outlet to allow a larger outflow: but the lake outlet is located in Italy, not in Switzerland! So all the Swiss government could do was proposing the excavation to the Italian government. In turn, Italy could not accept the proposal without considering its own interests in the Verbano system: the effluent of Lake Maggiore, the River Ticino, feeds a rich irrigated agricultural activity, producing 45% of Europe's rice, and several hydropower plants; the landscape around the river, a maze of ditches, canals, oxbow lakes, and resurgences, is one of the most interesting anthropic wetland environments in Europe, which has been protected with the creation of two riverine parks; and finally, around the lake itself a rich community has developed, which gets its life and its riches from an articulated and refined tourist industry.

The aim of this book

The search for a solution of the Verbano management problem is the apparent subject of this book. Apparent since it actually is a 'false target': the true aim of the book is to introduce the reader, through a practical approach, to the tools that are available to find solution to problems such as that and to show her/him how the research should be organized in order to create consensus around the solution.

This aim is strategic because the planning and management of water resources are today already key problems, but they will play an even more crucial role in the near future. Just

[1] Verbano is the name that the Lake Maggiore side dwellers use to refer to their lake. Since it is the lake's local name and at the same time shorter and more compact than 'Lake Maggiore', we will use it in this book.

as petroleum was a key resource in the XX century, water will be in the XXI. Global water demand has tripled in the last 50 years and it continues to rise every year, since every year the world population grows by 80 million people (Brown, 2001). Seventy percent of the water that is derived from rivers or pumped from aquifers is used for irrigation and, thanks to that, agricultural production provides 40% of the food that we consume. However, demand is increasing, and water availability remains constant in the best cases, when it doesn't actually decrease due to global phenomena, pollution and competition with industry. So the development of methodologies for a more careful use of the our precious water capital is a strategic factor for our future. In the last twenty years we have witnessed a flowering of new tools that try to meet this goal.

However, we know well that good tools are not sufficient: it is vital that interventions that are good in name be then actually realized. Many difficulties, however, interfere with the implementation of a project: the main one often being the dissatisfaction of those who were excluded from the benefits provided by the project, or who are, or believe to be, damaged by it. It is essential therefore to be able to evaluate the effects that a given intervention will produce not only on the sector that it has been conceived for, but also on all the sectors that it influences. It is for this reason that in the late 90s a new management paradigm was conceived and began to spread: Integrated Water Resources Management (IWRM) (GWP, 2003).[2]

Within this paradigm my research group developed a decision-making procedure that is *participatory*[3] and *integrated*.[4] It was called PIP (*Participated and Integrated Planning*) procedure and is presented from the theoretical standpoint in another book: *Integrated and Participatory Water Resources Management: Theory* (in the following referred to as THEORY). The present book is aimed at presenting the PIP procedure from a practical point of view, using the Verbano problem as an example and describing in details the actual development of the *Verbano Project*, which was launched in 1999 by the European Union, among the INTERREG-II projects devoted to transnational problems of the EU. Of course, other textbooks exist that give examples of projects for problems of this type, but unfortunately they are often focused only on normative and procedural aspects, or to strategic analysis, omitting all those details that actually are the essence of the project and that make the difference between a successful project and a failure. My ambition for this book is to fill this gap.

Methodological novelties

The PIP procedure contains several interesting methodological novelties. Environmental Impact Assessment and the methodologies of *Multi-Attribute Value Theory* (MAVT) have been employed until now mostly for the assessment of projects concerned with non-dynamical systems (or with systems that are given a non-dynamical description), where only *planning* decisions are considered, i.e. decisions that are made once and for all. This is a strong limitation since most of environmental systems actually are dynamical. For example, Lake Maggiore is dynamical: its state, the lake storage, varies in time and every day a decision

[2] See also www.gwpforum.org, last visited September 2006.

[3] Participation is not limited to providing the Stakeholders with information (*Informative Participation*), nor to just asking them for information (*Consultation*), but involves the Stakeholders in the design and evaluation of the alternatives (*Co-designing*) and possibly in the final choice (*Co-deciding*).

[4] In the sense that it considers all the interests in play, independently from their relative relevance.

must be made about the water volume to be released in the next twenty-four hours. Clearly, the release decision must take the available resource into account, namely consider what the lake storage is. Moreover, the decision influences the storage that will result in the following day, thus influencing also the subsequent release decisions: decisions are concatenated and as such they are *management* decisions. Such decisions can be rationally assumed by designing appropriate *management policies*. In the PIP procedure a methodological effort was made to extend the applicability of decision techniques to dynamical system and to the design of management policies.

A second innovative aspect concerns the decision-making procedure. A rich literature exists that deals with the case in which, even in the presence of conflicting interests, the final decision is the duty of a single, high-level decision maker. However, 60% of freshwater in the world is collected in transnational basins, where possible disputes for water use cannot be solved by a supervisor like a government or governmental agency, for the simple fact that such supervisor does not exist. This is the case of Verbano, which extends both in Swiss and Italian territory. In such cases, negotiations are the only possible way to reach an agreement. The PIP procedure was designed by explicitly considering these issues.

Nevertheless, this book is not and is not intended to be a theoretical work. The theory will be referred to only in the measure necessary and sufficient to allow the reader to understand what was done for the Verbano problem and, above all, what should be done in good decision-making practice. It involves issues from four major disciplines: Hydrology, System and Control Theory, Decision Theory, and Information Science. For deeper understanding in the theoretical aspects, the interested reader will find precise references to THEORY.

To whom this book is addressed

The natural users of this work are engineering firms, water agencies and university students.

As for university students, the book can be employed in courses that deal with management and planning of water resource (or with Environmental Impact Assessment) in both Environmental Engineering and Science Faculty, both at under-graduate and post-graduate levels. In first level courses, it could be opportune presenting only procedural and decisional aspects, and thus Chapters 6–8, which deal with modelling and control, might be skipped. Finally, this book can also be used as a source of examples for more theoretical courses that adopt THEORY as a textbook.

Please note that this book and THEORY share the same terminology and the same overall structure. Even more: they share Chapter 1 (*Making decisions: a difficult problem*) since, in a sense, the following chapters in the two books are none other than two parallel and alternative ways of developing the content of that first chapter: the theoretical oriented approach and the practically oriented one.

Three levels of reading

Too much attention to detail could easily produce a work that is boring and unreadable. Conscious of this risk, I have structured the book at several levels of reading.

The first, introductory level is supported by a video documentary and by the summary document. The documentary, which was created by Swiss Italian television with my advisory support, gives the reader a live introduction to the Verbano problem and the decision-making procedure adopted for its solution. It serves as both an audiovisual introduction to the entire book, as well as an index, since its captions link to the section of the book that describes the issues that are being dealt with in the video. The summary document (Chapter 15) is a document that summarizes the development and results of the Project and that is submitted to the final Decision Makers, who needs to catch the essential of the Project results without lingering on technical details. A synthetic document of this type is required by good practice and is helpful to improve participation and communication with the final Decision Makers. Part C of the book begins with this document, which provides the reader lacking time with an overall vision and allows the reader that has followed the work step by step to review the principal phases.

The second level, embodied in the volume that you have in your hands, describes the entire Verbano Project. It describes: the decision-making procedure used in the Project (Chapter 1); the setting of the Project (Part A), from the recognition of the system and the definition of the Project Goal, to the practical definition of the procedure to be used for identifying the most interesting alternatives (called *reasonable alternatives*); the implementation of that procedure and negotiations among Stakeholders (Part B); the summary document, considerations about the final negotiations among the Decision Makers and about the implementation of the best compromise alternative (Part C).

At the third level the description of the Verbano system is further detailed in its social, economic and environmental aspects. The presentation is made by Experts from each sector and enriched by images and videos. All this material is presented in a hypertext format on the DVD attached to the volume.

To facilitate the 'navigation' of the work, the text and the DVD share the same index. It is reproduced in both and the lighter shade of the title of a section and the icon indicate that the corresponding section is physically present only in the other support. The icon , inserted behind a section title printed in the normal shade signifies that additional material on the same topic is contained on the DVD. I hope that in this way we have helped the reader to form a single mental image of the work, even though it is physically based on two different media.

WEB site

To facilitate the updating of this work, a WEB site is available (www.elet.polimi.it/people/soncini) where the reader will find didactic material, updates and the errata–corrigenda prepared on the basis of the comments and suggestions provided by the readers.

Where to begin

I believe that I have set out all of the salient points. Now it is your turn. If you are a hurried reader with some knowledge of the matter, I suggest to begin with Chapter 15, the summary document of the Project: it will provide you with a quick panoramic view and direct links to the chapters that interest you the most. If you have interests in the methodology you can start from Chapter 1, which illustrates the decision-making procedure (PIP) that was developed for the study and that constitutes an annotated index of Part A. Those that have time, can just start from Chapter 1 and proceed with the subsequent chapters, following the phases of the PIP procedure step by step. But if you come from reading THEORY, you may omit Chapter 1, since it is analogous to Chapter 1 of that book. However, everyone is first invited to see the documentary *Verbano, let's get it together!* contained on the DVD (it only takes 30 minutes), the subtitles of which point to the part of the work in which the issues presented are dealt with.

Enjoy your reading!

I would be very grateful for comments or suggestions.

Politecnico di Milano, 10 December 2002

Rodolfo Soncini-Sessa
rodolfo.soncinisessa@polimi.it

Post-scriptum of June 2006

Once this work had been completed I prepared myself to create a work with a stronger theoretical scope, which would describe in details the methodological aspects that were developed *ad hoc* in the elaboration of the *Verbano Project*. This is how THEORY was born. In the meantime the application of the PIP procedure to other projects both national (*Iseo Project, Piave Project*) and international (MERIT Project EU-FP5) demonstrated that it had a much greater value than the one I had perceived at first. This conviction was confirmed by the adoption of PIP as one of the basis of the *Planning Framework* proposed[5] by the Harmoni-Ca Project (EU-FP6) for the development of the River Basin Management Plans required by the Water Framework Directive (WFD) of the European Union [Directive/2000/60/EC]. The experience gained through Harmoni-Ca shed light on the usefulness of the PIP procedure not only for the drawing up of participatory and integrated River Basin Plans, as required by the WFD, but also to make Plans more comparable.

The text is not systematically framed in the WFD, since it was born before it but the core concepts of the directive were correctly anticipated. For the publication of the English version, I thought about updating it, to make it coherent with the way that these concepts have matured two years after the completion of the first edition, but I felt that the effort would be useless: it would have delayed publication by at least one more year and in the meantime THEORY would also have aged. Knowledge advances very quickly these days, especially at the interface between society and technology in the so-called Information-Society, while the time it takes to write a book has not changed for centuries. This is the reason why the reader will not find in this book some ideas that do appear in THEORY, which we highly recommend to the interested reader.

[5] See Work Package 3 on http://www.harmoni-ca.info, last visited on September 2006.

Acknowledgements

The translation of the book was partially supported by the Project FONDAZIONE CARIPLO TWOLE-2004.

The work would never have been what it is without the enthusiasm, the passion and the mirth of Alessandro, Emanuele and Francesca, three students from my course that began the work as master theses and finished it once they had already got the master. They did a great amount of work and so their names deservedly appear on the chapters of this book. Nevertheless, without the tenaciousness of Enrico Weber, the computations would never have been completed. He created large part of the computer codes, launched optimizations and simulations for a total of about one hundred full days of computing time, and finally, heroic work, he restored the model's database after the disastrous surprise attack from a hacker in the Christmas vacation of 2001, that caused us enormous damage, especially because the last back up had been mistaken (and who says that Murphy's law is not scientific?!).

Many are the people to whom I am in debt for the help and the collaboration that they offered to us. A special acknowledgement goes to Bruno Bolognino, from the *Associazione Irrigazione Est Sesia* and to Nicola Patocchi, from the *Fondazione Bolle di Magadino*, for their willingness and availability to respond quickly to every one of my questions, even if at times they came at an un-orthodox hour. I am grateful to all of the representatives of the *Stakeholders*: Raffaele Babuscio, Doriana Bellani, Gerolamo Boffino, Lorenzo Del Felice, Andrè Engelhardt, Piero Ferrozzi, Gianfranco Gasparini, Francesco Grecchi, Claudio Peja, Ambrogio Piatti and Cesare Puzzi who actively participated in the negotiations meetings. I am even more grateful to those of them who were willing to enrich the DVD with a description of their sector, in which they encapsulated a great deal of experience. You will find their signatures on the texts that they produced. Thanks go to Andrea Rizzoli, coauthor of the first release of TWOLE, the MODSS used in the Project, who affectionately helped us in making several difficult modifications to it and to Luca Villa who created the first version of TWOLE. Final thanks go to Andrea Castelletti and Mariangela Crocetta who re-read the whole Italian manuscript with patience and care; to Guido Sagramoso who nursed, with his natural perfection, the adaptation of the videos that furnish the DVD, to Giulio Roggiero who designed it friendly interface and solved the problem of reading it with free software tools, and lastly (last just because in every list someone must be the last) to Marco Montanini who spent a great deal of energy to create the architecture and to assemble the DVD.

For the English edition a very, very warm thanks goes to Valerie Cogan for the care and love that she placed in translating and to Jonathan West for his careful, high professional revision of the manuscript.

A completely separate acknowledgement goes, at the very end, to the Swiss-Italian Television (*Televisione della Svizzera Italiana* (*TSI*)) and to its producer Enrico Pasotti, who agreed to create the documentary that describes our study and to enrich the DVD with other documentaries produced by them. They courageously opened the road to an activity which I believe will be a joint activity of publishing and television in the future: the creation of interactive video-texts for more incisive teaching.

Authors

The chapters of this volume are signed with acronyms that correspond to:

AC	Andrea Castelletti	DEI–Politecnico di Milano, Milano, Italy
AN	Andrea Nardini	CIRF, Mestre, Italy
AP	Alessandro Passoni	Environmental Engineer
EB	Emanuele Betti	Environmental Engineer
EW	Enrico Weber	DEI–Politecnico di Milano, Milano, Italy
FC	Francesca Cellina	POLIEDRA – DEI–Politecnico di Milano, Milano, Italy
FP	Francesca Pianosi	DEI–Politecnico di Milano, Milano, Italy
RSS	Rodolfo Soncini-Sessa	DEI–Politecnico di Milano, Milano, Italy

The DVD contains contributions also from:

Gerolamo Boffino	*Parco Naturale della Valle del Ticino*, Cameri (NO), Italy
Bruno Bolognino	*Associazione Irrigazione Est Sesia*, Novara, Italy
Gianni Del Felice	*(previously at) Consorzio di Bonifica Est Ticino-Villoresi*, Milano, Italy
Andrè Engelhardt	*Comune di Locarno*, Locarno, Ticino, Switzerland
Piero Ferrozzi	*Ente Gestione Governativa Navigazione Lago Maggiore*, Milano, Italy
Nicola Patocchi	*Fondazione Bolle di Magadino*, Magadino, Ticino, Switzerland
Claudio Peja	*Parco Lombardo della Valle del Ticino*, Pontevecchio di Magenta (MI), Italy
Ambrogio Piatti	ENEL *Produzione*, Sondrio, Italy
Cesare Mario Puzzi	*Gestione e Ricerca Ambientale Ittica Acque* (GRAIA), Varano Borghi (VA), Italy
Stefania Trasforini	*Gestione e Ricerca Ambientale Ittica Acque* (GRAIA), Varano Borghi (VA), Italy

Translator's note

In order to avoid boring forms like (s)he and his/her, which are used to acknowledge the fact that both men and women can be found performing all the roles described in this book, we chose to allocate each of our characters with a gender right from the beginning, and have been consistent throughout. Place names and names of institutions are for the most part in Italian. As much as possible we have maintained the original names, while substituting nouns such as ‘bridge’ (*Ponte Coperto* → *Coperto Bridge*) and ‘consortium’ (*Consorzio Est Sesia* → *East Sesia Consortium*) with the English word to give the reader a better understanding.

Valerie Cogan

Chapter 1

Making decisions: a difficult problem

AC, AN and RSS

The evolution of a natural system that is subject to anthropic pressure is well described by the DPSIR framework, proposed by the European Environmental Agency (EEA) and reproduced in Figure 1.1 (EEA, 1999, see also OECD, 1994 and UNCSD, 1996). The *Drivers* generate *Pressures* that change the *State* of the system. This variation produces *Impacts* on society, which reacts by devising and implementing *Responses*, which can be directed at the Drivers, as well as the Pressures, the State or the Impacts themselves.

The following example is useful to clarify the framework: consider an enchanting lake, surrounded by fields, forests, a fishing village and a few small hotels.[1] The Drivers are the agricultural, industrial and domestic practices. They produce a flow (Pressure) of nitrogenous substances that reaches the lake, through agricultural land run off, or through direct or indirect discharge from the sewage system. It follows that there is an increase in the trophic level of the lake, which induces algal blooms, anoxic conditions and mass fish death, and so, a variation in the State of the lake. In this way two Impacts are produced: a reduction in fishing activity and a loss of the lake's appeal to tourists. In order to respond to the fishermen's and hotel-keepers' discontent, the Environmental Agency (EA) must design an intervention (Response). It can choose among different forms: issue a regulation regarding the use of nitrogenous fertilizers in agriculture (arrow 1 in Figure 1.1), create a stage for the removal of phosphorous in the treatment plant that purifies the sewage prior to discharge (2), collect the algae when necessary or inject oxygen at a certain depth to prevent lake waters from becoming anoxic (3), or simply introduce a monetary compensation (4) for the damage. In general terms, the EA is not limited to choosing only one of these *interventions*, each of which can be realized in different forms and degrees: it can also select a combination of them, in an integrated and coordinated package, that we will call *alternative*.[2]

---

[1] In what follows we will refer to this example as the 'enchanting lake' example.

[2] A specification for readers who are familiar with negotiation theory: in the literature (see, for example, Raiffa et al., 2002) there is a distinction between the alternatives that a Party can pursue alone (i.e. without reaching a negotiated agreement), and those that are subject to negotiations, because they contain actions that can be carried out only after an agreement has been reached. The term 'alternative' is reserved for the first, while the term 'option' is used for the second. We will not make this distinction (except for Appendix A10 of THEORY) and we will use the term 'alternative' to designate both of them.

In some contexts the term 'programme of measures' is used instead of 'alternative'.

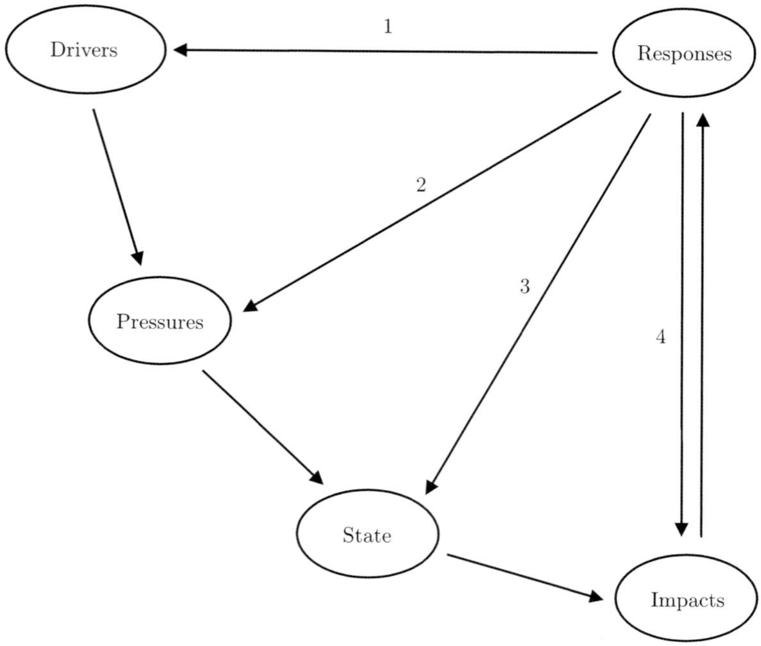

Figure 1.1: The DPSIR framework.

In practice, the same system will require different Responses as time goes by and the context changes. Therefore, very rarely the alternative is established once and for all, more often interventions occur in cycles, as in Figure 1.2. Each cycle is a sequence of events: a *planning* phase, in which an alternative is chosen, its *implementation*, and the *management* of the modified system for a period of time afterwards, during which the system behaviour,

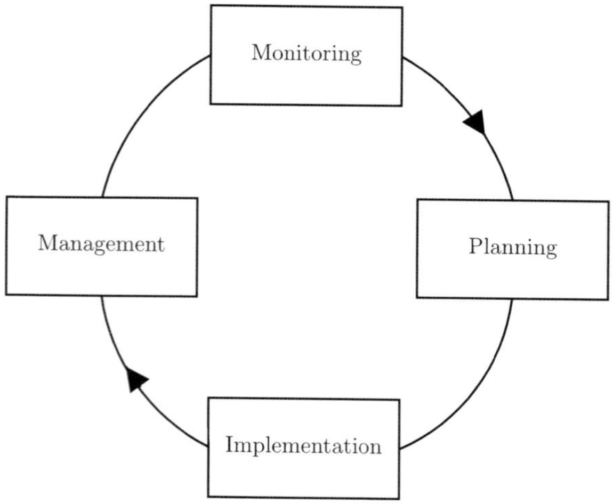

Figure 1.2: The intervention cycle.

i.e. the Impacts produced, is monitored (*monitoring*). When such impacts require a new response, a new cycle begins.

The choice of an alternative constitutes a *Planning Problem*, the solution of which requires a *procedure* to be established. This chapter is dedicated to the individuation of the latter, but before proceeding, an effort is made to clarify what kind of interventions we are talking about and what relationship they have with actions and decisions.

1.1 Interventions, actions and decisions

The set of interventions we are concerned with in this book often makes up a *Project*, i.e. the choice of an alternative that permits the achievement of given goals. But they could also make up a *Plan* (substantially a project, but with a broader range, that usually deals with a wider area and considers larger scale decisions, but at a lower level of detail), a *Policy*[3] (less detailed than a Plan and at a level that is more strategic and less operative) or a *Program* (a set of Projects, organized in time, through which a Plan or Policy is implemented). In practice, the boundaries among these terms are not well defined and are therefore subjective. For this reason, sometimes Projects may seem like Plans, or Policies may seem like Projects. Independently of the name that is adopted, the essential nature of the thing does not change.

We are now ready to give a more precise idea of the types of interventions that we are thinking about when we speak of Projects: these are the different approaches that can be used to reach the Goal that has been set. In practice, interventions can vary greatly from case to case and be quite case-specific. We give the following general guidelines for identifying possible interventions in any given case: do not exclude any intervention without first evaluating it, even if at first glance it is not very 'orthodox'; and avoid, as much as possible, interventions that for sure will preclude others in the future. As it would be impossible to make a list of all the possible interventions, we prefer to exemplify with two projects.

1.1.1 A first example: the Egyptian Water Plan[4]

The Egyptian system and the reasoning behind the Plan are described in the box on page 4. In brief, its aim is to improve the country's quality of life by targeting water availability, bearing in mind the need for environmental sustainability. The Plan covers a period of fifty years into the future.

Given the significant extension of the time horizon, it is necessary to consider all the activities that may affect the availability or quality of water resources in the long run. For example: the increase of urbanized areas, which follows from population growth and takes fertile land away from agriculture, and industrial development, which increases the demand for water and depletes the available resources through pollution. The 'quantity' of the resource is, in fact, strictly connected to its 'quality': for water to be usable, appropriate standards must be passed, which depend on its use. Not only must the quantity and quality of water in bodies of water be considered, but also the effects of water on soils (salinization, desertification), on crops (increase or loss of productivity, bioaccumulation of toxins), on human health (diffusion of water-borne pathogenic agents) and on the natural state and biodiversity of the environment. Finally, given that the largest portion of the demand is due to irrigation, among the interventions to be considered one should include all those which affect farmers' choices. These can be laws about land use, provisions for water

[3]Not to be confused with the meaning of the term that will be defined in Section 5.2.1.

[4]Following Nardini and Fahmy (2005), with a few modifications for didactic requirements.

The Egyptian Water Plan

System description For Egypt, water is the most important natural resource, and at the same time the one at the greatest risk. Egypt's groundwater reserves are, in fact, very limited. The water required for domestic use, industrial activities and irrigation comes from the Nile. This is regulated by the High Aswan Dam that creates Lake Nasser, which collects the flow from a vast catchment. It is expected, however, that in following years the flow will decrease, not only because of the climate change that is probably occurring, but mostly because many towns in the catchment area foresee an increase in the amount of water drawn from the Nile to meet the needs of a growing population. Problems come not only from the limited quantity of the resource (the entire volume supplied by Lake Nasser is consumed before it reaches the sea), but also from its quality, which is depleted by many sources of pollution. It is legitimate therefore to affirm that the quantitative and qualitative scarcity of water resources is threatening to become the principal factor limiting the future development of the country.

Project proposal The water resource problem in Egypt has been a subject of study for many years (Elarabawy et al., 1998; NAWQAM, 1999) and a great number of proposals have been put forward. They are, nevertheless, primarily sectorial and the positive effects that they produce in a given sector are often accompanied by negative effects in others. For example, using recycled (polluted or brackish) water for irrigation makes it possible to reallocate precious water volumes from agriculture to domestic uses. However, if this practice is abused the fertility of the land is reduced in the medium to long term. The development or intensification of agricultural practices in one zone can bring about a reduction in the availability of water in another. The introduction of more efficient irrigation techniques can slow the recharge rate of aquifers and therefore reduce their usability downstream or in the future. In order to consider this complicated tangle of factors, the Egyptian Government decided to prepare a National Water Plan, which defines the terms and timescale for the interventions to be carried out and the measures designed to guarantee the water resources that the country will need in the medium to long term. The Egyptian Government's objective is to improve quality of life for the country over the next 50 years. The Government also wants the plan to be economically efficient, environmentally sustainable, financially and politically feasible, and socially equitable so that it guarantees national security.

saving, economic incentives and disincentives, and fixing prices of production commodities and foodstuffs.

Below is a summary of the options for intervention that can be considered within Egypt itself, excluding political agreements that could be established with the States of the Nile basin to improve the quantity and quality of water that reaches Lake Nasser. This summary should not be considered either absolute or exhaustive. In a different geographic situation or with other aims it would be appropriate to eliminate some of the headings and add others.

- **Interventions to locally increase water availability**
  - constructing desalinization plants for brackish and marine water;
  - constructing structures for exploiting rainfall and flash floods;
  - constructing waste water recycling plants;
  - constructing water treatment plants to provide drinking water;
  - constructing pumping stations for superficial or deep aquifers;
  - installing pumps for lifting water from interceptor canals into the irrigation network;
  - improving irrigation network efficiency, e.g. by coating the earth canals;
  - improving drinking water distribution efficiency, e.g. by reducing seepage and leakage;
  - extending the irrigation network into lands reclaimed from the desert;
  - changing irrigation method: a shift from the flooding method, which is extremely water consumptive, to the sprinkler or drip methods.[5]

- **Interventions to safeguard the quality of the environment**
  - constructing domestic and industrial wastewater treatment plants;
  - improving the drainage network to counteract the salinization of soils;
  - enacting laws that establish quality standards for effluents or receiving bodies.

- **Regulation policies[6]**
  - defining regulation policies for Lake Nasser, the only surface reservoir that serves Egypt;
  - defining regulation policies for the aquifers (define the extractable volumes from each of them according to the month of the year and the condition of Lake Nasser);
  - defining distribution policies between land districts.

- **Interventions aimed at guiding behaviour**
  - sizing areas used for particular crops; imposing crop rotation in the first years, in lands reclaimed from the desert; imposing taxes/subsidies on some crops;

[5] One may ask why not simply decide to change over to the drip method at one fell swoop, which would drastically reduce consumption and might finally allow Egypt's thirst to be satisfied. There are several reasons. First of all, the high costs of the equipment, and the fact that not all soils are suited to or able to sustain the crops that are compatible with drip irrigation. In addition, the surplus water required by the flooding method is not all lost. A great deal filters into the water table, from where it evaporates less than from the canals, and from where it is then pumped to be reused downstream. Lastly, there is cultural inertia. The agricultural community is structured around the extremely dense network of irrigation and interceptor canals, the very structure of the flooding method, which delimits the plots of land. The people's way of life is linked to this structure. The modification of this state of affairs requires a cultural education programme which can be a slow process. This should be considered in the context of an integrated intervention.

[6] A regulation policy is a procedure that defines the rate of flow to release from a reservoir (see Section 5.2.1) or to divert at a diversion dam, or to pump from an aquifer as a function of the data acquired from the information system.

- setting limits to urban expansion;
- improving irrigation efficiency through education programs and economic incentives;
- setting taxes/subsidies for the use of chemical products;
- setting quotas, taxes/subsidies for importing/exporting certain foodstuffs;
- setting prices for products controlled by the government;
- defining tariff schemes[7] for water service users, both domestic and industrial;
- defining incentives for the settlement of farmers in reclaimed lands.

Note that all these interventions have (and must have) something in common: they can all be carried out and managed by the organization that commissions the Plan: the Egyptian Government.

1.1.2 A second example: controlling hydraulic risk[8]

The management of hydrogeological instability and especially the risk of flooding are of vital importance in many countries. Traditionally, the interventions that are considered are:

- **Structural interventions for hydraulic protection and regulation**
  Among the most common are the construction of flood detention areas,[9] detention basins,[10] dry dams,[11] dikes, riverbank defences, flood diversion canals and rectification of the river channel.

All these interventions have reduced the naturalness and beauty of the water courses and the landscape, they have impoverished biodiversity and, albeit surprising, they often have not reduced the risk of flooding. This is due to two reasons: the first is that in the 'safety' zones land use has intensified, increasing the potential damage. The second is that the interventions have increased the flood peaks in downstream transects, thus worsening the situation, because they have reduced the arrival time and the detention capacity of the river channel. Experience has demonstrated that these interventions are too often neither sufficient, nor effective, nor efficient, nor sustainable. As a result, in the last decade in many countries a new approach is emerging, which aims at a generalized renaturalization and is based on the following actions:

- **Interventions to reduce the potential damage**
  Avoiding and eliminating the presence of assets in flood risk zones, establishing binding building regulations and/or promoting the delocalization of settlements already present through regulations and economic incentives.

---

[7] The rules that establish how much each user must pay for the water used as a function of the volume of water withdrawn, of the withdrawal period, and of the category to which the user belongs.

[8] This section is based upon Nardini (2005).

[9] Portions of land that are dedicated to the temporary storing of the volumes of water that overflow from a water course, thus lowering the peak level of the flood and thus alleviating the flooding risk for areas downstream.

[10] These are flood control structures similar to detention areas, but artificially controlled: they typically include banks, which separate the river channel from the flood zone, a spillway that regulates the intake of water from the river channel, and a drainage system.

[11] A non-regulated dam that, in normal conditions, holds back no water and allows the river to flow freely. During periods of intense flow, which would otherwise cause flooding downstream, the dam temporary holds back the excess water, releasing it downstream at a controlled rate.

- **Interventions to increase storage capacity**

  Modifying the general situation of the territory: reafforesting; constructing retention basins for storm water in urban zones; reducing the impermeability of urban ground, creating draining surfaces in parking lots, in squares, and on the roofs of buildings (*green roofs*); and establishing incentives/disincentives to drive the actions of the private sector in this direction.

  Reestablishing space for rivers to flood, to modify morphology and to meander: moving or eliminating banking, rectifications, and riverbank protection where possible, particularly in the minor network; substituting the interventions to protect areas of a low intrinsic value (e.g. agricultural and treed areas) with mechanisms for damage compensation.

- **Interventions for recovering geomorphologic equilibrium**

  In the basin: afforestation to stabilize the slopes, monitoring and controlling fires and grazing to reduce erosion, identifying feeder basins for refurnishing the river beds with sediments, eliminating dams and dredging reservoirs (releasing the sediments into the river downstream).

  In agricultural zones: giving incentives for suitable agricultural practices and crops, and regulating land management.

  In the river channel: enhancing the riverbank vegetation, forbidding the extraction of gravel, and limiting the artificial protection of riverbanks from erosion as much as possible, given that the possibility to sediment and erode is key to maintaining the equilibrium of the river bed.

- **Interventions for living with risk**

  Interventions to raise responsibility: informing, sensitizing and educating the public; inducing people to participate in decision-making and assume responsibility.

  Planning the management of emergencies: setting up efficient warning systems for flooding events and appropriate emergency plans; equipping the zones to face these events.[12]

Note that also in this case all these interventions can be decided upon by the administration responsible: a river basin authority, a regional authority, or a local body.

1.1.3 Actions and decisions

Every option for intervention should be then broken down into one or more *actions*, which are characterized by the fact that each one of them can be completely and precisely identified through the specification of the values assumed by a set of attributes (parameters and/or functions). In the 'enchanting lake' example that opens the chapter, the normative intervention can be, for instance, specified by the maximum load [kg/ha/year] of nitrogen allowed in field fertilization, while the construction of the water treatment plant can be defined by its location and by the percentage of nitrogen removed. The *decisions* that the Project must take are concerned with the options for intervention to consider, the type of actions by which

[12] For example, construct buildings on piles; provide openings below flood level with watertight doors; construct retaining walls around buildings; make basements and ground levels floodable without incurring damage; avoid locating residences on the ground floor; provide the sewage system with one way valves; locate the electricity plants, telephone systems and heating above flood level.

to realize them and the values to assign to their attributes. The attributes must be defined in such a way that it is always possible to leave things as they are, i.e. to choose non-action.

As we have already said, an *alternative* is an integrated and coordinated package of actions. The purpose of the Project is to identify the alternative (or alternatives) which permits the achievement of the overall Project Goal among the alternatives being considered. The set of these must always include the *Alternative Zero* (denoted with $A0$), which is composed by non-actions and therefore is often described as *business-as-usual*.

1.1.4 Classifying actions

The actions, and as a consequence the decisions that are concerned with them, can be classified in various ways.

1.1.4.1 Structural and non-structural actions

The first distinction is between *structural actions* and *non-structural actions*: the first are concerned with physical modifications of the system, as, for example, the location and dimensions of the structures for the collection, transportation, distribution and use of the resource. The second either modify the system only functionally or they alter the effects that the system produces. Examples of structural actions are: the construction of a dam or a canal; the installation of an irrigation system; the construction of a waste water treatment plant; and the renaturalization of a river that has been rectified in the past. Examples of non-structural actions are: a regulation that introduces quality standards for effluents; setting tariffs for water services; an incentive programme for farmers to encourage 'virtuous' behaviour (e.g. adopting crops that need less water or planting woody buffer strips); and the regulation policy of a reservoir.

The assignment of an action to one class or the other is not always univocal: for example, the US Army Corps of Engineers classifies the action 'raising a building on piers' as non-structural. The reason is that such an action does not modify the functioning of the system (the river flooding), while it does influence the effects that it produces. Also note that a non-structural action can indirectly produce structural actions. For example, an incentive programme for farmers can encourage them to plant buffer strips along water courses, so modifying the flood regime. Therefore the border between the two classes of actions can be, at times, very elusive, but the classification is useful just the same.

1.1.4.2 Planning and management actions

A second distinction is made between *planning* actions and *management* actions. The discriminating factor is the time step with which the actions are decided. An action is a *planning* action when it is decided once and for all (a typical example is the construction of a dam). An action is a *management* action when it is decided upon frequently or periodically.

There are two outstanding characteristics of management actions. First: when one decides on the next action, up to date information about the system is available so that the evaluations that were carried out to take the previous decision can be updated. Even the decision-making method can be reviewed on the basis of this update. Second: the decision is *recursive* when the system is *dynamical*.[13] This means that every action will be decided by considering the decisions that will have to be taken in the future on the basis of the states that today's decision will have produced. In some ways it is more difficult to choose

[13]For the meaning of this term see Section 6.2 or Appendix A3 of THEORY for more details.

management actions than planning actions, because it is necessary to evaluate not only the current decision, but also all the decisions that will need to be taken in the future, while taking into account the effects that the first ones will have induced. However, a planning problem can be conceptually more difficult than a management problem, because it may incorporate management problems (think of the Egyptian Water Plan).

We can summarize the twofold classification that has been introduced above with a 'sample' of actions, which is a far from exhaustive list, accompanied by the principal attributes that define them:

- **Planning actions**
  - Structural actions: construction of
    - *Reservoirs*[14]: location, size of the dam, characteristics of the outlets;
    - *Curtains* for pathogenic control or fixing nutrients in natural lakes and reservoirs: location and maximum reachable depth;
    - *Aerators* in natural lakes in anoxic conditions or those requiring destratification: location, depth, power;
    - *Diversions*: location, regulability, maximum derivable flow;
    - *Canals*: location, layout, minimum and maximum flow, presence or absence of coating;
    - *Irrigation systems*: location/extension, irrigation/drainage technique;
    - *Pumping stations*: location, capacity, head;
    - *Aqueducts*: layout, average and maximum flows, losses;
    - *Hydropower plants*: location, intake and outlet points, maximum and minimum flows of the turbines.
  - Non-structural actions: definition of
    - *Management criteria for reservoirs and diversions*:
      - *minimum environmental flow* (MEF), i.e. the minimum flow that must be released to the river downstream from a reservoir whenever the inflow exceeds the MEF[15];
      - storage constraints (regulation range);
      - constraints on the dam operation;
    - *Land use regulations*: zoning, limits to expansion, urban regulation;
    - *Regulation and/or water distribution policies*[16];
    - *Economic instruments*: tariff schemes for water services, insurance plans against the risk of flooding;
    - *Information and education campaigns* to increase awareness: program, people involved, budget, means employed.

- **Management actions**
  - Structural actions
    - *Maintaining storage structures*: volume and location of sediments to be removed in reservoirs and diversion dams;

[14] In the text that follows we will use the term reservoir to refer to both artificial reservoirs and regulated lakes; in other words, we will use it as a synonym for regulated storage facilities.

[15] When the inflow is less than the MEF it is never compulsory to release more than the inflow. This does not exclude that a reservoir could be used for low-flow augmentation.

[16] See page 107.

  - *Planting woods*: number, type of plants, and frequency and zones of afforestation.
- Non-structural actions
  - *Releasing water from reservoirs*: flow rate;
  - *Aerating natural lakes and reservoirs*: intensity;
  - *Operating curtains*: depth;
  - *Reviewing economic tools* in the light of contingent conditions: degree of variation;
  - *Broadcasting alarms and pre-alarms for floods*: area involved.

Note finally that another attribute, which must be considered for all these actions, is the time at which its effects will come about: for example, the time a dam becomes operational or the time a regulation comes into action. This attribute is of particular interest not only for Programs, but also for Projects, when short-term (transient) effects are considered (see page 71).

1.1.4.3 Other classification criteria

Another classification criterion is based on the aim of the actions. This is how we have classified the actions in the two examples that open this paragraph. A further possibility is to classify the actions according to the Decision Maker (DM) that can take them, or according to the decision-making level at which they are established. For example, a public administrator, a single farmer and an individual citizen all operate at different levels: the first can decide on all the actions listed in Section 1.1.1, the second only on crop rotation, irrigation techniques and agricultural practices, while the third decides how to save water when brushing his teeth or taking a shower.

1.2 Difficulties and keys to their solutions

The projects of which we are speaking show two fundamental characteristics: they involve many individuals (even sometimes many DMs, as is the case of Verbano Project), and they require a decision from an Agency or a public body (the choice of the alternative to be implemented or simply the authorization to implement an alternative proposed by others). For this reason we talk about decisions in the public realm. Such decisions are usually taken by following the procedure described in the diagram in Figure 1.3. In the diagram the term *Decision Maker* (DM) refers both to the commissioner of the Project, who oversees the first phases (e.g. the Egyptian Ministry of Public Works that wants to produce the Water Plan), and the final DM, who must approve it (in the example the Government or Parliament). The *Analyst* is (s)he who actually conducts the necessary studies and draws up the Project on paper. It can be a technical office of the same administration, but, more often, it is a consultancy that has been entrusted with the job after putting in a tender. The *Stakeholders*[17] are either all those (people, institutions, organizations) that experience the effects of the Impacts for which a Response[18] is being sought, or those that could be influenced by the options for interventions considered for implementing the Response. The horizontal arrows indicate the moments in which the *actors* (DMs, Analyst, and Stakeholders) interact and their direction leads from those that pose the questions to those that must respond.

[17] Some prefer to call them 'rightholders', which is a broader category, according to those who support the idea, than Stakeholders. In actual fact, however, often there are Stakeholders whose rights are not recognized.

[18] According to the terminology of DPSIR scheme in Figure 1.1.

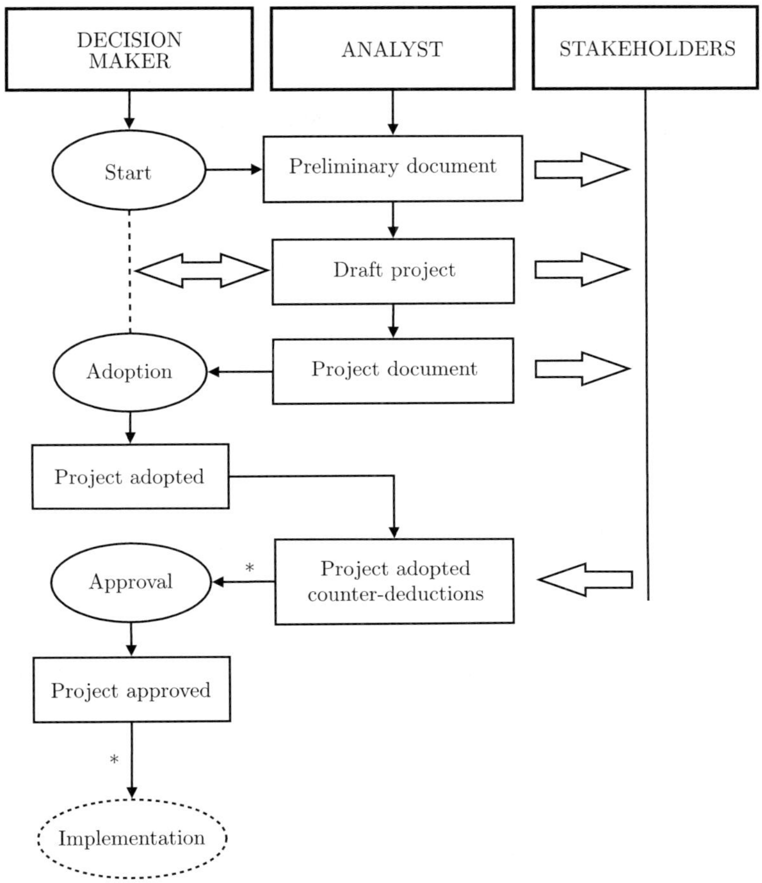

Figure 1.3: The standard procedure for a planning process.

Often the two steps that are indicated with an asterisk are protracted for an excessive amount of time, sometimes even for years, and, since they are not transparent, it is not always very clear what is happening. Informal negotiations, lobbying and political battles are likely to develop. There are two principal difficulties that provoke these delays:

- the *conflicting interests*[19] are often of great importance, especially economic importance and when not adequately managed, can lead to lobbying, opposition and boycotts;

- the *inadequacy of the approach* that is top–down, i.e. a Decision–Announcement–Defence approach. The administration decides, communicates the decision and defends it from the Stakeholders' reactions; the latter, having had very little influence on the decision, are almost always against it. Furthermore, the approach emphasizes the political component of the decision, which while it is certainly essential for ori-

[19] "Perceived divergence of interests, or belief that the Parties' current aspirations cannot be achieved simultaneously" (Rubin et al., 1994).

enting the evolution of the system in the long-term, by its nature does not provide the transparency, the credibility and the explanation of the choices that the Stakeholders demand. In addition, it risks being technically inefficient and unsustainable because the information-modelling and decision-support tools that are available today are too often not really integrated into the approach, but only juxtaposed, at times, acting purely as frills.

In order to achieve good planning and good management it is absolutely necessary to overcome these difficulties by conducting a *participatory, integrated and rational decision-making process*. Let us examine these three adjectives one by one.

1.2.1 Participation

1.2.1.1 Awareness

If we do not manage to awaken or instill profound human values, such as a sense of belonging to the territory, a love for the environment, and a responsibility for its evolution, we will not be able to get very far. Therefore, education is fundamental and must be structured at many levels, from children to administrators and politicians, including technicians and the general public.

1.2.1.2 The participatory process

The top–down approach of the standard planning procedure needs to be reversed, proposing and launching a participatory process (Renn et al., 1993; Renn, 1995; Budge, 1996; Delli Priscoli, 2004), that is 'bottom–up' and based on the management of participation: a process of participation that begins from the proposal of the Project, the recognition of the need for the process itself, and continues with the choice of the alternative to be implemented, right through to the monitoring of the effects after its implementation. This process should not be limited to providing the Stakeholders with information (*Informative Participation*), nor to just asking them for information (*Consultation*), but should also involve the Stakeholders in the design and evaluation of the alternatives (*Co-designing*) and ideally even in the final decision (*Co-deciding*)[20] (Mostert, 2003; Hare et al., 2003).

In this way, a process of *social learning* is created, in which the Stakeholders become aware of the problem, of the alternatives, and of the viewpoints of others; they take responsibility and together they develop the alternative to be carried out (Renn, 1995). The key ingredients are information, transparency, repeatability, and the active involvement of the Stakeholders. In order for the process to be effective, it must be supported by an open and flexible decision-making procedure to accommodate the unforeseen events and elements that are introduced by the Stakeholders. At the same time this process must be structured, so that it does not degenerate into a 'condominium meeting' (see Renn et al., 1995, for an evaluation of different approaches to participation). As it will be shown in Section 1.3, the

[20]With surprising speed legislation has already adapted to this need: the Århus Convention, signed in 1998 and in force since 2001, recognizes citizens' rights to "have access to information and be enabled to participate in the decision-making process with regard to the environment", based on the principle that only participation can make sustainable development possible. In escort to that convention, the Directive 2003/35/EC (European Commission, 2003b) was issued by the European Parliament which establishes that "the public concerned shall be given early and effective opportunities to participate in the environmental decision-making procedures", right from the initial phases, so that they have an effective possibility to influence the choices. The Water Framework Directive (WFD) [Directive/2000/60/EC (European Commission, 2000)] anticipated this position.

synthesis between these antithetic characters can be achieved with an accurate, explicit and shared definition of the phases of the process and professional guidance of the moments in which the interaction and negotiation between Stakeholders take place.

A participatory process takes time and it is this aspect in particular that often discourages its implementation. The final goal should be made clear: if it is the 'actual implementation of Responses', then a participatory process is almost always quicker and more efficient than a non-participatory one (the duration is measured between the beginning of the Project and the end of the implementation phase). If, instead, the goal is just to 'draft the Project', then the non-participatory process is faster, because it avoids many tiring phases of negotiations (in this case the duration is measured from the beginning to the approval of the Project). However, drawing up a Project which almost surely will not be implemented because it is not shared, is a real waste of public money.

Here are some *guidelines* to stick to when developing a participatory process:

1. Share the idea that everyone is working together to solve a problem and improve the quality of life.

2. Participation, not just communication: accept that the participation really influences the final decisions.

3. Always stick to what was agreed upon in the previous meetings.

4. Create responsibility through agreement.

5. Try to transform the difficulties into opportunities,[21] for example, looking for alternatives so that no one is, if possible, worse-off than before[22] (*win–win alternatives*).

6. Look for equity by identifying who bears the costs and who reaps the benefits; look for interventions that make those that impose costs on others responsible for their actions.[23]

7. Guarantee flexibility, but maintain rigor, to avoid 'houses of cards'. One should not proceed by taking decisions on a weak basis that can crumble afterwards. A well conducted process gains the confidence of Stakeholders, who will then be under pressure to participate for fear of being excluded from decision-making.

8. Look for agreement at every step, but at the same time accept that differences and uncertainty are integral and inevitable parts of the process. Do not force the Parties to rush at an agreement, but accept that, in order to reach one, time and interaction are required.

---

[21] Often it is only a matter of imagination: if, for example, a group of farmers are opposed to a project for a river park because they fear that they will have to change their activities and be dispossessed of their land, the opposition could be overcome by also including among the actions of the Project, to be developed with those same farmers, a reconversion action of cropping patterns: e.g., substituting the production of maize with biological crops or high value herbs that can be cultivated and certified, thanks to the very fact that the park exists. For this reason, also interventions for training and technical assistance have to be considered in the project, and potentially subsidies. Furthermore, instead of expropriating the lands, one can think about a form of contract to use land for reciprocal advantage.

[22] Excluding, clearly, those that started in an illegal condition.

[23] For example, by imposing a tax on dumping polluted waste.

9. Provide guarantees instead of demonstrations. It is useless to insist that there will not be undesirable effects; it is much more convincing to sign a commitment to a corrective action in the eventuality that the effects that the Stakeholders fear do occur.

10. Recognize that the public has a role to play in monitoring and carrying out the project.

11. Accept the existence of different points of view.

12. Distinguish facts from value judgements, which are the product of subjective preferences.

13. Keep the decision-making process transparent and repeatable, guaranteeing access to information.

14. Respect the role of the political DM, but make sure that she explicitly clarifies the reasons for her choices.

And here are a few *sine qua non* conditions that the Analyst must respect, if the participatory process is to be successful:

1. State the criteria and the rules to follow and respect them. The 'statute of participation' (Connor, 1997) is a useful tool: it is a document that states principles, intentions and rules of the decision-making process.

2. Define the decision space clearly, and clarify the 'power of the participation machine' and the relations with the Administration.

3. Have the DM participate in the meetings with the Stakeholders, ideally in person, or at least by sending a delegate.

4. Act in a way that gains the trust of all the Parties and never betrays that trust.

5. Be very careful to uncover any misunderstandings that are produced by the terminology and have the patience to take the time and energy to resolve them.

6. Evaluate often the state of the participatory process.

1.2.1.3 Evaluation for negotiations

It is almost always impossible to identify an alternative that produces the best possible effects for all the Stakeholders. Each alternative is a particular compromise between the interests[24] at stake and thus it is essential that its effects be evaluated from the viewpoint of each Stakeholder, so that each one can express his/her opinion about it and negotiate the *best compromise*.

1.2.2 Integration

The decision-making process must be founded in the *principle of integration*, which manifests itself at many levels:

[24]By the term *interest* we mean the needs, desires, worries and fears, concerns, and more generally, whatever reason that encourages a Party to negotiate.

- among the parts that compose the system;
- between rationality and emotionality;
- among Stakeholders and political DMs;
- among the Stakeholders themselves, particularly between those who benefit and those who incur damage;
- among the evaluation approaches: Cost Benefit Analysis (CBA), Cost Effectiveness Analysis (CEA), Multi-Attribute Value Theory (MAVT), Environmental Impact Assessment (EIA), Strategic Environmental Assessment (SEA);
- between environmental policies and sectorial policies (applying the *precautionary principle*[25]);
- between technical approaches to solution and decision-making techniques (*integrated technical approach*);
- between planning and implementation (plan/project; SEA/EIA) and between strategic and tactical scales (applying the *subsidiarity principle*[26]).

In our view, the keys to integration are the correct identification of impacts and the correct identification of the indicators that quantify them, and so two conditions must be respected:

(a) the Stakeholders whose interests will be, or might be, affected must be clearly identified: not only those who could be disadvantaged by the decisions being considered, but also those who may benefit;

(b) the values that the Stakeholders attribute to the impacts must be made explicit (Keeney, 1992).

In this way, the evaluation can focus on the reasons why an alternative is preferable to another, rather than fall into a sterile conflict of positions, in which some defend an alternative to the utmost while others attack it.

1.2.3 Rationalization

We have therefore understood that the decision-making process must be rationalized through the adoption of a precise *decision-making procedure* that allows the *best compromise alternative* to be identified. This is done by way of a participatory negotiation process that is structured and transparent, and whose core is an integrated evaluation that allows each Stakeholder to evaluate the effects that will result from each alternative, and explicate the political compromise between conflicting interests upon which every choice that is made is founded.

[25] Environmental phenomena are very complex and in many cases we do not know how to predict their effects, especially in the long-term. It is therefore advisable not to take actions whose effects we are unable to evaluate or, if it is really necessary to take such actions, to do so only with appropriate security measures.

[26] To each administrative level (e.g., Ministry of the Environment), leave only the decisions that cannot be taken at a lower level (e.g., Province) so that procedures are quicker and more responsibility is given to the lower levels.

The first, fundamental condition to identify an alternative that enjoys a wide agreement is that the very decision-making process get the consent of the Stakeholders. To this aim the sought-for procedure must:

- break the process down into phases and establish the sequence in which they are executed; and
- specify the aim of each phase and the technical means (algorithms and procedures) by which it will be achieved.

In addition, it must give a concrete form to principles 11–14 stated above. More specifically:

- accepting different points of view translates into the fact that each phase (except for the last) can close with a plurality of outputs, all equally adequate;
- maintaining the distinction between facts and value judgements does not mean that subjective preferences are suppressed, because they too guide the choices, but that they are kept distinct from that which is 'objective';
- the transparency and repeatability are realized by making the information available to all the Stakeholders and the DMs, documenting and distributing the results of all the phases;
- and lastly, respecting the role of the final DM means that the final decision (and therefore the last phase of the procedure) is reserved for her and she is guaranteed the right not to choose one of the alternatives that have emerged from the decision-making process. However, the procedure must ask her to justify the choice with the same instruments (indicators and criteria) that were defined by the Stakeholders during the course of the process. If the process was well conducted, the DM's different choices should be justified only by the different relative importance she gave to the evaluation criteria.

The definition of the decision-making procedure is the subject of the next section.

1.3 Planning: the PIP procedure

Sometimes there is more than one DM, as would be the case if the 'enchanting lake', in the example that opens this chapter, and its inlet were to define the border between two countries. In that case, neither of the DMs (the Environmental Agencies of the two countries) could assume efficient decisions autonomously. Even when only one DM is concerned, we have seen (Section 1.2.1) that it is a good idea to choose the best compromise alternative by taking into account the Stakeholders' viewpoints. For this reason, the decision-making procedure should not be limited to considering information collected from the Stakeholders (*Consultation*), but they should, instead, be treated as if they were DMs that must negotiate a compromise alternative (*Co-deciding*). The goal of the decision-making process is to reach an agreement that is acceptable to them all, to which they remain committed, and which is actually implemented. Only the last, decisive step of the procedure (the formal choice of the alternative that is to be implemented) is in most cases the reserved responsibility of the DM (or DMs) that has (have) the institutional power and responsibility to make the choice.

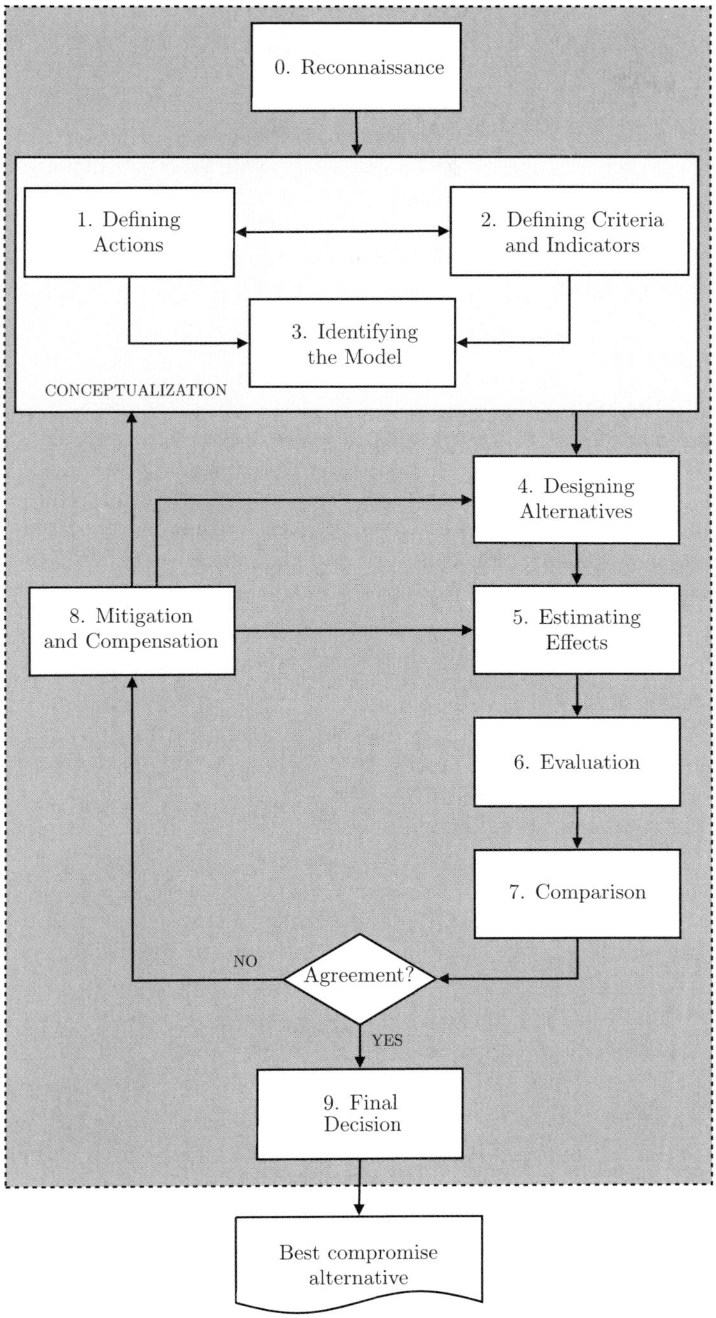

Figure 1.4: The phases of the PIP procedure.

In Figure 1.4 the flow diagram of the procedure that we are proposing is shown. We call it PIP procedure, for Participatory and Integrated Planning procedure.[27] The diagram is also a conceptual map of the modules of the software systems used to implement the decision-making procedure, called *Multiple Objective Decision Support Systems* (MODSS, see Section 1.5). This structure is also reflected in the succession of the chapters in Part A of this book. The description of the component phases is only a preview of what will be presented in those chapters, applied to the Verbano Project; a theoretical presentation of each phase can be found in THEORY. We are going to introduce many new concepts and some may, inevitably, seem obscure to the reader. If this is so, (s)he should not be discouraged, because such doubts are completely normal and will be dispelled further on. What we wish him/her to capture is just a view of the whole procedure.

1.3.1 The phases

Phase 0 – Reconnaissance

Statisticians consider two types of errors: type I error (rejecting a true hypothesis) and type II errors (accepting a false hypothesis). The aim of *Reconnaissance* phase is to avoid "type III errors": solving the wrong problem (Raiffa et al., 2002). The work is concentrated on defining the *Project Goal*, the (spatial and temporal) boundaries of the system being considered,[28] the normative and planning context in which the procedure operates, the data available, and the information that needs to be collected.

One must start off from the identification of the Stakeholders involved and their needs, expectations, fears and perceptions, in a word, their *interests*.[29] In fact, the definition of the Project Goal closely depends on the interests being considered and on the hopes and expectations that one wants to fulfil.

Then the PIP procedure has to be explained to and accepted by, or, if necessary, negotiated among all the actors (Stakeholders and DMs).

At this point it becomes possible to define the *Goal* that the Project must pursue. It is derived from the DM's strategic goals, from the Stakeholders' interests and from the regulatory and planning context. In the case of the Egyptian Water Plan the specific Goal of the Plan ("to improve the quality of life for Egyptians by targeting water availability") was derived from the government's strategic goal "to improve the quality of life for Egyptians". It is useful to translate the Goal for each Stakeholder into a *vision* that visualizes, with words, or better still, with a picture, the condition that the Project aims for. For example, the vision for the civil users of the Egyptian Water Plan could be: "no longer water only in the evening from 17:00 to 20:00 one day in three, but a continuous reliable supply, which is not too expensive ... "; while a project for upgrading a river system could be expressed by the pictures in Figure 1.5. The choice of a good vision is important when the Stakeholders are not very motivated to take part in the decision-making process or resistant, or unable to express their own goals.

---

[27] Even though it was devised autonomously, it can be interpreted as a variation of the PROACT scheme proposed by Hammond et al. (1999), which has been suitably modified to take into account that, in the case we here examine, the decision-making process is targeted at consensus building and actions include management, i.e. recursive decisions.

[28] These two points are often referred to as *scoping*.

[29] "It is crucial for the legitimacy of a planning process to start dialogue as early as possible in the phase of problem definition" (European Commission, 2003b).

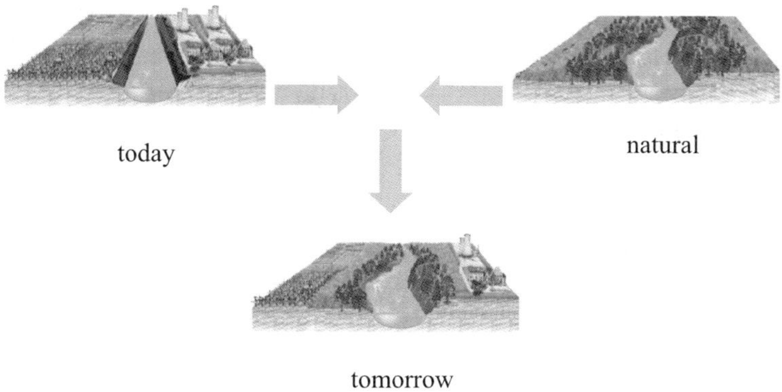

Figure 1.5: The *vision* of a river upgrade project (from CIRF, 2004).

An extensive knowledge of the system is the fundamental support to all these activities and it is acquired by: analysing the regulatory and planning context of the Project; collecting the information and data available; identifying the missing information; and finally, filling the information gaps by conducting hydrological, economic and social surveys. The actors should share all the available information, agree upon its validity (this is crucial!) and the potential need for further investigations. In other words, when necessary, even the validity and availability of the information must be negotiated.

Phase 0 of the Verbano Project is presented in Chapter 2.

Phase 1 – Defining Actions

In this phase, the options for intervention that are supposed to achieve the Project Goal must be identified, bearing in mind the interests of the different Stakeholders. This is not a simple operation because the opinions can be very discordant. For example, for some the obvious solution for the 'high water' problem in Venice would be the construction of the MOSE[30] at the mouths of the lagoon, for others it is the construction of gate-ships,[31] which are less complicated to construct and which would be more adaptable to the bradyseism of the lagoon bottom; others say that the only sensible option would be the reduction in the green house gas emissions that are responsible for the eustatic sea[32] in front of the lagoon. From this initial, decidedly disorganized, collection of ideas, which are in part silly, in part gifted with incredible wisdom, good ones always emerge. It may seem strange to begin with brain storming; however, it is essential to promote creative decision-making that considers more than just a set of interventions given a priori, and is

[30] A system of submerged hollow steel gates, hinged at the bottom of the lagoon and installed at each of the lagoon's three openings to the Adriatic sea. When 'high water' is foreseen, they can be raised by pumping compressed air into them and creating a sea barrier.

[31] Two ships, whose length is about half the width of the mouth of the lagoon, which are hinged to the two offshore piers that mark the mouth's boundaries. The free end of each ship is equipped with a propeller that allows them to position themselves across the mouth opening. When the ships are in this position they flood the compartments that make up their hulls so that they sink, creating an insurmountable barrier for the sea. When the high water event is over, the mouth of the lagoon is reopened by reversing the operation. The major advantage with respect to the MOSE is that the construction and the maintenance are done on dry land and the mouth of the lagoon would need to be modified only slightly.

[32] The phenomenon of sea level rise in the long-term in response to geological and climatic changes, such as, for example, the melting of the polar ice caps, produced by the green house effect.

able to open new perspectives and discover unexpected alternatives. If an intervention is really useful it will emerge in the following steps, and if all the Stakeholders' suggestions are considered and processed, they are more prone to collaborate since they feel they are being "taken seriously" (WFD, Annex VII, point A9, see European Commission, 2000). Moreover, how can there be a participatory process if one does not listen carefully to the ideas and proposals of the actors?

This first, creative phase must necessarily be followed by a phase of 'sedimentation', in which we separate what can be decided and what cannot, and thus, the effective decision space becomes apparent. It is, in fact, useless passionately to debate aspects that cannot be decided in the Project. But note carefully that this is not the time to discriminate the useful interventions from the useless ones, or the sensible ones from the less sensible. This will be a task for the next phases. Here, we only aim at obtaining a list of the interventions 'that can be decided upon' in the context of the Project, e.g. in the case of the Egyptian Water Plan one obtains an extract from the long list of actions in Section 1.1.1.

Each intervention is finally broken down into one or more *(meta-)actions*, i.e. into elementary interventions that can be fully and easily defined by specifying the values of their attributes (see Section 1.1.3), that is by specifying who is doing what, how and when. In this way a meta-action is transformed in an *instantiated* action. Technically, this transformation is carried out by assigning values to the parameters and/or the functions that describe the attributes of the meta-action. The specification of these values is a matter for a future phase (Phase 4: *Designing Alternatives*), but their feasibility sets have to be defined in this phase, thus fixing the meta-actions to be considered. The instantiated actions are the 'building blocks' from which the alternatives will be constructed later. In the following, as we have done so far, we will use the term 'action' to denote both the meta-actions and the instantiated actions. The actual meaning will be clear from the context.

The development and results of this phase are presented in Chapters 3 and 5.

Phase 2 – Defining Criteria and Indicators

To evaluate and compare the effects of the alternatives on the system it is necessary to identify, together with the Stakeholders, a set of *evaluation criteria* that reflect the characteristics of the problem and the values that are at the base of the judgements that the Stakeholders express. The criteria do not have to pertain only to the Project Goal, but to all the positive and negative effects that the Stakeholders hope for or fear: in other words they must express their interests.[33] In particular, the criteria for sustainable development will be proposed by the Agencies and by the Environmental Associations which must always be included in the Stakeholder group.

Not every evaluation criterion is necessarily expressed in an operative way, i.e. it may not spontaneously define a procedure that allows us to ascertain how much a given alternative satisfies that criterion. This is why an *index* must be defined, i.e. a *procedure* which associates the criterion with a value that expresses its satisfaction. This is done through the identification of relationships between the evaluation criterion and the variables (e.g. lake level and water release, river and channel flows) that describe the system condition. In practice, one proceeds by first splitting the evaluation criterion into lower level criteria

[33] Objectives are the same as interest: unfortunately the negotiation theorists and the decision-making theorists have not agreed on a common term. The first talk about 'interests' and the second about 'criteria and objectives'; but they mean the same thing. We will adopt the jargon of the second group.

and, in turn, splitting those into even lower level criteria, until it is possible to associate each one of the criteria at the lowest level (*leaf criteria*) with an *indicator*, i.e. a function of the trajectories of the variables describing the system condition. In this way a hierarchy of criteria is obtained for each evaluation criterion. The definition of a criterion and of its hierarchy ought to encompass either thresholds (e.g. minimum environmental flow or the level above which a flood occurs) or a Stakeholder's wish (*leitbild*, see Egger et al., 2003), which is often related to the performance level (s)he demands (e.g. a preferred flow).

It is necessary to dedicate a great deal of time and attention to interactions with the Stakeholders and to studying their points of view, because it is essential that each Stakeholder sees that his/her interests are expressed in at least one of the indicators. If this does not happen, negotiations in Phase 7 will inevitably fail.

In the next phase we will see that very often the system is affected by *random disturbances* (either *stochastic* or *uncertain*). It follows that also the values assumed by the indicators are not generally deterministic. When this occurs, it is necessary to take account of the *risk aversion* that the DMs and Stakeholders may have. This can be expressed through the classical approach of utility functions, proposed by Keeney and Raiffa (1976), but more often it is translated through *criteria*[34] from which the most frequently adopted are the *Laplace criterion* (expected value) and the *Wald criterion* (worst case) (French, 1988). We will return to these issues in Sections 7.6.1 and 7.6.1; for more details, see also Chapter 9 of THEORY.

Like all the phases of the PIP procedure this one must also be participatory: the evaluation criteria should be forthcoming from the Stakeholders and the definition of the indicators must have their contribution and approval. This last step is, however, often very technical and so, as for the technical steps in the following phases, the Stakeholders can be supported by *Experts*.

We will deal with this phase for the Verbano Project in Chapter 4.

Phase 3 – Identifying the Model

In order to quantify the effects that the different alternatives would produce on the different indicators if they were to be implemented, it is necessary to provide a *model* that describes the cause–effect relationships within the system. Such a model can take the form of an Expert, who, on the basis of his experience, is able to estimate the effects that each alternative will produce (see for instance the MÖLL Project (Muhar and Schwarz, 2000; Jungwirth et al., 2000)). Alternatively, it can take the form of a mathematical model, which is the type of model used in the Verbano Project. The choice of the level of detail in which the model must describe the phenomena is strictly connected to the indicators defined in Phase 2 and to the actions that are considered. In the case of the 'enchanting lake' described in the introduction, the regulations could be planned by describing the system by a set of algebraic equations, while an executive project for artificial aeration would require the system to be described by differential equations.

The input variables of the model must include the parameters that quantify the attributes of the actions (e.g. the maximum nitrogen supply allowed in agricultural practices and

[34]Take care not to confuse this meaning of the term criterion (Stakeholders' attitude towards risk) with the one previously introduced (judgement category). We will encounter semantic ambiguities of this type on other occasions: they occur when the subject of discourse is on the frontier of different disciplines, each of which has independently developed its own jargon.

the nitrogen removal efficiency of the treatment plant) as well as all the variables that allow the future conditions of the system to be described (e.g. the precipitation in the catchment area and the users' water demand). The choice of values to attribute to the former constitutes the subject of the Project, while the values assigned to the latter describe the context within which the alternatives are evaluated and are therefore called the *scenario*.[35] Both alternatives and scenarios have to be quantitatively specified before the model can be run. Note that there can be more than one scenario: in the 'enchanting lake' example we might be interested in evaluating what would occur in a 'high' or in a 'low' rainfall scenario). Moreover, the scenario is not compulsorily deterministic: very often it can be random.

The scenario(s) may be chosen by Experts or it (they) can be obtained by running models, if they are available, that describe the processes that produce the driving forces. In the 'enchanting lake' example the future scenario of rainfall can be generated by a climate change model, while the future scenario of agricultural practices can be suggested by an Expert. When all the processes are stationary, the *historical scenario*, i.e. the trajectories registered in the past, is often adopted for the reasons that will become apparent in Phase 5. In any case, the time horizon of the scenario should be sufficiently long to capture all the types of significant events the system may face. It is a common practice to adopt different scenarios for the design of the alternatives (*design scenario*) and for the estimate of the effects (*evaluation scenario*, sometimes also called *baseline scenario*), which are generally fixed in the corresponding phases.

To facilitate a *social learning* process and help the Stakeholders share a quantitative understanding of the system it is important that they go through the same thinking process and be exposed to the same information and arguments as the Analyst. For this reason, the implicit assumptions of the models should be made explicit and the modelling activity should be supported by an MODSS that must be flexible enough to identify models through a participatory process. Only in that way can the Stakeholders share a common interpretation of the system behaviour (model), which is necessary for them to be able to trust the effects that are estimated with the model. Agreeing on the same model does not prevent them from having different perceptions (indicators) of these effects.

In Chapter 6 we will see the application of these concepts to the Verbano water system.

The set of Phases 1–3 constitutes the *conceptualization* of the Project.

Phase 4 – Designing Alternatives

Unfortunately, very often, it is common practice to consider only the alternatives that are prompted by the Analyst's experience and the suggestions from the Stakeholders. These alternatives make a good starting point, but we believe that it would be a mistake to limit the choice to them alone. More correctly, remembering that an alternative is an integrated package of actions, all the alternatives that can be obtained by combining the actions identified in Phase 1 in all possible ways should be considered. Often the number of alternatives that follows is so high that it would be impossible to examine them all in the following phases and so it is necessary to select only the 'most interesting' ones.

[35]Dictionaries give the term scenario the following definition: "a possible set of future events". Brought into our context the term lends itself to three different possible meanings: (1) synonymous to alternative (e.g. *business-as-usual scenario* means *Alternative Zero*); (2) the set of effects that an alternative produces; (3) the time series of input variables that are not controlled by the DM. We will strictly adhere to this last meaning.

However, these must be chosen still following the Stakeholders' criteria, which were identified in Phase 2, rather than the Analyst's preferences.

In more complicated projects, such as the Verbano Project, where a higher level of mathematical formalization is required, to identify the 'most interesting' alternatives it is necessary to define a mathematical problem (a Mathematical Programming Problem or an Optimal Control Problem), called *Design Problem*, which selects the *efficient alternatives* with respect to *objectives*, accordingly called *design objectives*. These are defined on a *design horizon*, given the *design scenario* and taking into account only a subset of the evaluation indicators: the *design indicators*. This simplification is introduced when the considering of all the evaluation indicators would make the Design Problem unsolvable in acceptable computing times; it does not excessively polarize the result if the design indicators are carefully chosen, since in Phases 5 and 6 the alternatives will be evaluated with respect to the complete set of indicators.

The characteristics of the system appear in the Design Problem as *constraints*, while other elements that define the design scenario (e.g. user demands, produce prices) contribute, along with the structural and normative actions, to determine the value of parameters that appear in the constraints and in the objectives of the Problem. Solving the Problem through an appropriate *algorithm*, provides the set of alternatives that will be examined in successive phases. To these the *Alternative Zero* ($A0$) is always added, that is the alternative that assumes that nothing is done and everything remains the way it is (business as usual).

The formulation of the Design Problem will be outlined in Chapter 5 and completed in Chapter 7, where the algorithm used to solve it is described.

Phase 5 – Estimating Effects

Once the alternatives have been identified, the *effects* that each produces must be estimated: in other words, it is necessary to compute the values that the indicators assume as a result of each of the alternatives being implemented. When the system is not dynamical, the evaluation is immediate. When the system is dynamical this estimation requires that each alternative be simulated over a time horizon (*evaluation horizon*) long enough to make extreme events (e.g. droughts or floods) likely to occur, in order to avoid the risk of estimating the effects in 'average conditions' only. In both cases it is necessary to feed the model with an appropriate input: the actions of the alternative considered and one or more *evaluation scenarios*. The alternatives will be compared to single out the 'best' one on the basis of the effects estimated in correspondence with one of these scenarios, the most probable for example; the effects estimated with the other scenarios will be useful for evaluating what would happen if the selected scenario did not occur and in order to adopt a precautionary viewpoint. The choice of the scenario(s) to adopt can be critical, and all DMs and Stakeholders must agree upon it, otherwise, the following phases would fail.

The adoption of an *historical scenario* (i.e. of a situation that was historically recorded) has an advantage in that it allows the comparison between what happened and what would have happened if the alternative had been implemented at the beginning of the historical horizon being considered. This information has a heightened significance for the Stakeholders and DMs because it provides a more immediate perception of the effects of a given alternative, when they have, as they often do, a direct memory of those events. If the historical horizon is too short, artificially generated scenarios can be used, provided that they are as probable as the historical one. By doing so, the estimate of the effects is statistically more reliable, but the psychological significance is lost. Both of these ways to proceed are meaningful only when one can reasonably assume that the processes that generate the scenario (e.g. the meteorological system and land use) remain unchanged into the future. If not, the scenarios have to be generated with models that describe the expected changes.

At the end of this phase the values that have been obtained for the indicators are organized in a matrix, called *Matrix of the Effects*, whose columns correspond to the alternatives and whose rows to the indicators.

We will deal with this phase in Chapter 8.

Phase 6 – Evaluation

An indicator measures, in physical units, the effect produced by an alternative on a particular leaf criterion (see page 21). Nevertheless, the 'value' that the Stakeholders attribute to an alternative, in other words the satisfaction that they get from it, is not always directly proportional to the value assumed by the *evaluation indicator*. In the 'enchanting lake' example, the 'value' that the fishermen attribute to the catch grows very rapidly for low catch yield, but very slowly at high catch yield, i.e. it 'saturates' when the fishermen feel satisfied. To account for this effect, it is necessary to translate each indicator (sometimes a group of indicators) into the 'value' assigned by the Stakeholders. This can be done by means of a *partial value function* that has to be identified through interviews with the Stakeholders. Once all the indicators are transformed into 'values', a Stakeholder (or a DM) can express the overall satisfaction that (s)he assigns to an alternative through a dimensionless *index*, whose value can be computed from the attained 'values'. Therefore, it is possible to sort the alternatives by decreasing values of the index, thus identifying the alternative that the Stakeholder (or the DM) prefers (the first alternative in the ranking).

If there is only one Stakeholder (or DM), the *optimal alternative* is found and the decision-making process is concluded. We will describe this phase in Chapter 9.

When, instead, as is almost always the case, there is more than one Stakeholder (or DM), by working in the aforesaid manner a different ranking is obtained for each one of them. The choice of an alternative then requires the expression of a judgement about the relative importance of the involved Parties (Stakeholders or DMs), i.e. requires that the Parties negotiate between themselves or that a DM (or a Super-DM) expresses her preferences among the Parties (Stakeholders or DMs). Since, however, preferences and negotiations concern subjective aspects, dealing with them is postponed to the successive phases to maintain the distinction between facts and value judgements, as required by guideline 12 in Section 1.2.1.

Phase 7 – Comparison

The aim of this phase is the identification of an alternative that is judged to be an *acceptable compromise* by all the Parties and so does not encounter opposition from anyone. Clearly a *win–win* alternative, i.e. an alternative that improves all the Parties' indices with respect to the Alternative Zero, would be the ideal solution for the decision-making process. Unfortunately, such an alternative does not always exist. In a case of irresolvable conflict between the interests of different Parties, the phase concludes with the identification of the alternatives that obtain wide *agreement* from them and listing the supporting and opposing Parties for each of them. We call these alternatives *reasonable alternatives*[36] (or *compromise alternatives*[37]). Thus, with this term we refer to the alternatives that are supported by at least one Party, are admissible (because they satisfy physical, technical and legal constraints), are economically feasible and are Pareto-efficient, i.e. they are such that it is impossible to improve the satisfaction of one Party without worsening that of another (see Section 5.2.1 and Chapter 18 of THEORY for the definition of Pareto Efficiency).

To achieve this result, first of all a series of activities is promoted, which help each of the Parties to know and understand the others' points of view, and, if such exist, the negative effects that the alternative (s)he prefers produce for the others. Once this information has been shared, the heart of the phase is the search for a compromise through *negotiations* among the Parties.

The negotiation process can take place with different procedures (see Chapter 21 of THEORY); the one used in the Verbano Project is described in Chapter 10. Sometimes it is necessary to suspend negotiations and move back to Phase 4 to design other alternatives, in view of what has been understood of the needs, aspirations and requests of the Parties; the effects of the new alternatives should be then estimated (Phase 5), evaluated (6) and brought to negotiations (7). In this way an iteration between Phases 4–5–6–7 is established. As we will see (Section 5.4), this is precisely what happened in the Verbano Project: Part B of this book is completely devoted to the description of this sequential procedure for identifying the reasonable alternatives.

Phase 8 – Mitigation and Compensation

If an alternative enjoys the agreement of the majority of the Parties, but not all them, it is important to explore whether or not it is possible to enlarge the agreement and satisfy some of the unsatisfied Parties through *measures* (meta-actions) of mitigation or compensation. To do this it is necessary to identify new (meta-)actions to include in the alternative, which act specifically on the criteria of unsatisfied Parties. Once these (meta-)actions have been identified, they must be instantiated into actions (Phase 4) and their effects estimated (5); then they must be evaluated (6) and compared (7) with the reasonable alternatives previously identified, in order to see if they actually broad the agreement. In this way one obtains a new set of reasonable alternatives, which could be examined in their turn to find new mitigation measures. Mitigation of some of the reasonable alternatives of the Verbano Project will be described in Chapter 14.

Here a recursion is established between Phases 4–5–6–7–8 (Figure 1.4), which sometimes also includes Phases 1–2–3, during which the whole set of alternatives is 'sifted' in order to single out the reasonable alternatives. Sifting ends when a reasonable alternative is identified, which is accepted by all Parties; or when it is no longer possible to identify mitigation measures or new measures that make it possible to enlarge the agreement; or simply when the time available for the decision-making process has run out. By construction, each alternative obtained in this way has the support of at least one of the Parties. All of them are presented in the summary document of the study (Chapter 15), which sums up the entire

[36]This term is taken from art. 5 of the Directive 2001/42/EC (European Commission, 2001) about Strategic Environmental Assessment (SEA).

[37]They are given this name because they emerge from a process of negotiation in which an attempt is made to find a compromise among different points of view. However, it is not necessarily possible to achieve this, so the term seems equivocal to us and we prefer the first.

development of the Project and its results. This document is the material needed to begin the next and last phase.

Phase 9 – Final Decision

This phase is put into practice only when there are one or more DMs, at a higher level than the Parties who sifted the alternatives, who are responsible for the final decision about which alternative will be implemented. It is therefore up to these DMs to choose the *best compromise* alternative from the reasonable alternatives, where 'best compromise' means the alternative that best reconciles the different interests, or simply the one upon which they manage to agree. In many cases, this phase is simply a comparison (if there is only one DM) or a negotiation process (if there is more than one) of the reasonable alternatives, which is often carried out with less formalized methods than those used in Phase 7, taking account of the customs and local culture. Sometimes, however, the DM(s) feels the need to explore new alternatives or introduce new criteria. In that case the phase is transformed into a new cycle of Phases 1–8.

In Chapter 16 we outline the way in which the negotiations between the two DMs (the Italian and Swiss governments) could take place, even if the Final Decision has not yen been taken.

1.3.2 Remarks

Often the importance of phases that have an engineering or modelling character (in particular *Identifying the Model* (Phase 3) or *Designing Alternatives* (Phase 4)) is emphasized, at the expense of more socio-political phases, like *Defining Criteria and Indicators*, *Evaluation* and *Comparison*. This is a mistake, since a correct decision can be taken only when the expectations, desires, images, knowledge, problems and fears of the Stakeholders are as well described and understood as the physical, technical and economic aspects of the system. Therefore, not only are Phases 3 and 4 of equal importance to the others, and must be considered as such, but the participation of Stakeholders should be full and continuous in all the phases, because only in this way will negotiations in Phase 7 be successful. We will never tire of repeating that if the Stakeholders do not believe in the index values that are shown them in that phase, they will never be willing to negotiate. Actually, they will probably not decline to participate in negotiations, but these latter will develop laboriously, with Stakeholders that listen passively or react aggressively and the result, even if formally it can be achieved, will not really change anything in the existing conflict.

Not all the phases are always necessary. If, for example, in Phase 2 only one criterion is identified, the decision-making process concludes with Phase 4 (or at best with Phase 5). If, instead, there is only one DM and she does not intend to activate a participatory decision-making procedure, the process concludes with Phase 6. If there is no DM above the Parties that participated in negotiations, it makes no sense to go through Phase 9.

It is important to underline that the real development of the decision-making process is not serial as Figure 1.4 might lead one to think. Besides the recursion between Phases 4–5–6–7–8, which is explicitly highlighted in the figure, many others can appear. For example, the criteria cannot actually be correctly identified if one does not known the actions being considered, since these latter produce the effects that the Stakeholders endure. On the other hand, it is not possible to identify the actions without knowing the interests at stake, and therefore the criteria. The presence of recursions is essentially due to the fact that in carrying out the decision-making process new information is produced, because it is a process of *social learning* (Renn, 1995). In view of the new information that is acquired, it is then necessary to re-examine the conclusions of the phases that were considered to be already finished and, when necessary, to modify them. In some sense the aim of the decision-making process is to increase the actors' understanding about the Project, so that they can formulate more and more precise requests and justified opinions.

The PIP procedure has to be supported: except for the phase of *Recognition*, all the phases must be handled by an appropriate set (*toolbox*) of ICT (Information and Communication Technologies) tools, which its users must perceive as part of a unique and coherent system, i.e. a Multi-Objective Decision Support System (MODSS[38]). In the literature this term is sometimes used only in relation to Phases 6 and 7, but we consider this to be inappropriately restricted.

The PIP procedure has to be managed: in all the phases the Stakeholders must be assisted by the Analyst, for the more technical aspects, and helped by a Facilitator during the negotiations and for all organizational and relational aspects (see Appendix A10 of THEORY).

Finally, it is important to underline that there is a difference between a phase and the method used to implement it in a given context. The phase defines a methodology, i.e. a set of methods and the rules to choose between them. Thus there is not a one-to-one, but a one-to-many relationship between phases and methods. The aim of THEORY is to illustrate these very relationships from the theoretical standpoint, the aim of this book is to describe the application of some of those methods to the planning of Verbano Water System.

1.3.2.1 The FOTE paradigm

Negotiations and relations among the Parties are easy or difficult in relation to the degree to which they share and exchange information about the problem, the system and their own personal interests; that is the degree to which they adopt a paradigm of *Full Open Truthful Exchange* (FOTE) (Raiffa et al., 2002). This is why the PIP procedure does not begin with negotiating the alternatives, but instead with an information exchange (Phase 0). It goes on with a participatory definition of the actions (Phase 1), the enunciation of the interests (Phase 2), the identification of a shared model of the system (Phase 3). Not always the Parties accept to adopt the FOTE paradigm, but the Analyst must always suggest its adoption, because, in the opposite case, the Parties might not make the most of their potential synergies. This is the approach we followed in the Verbano Project, where the FOTE paradigm has been adopted by the Stakeholders.

1.3.2.2 Data

Do not underestimate the essential role that the data and the methodologies for their acquisition play in the decision-making process. The data are used, in a qualitative way, in the phases of *Recognition* and *Defining Actions*, and in a quantitative form in Phases 2, 3, 4, 5 and 8. Their quantity, availability and accuracy are essential to the success of the decision-making process, but even more important is that all the Parties believe that the data are valid and meaningful. The credibility of the model, which is the basis for the credibility of the evaluation of the alternatives, is founded upon this belief, without which the negotiation process is a fruitless exercise. In simple terms: the entire decision-making process depends on the social acceptability of the data.

[38]This is the acronym we have adopted for the DSSs (Decision Support Systems) that are used to take decisions in a way that is coherent with the IWRM paradigm.

1.3.2.3 Uncertainty

Ignorance is being unaware that our knowledge is imperfect. An imperfect knowledge implies uncertainty and uncertainty generates apprehension. For this reason, DMs often have the tendency to remove the problem of uncertainty: they want scenarios to be deterministic and models to provide exact estimates, so that their evaluations will be perfect. However, hiding uncertainty is none other than a form of ignorance. Thus, in many phases of the PIP procedure the problem of treating *uncertainty* arises. Uncertainty is produced by corrupt, insufficient or scarce information, and by the errors that are committed unknowingly. We will see in the following that all these causes can be represented as the effects of *disturbances*, which can assume different forms: disturbances are *stochastic* when we know or we can estimate their probability distributions; they are *uncertain* when we know only the set of values that we guess they might assume. The form of a disturbance depends upon the source that generates it. For example data collection generates the most common uncertainty: *measurement errors*, which are always described as stochastic and afflict all the phases in which the data are used. In Phase 1 uncertainty appears also in the description of the actions, since the way in which they will actually be implemented is not always certain (implementation uncertainty); this uncertainty is not only due to implementation aspects, but also to institutional inertia. In Phases 3 and 6 one must account for the disturbances that can make the design and evaluation scenarios uncertain: it is when confronted with this type of uncertainty about the future that DMs and Stakeholders reveal their aversion to risk, which we discussed in the description of Phase 2. In Phase 3 it is necessary to take into account the so-called *process errors*, i.e. the eventuality that the model does not perfectly describe reality.

The effects produced by all these disturbances combine to generate the uncertainty that afflicts the indicator values that make up the Matrix of the Effects. One must keep this fact in mind in Phases 6–9.

1.4 The decision-making levels

Once the best compromise alternative has been selected (Phase 9 of the PIP procedure), it has to be implemented (see Figure 1.2): this is achieved by implementing the structural and normative actions it includes, and applying the regulation policy, if this has been designed, at the scheduled time instants. Since this policy may leave some degrees of freedom to the Regulator (see Section 5.2.1.1 on *set-valued policies*), a decision-making problem must be formulated at the *management level* as well, though the degree of freedom that the policy allows is much less than the degree that the DM had when the problem was formulated at the *planning level*, i.e. at the level that we have been considering until now. The best compromise alternative is therefore at the same time the conclusion of the decision-making process at the planning level and the starting point for the decision-making process at the management level, which is renewed periodically, often daily, on the basis of the new information which is obtained as the time goes on. But the practical implementation of the decision may require some more technical and operation decisions, i.e. establishing which actions to undertake in enacting the decisions taken at the management level. For example, one must establish how to operate the control gates in the course of the day, so that at the end of that period the volume determined at the management level has actually been released. This is the task of the *operational control level*.

In conclusion, three are the levels at which decisions are taken in a water system: the planning level, the management level and the operational control level (Figure 1.6). Moving

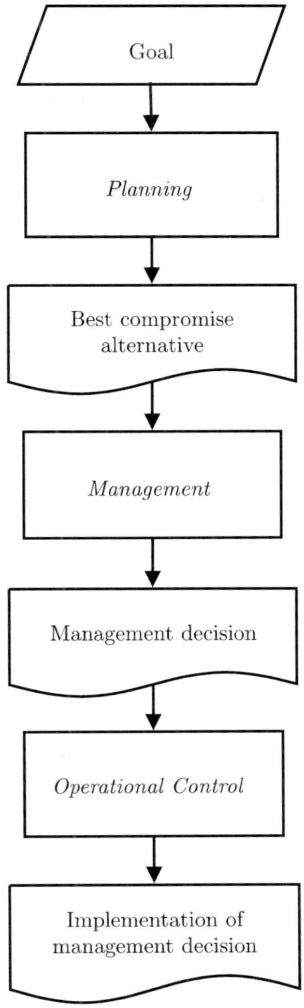

Figure 1.6: The decision-making levels in a water system.

from one level to another changes not only the object of the decision but also the time horizon that it is concerned with, and so the horizon over which the system is observed and objectives are defined. The Verbano Project deals with the planning level, but the formulation of the decision-making process at the management level is outlined in Chapter 17. The operational control level does not fall within the scope of this book.

All three levels must have input from a monitoring and data collection system, which includes the validation and pre-treatment processes of the data. In particular, data availability is crucial at the management level to take daily decisions through the regulation policy.

1.5 A multi-level MODSS: TWOLE

As we have already said, at the first decision level, i.e. at the planning level, the decision-making process is carried out through the PIP procedure, which is supported by a software

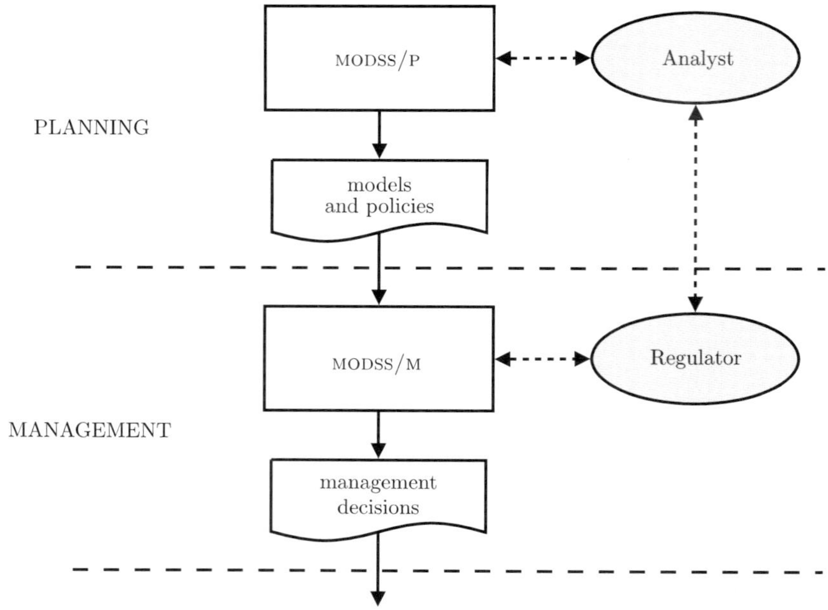

Figure 1.7: The two decision-making levels of TWOLE (Soncini-Sessa et al., 1999).

system called MODSS. A similar tool is useful for supporting the decision-making process at the management level, too. Two MODSS are thus necessary: one for the planning level (MODSS/P), whose user is the Analyst, and one for the management level (MODSS/M), whose user is the Regulator.

However, from the technical point of view, it is not advisable that the two MODSS be independent monads because the policies produced and the models identified with the first (MODSS/P) have to be utilized in the second (MODSS/M). This is why one ought to think of an MODSS on two levels (Soncini-Sessa et al., 1990). The acronym TWOLE (TWO LEvels) given to the system we have developed comes from this idea (Figure 1.7). A system of this type has another advantage in that the models and data are used daily, so that in the future, if it is necessary to return to the planning level to take new decisions, they will be up to date and available.

The prototype version of TWOLE/P was used to support the development of the Verbano Project. Its architecture is not described in this book, the interested reader can find a detailed description in Chapter 24 of THEORY.

1.6 Monitoring

Once the best compromise alternative has been implemented and management has begun, the effects that are produced should be monitored (see Figure 1.2) continuously over time, in order to make sure that the real effects are actually those that were foreseen. Should they not be, it is necessary to open a new cycle of intervention. For a general treatment of this issue, see UNECE (1990) and Wiersma (2004), as for the monitoring in Verbano Project see Section 15.5.

PART A

SETTING THE VERBANO PROJECT

Chapter 2

Reconnaissance: territory, Stakeholders and regulation

EB, FC, AP and RSS

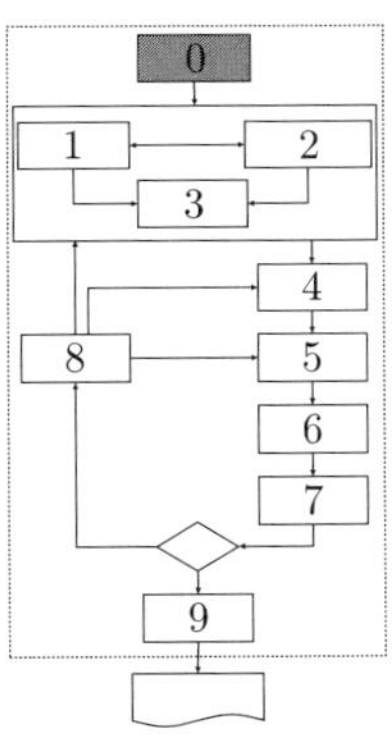

Figure 2.1: The phases of the PIP procedure. The *Reconnaissance* phase, described in this chapter, is highlighted.

Verbano (often called Lake Maggiore) is outstandingly important, not only from a socio-economic, but also from an environmental point of view. Over time there has been an increase and diversification of the ways its waters are used. The first major projects – the construction of the Sforzesco and Langosco waterways off the Ticino effluent in the 15th century and 17th century respectively – were carried out to favour navigation, especially for the transport of goods and raw materials. Afterwards, interest shifted to agricultural and industrial production and as a result, new canals and hydropower plants were constructed. The use of the lake's waters for irrigation and hydropower led to a need to make the availability of the required flows as reliable as possible, something that is difficult to achieve in a natural lake. The dynamics of the outflow from a natural lake is driven by the inflow, which is uncertain by nature. For this reason, as early as the beginning of the 1800s there was thought of transforming Verbano into a regulated lake, with the construction of a *dam* across the Ticino effluent. The first studies for the creation of the structure date back to 1865, with the Villoresi–Meraviglia project, but not until 1902 was the first complete design drawn up. The construction was postponed even further, and so the regulation dam, called *the Miorina dam* from the name of the place in which it is situated, did not officially begin operating until 1st January 1943. From that date on Verbano has been a *regulated lake*.

The use of Verbano's waters, and the related interests, go beyond those that are mentioned above. To appreciate the context of the Project presented in this book, it is essential to consider the needs of the lakeside populations, who are worried about the lake flooding, and the safeguarding of natural environments, whose perceived importance has increased with the passage of time, so that today it would be impossible to ignore them.

One might well suppose that this is a complex situation with numerous points of conflict, the most important and evident of which is certainly the conflict between the lakeside populations, on the one side, and the irrigators and hydropower producers, on the other side. In fact, the transformation of Verbano into a regulated lake generates an increase in lake levels just in the period when the inflow is most abundant and the risk of *flooding* is greater.

The conflict of interests motivated some of the involved people to formulate intervention proposals, mostly aimed at resolving the problems closest to their own interests. The evaluation of the effects of these proposals was usually carried out by the proponents themselves, with the result that the other interested parties, fearing that the evaluation was partisan, systematically discounted them.

The first tentative definition of the *Project Goal* can be stated as follows: find an intervention option that takes care of all the issues mentioned so far, thus increasing the satisfaction of all the Stakeholders.

The *Reconnaissance* phase starts with a description of the characteristics of the territory and the identification of the *Stakeholders* involved, using the 1999 situation as a reference point (the year that the Project began).

Since the Miorina dam is an important physical division, we keep the upstream and downstream parts of the system distinct from each other in our description. Therefore, in the following the terms *upstream* and *downstream* will be used with respect to the Miorina dam.

2.1 The territory

2.1.1 The catchment

Verbano has a surface area of about 211 km^2, 170 of which are in Italian territory while the rest are in Swiss territory. Its *catchment*, whose outlet is the mouth of the lake at Sesto Calende, extends for 6598 km^2 (see Figure 2.2) and is equally divided between Italy (3229 km^2) and Switzerland (3369 km^2). The ratio between the surface area of the catchment and that of the lake is 31 to 1, the highest ratio of all the great Alpine lakes[1]: it follows that there can be large *floods* and very rapid excursions in the lake level.

The territory in the catchment is Alpine, characterized by steeply sloping mountainsides and an average elevation of 1270 m a.s.l.[2] The average elevation of the lake is about 194 m a.s.l.

The catchment can be subdivided into 18 *tributary catchments*: the principal ones are the upstream Ticino, the Maggia-Melazza, the Toce and the Tresa. They include other natural lakes – the largest of which are Lugano, Orta and Varese – and numerous artificial reservoirs for hydropower production, both in Italy and Switzerland.

The pluviometric regime in the catchment is typical of sublittoral Alpine zones. In fact, the *trajectory*, i.e. the time path, of the (median) *inflow* to the lake, shown in Figure 2.3 for the years 1974–1998, has a minimum in winter and peaks in autumn and late spring. Precipitation in general is high and rather unevenly distributed over the tributary catchments:

[1] For example, Lake Garda has a 6 to 1 ratio, Lake Como 31 to 1, Lake Lugano 13 to 1, and Lake Geneva 14 to 1.

[2] The reference assumed for the heights is the Italian one. This distinction is necessary since the Italian and Swiss reference *ellipsoids* differ by 0.352 m in the zone of Verbano.

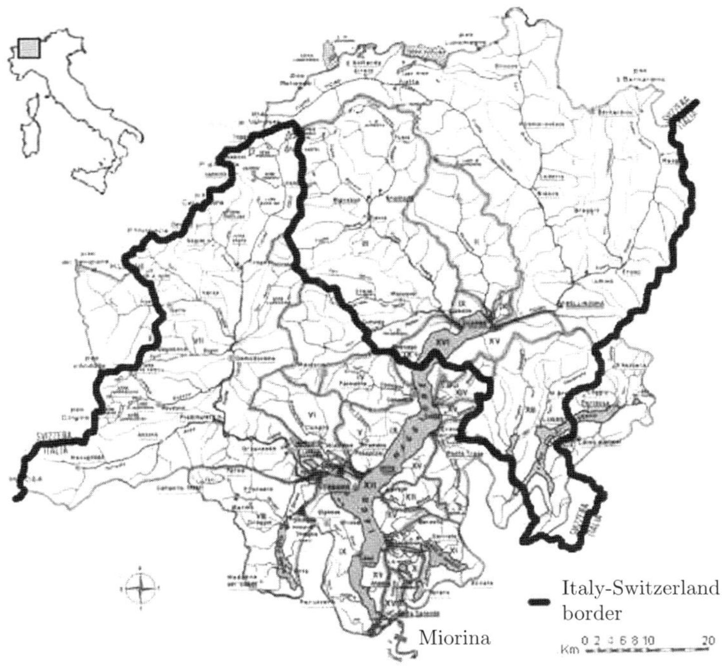

Figure 2.2: The Verbano catchment. The black line shows the boundary between Italy and Switzerland, while the boundaries of the tributary catchments are shown in grey.

the average annual precipitation over the entire basin is 1692 mm, but the Toce valley contributes with more than 2000 mm (the highest in the entire River Po catchment), while the upper Ticino valley contributes with only 700÷1000 mm). Most of the inflow to the lake comes from the rivers Toce, Maggia-Melezza, Ticino, Verzasca and Tresa, which together drain 77% of the entire catchment.

However, the information contained in Figure 2.3 is not sufficient to understand the nature of the floods. One can only deduce that the peaks reached in late spring are on average higher than those reached in autumn. But the spring floods are produced essentially by melting of snow pack, and thus they are slow and develop over many weeks. Even when precipitation contributes to them, the runoff towards the lake is slowed by the presence of snow on the ground in a significant part of the catchment. On the other hand, the autumn floods, which are completely caused by precipitation, are fast and sudden, particularly if they come after some days of rainfall, when the soil is saturated. This is well demonstrated by the annual trajectory of the standard deviation of the inflow, shown in Figure 2.4. It can be seen that the autumn floods are much more dangerous than the spring ones and that the menace of autumn floods is even more acute because, while spring floods can be easily forecasted (from the moment in which the air temperature rises, it takes about three or four days before the snow pack is saturated and the runoff begins to reach the lake), the autumn ones are fast and unpredictable.

The trajectory of the inflow does not depend only on the rainfall regime, but also on the presence of *reservoirs* for hydropower production in the lake catchment. This point is important, so we will focus on it in the rest of the section.

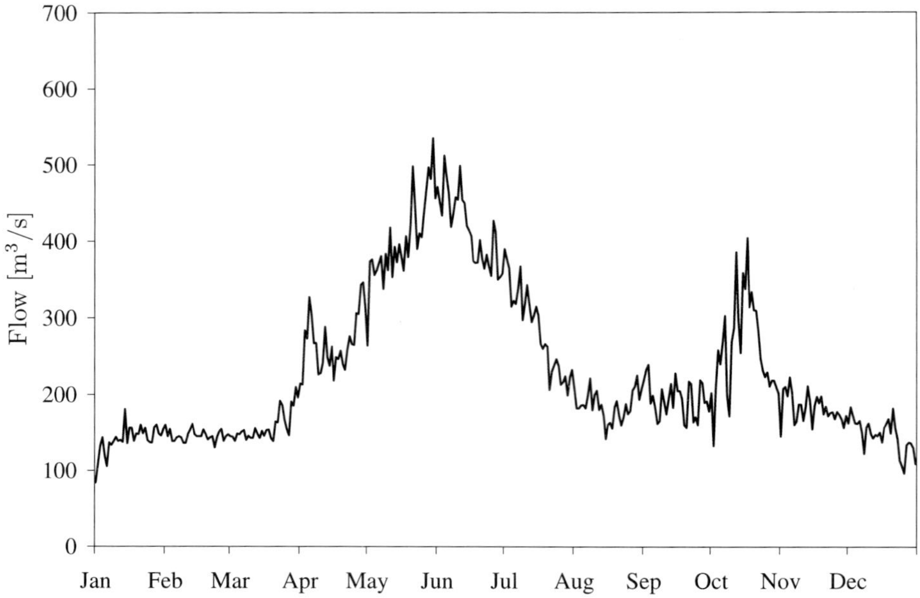

Figure 2.3: The annual trajectory of the median inflow to Verbano (years 1974–1998).

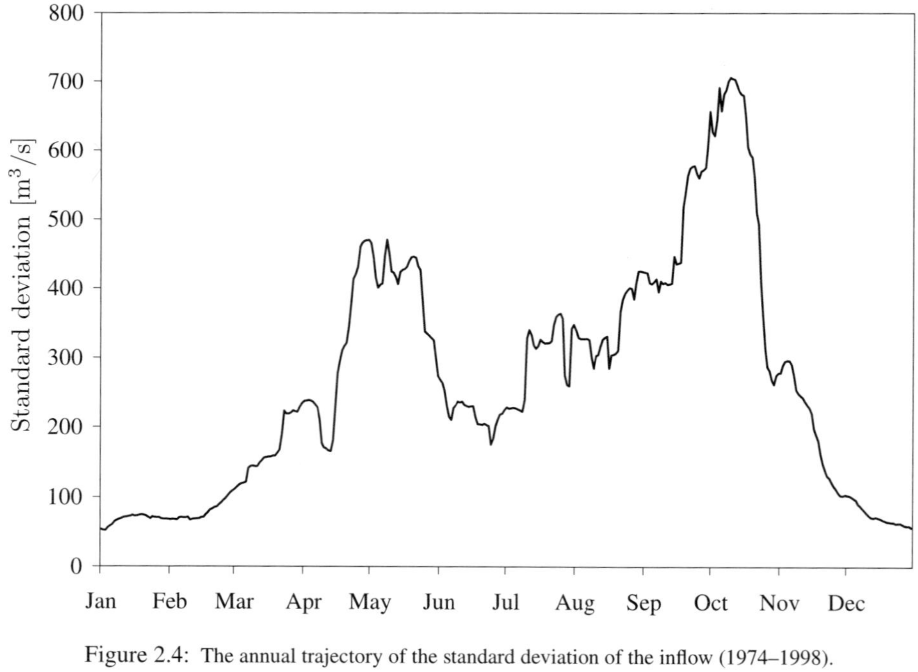

Figure 2.4: The annual trajectory of the standard deviation of the inflow (1974–1998).

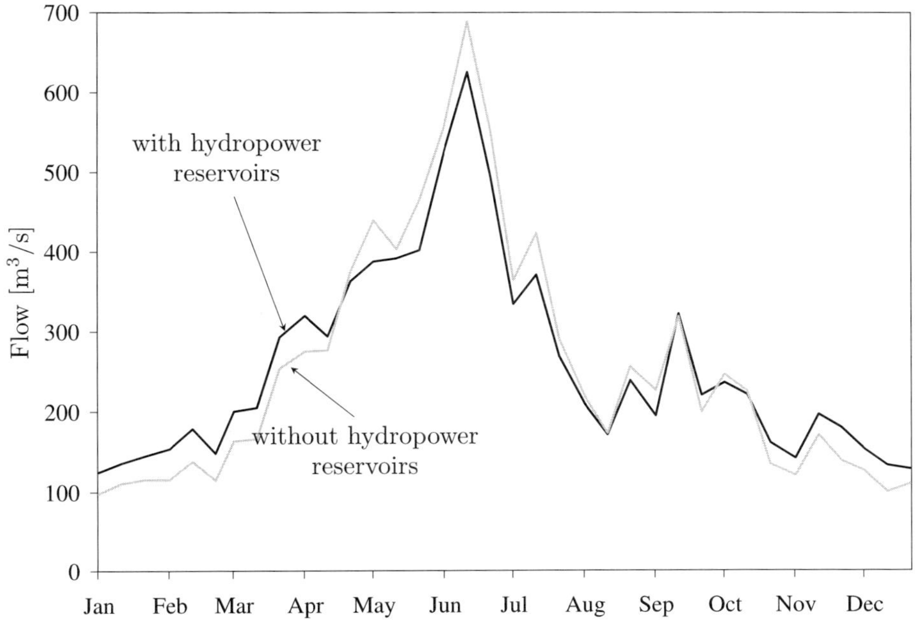

Figure 2.5: A comparison of the historical inflow, influenced by the presence of hydropower reservoirs, with the inflow that would have occurred in their absence (median of ten-day inflow, 1970–1978). Re-elaboration from Citrini and Cozzo (1981).

The construction of water reservoirs for the production of hydropower began in 1911 with the creation of Lake Delio, situated in Italian territory, and it continued until the beginning of the 70s. The last reservoir came into service in 1973; only since that year has the configuration of the catchment been stable and this explains why only data after 1974 was considered to produce the figures reported in this chapter. Today the total area subtended by reservoirs is 1075 km^2, equal to 16% of the area of the entire basin. Their overall storage amounts to more than 600 million cubic metres, of which more than two thirds is located in Swiss territory. To appreciate the significance of this number, imagine that if this whole volume of water were released and the lake outlet were closed, the level of Verbano would rise more than 2.5 m (Citrini and Cozzo, 1981). It is thus easy to understand how the regulation of these reservoirs can significantly influence the inflow regime.

The effects of regulation can be observed on different time scales, first of all on the annual level. In Figure 2.5 the trajectory of the median of the ten-day inflow, recorded in the years 1970–1978,[3] is compared with the inflow that would have occurred if there had been no hydropower reservoirs. This hypothetical inflow was reconstructed by adding (or subtracting) the total volume that was stored (or released) by the reservoirs over a ten-day period to the measured inflow in the same period (the *concentration and travel times* are less than 24 hours). One may observe that the effect of the reservoirs' regulation is to

[3]The choice of this years is not coherent with what was stated previously. The reason is that the time series of the levels of the hydropower reservoirs, and therefore those of the stored/released volumes, do not go beyond 1978. Thus we decided to consider data from as early as 1970, rather than beginning with 1974 as would have been more correct, so that the series would not be too short. The variation in the overall storage capacity of the reservoirs between 1970 and 1974 is only a few percent.

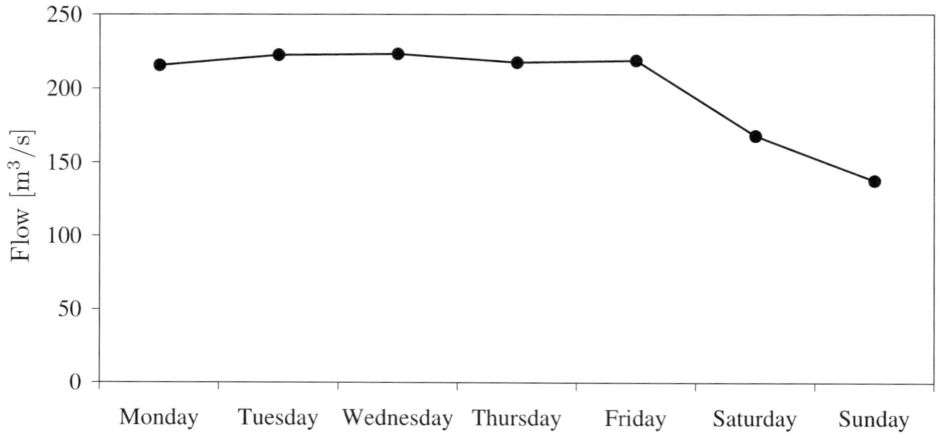

Figure 2.6: Weekly pattern of the median inflow to Verbano in the October–April period (years 1974–1998).

increase the inflow in the months from October to April and to reduce them in the other months. This is because the demand for electricity, and thus its price, is higher in the winter period, so that the reservoirs are regulated in order to transfer water volumes from the period of the snow melts (spring–summer) to the winter time.

In the October–April period, in which, as we have seen, the reservoirs release water, the inflow to Verbano is periodic of a weekly period, with a reduction in the flow on Saturdays and Sundays (Figure 2.6). This is due to the weekly pattern of the electric power demand, which diminishes on weekends. The same effect occurs also in other months of the year, even if it is less evident.

2.1.2 The downstream territory

Verbano's effluent is the Ticino (to be distinguished from the river of the same name that enters the lake in the upper north corner), which flows into the River Po a little downstream from the city of Pavia, after travelling about 100 km and dropping 130 m. On the banks of the Ticino there are important urban centres, including Pavia and Vigevano, and zones with a great environmental value. The river feeds a dense network of irrigation ditches and canals (Figure 2.7), the most important of which are the Regina Elena Canal, which departs from the Ticino a little upstream of the Porto della Torre dam, and the Villoresi and Industrial Canals, which branch off from the Panperduto diversion dam. A fraction of the flow in the Industrial Canal feeds the Naviglio Grande waterway, which once derived directly from the Ticino, downstream from the Busto–Novara railway bridge, near Oleggio. The remaining flow is returned to the Ticino through several different drains, according to the conditions at the time. This water, along with the water that has remained in the *riverbed* of the Ticino after passing Panperduto and which has been enriched in the meantime by drains and *resurgences*,[4] is diverted[5] into the Langosco and Sforzesco waterways.[6] With

[4]Distributed underground flow that reaches the surface in the Ticino valley; on average it increases the Ticino flow by 1 m^3/s per kilometer.

[5]Through the so-called *filarole*, which are temporary dams of sand and gravel that deviate part of the flow into the canal.

[6]Both these canals and the Naviglio Grande have been constructed as waterways but are not used for that purpose anymore.

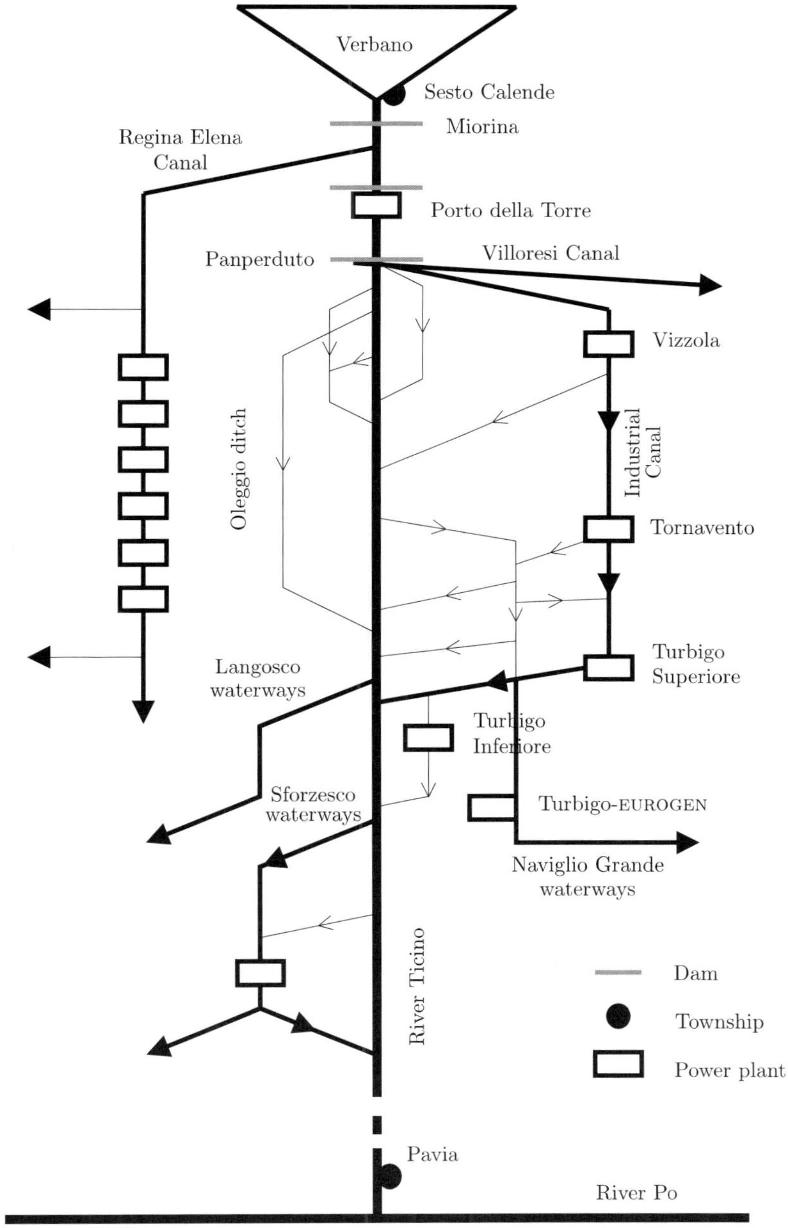

Figure 2.7: The network of canals that is fed by the Ticino effluent.

the term *downstream territory* we mean the whole territory served by the Ticino and the network of its canals.

The presence of these canals has contributed over the centuries to the development of a rich agricultural system, based on the production of rice, corn and wheat in the vast regions of Lomellina, Novarese, Vercellese and Pavese. A new agricultural-irrigation ecosystem has thus substituted the original natural ecosystem.

The waters of the Ticino are not used only for irrigation but also used to produce hydropower, through run-off-river power plants. The Industrial Canal, mentioned above, was constructed at the end of the 1800s by the Vizzola Co., to feed the power plant of the same name, which came into service in 1899. Subsequently, two other plants were constructed along the same canal (Tornavento and Turbigo Superiore) and, along the Ticino, the Porto della Torre plant was built, which exploits the hydraulic head between Miorina and Panperduto, through a dam constructed upstream of the latter. In the 90s, six small plants were built on the Regina Elena Canal.

Diverting water from the River Ticino to the canals not only diminishes the flow in the river downstream from the canals' intake but, a few weeks after the start of the extraction, it enriches the river flow: part of the water that was diverted, in fact, returns directly to the river through the drainage system of the irrigation network, and part of it feeds the water table, which in turn is drained by the river.

The latter effect is due to the fact that, except in particularly dry years, the ground water level is higher than the surface level of the river, since the Ticino is cut deeply into the surrounding terrain, especially in the first stretch. This is also the reason why residual water from the irrigated lands near the Ticino drains directly into the river itself.

The Ticino finally reaches Pavia and, soon afterwards, flows into the Po at the Becca Bridge. The level of the Ticino at Pavia is usually higher than the level of the Po by about one and a half metres, but the difference can reduce to fifty centimeters when the Po is in flood. In such conditions, in order to flow out into the Po, the Ticino at Pavia must reach a level that is much higher than the level that the same Ticino flow would reach in normal conditions: the Po flows back into the Ticino (*backflow effect*). Therefore a particularly dangerous case is when the peak of a flood, flowing down the Ticino, reaches the confluence at the same time as a flood wave travelling down the Po. When this occurs, the river levels at Pavia are greatly elevated and the united wave from the two rivers increases the risk of flooding along the Po valley, particularly in the Polesine region.[7]

2.2 The Stakeholders

By the term *Stakeholders* we mean those people whose interests (i.e. needs, desires, worries and fears, concerns) are influenced, either positively or negatively, by the level of Verbano or by the flow released from the Miorina dam. As with the territory, we divide them into upstream and downstream Stakeholders.

[7] The low-lying region of the Po delta.

2.2.1 Upstream Stakeholders

2.2.1.1 The socio-economic system

Among the upstream Stakeholders there are the shoreline inhabitants, the *laghée*,[8] who own property that could be damaged by water during lake *floods* or whose activities could be negatively influenced by the effects induced by high lake levels (for example, because their shops become inaccessible or because of traffic congestion caused by flooded roads). All the tourist operators are also shoreline Stakeholders (hotel and restaurant owners and people who run recreational activities, for instance) because they use the lake environment as a tourist attraction and the lake surface as a place to organize water sports. We assume that these Stakeholders can be represented by lakeside townships[9]: 44 in all, of which 14 are in the Ticino Canton in Switzerland, and 30 in Italy, some in Lombardy and some in Piedmont, the two Italian Regions with Verbano shores. There are 34 529 inhabitants overall in the Swiss townships, and 151 388 in the Italian ones.[10] The most important centres are Locarno (14 506 inhab.), in Swizerland, and Verbania (30 517 inhab.), Arona (15 543 inhab.) and Luino (14 883 inhab.), in Italy.

The lake is also used by professional fishermen and by the Lake Navigation Company to transport goods and people, with boats and ferries. Navigation would get a big boost if a route were opened (or reopened, since in the past it existed and thrived) between Locarno and Venice, via Lake Maggiore, the River Ticino and the River Po.

2.2.1.2 The environment

The lake significantly influences the climate and the vegetation in the neighboring areas and supports the survival of unique biotopes. The Fondotoce Natural Reserve includes the most extensive reed bed on the Piedmont side of Lake Maggiore; the Canneti di Dormelletto Natural Reserve, situated on the south-west bank of Lake Maggiore, is important for its fauna; and the zone called the Bolle di Magadino is a natural area of international interest located on the delta of the rivers Ticino and Verzasca, in Switzerland, which is managed by the Bolle di Magadino Foundation.

2.2.2 Downstream Stakeholders

2.2.2.1 The socio-economic system

Just as in the upstream area, downstream inhabitants who own property that could be damaged by floods in the Ticino effluent or whose activities could be influenced by the effects induced by the variations of river flow are counted among the Stakeholders. The most expensive flooding damage occurs in the townships of Vigevano and Pavia. In Pavia the floods are not caused only on the flow released from the lake, which travels down the Ticino, but also on the state of the Po.

Among the downstream Stakeholders the so-called *water users* are also included, i.e. the associations, companies, and public bodies that use the water that is released from the Miorina dam. Among them, it is important to distinguish *irrigators* and *hydropower producers*, because they use the water in different ways.

[8] From time immemorial, in local dialects this term has been used to refer to the people who live around Verbano.

[9] In this work the term is used to mean an administrative unit of population.

[10] Data from 1990, source UST (Berna) and ISTAT (Roma).

The irrigators are subdivided into:

- The *Old Users*. The term denotes all those users who gained a right to extract water to irrigate vast territories, on both sides of the river, in Lombardy and Piedmont earlier than 1884. They are not united in a Consortium and thus do not have elected representatives. Originally they derived water through the Grande, Langosco and Sforzesco waterways, the Oleggio irrigation ditch and numerous other minor derivations, but nowadays almost all the water they use is derived through the Industrial Canal, which directly feeds the Grande waterway and indirectly the Langosco and Sforzesco waterways. The first waterway is managed by the East Ticino Villoresi Consortium, while the East Sesia Irrigation Association is responsible for the others. The Oleggio ditch and the other minor irrigation ditches take water directly from the Ticino; their demand is very modest (25 m^3/s on average), and is almost always satisfied by *resurgences* into the Ticino riverbed. For this reason, the irrigators that take water from these minor ditches will not be considered in the Project.

- The *East Ticino Villoresi Consortium* (in the following indicated as the *East Ticino Consortium* for short) brings together the irrigators on the Lombardy side of the Ticino. It was founded recently (1999) from the fusion of the pre-existing Eugenio Villoresi Consortium and the Basso Pavese Consortium.

- The *East Sesia Irrigation Association* (in the following indicated as *East Sesia Consortium* for short) includes all the users of state waters for irrigation purposes, who are located in the district that is delimited by the rivers Sesia, Ticino and Po. The Consortium uses several different sources for irrigation, but only the Regina Elena Canal is of interest to us, since it is the only one that is fed by waters from Verbano.

There are two hydropower producers:

- The National Electricity Board (called ENEL, an acronym that stands for *Ente Nazionale per l'Energia Elettrica*): owns the plant at Porto della Torre, immediately downstream from the Miorina dam, and uses the Industrial Canal to feed the hydropower plants at Vizzola, Tornavento and Turbigo Superiore (for an overall capacity of about 70 MW) and to cool the important Turbigo-EUROGEN thermoelectrical plant (1230 MW);

- The *East Sesia Consortium*: built six small powerplants on the Regina Elena Canal, with which, since 1996, it has been supplementing agricultural activities through the production of hydropower.

The East Ticino Consortium would also like to build some small powerplants to exploit the difference in level along the course of the Villoresi Canal. Nevertheless, today this is still more of a wish than a project, and so we will not take account of this possibility. Clearly, the management of the lake would have to be reviewed in an appropriate way on the day that these plants were built.

Both types of water users, irrigators and hydropower producers, are characterized by their water *demand*, which varies in time and which, as a fairly good approximation, we can consider periodic of one year. Defining their demand values is not a trivial task, as one might think at first.

From a legal point of view, the flow that may be derived by each user is defined by an *Abstraction Licence*. The form of the licence is not the same for all the users: for some it is defined by the maximum amount of water that can be diverted in a given time interval, for others it is an average flow that can be diverted during certain periods. Nevertheless, the flow that is actually abstracted from the canal is not always what is stipulated by the licence: the procedure for licence approval is very long (even scores of years), and so sometimes water is abstracted on the basis of a licence application that has been put forward, but not yet approved, while in other cases the user may have a provisional licence. This is the case for ENEL: its licence for the Industrial Canal is 120 m^3/s, but, since it was temporarily raised to 138.5 m^3/s during the years 1977–1998 (with short interruptions), ENEL has continued to extract the higher amount, having applied for a definitive licence in 1999.

Thus, it is not possible to define the users' demand by referring to existing licences alone. Note that this would not serve our purposes, anyway: our task is to evaluate alternatives which, if approved, would be enacted in the future; it is thus essential to estimate the flows that the various users might demand on that future date. To estimate these future flows, we used the water demand declared by each user, provided that it was accepted by all the other water users and coherent with the maximum flows that the canals can physically carry and that the plants can run through their turbines. The patterns of demand determined in this way are shown in Figure 2.8. Each demand is distinguished by the acronym of the canal through which the water is derived and by the suffix *irr* (irrigation) or *hydro* (hydropower), according to the use it is destined for. In this way, the Old Users' water demand, for example, which is satisfied through the Industrial Canal, is individuated with the symbol *IC,irr*.

It is important to bear in mind that, both for the Industrial Canal and for the Regina Elena Canal, at every time instant the total water demand is the maximum between the irrigation demand and the hydropower demand: the reason is that the flow diverted into the canal is first sent to the plants, to be run through the turbines, and is then delivered to the irrigators. Instead, the Villoresi Canal has only irrigation demand.

From Figure 2.8 it can be noticed that the Regina Elena and Villoresi Canal have two dry periods, which mark the beginning and the end of the irrigation season. In the territory served by the Regina Elena Canal, rice is the main crop and the season begins with the flooding of the rice paddies at the end of March, while in the territory served by the Villoresi Canal the season opens around mid-April. In this district principally corn is grown and so the irrigation does not terminate until the beginning of October, a little later than in East Sesia.

2.2.2.2 The environment

The Ticino valley is scored by a complex network of canals, oxbow lakes and resurgences which, along with the river, constitute a unique landscape specimen that has great environmental value because it is a natural environment molded by thousands of years of human activity. This is the reason why it has been protected with the institution of two regional parks, one in Lombardy and the other in Piedmont (*Parco Lombardo del Ticino* and *Parco Naturale della Valle del Ticino*), which will be simply named *Ticino Parks* in the following.

A detailed description of all the Stakeholders is contained in Section 2.2 of the DVD.

2.3 The regulation of the lake

Before the construction of the Miorina dam the waters flowed freely from the lake. We must often refer to this condition and so it is convenient to give it a name: we call it the *natural regime* and the corresponding lake is called the *natural lake*. In a natural lake, at every time

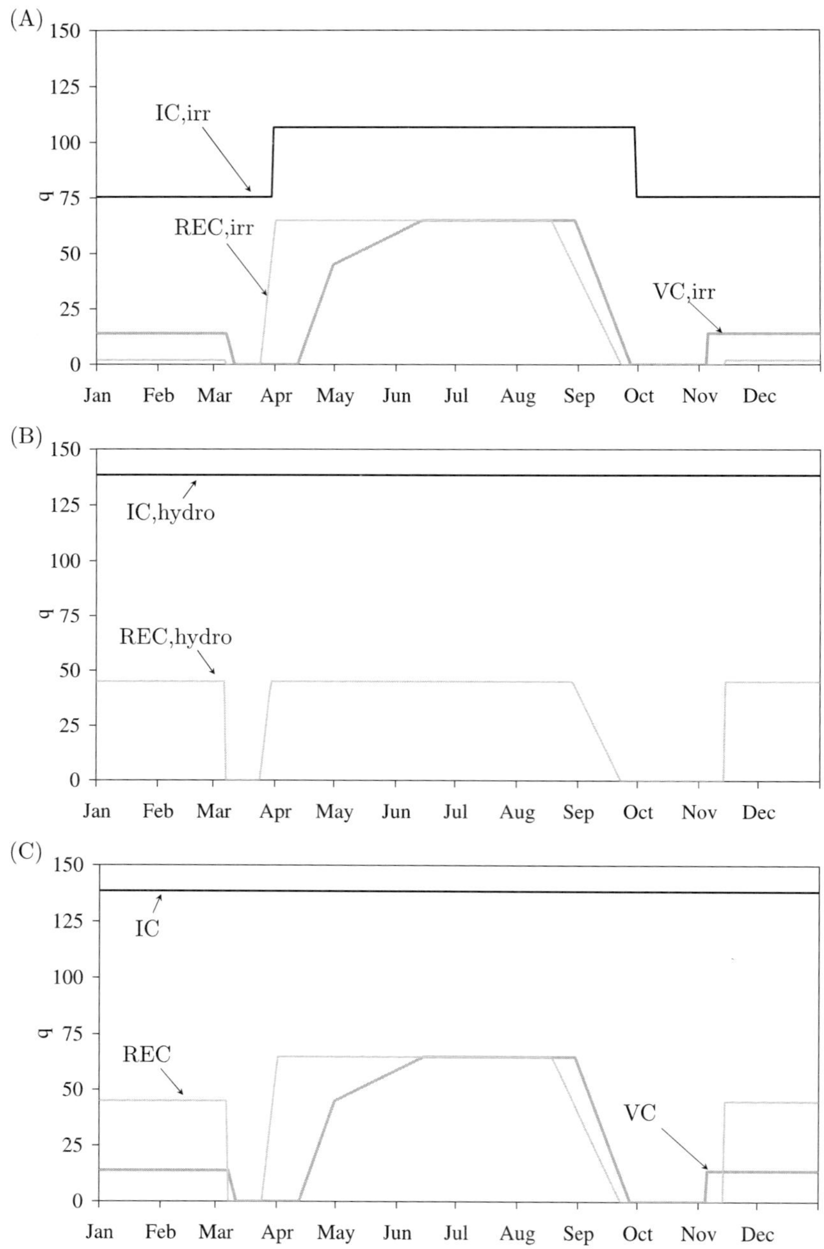

Figure 2.8: Irrigation demand (irr) (A), hydropower demand (hydro) (B) and total demand (C) patterns for the Regina Elena (REC), Villoresi (VC) and Industrial (IC) Canals.

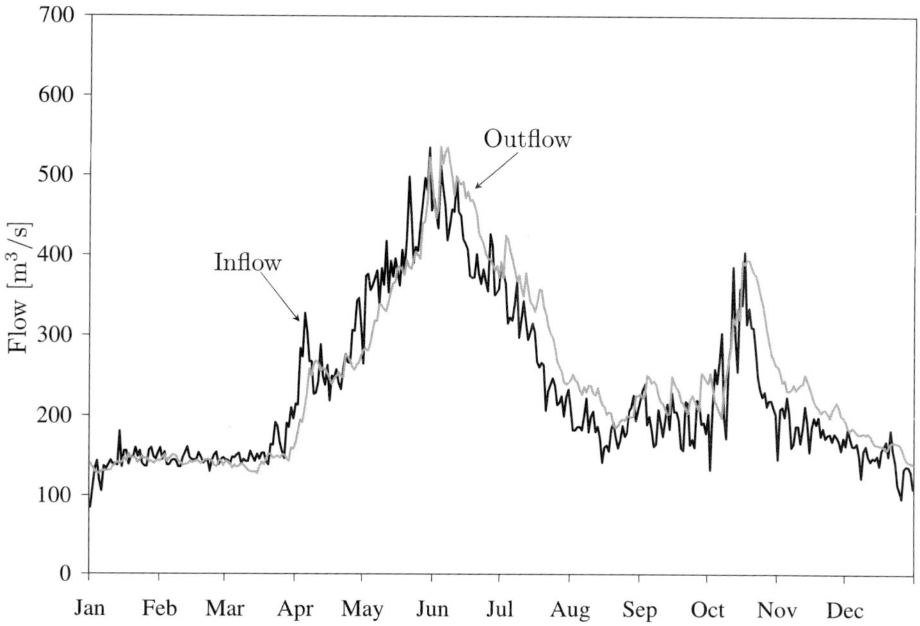

Figure 2.9: Median inflow and outflow in natural regime (years 1974–1998).

instant, the outflow is related to the lake level through a relation called the *natural stage–discharge relation*. From a theoretic standpoint, the stage–discharge relation can be defined only under conditions of *permanent flow*, but generally only modest errors occur if it is used also in conditions of *non-permanent flow*, provided that the variation of the lake level is relatively slow.

In natural regime, the dynamics of the level and the outflow[11] is driven exclusively by the inflow. Figure 2.9 shows that the trajectory of the outflow basically follows that of the inflow: more precisely, the outflow is a little lower than the inflow, when the latter is increasing, and it is a little higher in the opposite case. This delay is provoked by the *detention* capacity of the lake, a phenomenon which can be understood by observing that an increase of the inflow produces an increase of the outflow only after it has produced an increase of the lake level. This means that rapid variations of the inflow are not reproduced in the outflow; the trajectory of the outflow is thus smoother than the trajectory of the inflow.

In spite of the lake's detention capacity, the water users felt that the variability of the natural outflow was too great and wished for a more reliable supply. Furthermore, with the construction of larger canals (Villoresi in 1884, Industrial in 1897), it became clear that the natural flow was not able to satisfy even the average users' needs. Figure 2.10 shows that in three periods (winter, the submersion of the rice paddies, and summer) the median of the outflow is lower than the total water demand (irrigation demand plus hydropower demand). If follows that for more than 50% of the time, the demand was not satisfied, causing a *deficit* or *water failure*.

[11]The series of measures of the natural level and outflow end in 1943, when the dam came into service. From that year on, it has been the task of the Ticino Consortium, the body given responsibility for the regulation of the lake, to reconstruct the level and the outflow that would have occurred in the natural regime each day.

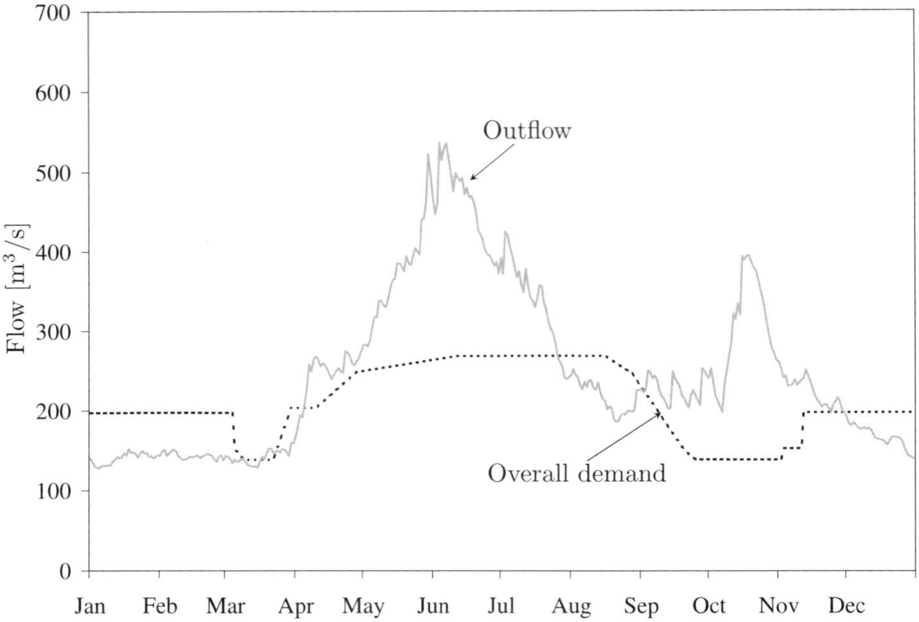

Figure 2.10: A comparison of median natural outflow (years 1974–1998) and overall water demand (from irrigators and hydropower producers). Note the failures in the winter period, during the submersion of the rice paddies (March–April) and in summer.

To solve the problem, the construction of a dam across the Ticino effluent was planned, with the aim of storing water in the lake in the periods when the inflow is high, and using it in the following deficit periods. Note, in fact, that in Figure 2.10 the deficit volume (the area between the demand and the outflow, when the first is higher than the second) are clearly inferior to the surplus volume (the area between the outflow and the demand, when the first is greater than the second). Thus, if the surplus volume could be stored in the lake and delivered when a deficit occurred, the demand could be totally satisfied; to achieve this, however, a great storage volume would have to be available.

The lakeside dwellers, both the Italians and the Swiss, were opposed to the idea of regulation because they feared that an increase in the lake storage, which would occur in the very periods of high inflow, might cause an increase in the frequency of lake floods. The opposition was overcome with a compromise: on the one hand, the construction of a dam was authorized, and on the other it was established that it would have to be completely opened whenever the lake level exceeded a pre-established level, to allow the waters of the lake to flow out rapidly. In order that the outflow be faster than in natural regime, the lake mouth was excavated before the construction of the dam. This and the following management of the lake's waters were entrusted to a body that was constituted for that purpose: the Ticino Consortium. In 1940 an agreement was signed between Italy and Switzerland on the basis of which the regulation methods were fixed, and embodied in a *Regulation Licence*; in addition, an *Italian–Swiss International Commission* was constituted, which was given the responsibility for making sure the license was followed.

The Miorina dam (Figure 2.11) is 200 m wide and made up of 120 metal gates which hinge on the river bottom. The gates can be completely lowered, and their position is regu-

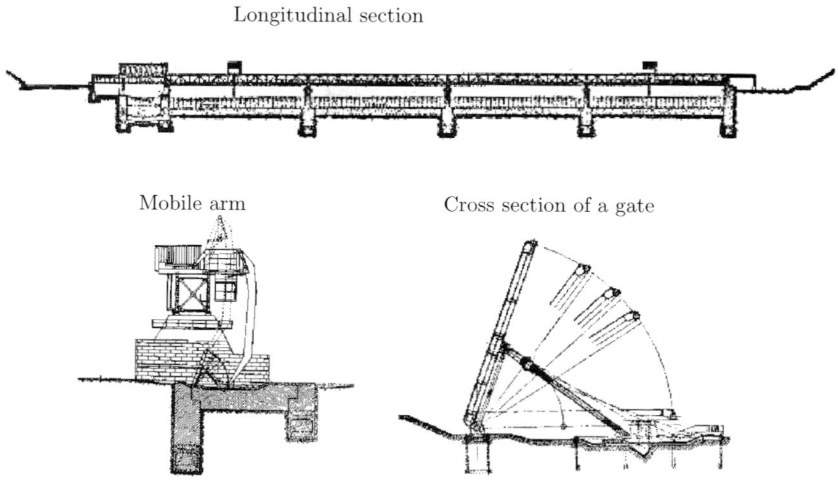

Figure 2.11: The Morina dam.

lated by mobile arms, which are fixed to bridge cranes. The dam came into service on 1st January, 1943.

2.3.1 How did regulation come about

2.3.2 The dam

2.3.3 The Regulation Licence

The Regulation Licence establishes a *regulation range* which is the interval of lake levels within which the *Regulator* – the director of the Ticino Consortium – can freely decide the release. Unlike the natural regime, in this new condition, called the *regulated regime*, a one-to-one relationship between levels and flows no longer exists. The extremes of the regulation range are seasonally variable and have changed over time. At the beginning of the Verbano Project (1999) they had the following values:

- in the summer (16th March – 31st October) the regulation range is defined by the limits -0.50 m and $+1.00$ m (above zero at Sesto Calende, which is fixed at 193.016 m a.s.l.);
- in the winter (1st November – 15th March) it is defined by -0.50 m and $+1.50$ m.

The transition from the winter value of the upper extreme to the summer value does not occur from one day to the next: the 50 cm decrease is staged over a period of 15 days, as shown in Figure 2.12. The inverse transition, namely the passage from the summer value to the winter one, is carried out in a similar way.

When the lake level reaches the lower extreme of the regulation range, the Regulator is obliged to release a flow no higher than the inflow, so that the lake level cannot fall any further. In this way the socio-economic activities, such as tourism and navigation, which are negatively affected by lake levels below a certain threshold, are safeguarded. Conversely, when the lake reaches the upper extreme of the regulation range, the Regulator must open

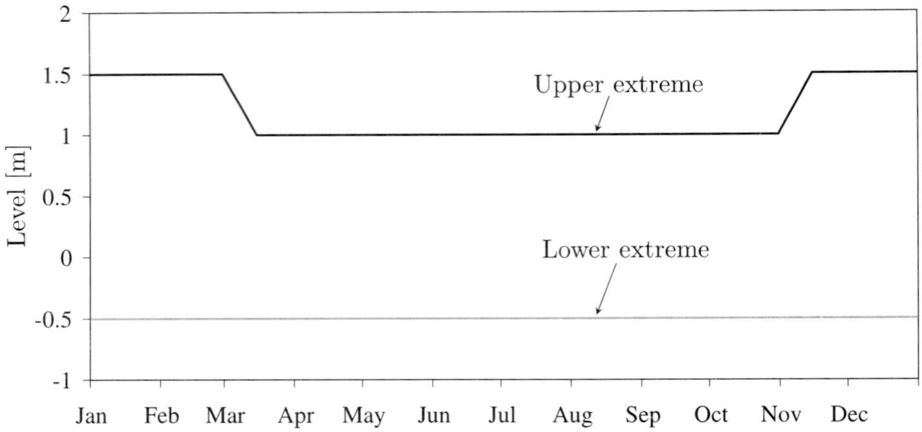

Figure 2.12: The time pattern of the regulation range.

the gates completely, to achieve the greatest possible outflow. In this way the interests of the lakeside population are safeguarded, by minimizing the risk of flooding around the lake. Once the gates are opened, the outflow is left to evolve without control and the lake is said to be in *free regime*. In these conditions the lake behaves as if it were the natural lake, with the only difference that now the section of the outlet is greater and so the outflow is faster.

2.3.4 The effects of regulation

In the regulated regime the outflow, usually called *release*, depends not only on the level of the lake, but also on the Regulator's decision. Let us see how he has worked so far. In Figure 2.13 the trajectories of the (median) levels in the regulated regime are compared with those that would have occurred if the lake had been in its natural regime. Observe that the first are substantially different from the second, particularly in autumn and winter. As

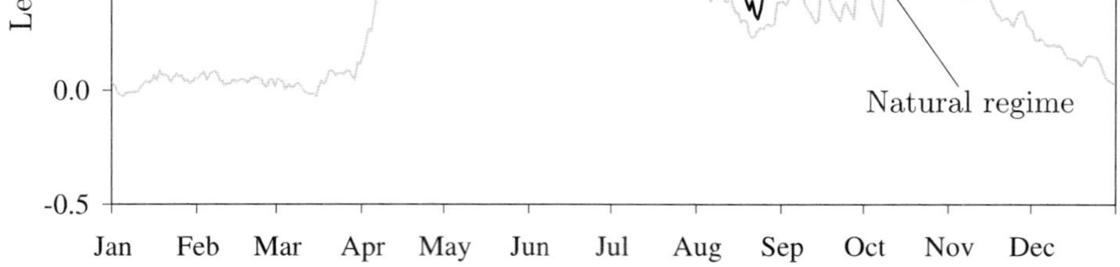

Figure 2.13: Verbano's median levels in the regulated and natural regimes (years 1974–1998).

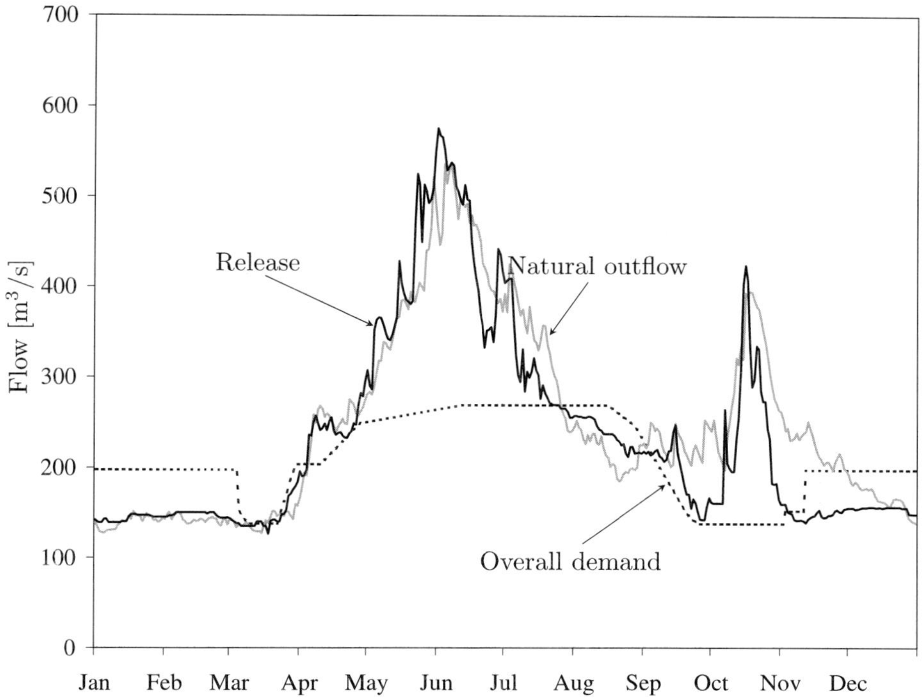

Figure 2.14: Median natural outflow and release (years 1974–1998) compared with the overall water demand.

planned in the dam project, with the exception of rare occasions, the lake has been storing water in spring and autumn. Water is released in the summer period, principally for irrigation needs, and in winter, almost exclusively for hydropower production.

In Figure 2.14 the trajectories of the release and the natural outflow are compared. Note that the first is higher than the second when the natural outflow is less than the total water demand. The release is thus better for the water users' needs. On the other hand, in periods when the natural outflow is higher than the demand, the trajectory of release shows steep short waves; they are produced by the complete opening of the dam, after the upper extreme of the regulation range has been exceeded. In such moments the release is higher than the natural outflow, and this could be surprising because one might think that the storage would not be able to form. Actually, the storage is formed in the phases in which the inflow is decreasing and the probability of flooding is lower. In these phases, in fact, the release is lower than the natural outflow. The autumn deficit is greater, because in this season the aim is to create the lake storage and to maintain it until spring. This is why in winter the release is only slightly superior to the natural outflow. The reader might be surprised that the reduction in the spring deficit is inferior to the increase in the autumn deficit, and the transfer of water volumes could appear to be very modest. However, (s)he must note that the trajectories shown are those of the medians: it is not possible to draw conclusions about the mass balance. Furthermore, the real stochasticity of the system is hidden. In rainy years the natural outflow is sufficient to satisfy the total water demand, so it is not necessary to modify the natural regime through regulation; on the contrary, in drought years the transfer of water volumes due to regulation is very evident.

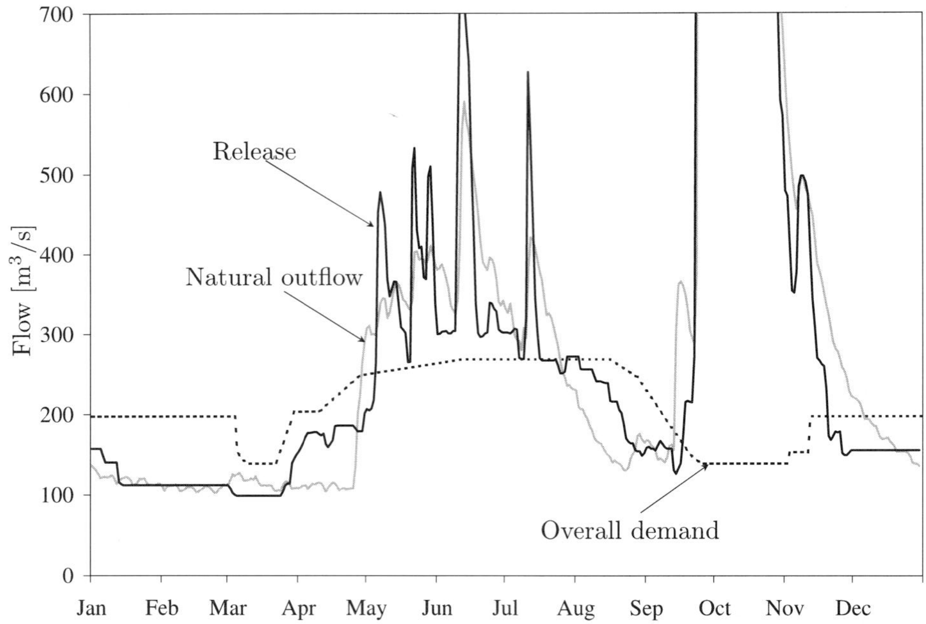

Figure 2.15: A comparison of natural outflow, release and overall water demand (1993 data).

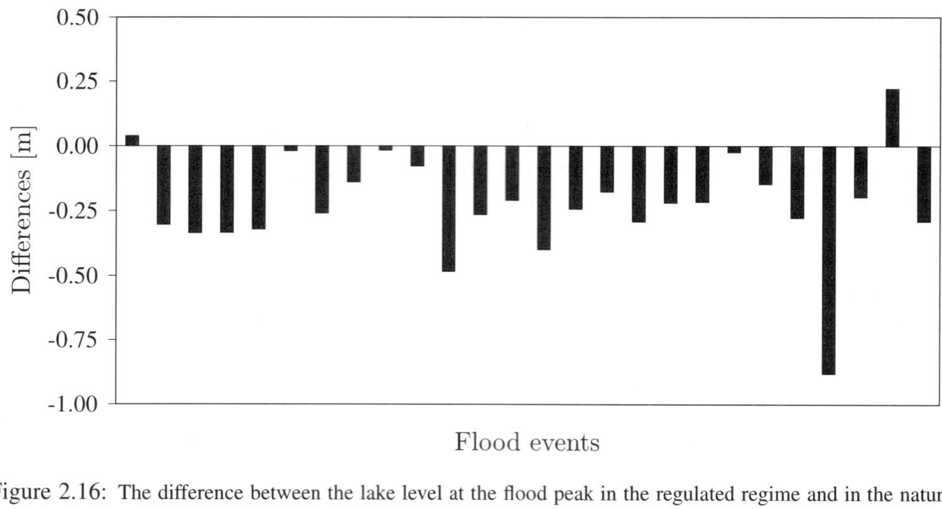

Figure 2.16: The difference between the lake level at the flood peak in the regulated regime and in the natural regime for the flood events between 1974 and 1998.

This can be demonstrated by considering the release trajectory over one year, for example 1993, in which the inflow was scarce. Figure 2.15 shows that, thanks to regulation, the irrigation deficit was halved with respect to the value that would have been registered in natural regime, both in the period when the rice paddies are submerged and in the month of August. On the other hand, the autumn deficit increased for the reasons explained above.

The regulation of the lake has produced a reduction in the average annual deficit, by providing on average 208 million m^3 of *new water* each year. New water is the surplus water volume made available to the users with respect to the volumes they would have used in the natural regime. Even if the amount of new water might appear modest, with respect to the average volume that is released from the lake in a year (8800 million m^3), this new water is very important, because it often makes the difference that allows irrigators and hydropower producers to achieve their aims in periods of drought, as was the case in 1993.

It is thanks to regulation that it was possible to create new water without increasing flooding. For each flood event between 1974 and 1998, Figure 2.16 shows the difference between the peak level in the regulated regime and that in the natural regime: observe that regulation has reduced the peaks of 24 events and worsened only 2.

2.4 Dissatisfaction

The regulation of the lake has thus brought advantages, both upstream and downstream, with respect to the natural regime. Nonetheless, the Stakeholders are not satisfied because they perceive only the damage they have incurred, while they have no way of appreciating the damage that has been avoided. The lakeside population believes that flooding is still too frequent, while the irrigators and hydropower producers complain of the deficits that reduce their production. For both groups the expression "the more you have, the more you want" holds true. The reduction in flooding has resulted in the fact that areas that were once designated for the expansion of the lake have become urbanized, so that floods which once would have produced very modest damage today produce very significant ones. Agriculture has partially extended, especially in Novarese, where water demand from irrigators is higher today than in the past. With the passage of time issues such as environmental protection have assumed an importance which before was ignored or neglected. The Bolle di Magadino Foundation, for example, is dissatisfied with the regulated regime, since it damages the habitat of important species (Section 2.2.1.6 of the DVD); the fishermen complain, because sometimes a sudden increase in the lake's outflow compromises the reproductive cycle of the bleak and the white fish (Section 2.2.1.5 of the DVD). It is no longer acceptable that no attention is paid to these problems and this is also true for those linked to tourist enjoyment and navigation on the lake.

Then there are other problems, which may seem smaller, but which have great importance for the people directly involved. For example, during floods hydrostatic pressure can make the oil tanks rise out of the ground if they are not properly ballasted, and this can cause oil spillage and serious environmental damage. If water reaches electricity boxes during floods, it can create a very dangerous situation but, if electricity supply is suspended, emergency services must work in the dark and the people who are already in great difficulty lose the comfort of illumination. All these problems require solutions, but in which frame of reference?

With their own interests at heart, a number of Stakeholders proposed changes to the structure of the system or to the way that regulation is conducted. In judging those proposals, however, we must be careful that solving one problem does not worsen another.

This plurality of expectations and desires demanded a response and so the Verbano Project was launched: a study financed by the European Commission under the INTERREG II Program.

In the next chapter we will illustrate the *intervention* proposals that were put forward by the Stakeholders, and we will consider how to find a solution that would favour all of them if possible, and thus be accepted by all. Before proceeding, however, there are still two points to be clarified: who makes the final decision and what the Goal of the Verbano Project is.

2.5 The Decision Makers

The transformation of Verbano from a natural lake into a regulated lake was the result of a long process (described in Section 2.3.1 of the DVD) with which the hostility of the lake-side population was overcome. Remember that these people are citizens of two different States. To favour dialogue between the parties, the governments of these States created an *Italy–Switzerland International Commission*, to which they assigned the tasks of screening the regulation projects, expressing an opinion about their feasibility and their expected effects and, once the regulation had begun, of supervising its progress and putting forward proposals for its improvement. The Commission is active to this day and its role remains unchanged.

The Stakeholders can express their needs and put forward proposals; they can also influence decisions with their social and economic weight, but they cannot take decisions directly. Final decisions are the responsibility of others: the proposals must firstly pass the screening of the *Italy–Switzerland International Commission*, and then be submitted to the two governments. Therefore, the *Decision Makers* are the Italian and Swiss governments.

We know that the management of the Miorina dam was assigned to the Ticino Consortium. The Consortium is thus a third Decision Maker, even if at a lower level of the hierarchy than the two governments, since its role and its responsibilities are defined by them through the Regulation Licence.

2.6 The Goal

Lastly, let us define the *Goal* of our Project: to evaluate the interventions proposed by the various Stakeholders in an integrated way, i.e. in such a way as to analyse the effects on the entire system and to identify some alternatives to submit to the final Decision Makers, with indications about which Stakeholders are in favour and which are opposed.

The starting point is thus the interventions that have already been proposed and that we will describe in the next chapter. As we will see, it is advisable that these interventions be made more coherent by adding linking actions; but the latter must be such that they do not question the current structure of the system (with the exception of the mouth of the lake, for which a modification has already been proposed) or the existing water rights. From this it follows, for example, that we cannot consider among our intervention options a modification in the ditch irrigation method, or a variation in the water demand that the canals currently serve (see Section 2.2.2).

Chapter 3

The options for interventions

EB, FC, AP and RSS

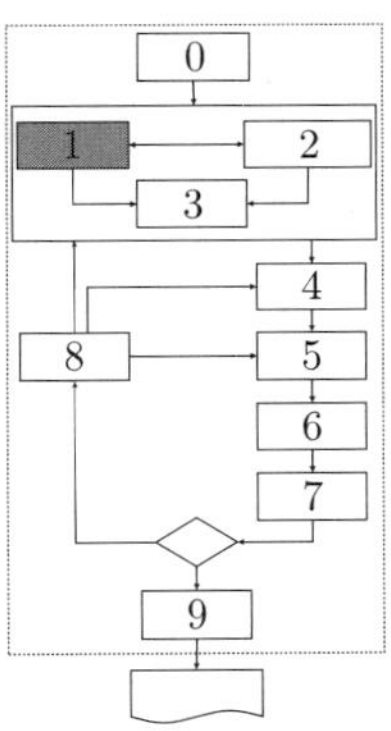

Figure 3.1: The phases of the PIP procedure. Phase 1, which is described in this chapter, is highlighted.

The disastrous flood events of 1993 and 2000, increasing awareness about environmental issues and the persistent interest in irrigation and hydropower production have motivated several groups to propose *interventions* with the aim of improving the performance of the Verbano system. These proposals are presented in this chapter, in order to identify the options for intervention to consider for achieving the Goal of our Project. It will be an initial collection of ideas. It may seem strange to begin with brainstorming, but in the PIP procedure (see Figure 3.1) it is essential to promote creative decision-making that is able to discover unexpected synergic possibilities among the options for interventions and to consider all the Stakeholders' suggestions. Then they are more disposed to collaborate since they feel they are being 'taken seriously'.

As we learned in Chapter 2, the first creative phase has to be followed by a phase of 'sedimentation', in which one separates what can be decided and what cannot: in fact, it would be useless to debate aspects that cannot be decided in the Project. This is the reason why we will critically analyse each one of the options, even if for the moment we will not try to discriminate between the useful interventions and the useless ones. It is not yet the time for that discrimination, which will come in the following phases. In this chapter, we only aim at obtaining a list of the interventions 'that can be decided upon' in the context of the Project.

3.1 Modifying the outlet

The problem of lake flooding is certainly among those most felt by people. When the lake level reaches the upper extreme of the regulation range, the Regulation Licence establishes that the Regulator must completely lower the gates at the Miorina dam, so as to allow the maximum possible outflow. The outflow rate is determined by the characteristics of the

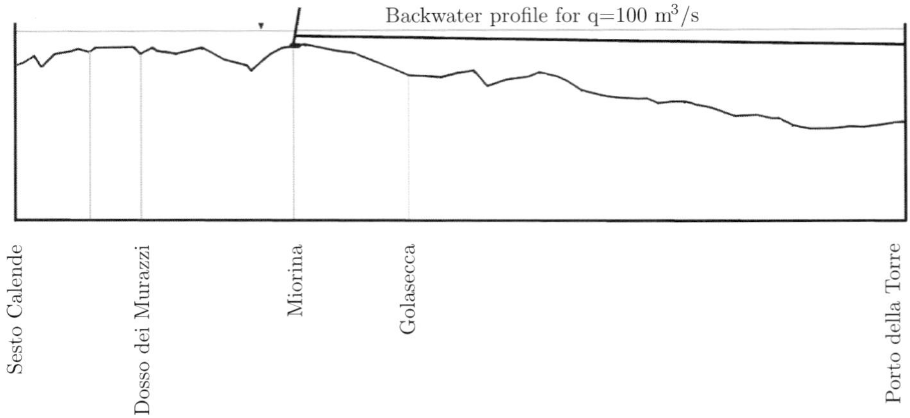

Figure 3.2: Longitudinal water-surface profile of the River Ticino from the hydrometric station at Sesto Calende to the dam at Porto della Torre, after the dredging at Dosso dei Murazzi (re-elaborated from Marchetti and Raffa, 1965).

outlet: the elevation of the submerged threshold and the form of the cross section. The lower the threshold and the wider the cross section, the greater the volume of water that can be discharged in a given time period and consequently, for the same inflow, the lower the increase in the lake level. This information is quantified in the so-called *stage–discharge relation*. The dream of the lakeside population is thus to modify the outlet so that for the same lake level, the outflow in free regime is increased.

When the Miorina dam was constructed (1943) the lake outlet was modified, with the aim of increasing the outflow capacity, for both high and low lake levels. Afterwards, over the years 1954–1963, the bottom of the river channel was dredged at Dosso dei Murazzi, which is located 1.5 km upstream of the Miorina dam (see Figure 3.2 and Section 2.3.1 of the DVD). Thus the stage–discharge relation has already been modified with respect to the natural regime. Other small variations were produced later by particularly extreme flood events, which eroded the mouth of the lake.[1]

Thus, over time Verbano's stage–discharge relation has continued to evolve, and new calibrations have become necessary. Therefore, a number of relations are available (see Figure 3.3), which should be thought of as frames of a film:

- *Natural stage–discharge relation* (Marchetti and Raffa, 1965), estimated by the Italian Hydrographic Service on the basis of measurements taken over the years 1938–1939. It describes the outflow capacity of Verbano before the construction of the Miorina dam; it is used to reconstruct the natural regime.

- *1946 stage–discharge relation* (Marchetti and Raffa, 1965), estimated by the Italian Hydrographic Service on the basis of measurements taken over the years 1946–1953. It shows the increase in outflow capacity produced by changes to the river channel and by lowering the threshold at the Miorina when the dam was built. It can be considered valid from 1943 to 1963.

[1] Phenomena of material sedimentation that reduce the outflow capacity have never been registered.

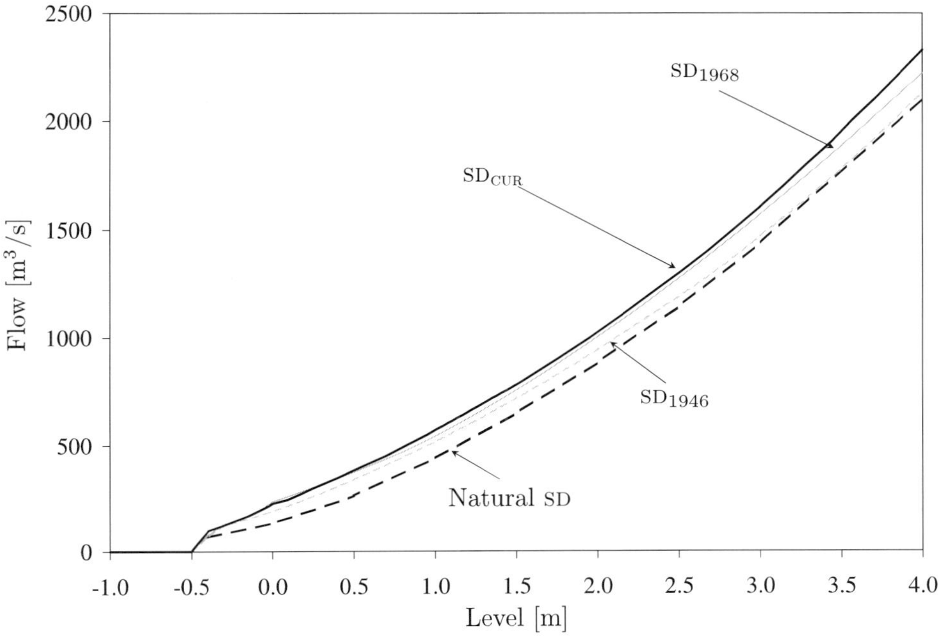

Figure 3.3: Verbano's stage–discharge (SD) relations over time.

- *1968 stage–discharge relation* (Citrini and Cozzo, 1982a), estimated by the Italian Hydrographic Service on the basis of measurements taken over the years 1968–1973. It allows us to measure the effects of dredging at Dosso dei Murazzi: the comparison with the 1946 scale shows an increase in the flow that can be delivered both in flood and drought conditions. It can be considered valid from 1964 to 1990.

- *Current stage–discharge relation* (Maione and Mignosa, 1995), denoted in the following as SD_{CUR}, was estimated by the Hydraulics Department (DIIAR) at the Politecnico di Milano on the basis of measures taken during the flood events of 1991 and 1993. It reveals that, with respect to the 1968 scale, there has been a further increase in the outflow capacity of the outlet in flood conditions.

Switzerland proposed two different interventions for excavating the outlet, which aim at producing an increase in the outflow in the free regime of 300 m^3/s (Anastasi, 1992) and 600 m^3/s (Anastasi, 1994) respectively, in correspondence to the level 2.78 m at Sesto Calende hydrometric station. In practice, the effective increases that would be generated are less, but it became customary to call the stage–discharge relations resulting from the two interventions as '(stage–discharge) relation +300' and '(stage–discharge) relation +600'; in the following we will denote them with the acronyms SD_{+300} and SD_{+600}. They are shown in Figure 3.4 along with SD_{CUR}. Note that the differences between them and the current stage–discharge relation are decisively greater than the difference between this latter and the natural stage–discharge relation. The dynamics of the lake level and of outflow would thus be radically modified if either one of them were adopted, and this would have important effects (positive or negative according to one's point of view) both upstream and

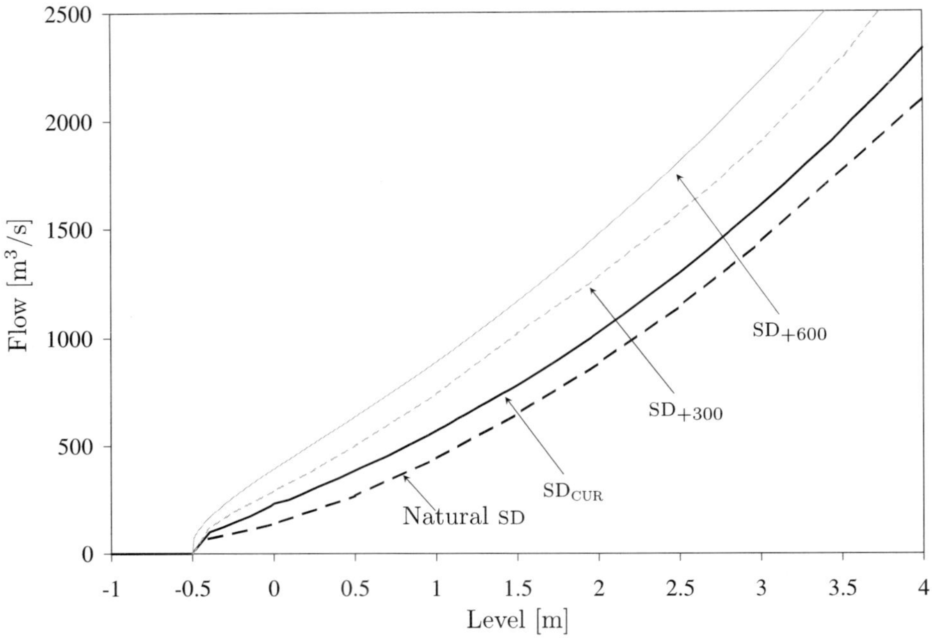

Figure 3.4: A comparison between the current stage–discharge relation (SD_{CUR}), those proposed by Switzerland (SD_{+300} and SD_{+600}) and the natural one (re-elaborated from Anastasi, 1992).

downstream from the Miorina dam. In particular, it is expected that increased outflow capacity would diminish upstream flooding (this is not certain, however, because, as we explain in Section 3.5, the effect would also depend on the regulation adopted once the excavation had been completed). At the same time, it is feared that the downstream flood regime would be worsened. This fear explains why the excavation proposals were put forward only by Switzerland.

Let us examine the two proposed relations in more detail.

3.1.1 The relation +300

Anastasi's study (Anastasi, 1992) started by considering two alternative ways for increasing the outflow by about 300 m^3/s:

- excavating the river bed and removing about 570 000 m^3 of material;
- constructing a by-pass tunnel (4000 m long and with a diameter of 7.5 m), which would capture the waters upstream of the Miorina and channel them into the Ticino, upstream of the diversion dam at Panperduto. Its path is shown by the continuous line in Figure 3.5.

The first possibility would cost next to nothing: the excavation costs could be covered by selling the gravel that was excavated. This is not true for the second possibility, which is extremely onerous. Even from the hydraulic standpoint, excavating is more efficient: the effects of the tunnel would be felt only downstream from its intake (see the continuous line in Figure 3.5), while upstream the current outflow limit would persist. To remedy this,

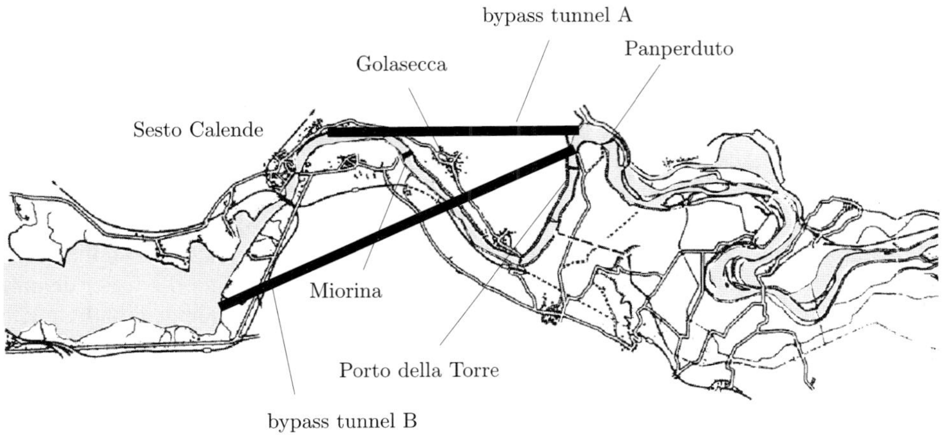

Figure 3.5: Planimetry of the Miorina area and two alternative paths for the *bypass* tunnel.

it would be necessary to construct a tunnel longer than the one shown in Figure 3.5, by placing its intake in the lake itself, but then the tunnel would have to pass under the River Ticino (see the dashed line in Figure 3.5), and its construction would be even more difficult and expensive. Anastasi thus concluded that the possibility of constructing a tunnel should be discarded and that only the excavation of the river bed was worth considering.

3.1.2 The relation +600

In 1999, when the Verbano Project started, on the basis of Anastasi's study (Anastasi, 1994) it was believed that to obtain an increase in flow of about 600 m^3/s (550 m^3/s more precisely) it would be necessary to excavate the river bed (with an overall extraction of 1 170 000 m^3 of material) and lower the threshold of the Miorina dam by 50 cm. It was estimated that the sale of extracted material would cover a good part of the intervention costs, but there were serious doubts about the costs of modifying the dam.

The results of this study were substantially confirmed by Mignosa, who made small changes to the stage–discharge relation presented in Anastasi's study, to account for the backflow effect induced by the Porto della Torre dam on the Miorina level at low flow rates. The relation elaborated by Mignosa[2] is the one shown in Figure 3.4 as SD_{+600}.

After the end of our Project, the results of another study were published (Maione and Mignosa, 2001): they state that it is possible to obtain an increase of 520 m^3/s in the outflow without modifying the dam and this would require only 700 000 m^3 of material to be removed. These indications make the intervention appear much simpler and the hypothesis that the overall cost would be close to zero seems more reliable. The differences between the stage–discharge relation that the latter intervention would create and that of the Anastasi's +600 proposal (SD_{+600}) are quite small, so that it is reasonable to consider the new proposal as a different way of realizing the previous one.

[2]Private communication.

3.1.3 The effects on the free regime

3.2 Norms for hydropower reservoirs

In Section 2.1.1 we showed that the activity of the hydropower reservoirs in the lake's catchment significantly modifies the trajectory of the inflows to the lake. Figure 2.5 shows that, on average, the reservoirs cause an increase of the inflow from October to April and a decrease in the rest of the year. The effect of this variation on the downstream users is not easy to guess: on the one hand, the increase of the inflow from October to April facilitates the formation of the lake storage in spring, which is essential for satisfying the water demand in summer; on the other hand, the reduction of the inflow in the summer months makes the resource scarce in the very period that it is most necessary. Two studies dedicated to the problem (Gherardelli, 1964; Citrini and Cozzo, 1981) have shown that the management of these reservoirs has not produced, on average, unfavourable effects for the downstream users, except in the summers of 1962, 1965 and 1976 (note that the most recent study is from 1981). However, these are just the years in which water failure was most severe, and in these conditions the activity of the reservoirs further worsened the situation.

One might consider introducing a norm that would establish when and how much the reservoirs must release in favour of the downstream users, or, more precisely, in favour of the irrigators. In fact, the downstream hydropower producers do not have any interest in this, because electrical energy has a higher price in winter and so they are quite happy that the upstream reservoirs reduce the inflow to the lake in summer and increase it in winter.

The hydropower reservoirs are all private property and they operate on the basis of abstraction licences that were granted to them by the State (Italy or Switzerland). Any norm that would force them to release water would have to be analysed firstly from a legal point of view, to see if it would be acceptable, given the nature of the licence agreements. The legal problem seems even more complex when we note that most of the reservoirs are in Swiss territory and so the State that grants the licences is different to the State in which irrigation is being practised. This is why this type of proposal has never been put forward and only once, in the summer of 1976, during the most serious drought of the last century, did the Ticino Consortium negotiate an agreement with ENEL so that a water volume sufficient to prevent the total loss of the harvest was released.

A theoretically irreproachable Project would have to analyse the possibility of a norm for regulating the releases from the reservoirs in the system, but since this has not actually been proposed by any of the Stakeholders, the possibility was not considered in the Verbano Project. Actually, we did consider other interventions that had not been proposed by anyone, but only when they were relatively easy and inexpensive to analyse. This one proved to be impossible to study: the analysis would require a very complex model because, other than the lake, 15 reservoirs would have to be considered. Even if one were only to consider a single equivalent reservoir,[3] modelling it would require the collection of data such as the time series of storages and releases for each of the 15 reservoirs, a difficult and costly task. Unfortunately limits of this type are not rare in environmental impact assessment, and when they do exist they should always be made explicit.

Nevertheless, this information gap is not excessively serious, because resorting to regulating the releases from the hydropower reservoirs would be an extreme measure, to be implemented only in extreme cases. It makes sense to plan other interventions first, and to

[3] This choice would also generate other problems, since the owner of each reservoir would obviously want to know what would happen to his reservoir following the application of such a norm.

consider this option only as an exceptional *mitigation* measure[4] which can be evaluated in a successive Project.

One might also think about using the reservoirs to buffer the floods; a recent work (Piatti and Martini, 2001) showed, however, that the effect would be minimal (see Section 3.6.7 of the DVD).

3.3 Modifying the regulation range

The introduction of a range within which the regulation exercise must be limited has the aim of decoupling the irrigators/hydropower-producers' interests from those of the other sectors: when the lake level is within the regulation range, the interests of the first pair are given priority, while outside the range the interests of the other sectors are protected. As mentioned in Section 2.3.3, the lower extreme of the regulation range serves to safeguard tourist activities and navigation, the upper extreme to protect lakeside populations from floods. Within the range, regulation can be freely established by the Regulator. The statute of the Ticino Consortium (1968) affirms that the Regulator has the job of "coordinating and regulating the exercise of using the water for the general interest" and of reaching "the best and most integrated use of the water in the interests of the Consortium members and to the advantage of agriculture and industry". In this framework, widening the range means favouring the downstream irrigators and hydropower producers, to the detriment of the upstream Stakeholders; conversely, reducing the range would be advantageous for the latter. The conflict between upstream and downstream interests can thus be controlled just by modifying the regulation range.

The lower extreme of the range has never been questioned, while the upper extreme has been the object of much discussion. Italy, particularly interested in agricultural and industrial production, has always pressed to raise this limit, while Switzerland pushes to lower it, with the aim of reducing flooding around the lake. Thus the range has varied over time. When the dam was constructed, the upper extreme was fixed at 1.00 m (at the Sesto Calende hydrometric station) for the whole year, but very soon afterwards Italy requested and obtained from Switzerland an increase in the winter period of up to 1.50 m. In winter, in fact, the risk of flooding is decisively less than in spring and autumn: therefore, storing water in winter, even at levels higher than 1.00 m, does not increase the risk for lakeside populations. The range that is in force today is thus time-variant[5] (Figure 3.6).

Since the Ticino Consortium did not consider the compromise that was reached to be completely satisfactory, other proposals were elaborated, and Switzerland responded with counterproposals (Figure 3.7):

- in 1961, Ticino Consortium put forward the proposal to fix the upper extreme at 1.50 m also between June and September. The proposal was based on a study of the sensitivity of the lake level's duration curve with respect to the regulation range; the study was prepared by De Marchi (1950a, 1950b);

- the Swiss counterproposal, put forward in 1973, was to prolong the period that the upper extreme is set at 1.50 m until mid-April and bring it back to 1.10 m by 1st July, in exchange for a reduction of the upper extreme down to 0.90 m from mid-April to 1st June and to 0.70 m from 1st October to 15th November;

[4] For the definition of the term see Chapter 14.

[5] The upper extreme of the current range is fixed at 1.00 m from 15/03 to 31/10 and at 1.50 m for the rest of the year; the lower extreme of the range is constant at −0.50 m all year round.

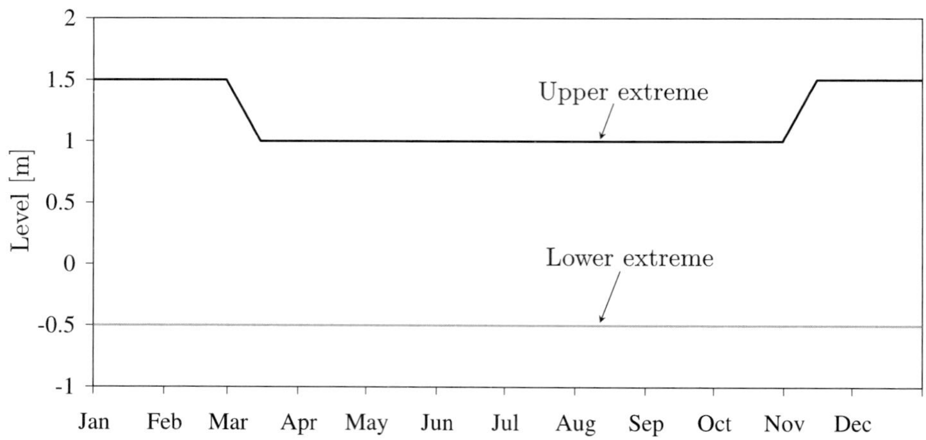

Figure 3.6: The time pattern of the current regulation range.

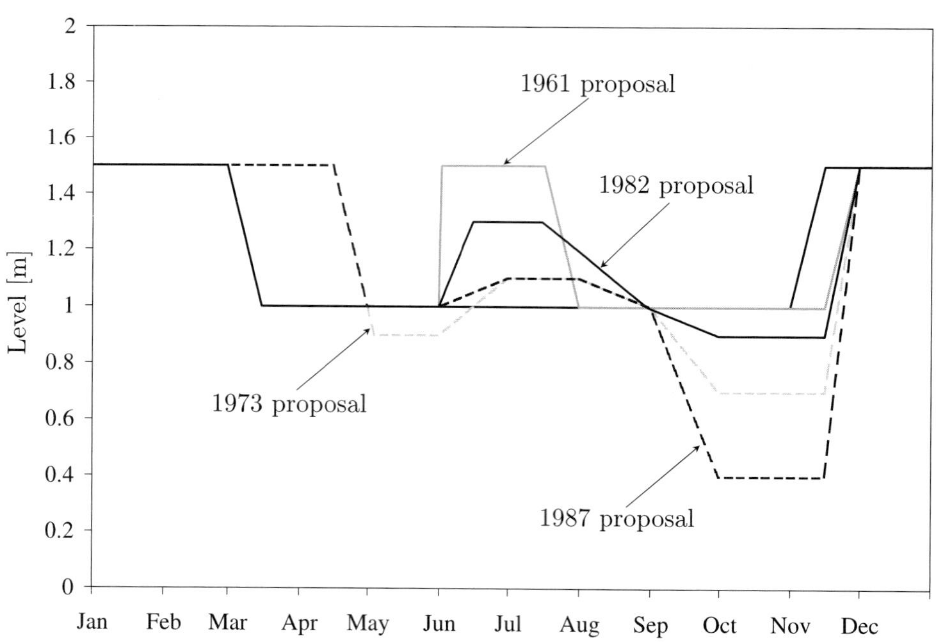

Figure 3.7: The trajectory of the upper extreme of the regulation range in the different proposals.

- in 1982 the Ticino Consortium, referring to a study by Citrini and Cozzo (1982b) which was reviewed by Baratti, put forward another proposal: a compromise between the 1961 proposal and the 1973 counterproposal;
- finally, in 1987 Guariso, Laniado and Rinaldi put forward yet another proposal for modifying the regulation range, which was characterized by the lowering of the upper extreme significantly in the autumn period.

These proposals were elaborated in different ways, which are described in Sections 3.3.1–3.3.5 of the DVD. Here we are only interested in underlining that in all these studies the change in the regulation range was considered as an isolated intervention. Today the situation is different: modifying the regulation range can be considered in a close relationship with the Swiss proposal to excavate the outlet. The latter is aimed at reducing the risk of flooding upstream and is of no interest to the downstream users, so they could propose that Italy request that the upper extreme of the regulation range be raised, taking it up to 1.50 m all year, as compensation.

3.4 Minimum environmental flow

The interventions examined until now are all inspired by the desire to solve the upstream–downstream conflict. We described and discussed them looking only at the lake levels and releases, and we did not consider how the releases are distributed in the downstream water network. In doing so, we implicitly assumed that the distribution policies for allocating the flows among the downstream canals[6] could not be modified. This hypothesis allows us to reason in terms of releases, and ignore the fact that the satisfaction of each irrigator or hydropower producer depends on the flow that (s)he actually receives. But is this hypothesis sound?

Actually, no *distribution policy* for allocating the flows among the canals and the Ticino has ever been codified. Distribution is carried out by the Ticino Consortium Regulation Committee, which is composed of the Director of the Ticino Consortium and representatives from the irrigation consortiums and from ENEL; the Committee decides how to act on a case-by-case basis. Interviews with members of the Committee allowed us to identify a policy that, given the time series of the releases at the Miorina, can reconstruct the series of the flows delivered to the canals and to the Ticino (see Section 6.7.3 and its extension on the DVD).

Before describing the characteristics of the policy that we identified, we must specify that by the term 'flow to the Ticino' we mean the flow left to the river at the Panperduto diversion dam (Figure 3.8), whose value remains practically unchanged until the Oleggio bridge. Downstream from this bridge, the flow progressively increases due to the effect of *resurgences* and, further downstream, also because part of the flow channelled into the Industrial Canal is returned to the river (see Section 2.1.2). The contribution from the resurgences is very important: during the big drought in the winter of 1993, measurements showed that the flow rate of 3 m^3/s at Panperduto had increased to 15 m^3/s about ten kilometers downstream from Oleggio. The flow rate at Panperduto is thus the lowest flow value along the whole course of the Ticino.

The distribution policy that we identified is articulated as follows:

[6] The canal mentioned in this book are all artificial watercourses not used for navigation.

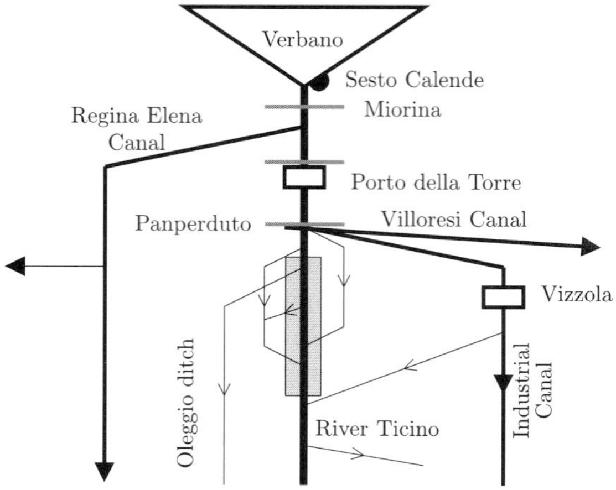

Figure 3.8: Diagram of the downstream water network. The grey rectangle identifies the stretch where the flow in the Ticino is lowest.

- The satisfaction of irrigators' and hydropower producers' needs takes priority over the environmental quality of the Ticino. The river is provided only with the *minimum environmental flow* (MEF) except for when water demand is completely satisfied and the lake is close to *spilling*. The (MEF) value is presently officially set at 3 m^3/s, but since 1997 the experimental value of 13 m^3/s has been used.
- When it is not possible to satisfy all the water users, the flows delivered to the hydropower plants are first reduced; then, if there is still a deficit, it is shared out among the irrigators in proportions that depend upon the season.

The policy is shared and accepted by all the water users, so there is no conflict between these Stakeholders. This explains why it is reasonable to group irrigators and hydropower producers together, as we have done up until this point. However, this policy is not accepted by the Ticino Parks (either in Lombardy, or in Piedmont), who do not agree with leaving the river in a subordinate position. Thus there is conflict about this point and the Parks put forward a dual request:

- the flow rate of the river should be equal to the MEF only in emergency conditions and not in normal ones. The minimum environmental flow, they argue, is a legal constraint whose function is to prevent the creation of conditions that put the survival of river ecosystems at risk. Consequently, the minimum flow should only last for brief periods of time.
- a MEF of 3 m^3/s is too low and its value should be raised. This value was defined in 1980[7] and it has been kept as a reference point, despite the fact that in the last 20 years new laws passed on the subject. On the basis of these,[8] in 1995 the two

[7] In the Enactment Norms of the Territorial Coordination Plan for the Parco Lombardo del Ticino, which in art. 20 establishes: "*downstream from every outtake along the River Ticino a minimum flow of* 3 m^3/s *should be guaranteed*".

[8] Particularly on the basis of the Law 36/94, so-called *Galli Law*, which in art. 25 establishes: "*in the context of national and regional protected areas, the management Agency for the protected area, after consulting the River Basin Authority, defines the spring water, running water and ground water necessary for the conservation of ecosystems, which cannot be tapped*".

Parks jointly decided that the value of the MEF should be set at 12–13 m^3/s for that year and that afterwards it should be gradually increased from year to year, to reach a value between 24 and 27 m^3/s in 2005. The Po River Basin Authority was asked to give its opinion about this decision, and the Authority approved the methodology that was used and confirmed the values that had been computed. Nevertheless the MEF established by the Parks was not applied because the users did not consider the decision to be binding as long as their Abstraction Licences had not been modified; this has not yet happened. Only after 1997 was the experimental value of 13 m^3/s adopted.

In addition to the study carried out by the Parks to support their decision, various other studies were conducted to determine the value of the MEF that is most suitable for the river and to evaluate its economic effects. Of particular interest is the study made by ENEL (1999), which analyses the impact of the MEF on hydropower production. As we explained in Section 2.2.2, the water demand for the Industrial Canal is 138.5 m^3/s, and thus equal to the maximum flow that the canal can divert and that the powerplants at Vizzola, Tornavento and Turbigo Superiore can run through the turbines. The Abstraction Licence is less, however: 120 m^3/s. The request made by ENEL is to bring it up to 138.5 m^3/s, in order to exploit the residual capacity of the power plants. In this way the flow run through the turbines could be increased when normal conditions hold. In return, in dry periods ENEL would give back to the river the equivalent of the surplus volume so obtained, thus producing an increase of minimum environmental flow, without reducing the overall power generation. In other words, ENEL's proposal is to keep the average flow derived from the Industrial Canal unchanged, but to redistribute it through time, in order to reduce the conflict between water users and Parks. On the basis of the simulations that were run, ENEL holds that it "*could sustain ... an increase in the minimum environmental flow of* 10 m^3/s *downstream from the diversion dam at Panperduto, reaching a constant* MEF *of* 13 m^3/s" (ENEL, 1999). The proposal from ENEL is thus different from the Parks proposal, and thus the problem of setting the MEF value, and the frequency with which the river flow is reduced to it, is still an open problem.

3.5 Designing a different regulation

We have seen that the Verbano's regulation can be freely established by the Regulator when the lake level is within the regulation range. Therefore, raising the upper extreme of the regulation range in several periods means giving the Regulator new decision possibilities. To estimate the effects of such a change, we must explicitly consider the regulation modes that could be adopted. A similar line of reasoning has to be applied to the proposal to excavate the lake outlet, which was put forward thinking that this might mitigate the floods around the lake. Actually, it is not certain that the excavation in itself could guarantee that, with equal inflows, levels lower than the historical records would be reached. Today, in fact, the Regulator tends to open the Miorina dam completely even before the upper extreme of the range is reached, to avoid flooding. Excavating the outlet could induce him to modify his behaviour: trusting in the increased outflow capacity, the Regulator could decide to wait until the upper extreme of the range was reached before opening the dam completely. During rapidly developing events, in which the lake level at the beginning of the flood plays a determinant role, this behaviour might produce higher floods than the ones that would have occurred in the previous situation.

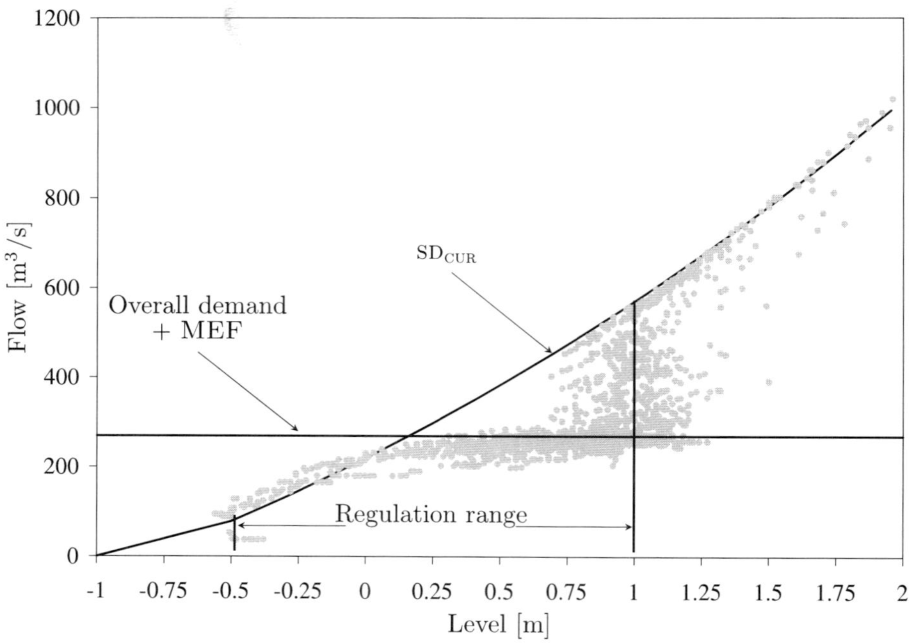

Figure 3.9: Level–release recorded pairs on the days between June 21st and August 20th in the years 1974–1998. Remember that the stage–discharge relation links level and instantaneous flow, while in this diagram the level value is the one measured at 8:00 am and the release value is the average release in the following 24 hours. This explains why, in free regime conditions, there are pairs that do not exactly lie on the stage–discharge function.

To evaluate the effects of any proposal, it is thus essential to explicitly consider the regulation of the lake. Furthermore, note that the analysis of the different regulation modes is interesting in and of itself, because it is not given a priori that the current mode of regulation uses the lake in an efficient way.

In normal conditions, the *release decision* is taken once a day (the *decision time step* is thus one day) on the basis of the lake level and, if necessary, of other information about the catchment (such as the volume of the snow pack, the overall storage in the hydropower reservoirs and the weather forecast) and about the downstream district (for example, the condition of the crops and electricity demand). Observe Figure 3.9 in which the level–release recorded pairs are reported for the days between June 21st and August 20th[9] for the years 1974–1998. For levels lower than 0.75 m, the correlation between level and outflow is very high: once a level has been fixed, the vertical dispersion of the points is reasonably small and can be explained by observing that, depending on contingent events (such as seasonal trends, the presence of precipitation in the irrigation district, or the closure of a

[9]This period was chosen for the following reasons:

(1) irrigation demand is constant in the whole period and this simplifies the interpretation of the diagram;

(2) it is a critical period for irrigation, given that in August inflows are scarce: this allows us to observe the regulation behaviour for low lake levels;

(3) the inflows in June can be particularly intense and so the risk of floods is high: this allows us to observe the regulation behaviour for high lake levels.

power plant), the total water demand can vary little from one day to another. Therefore, for levels lower than 0.75 m, the release decision is essentially a function of the lake level only. For levels higher than 0.75 m, the dispersion increases because the decision is made also according to other information. In these conditions, in fact, being close to the upper extreme of the regulation range (1.00 m), the Regulator begins to worry about flooding and would like to open the Miorina dam completely, in order to reduce the risk. By doing so, however, he might lose water volumes that are essential for irrigation, if no flood occurs. For this reason he tries to anticipate the flood arrival, using all of the available information.

The analysis of the other periods of the year leads to analogous considerations. Overall, we can thus consider that the behaviour of the regulator, on every day of the year, could be well approximated by a rule that associates a release value to every level value. A sequence of rules of this type is called a *regulation policy*. Since a priori one year is not different from another, we expect the regulation policy to be periodic of a one year period.

We will try to guess other characteristics. From Figure 3.9 it can be noted that, for levels lower than 0.75 m, the Regulator releases an average outflow that is at first almost equal to the water demand plus the MEF, but that, as the level decreases, becomes slightly inferior. This is rational, because small deficits are much less damaging than large deficits and so, when the resource begins to get scarce, it is advisable to reduce the release below the demand much sooner than one would be forced to do. Below 0.15 m, when the lake level is so low that it is no longer possible to satisfy the water demand, the Regulator tends to open the dam completely to provide the maximum flow possible; but even here, at higher levels one notes a tendency towards conservation, which disappears at lower levels, when water saving no longer makes sense, because it would mean damaging the crops. In those conditions it is better to release as much water as possible and hope that it rains. But how much and for how long is it advisable to save? It is necessary to analyse the problem in a quantitative way to respond and to evaluate if what has been done so far on the basis of an educated guess is the best that is possible. Similarly, it would be interesting and advisable to explore regulation policies that do not take account only of the needs of irrigators and hydropower producers. The reduction of flooding upstream, and the improvement of the quality of the river environment downstream, could be considered not only through normative interventions, but also with regulation policies that aim at realizing a compromise that is closer to meeting these interests than the current regulation policy is.

The current regulation policy is the fruit of a long learning process, based on experiments that were carried out in the field. The experiments that can be conducted in this way are, however, extremely limited, because there are high risks connected to mistaken decisions; to identify the best policy it is necessary to experiment more freely. Random attempts, however, will not necessarily be successful in identifying an efficient policy. Therefore, it is advisable to adopt a completely different approach: we must construct a model of the system and conduct experiments with it, with a search technique that guarantees identifying efficient regulation policies. By operating on a model it is possible to conduct all the experiments that we want, because they pose neither risks nor costs.

By proceeding in this way, it is also possible to analyse extreme regulation policies, i.e. the ones that take into account just one single interest and ignore all the others; not because these policies have any probability of being adopted, but because their analysis is very useful for negotiations between the Stakeholders. To understand why, let us dwell, for example, upon the lake floods. The regulation policy that minimizes the floods is obvious: always keep the gates of the Miorina dam lowered, as if the lake were not regulated. Certainly this policy would never be adopted, but estimating its effects lets us to know if it would be possible to completely eliminate the lake floods, or, if that were not possible, how many floods would be unavoidable. This information is very important because it could modify the expectations of the lakeside population and thus the degree to which it is willing to compromise with the other Stakeholders. For similar reasons it is useful to know the minimum and maximum satisfaction of each Stakeholder, because only with this information can they formulate their own requirements in a conscious way.

The search for and the analysis of the various regulation policies is therefore the key element of the entire Project.

3.6 Other interventions

3.7 Actions and alternatives

The considerations developed in the previous sections show that the various possible interventions can be synergistic. It is therefore evident that the Verbano problem will probably not be solved with one single intervention, for example by excavating the lake outlet alone. It is more likely that an 'opportune' combination of interventions will be adopted, for example the excavation of the outlet combined with a modification of the regulation policy. The adjective 'opportune' indicates that the combination must be chosen by the Decision Makers, on the basis of the indications provided by the Stakeholders, within a set of possible combinations, which we call *alternatives*. The aim of our Project will thus be to individuate all the feasible alternatives and to provide support to the decision-making process, in order to arrive at a conscious and shared choice of the most 'opportune' alternative.

In order to identify the set of the alternatives to consider, each possible intervention must be broken down into one or more *(meta-)actions*, i.e. into elementary interventions that can be fully and easily defined by specifying the values of their attributes, that is by specifying who is doing what, how and when. By fixing its attributes, a meta-action is transformed in an *instantiated* action. Technically, this transformation is carried out by assigning values to the parameters and/or the functions that describe the attributes of the meta-action. The specification of these values is a matter for a future phase (Phase 4: *Designing Alternatives*, see Chapter 7), but their feasibility sets have to be defined in Phase 1, thus fixing the meta-actions to be considered. The instantiated actions are the 'building blocks' from which the alternatives are constructed. As already stated, in the following we will use the term 'action' to denote both the meta-actions and the instantiated actions. The actual meaning will be clear from the context.

We postpone the identification of the actions to Chapter 5, even if this step constitutes the natural conclusion of Phase 1 of the PIP procedure. We have already noticed that the real development of the decision-making process is not serial, as Figure 1.4 might lead one to think. Indeed, the criteria cannot actually be correctly identified if one does not known the actions being considered, since these latter produce the effects that the Stakeholders endure. On the other hand, it is not possible to identify the actions without knowing the interests at stake, and therefore the criteria. This is the reason why we proceed with criteria identification in the next chapter and postpone action identification to Chapter 5.

3.7.1 Alternative *A*0

All the alternatives depend on the actions that have been devised, but one: the so-called *Alternative Zero* (*A*0), i.e. the alternative in which it is assumed that nothing is changed and everything remains the way it is. When we described the phase of *Designing Alternatives* (Section 1.3), we said that it is essential to always add this alternative in the set of alternatives that will be examined. Now, let us detail this idea more carefully.

Just as for the other alternatives, identifying *A*0 means defining its stage–discharge relation, regulation range, MEF value, and its regulation policy. The starting point for their definition is the analysis of the situation that existed until the moment when the Verbano Project started, i.e. the analysis that we carried out in this chapter; we will use the adjective *historical* to denote that situation. Observe, however, that, since the Verbano Project concerns the future management of the Verbano system, *A*0 will be defined on the basis of history, but it will not describe exactly the historical situation. The identification of *A*0 should in fact take into account the *scenario* (termed *evaluation scenario* in the following, see Section 8.3) which is believed to be the most probable in the future. Therefore, before assuming the existing rights and constraints, we should consider if each of them is likely to be modified in the future, independently from the conclusions of the Verbano Project. For example, in the definition of *A*0 the current stage–discharge relation and regulation range should clearly be considered, while the MEF value, according to what was said in Section 3.4, should not be assumed equal to 3 m^3/s (the current value) but equal to 13 m^3/s, which is the value that has been experimentally adopted from 1997: in fact, it is almost certain that the future MEF value will not be lower than that.

Also the regulation policy and the distribution policy must be defined, in other words we must specify how the release from the Miorina dam and the allotment to the water users and the Ticino will be decided day by day, if everything remains the same. In Section 3.5 we said that the Regulator, especially when the lake level is rather high, makes his decisions based on a number of factors. For *A*0 we can assume that, if the historical trajectory of the lake inflow occurred again in the future, the Regulator would adopt the same historical decisions, even if the water demands that he was facing when deciding in the past are not exactly the same as those in the evaluation scenario. However, the same hypothesis cannot be adopted for distributing water among the canals. In this case, in fact, there is strict dependence between the allotment decision and the water demand values. Then, we assumed that the water is allocated according to *allotment rules* (Sections 3.4 and 6.7.3) in which the future expected values of the water demands are considered rather than the historical demands. In Section 8.3.1 of the DVD the reader can find a comparison between the effects of *A*0 and those recorded in the past.

It should now be clear why it is essential to consider the Alternative Zero among the alternatives to evaluate and compare. Given that the Verbano system is characterized by conflicting interests and that the exploitation of water is very high, alternatives that *dominate A*0, namely alternatives that would be preferred to *A*0 by *all* the Stakeholders, are not likely to be found. In fact, even if from the standpoint of flood prevention, water supply and environmental protection the current situation does not completely satisfy some of them (see Section 2.4), it must be remembered that other Stakeholders based their activities on these very conditions. This is the case of tourism structures for example (such as wharves, beaches and camping grounds), which were built taking into account the lake level regime produced by the current regulation policy (see Section 2.2.1.3 of the DVD). A change of the policy, and thus the establishment of a new lake regime, most probably would require the modification of those structures and this would not be much appreciated. It is thus essential, when deciding if and how the Verbano system is to be modified, to assess how much each Stakeholder gains or loses in the change. Therefore alternative $A0$ provides the essential point of reference in all the phases of the decision-making process.

Chapter 4

Criteria and indicators*

AC, FC, FP and RSS

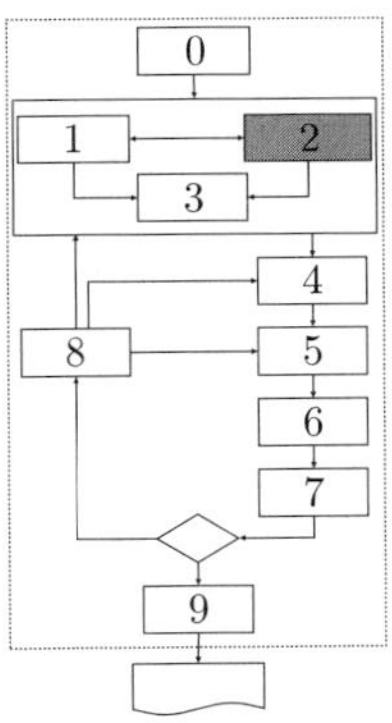

Figure 4.1: The phases of the PIP procedure. Phase 2, which is analysed in this chapter, is highlighted.

The context in which the decision-making process is carried out is defined by its Goal. The Goal of our Project was defined in Section 2.6. The options for intervention were identified in Chapter 3 and will be transformed into *actions* in Chapter 5, thus completing the description of Phase 1 of the PIP procedure. To do that, we first need to identify the *evaluation criteria* and the (*evaluation*) *indicators*, which are the subject of Phase 2 of the PIP procedure (Figure 4.1).

At this point, we should introduce a number of technical terms with formal definitions very abstract, but we would run the risk of disorienting the reader encountering them for the first time. Thus we prefer to proceed gradually, first by introducing these terms in an intuitive way and using them in the development of the chapter. Then we will return to their definitions in the last section of this chapter, to reflect about their nature and their use. In Chapter 3 of THEORY the interested reader will find the formal definitions and a theoretical foundation of the approach presented here.

Some definitions An *evaluation criterion* is an attribute, or a factor, through which the performance of an alternative can be judged with respect to a specific interest. For example, the hydropower producers assess the performance of the Verbano system on the basis of the *Power production*, while the Parks judge them from the *Environmental quality* standpoint.

A criterion in itself is not quantitative, i.e. it does not define a way to measure the satisfaction associated with an alternative. To do this, we must define an *index*, namely a *procedure*, which associates the criterion with a value expressing the degree to which it is satisfied by a given alternative. The value of the index can depend on a number of hydrological variables (level of the lake, release from the Miorina dam, flow diverted into the canals, . . .), or better, on the sequence of values that such variables assume over a given time horizon, under the effect of a given alternative. We term *trajectory* the sequence in

*This chapter is based on Chapter 4 (pp. 59–97) and Chapter 11 of *Modellistica Integrata e decisione partecipata in Pratica: Il Progetto Verbano*, by R. Soncini-Sessa, McGraw-Hill, 2004.

time of variable's values, and *evaluation horizon* the time period over which the alternatives are evaluated. The definition of an index directly from the trajectories of the hydrological variables over the evaluation horizon may prove to be difficult. To simplify this task, the passage from the trajectories to the index is broken down into two steps by introducing an (*evaluation*) *indicator*, i.e. an 'intermediate' quantity between the trajectories and the index: the indicator measures the effects produced by the alternative on the system (in some sense, it synthesizes the information contained in the variables' trajectories), while the index measures the satisfaction associated with those effects and thus with the alternative.

In any case, the definition of the index can still be difficult, but it may become easier if the criterion is broken down into a set of *sub-criteria*. For example, the *Environmental quality* can be articulated in a number of different factors, like the *Erosion of the lake's reed beds*, the *Possibilities for migratory birds to nest*, etc. If it is possible to associate an indicator to each of the new criteria, one can adopt the set of criteria in place of the original evaluation criterion. Otherwise, it is necessary to proceed with a new breakdown, i.e. subdivide each of the new criteria into a new subset. This procedure goes on until one gets to a set of criteria to which indicators are easily associated. This is a reductionist approach: the original problem of the index definition is broken down into a sequence of problems, each of which is easier to deal with. The above procedure can be represented through a *hierarchy of criteria*: in the hierarchy, criteria are ordered according to different hierarchical levels, beginning with the criterion that was originally proposed (called a *root criterion*) and working down to those criteria (called *leaf criteria*) to which the indicators are associated.

In Section 4.2 we will describe the hierarchies that have been defined in the Verbano Project, while in Section 4.3 some indicators will be presented as examples to illustrate how to define them and some of the difficulties that may be encountered. All the indicators that were used in the Verbano Project are defined in Section 4.5 of the DVD.

Once all the evaluation criteria have been defined, the performance of an alternative is described by a vector of indices. In order to evaluate the 'global satisfaction' associated to the alternative, the values of those indices must be aggregated, but this requires comparing the criteria and judging their relative importance. This operation is subjective: each Stakeholder, in fact, tends to attribute a high importance to those criteria which reflect his/her own interest and a low one to the others. An agreement can usually be found about the relevance of criteria concerning homogeneous interests (like, for example, all the criteria which contribute to assess the environmental quality), while conflict emerges when comparing high-level criteria (like, for example, environmental quality vs hydropower production). When an agreement involving a subset of evaluation criteria is reached, a *sector* is formed, which groups the subset of criteria and structures it according to a hierarchy. The root criterion of this hierarchy is called *sector criterion* and its satisfaction can be measured by a single index, called *sector index*, which is obtained by aggregating the indices associated to the criteria in the subset, through a shared procedure. After this definition, the remaining conflict is about the comparison between sectors indices, which is matter for negotiation and political choice. In Section 4.1.2 we will describe the sectors of the Verbano Project and outline the framework according to which they will be compared in Phase 7 of the PIP procedure.

Note that the structure of the Project in terms of indices and indicators is directly linked to the method used for the evaluation of the alternatives. We anticipate that our choice will fall on the Multi-Attribute Value Theory (see Chapter 9), which requires that, given the indicator values, one moves up the hierarchy of criteria until a numerical value is associated to each sector criterion. In Chapter 9 we will show how this aim can be achieved by defining a *partial value function* for each indicator and a *global value function* for each sector, as a weighted sum of the partial value functions. Since the global value is just the sector index that we are looking for, from Chapter 9 onwards, we will use the terms *global value* and *index* as synonyms.

Transient or steady-state? We have seen that the definition of a criterion helps us to evaluate the effects of an alternative with respect to a specific interest, but we have not specified of which effects we are speaking. Two different types can be considered:

- the effects in the construction phase, which means in the period in which the alternative is implemented, and the effects in the *transient* period, during which the system adjusts to the new condition;
- the *steady-state* effects, i.e. those that occur in the period (which is often considered to be never-ending) following the transient.

Consider, for example, the excavation work that would be necessary to excavate the mouth of the lake. During the excavation there would be an increase in suspended solids in the water, which would have a negative impact on the flora and the riverine populations downstream; a certain amount of noise would be produced by the dredges; and traffic would be disturbed by the trucks that transport the material removed. All of these effects would probably cause some discomfort to the population of Sesto Calende. After the excavation work had been finished, the lake level regime would be different and the lake ecosystems would need some time to acclimatize. For example, waves might break more frequently on the reed bed at Bolle di Magadino, causing its slow regression, which in turn would change the nesting possibilities for migratory birds. After a decade or so, the wetlands would probably reach a new state of equilibrium and the bird life would adapt to it. Several species would probably change their life cycle, others would abandon the area. The steady-state effects are those that appear in these new conditions.

Interest in one or the other or in both types of effects depends on the objectives of the Project. When interventions are planned that have effects that extend to the long-term future, as in our case, more weight is usually given to the steady-state effects, since those are the ones that last the longest. Clearly, this is reasonable only if the transient effects are not so devastating as to make the steady-state effects completely uninteresting. Furthermore, evaluating the transient effects requires a more detailed analysis than the evaluation of the steady-state effects. For example, to identify the steady-state effects it is sufficient to say that the excavation will result in a stage–discharge relation of +600 but, to evaluate the transient effects, we must specify the details of the excavation work such as the machinery that will be used; the season when the excavation will take place; the place where the removed material will be dumped; the number of trucks that will be used to remove it; its final destination; the devices that will be used to reduce the diffusion of suspended solids; and so forth. Without these specifications, the effects in the construction phase cannot be estimated. The time and costs of this analysis impose a limit on the study of the effects in the transient period of only a few of the alternatives. To identify the alternatives that should be studied in detail, all the alternatives should be compared with consideration for the steady-state effects alone, if one is reasonably certain that they do not produce transient effects that are largely superior to the steady-state effects. The alternatives that emerge from this first selection should then be analysed deeply to specify the details of the construction phase. This procedure usually increases the number of alternatives because there is more than one way to carry out the interventions. All the resulting alternatives should be subjected to a new decision process in which both the transient effects and the steady-state effects are taken into account. This is the work schedule that we adopted in our Project.

4.1 Sectors and criteria

Before illustrating the sectors defined in the Verbano Project and their evaluation criteria, let us examine the properties that the set of the evaluation criteria must have. As we have just said, the identification of the evaluation criteria can be carried out by defining a number of hierarchies, which can been seen as the branches of a overall hierarchy (the *Project hierarchy*) that structures the whole set of the Project's evaluation criteria.

4.1.1 Properties of the criteria

Keeney and Raiffa (1976) suggest that the criteria in the Project hierarchy should respect the following properties:

- *Completeness*: at a given hierarchical level, a set of criteria is complete if it is adequate for evaluating the extent to which the criteria at the next higher level of the hierarchy have been satisfied.

  To determine a complete set, one can recursively apply the *importance test* suggested by Ellis (1970): a criterion is *important* if the Experts believe that its presence can influence the choice of the best compromise alternative. When the Experts are no longer able to suggest important criteria, the set can be considered complete. In particular, completeness requires that for each action at least one criterion exists by means of which its effects can be evaluated; and that for each Stakeholder there is at least one criterion that expresses his/her viewpoint.

- *Operability*: the criteria must be chosen so that the evaluation of an alternative is easy and comprehensible to the Experts and the Stakeholders. It is thus necessary that these individuals collaborate with the Analyst to define the criteria.

- *Decomposability*: the set of criteria must be decomposable into a number of subsets, such that the evaluation of the criteria of each subset can be carried out independently from the others. If this condition does not hold, decision-making problems in which the number q of criteria is even modestly high (e.g. $q = 5$) become extremely difficult to handle.

- *Non-Redundancy*: the same effect should not be considered twice, i.e. aspects, which are actually the results of the same effect, should not be considered separately.

- *Minimum size*: to avoid increasing the complexity of the Project unnecessarily, the number of criteria must be kept as low as possible, clearly with respect for the Properties of Completeness and Operability.

4.1.2 Sectors

Now, let us see which are the sectors in the Verbano Project. In the first section of this chapter we have given a quite intuitive and approximate definition of this concept; a more

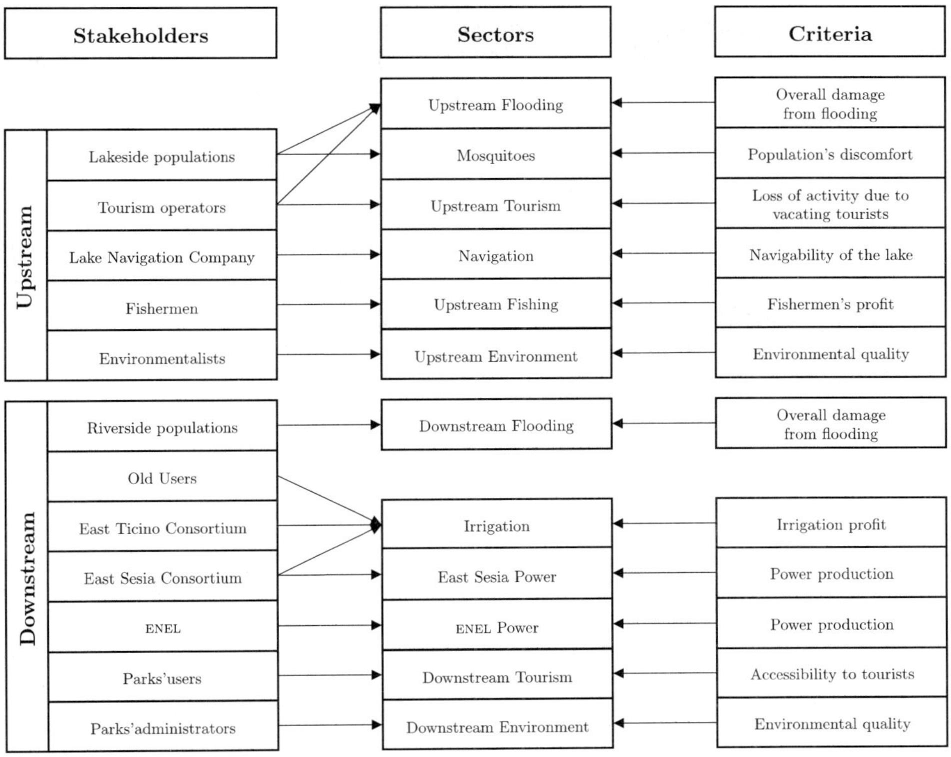

Figure 4.2: The Stakeholders, the sectors, and the corresponding criteria in the Verbano Project.

rigorous one can be found in the last section of Chapter 3 of THEORY. From these definitions it follows that the sector should be identified when the evaluation criteria are compared and their relative importance is determined, a task that is tackled in the moment when the alternatives are evaluated and compared (Phases 6 and 7 of PIP procedure). It is at this very moment, in fact, that the Stakeholders can appreciate the effects of the different alternatives and express their preferences. In practice, however, the sectors can often be determined, at least tentatively, right from the *Reconnaissance* phase, by individuating homogenous interest groups for which one can presume that the definition of a sector index will not generate conflicts during the evaluation. This is the way in which we proceeded in the Verbano Project. Figure 4.2 shows the set of sectors that have been defined, the corresponding sector indices and the Stakeholders interested in each of them.

The set of sectors constitutes a partition of the Stakeholders' interests in the system. This partition must be such that the same interest is not considered in more than one sector. This requirement is a consequence of the Decomposability and Non-Redundancy Properties (see previous section), which must hold at every level of the hierarchy, and thus also at the level where the sector criteria appear. In the following, we will check that this requirement holds for our set of sectors.

As for the upstream sectors, there is no overlap between *Environment* and *Fishing*, because the second represents the economic interests of the fishermen and is not concerned with safeguarding the lake and river environment, which are concerns for the first. There is no overlapping among these sectors and the others either. It might seem that there would be some overlap among the sectors *Flooding*, *Navigation* and *Tourism*, because flooding has a big influence on both navigation and tourism. Actually, double counting can be avoided by defining each sector with care.

In our definition of the *Navigation* sector, the damage caused by high lake levels is due to the fact that, when at one of the ports the water level gets close to or exceeds the level of the wharf, raised walkways must be provided to allow passengers and goods to board transport vessels. At particularly high levels, the service must be interrupted. The damage is due to the cost of putting the raised walkways up and to the lost revenue due to the decrease or interruption of the activity. Damage is incurred by the *Navigation* sector also when the lake level is very low and the Intra-Pallanza ferry must limit loading heavy goods to avoid hitting the bottom. Among the interests of this sector we do not include the damage to the ticket office or to the offices of the Lake Navigation Company, which are produced by flooding. They can be assimilated into the damage that all the other structures in the flooded area incur and so they are considered in the *Flooding* sector.

In general, tourists begin to abandon the riverine locations before the lake floods because the beaches are no longer usable. They may also decide to leave when the level of the lake is too low and long muddy beaches are exposed, or because mosquitoes begin to swarm. Since all these phenomena produce the same effect (the reduction of tourist numbers) it is logical to consider them jointly. It is these effects that influence the *Upstream Tourism* sector. Among the interests of this sector we do not include damage to accommodation structures that occur when the lake floods. They are considered in the *Flooding* sector.

The concerns about flooding could have been grouped with those involving the proliferation of mosquitoes, to form a *Lakeside Population* sector, but we preferred not to do so. Several observations led us to take this decision, since the two phenomena are very different, as are the interests connected to them. Firstly, the concern about flooding does not only involve the lakeside population but is shared by local business people as well (and to take account of this we would have had to create a separate sector *Business*). We also considered the fact that the problem of mosquitoes affects a specific location: the area of Bolle di Magadino.

Similar considerations are valid for the downstream sectors, and are not worth repeating in detail, but it is interesting to comment on the presence of two different *Power* sectors. The East Sesia Consortium interests as far as irrigation is concerned are equivalent to those of the East Ticino Consortium and the Old Users: they all pursue the maximization of irrigation profit. Since there is no conflict among these three groups (in Section 3.4 we said that they do not question the distribution policy with which the water is distributed between them, see also Chapter 6), it seems reasonable to group their interests into a single *Irrigation* sector. This means, as we will see, that all the irrigators will use the same sector index to evaluate the performances of the alternatives with respect to irrigation. On the other hand, the East Sesia Consortium is also interested in hydropower production, since it owns six small hydropower plants along the Regina Elena Canal. Its interests in hydropower production, however, cannot be grouped with those of ENEL, because the latter uses its abstraction not only for hydropower production but also for cooling the Turbigo-EUROGEN thermoelectrical plant (see Section 2.2.2), and thus the satisfaction of the ENEL *Power* sector and that of *East Sesia Power* cannot be defined in the same way, as will become clearer in the following section.

4.1.2.1 Comparing the sectors

The comparison of the sector criteria and the assessment of their relative importance is the object of the *Comparison* phase and will be discussed in Chapter 12. Here we just want to outline the way in which this comparison will be carried out. To do this, remember that, according to the participatory nature of the PIP procedure, the Stakeholders will play an active role during that phase too, so they have to choose the framework in which the comparison will take place. For example, they might want to discuss and compare the alternatives with the other Stakeholders from their own country first, and then with the Stakeholders from the other country. This approach groups the Italian interests on one side (encompassing all the sectors, both upstream and downstream), and the Swiss interests on the other (which consider only the upstream sectors), allowing for the introduction into the Project hierarchy of an intermediate level between the Project Goal and the sector criteria, composed of the two criteria *Italian interests* and *Swiss interests*. The choice of such an approach would reveal that the Stakeholders from each country are only sensitive to the issues that are present within the confines of their own territory and would inevitably lead to accentuating the conflicts between Italy and Switzerland. The adoption of this framework runs the risk of making fundamentally opposing positions emerge, diminishing the possibility to identify a compromise, and this is why the Stakeholders did not want to proceed in that direction. For the same reason, they refused to group the sectors according to an upstream and downstream division (i.e. they rejected a Project hierarchy with an intermediate level composed of the criteria *Upstream interests* and *Downstream interests*). Instead, they opted for a comparison of all the sectors at the same time, with no grouping at a higher level.

4.2 Hierarchies of criteria

Once the interests involved in the Project have been determined and the evaluation criteria have been listed, the corresponding indices must be defined. As already anticipated, the definition of the index associated to an evaluation criteria can be difficult when one tries to achieve it in just one step, but it can become easier if the criterion is recursively broken down into sets of sub-criteria, through a hierarchical procedure, until a set of criteria to which indicators are easily associated is found.

In the Verbano Project, we started by grouping the interests into sectors and defining a sector criterion for each of these (see previous section). Then, for each sector criterion, the above procedure was carried out, a sector hierarchy was determined, and the indicators associated to its leaf criteria were defined. To do this, the Analyst interviewed only the Stakeholders interested in the sector being considered. Since in this phase the Stakeholders are requested to collaborate in very technical operations, such as the definition of the indicators associated to the leaf criteria, it is suggested that they be accompanied by *Experts*. As we will see in Chapter 9, we will interact with these Experts in the *Evaluation* phase as well, using the hierarchies identified in the present phase. In Figure 4.3 the names of the Experts that collaborated in the Verbano Project are listed.

4.2.1 Upstream sectors

4.2.1.1 Upstream Flooding

The sector criterion is the *Overall damage from flooding*, whose intensity depends on the trajectory of the lake levels produced by the alternative under examination. A precise esti-

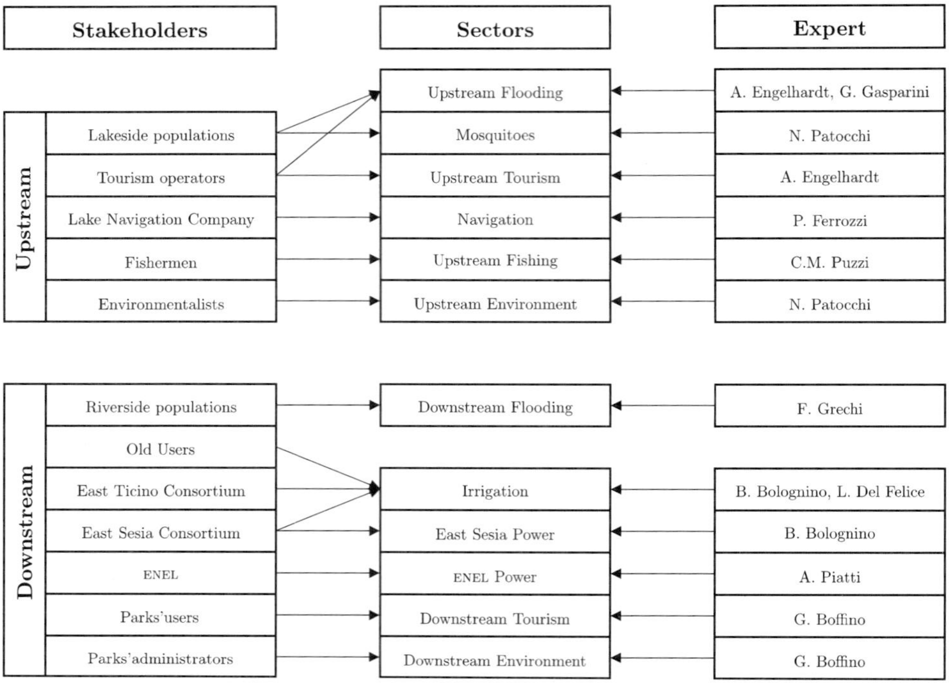

Figure 4.3: The Experts in the Verbano Project and their sectors of competence.

mation of the overall damage would require the assessment of the damage at every point along the lake's perimeter, but such an assessment would take too much time. The Experts believed that an estimate for each township would be sufficient (Figure 4.4). In each township the damage can be divided into three categories: the *Damage to structures* (including machinery), the *Damage to economic activities* and the *Damage caused by interruptions to*

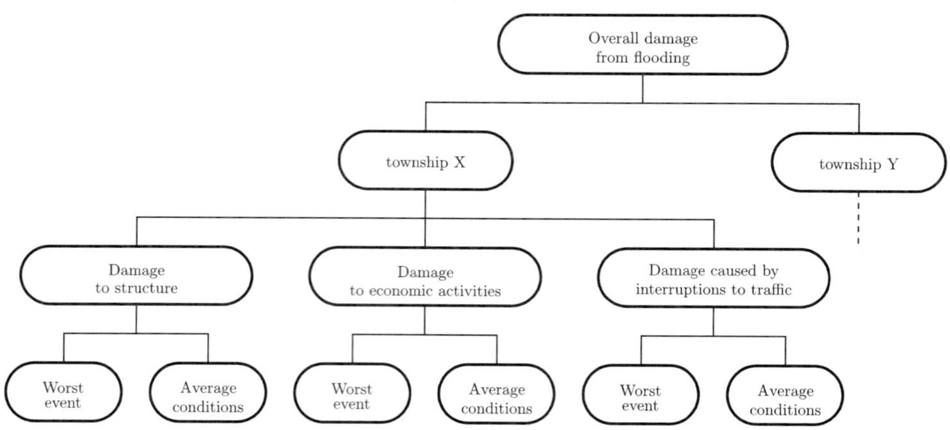

Figure 4.4: The hierarchy for the *Upstream Flooding* sector.

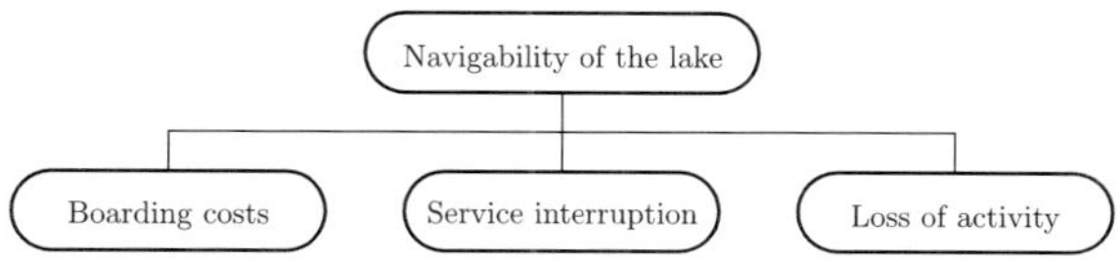

Figure 4.5: The hierarchy for the *Navigation* sector.

traffic. The first can be estimated according to the cost of repairing the structures; the second on the lost revenue due to the interruption of economic activity during the flooding and in the period required for the repairs to structures and machinery; and the last on the length of time that the streets are flooded. Note that the *Damage to economic activities* accounts both for the damage to the activities that are directly hit by the flooding and for the damage that is indirectly caused to activities that are related to them.

There are, then, three second-level criteria. Each of them is further subdivided into two third-level criteria, which take account of the Stakeholders' aversion to risk. They are, in fact, sensitive to the *Average conditions*, but look fearfully at the *Worst event*. For more details see Sections 2.2.1.1 and 4.5.5 of the DVD.

4.2.1.2 Navigation

The sector criterion is the *Navigability of the lake*, which is made difficult or impeded both when the water level is too high, and when it is too low (Figure 4.5). When at any port the water level reaches or exceeds the level of the wharf, the Lake Navigation Company must put up raised walkways to allow passengers to board. It is therefore subject to costs which are quantified with the leaf criterion *Boarding costs*. At very high levels the walkways are no longer safe and the *service* at that port must be *interrupted*. At *low levels* the ferry on the Intra-Laveno line risks touching the bottom. To avoid this, it is necessary to limit the amount of weight that can be transported which results in a *Loss of activity*. For more details see Sections 2.2.1.2 and 4.5.8 of the DVD.

4.2.1.3 Upstream Tourism

The sector criterion is the *Loss of activity due to vacating tourists* (Figure 4.6), which is due to *Reduced landscape aesthetics*, when the water level is low enough to expose the muddy lake bed, or to *Reduced access to beaches*, when the lake levels are very high, and also to *Discomfort produced by mosquitoes* (for details see Sections 2.2.1.3 and 4.5.10 of the DVD).

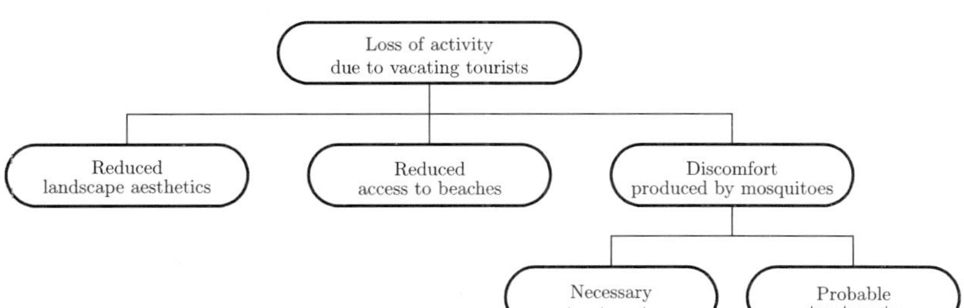

Figure 4.6: The hierarchy for the *Upstream Tourism* sector.

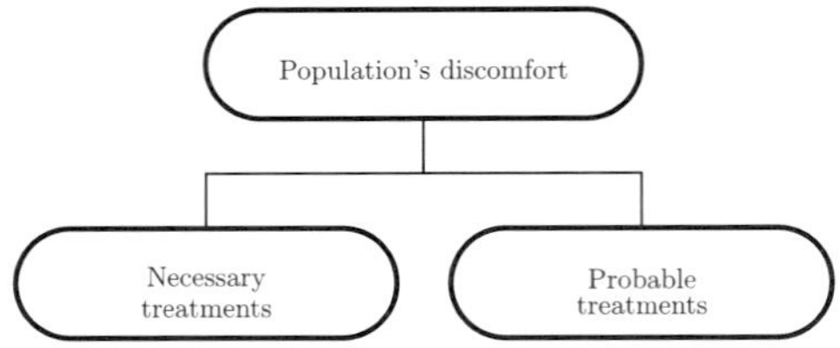

Figure 4.7: The hierarchy for the *Mosquitoes* sector.

The latter depends in turn on the leaf criteria that we will describe in discussing the sector *Mosquitoes*.

4.2.1.4 Mosquitoes

The sector criterion is the *Population's discomfort* produced by the swarming of mosquitoes.[1] The mosquitoes reproduce in the wetlands when the ground in which the eggs were laid is temporarily submerged in water (for details see Sections 2.2.1.4 and 4.5.12 of the DVD). In order to contain their proliferation in the Bolle di Magadino area, which is highly subject to the phenomenon, the Bolle di Magadino Foundation treats the reproductive areas, and the number of treatments depends on the frequency and the length of the periods they are flooded. It is possible to distinguish situations in which it is absolutely necessary to treat the areas and others in which it would just be advisable. In this second event, the course of action is evaluated on a case-by-case basis depending on the information available at the time. Since this cannot be known in advance, we are not able to estimate the exact number of treatments and so we make the distinction between *Necessary treatments* and *Probable treatments* (Figure 4.7).

4.2.1.5 Upstream Fishing

The *Upstream Fishing* sector includes the economic interests of Verbano's fishermen (see Sections 2.2.1.5 and 4.5.9 of the DVD) and is not concerned with environmental protection, which is included in the *Upstream Environment* sector. The sector criterion is the *Profit* of fishing operators, called *fishermen* for short (Figure 4.8). An unsuitable regime of lake levels can affect the reproduction of fish species that breed in coastal areas, with a consequent negative impact on the size of the catch and the fishermen's profit. This can happen for two reasons, depending on the species being considered. For some species (the bleak and white

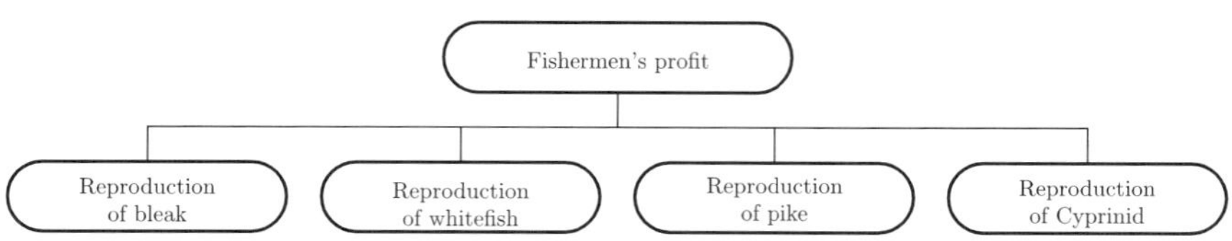

Figure 4.8: The hierarchy for the *Upstream Fishing* sector.

[1]Considering the discomfort produced by mosquitoes both for tourists and for the resident population does not constitute a violation of the Non-Redundancy Property because we are dealing with two distinct effects. However, these effects can still be traced back to the same leaf criterion since the cause is the same.

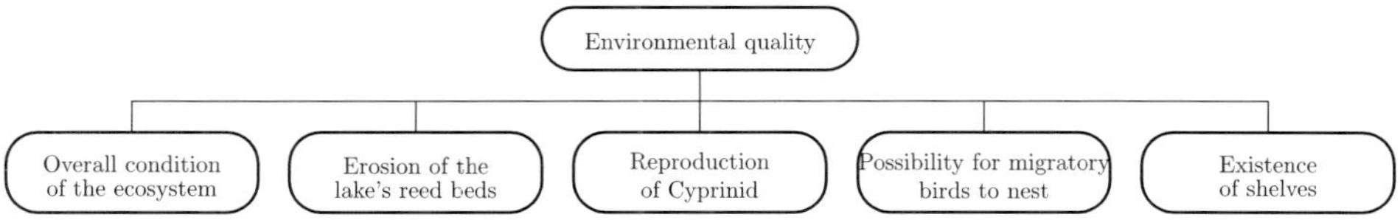

Figure 4.9: The hierarchy for the *Upstream Environment* sector.

fish that breed along the shoreline) there are problems when, in the breeding season, the water level in the lake decreases below the level where the eggs were laid, which dry out and die. For other species (the pike and Cyprinid) water levels that are too low can inhibit them from reaching the zones of reed beds where they lay their eggs. These are the species that are most sensitive to the lake level regime and they are also the most profitable ones for fishing. The leaf criteria thus refer to the bleak, whitefish, pike and the Cyprinid.

4.2.1.6 Upstream Environment

The sector criterion is the *Environmental quality*, which depends on the lake level regime, both because it influences the *Overall condition of the ecosystem* and because it affects the evolution of several specific phenomena that can compromise the functioning of the ecosystem (Figure 4.9). Specifically, we are referring to the *Erosion of the lake's reed beds*, to problems connected with the *Reproduction of Cyprinid*,[2] to the *Possibilities for migratory birds to nest* in the reed beds and, lastly, to the *Existence of shelves*, which are used both by migratory birds (*Waders*) as a resting place, and by endangered plants (such as *Nanocyperion*). Each one of these phenomena, which are described in detail in Section 2.2.1.6 of the DVD, is accounted for through a specific leaf criterion (see also Section 4.5.1 of the DVD).

4.2.2 Downstream sectors

4.2.2.1 Downstream Flooding

For the *Downstream Flooding* sector it was decided to adopt the same criteria that were identified for the *Upstream Flooding* sector, even if here it is the flooding of the River Ticino that is being considered (see Sections 2.2.2.1 and 4.5.6 of the DVD). The sector criterion is the *Overall damage from flooding* (Figure 4.10).

The definition of the criterion requires the clear specification of the term *downstream* when it is used in reference to flooding. In Section 2.1.2 we defined the *downstream territory* as the territory that is served by the Ticino and its network of canals, but a quick reflection shows that this definition is insufficient for defining the territory that could be impacted by a flood wave coming from the lake. We must return to that definition and make it more appropriate (this is a typical example of recursion within the decision-making procedure). There are three possibilities (Figure 4.11):

(1) consider only the stretch of the River Ticino upstream of Pavia;

(2) include Pavia, taking into account the *backflow* of the River Po;

(3) consider the whole stretch of the Po, right up to its delta at Pontelagoscuro.

[2] We consider this fish family here because it is at risk of extinction, not because of a reduction in fishing, an effect that has already been considered in the *Upstream Fishing* sector.

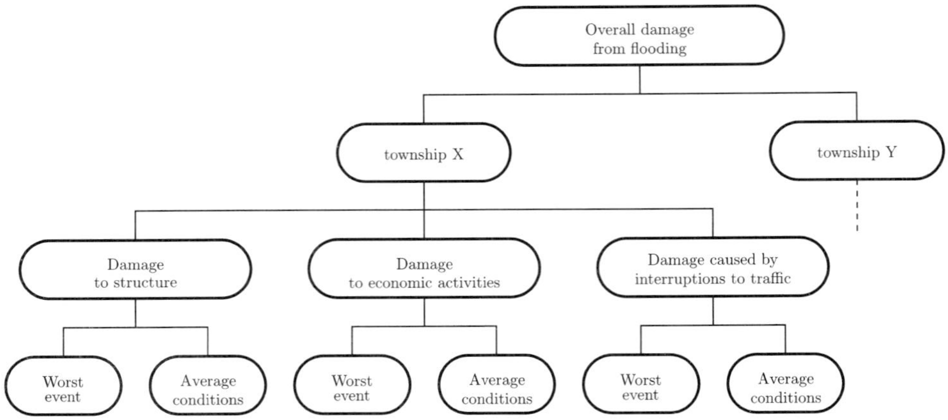

Figure 4.10: The hierarchy for the *Downstream Flooding* sector.

We rejected the first possibility because the Experts believe that the flooding at Pavia is a problem that the Verbano Project could not ignore. The third, which might seem the most thorough, is not right either: the regulation of Verbano lake would be given the responsibility of dealing with the flooding damage all along the Po, without taking into account that the damage is the result of the synergic effect of the release from all the regulated lakes (Lake Como, Lake Iseo and Lake Garda), whose effluents are tributaries of the Po, and that these lakes should be managed by an integrated control system. Nowadays, in normal conditions the management of the Po is delegated to the Agencies in charge of the regulation of the individual tributaries, and only during big flood events is control assumed by the 'Magistrato

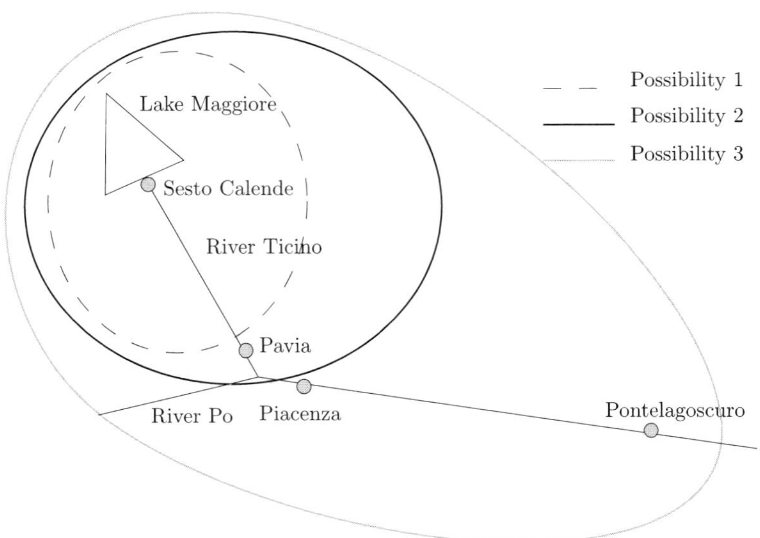

Figure 4.11: Three possibilities for the definition of the territory in which to evaluate the effects of downstream flooding.

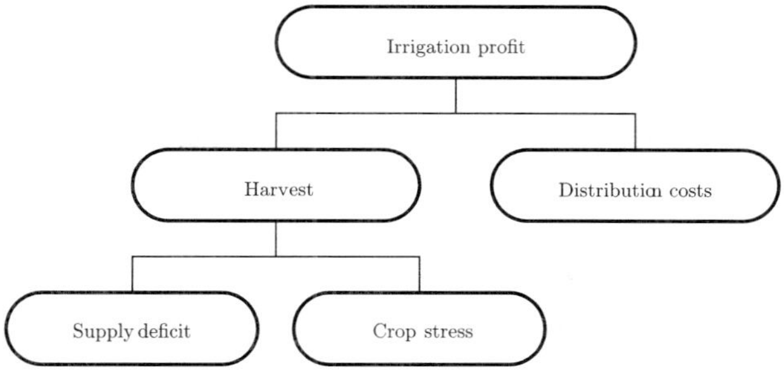

Figure 4.12: The hierarchy for the *Irrigation* sector.

per il Po', the Agency concerned with Po floods. From this perspective, the management system for each tributary can be designed considering that tributary as standing alone; then, when all the individual systems have been defined, the integrated system can be studied for the management of exceptional floods. The design of such an integrated system can thus be considered as a *mitigation measure* (Chapter 14) in the context of the Verbano Project. For these reasons, we adopted the second possibility. Note that from the point of view of protecting the downstream population, this choice provides a stronger guarantee than the one actually provided by the Verbano Regulation Licence, which does not consider the interaction between the Ticino and the Po. In fact, it does not require that the hydraulic conditions of the Po be considered in regulating Verbano, in order to avoid the flood peak of the Ticino arriving at Pavia synchronously with the peak of the Po. When this happens, the sum of the two floods can produce disastrous effects in the lower Polesine region, and, in fact, this is precisely what happened in the dramatic flood of 1951.

4.2.2.2 Irrigation

The sector criterion is the *Irrigation profit*, which increases with a greater *Harvest* and a reduction in the *Distribution costs* of the water in the canals, two quantities that are influenced by the regime of releases at Miorina (Figure 4.12).

Let us consider the first criterion. The harvest, i.e. the biomass of a crop at the time of harvest, could be computed directly with a crop (growth) model. In that case, the *Harvest* would be a leaf criterion and the *Average annual biomass produced* the associated indicator. Unfortunately, dynamic crop models have a very large *state*, because they describe the physiological processes of the plants in detail. They are therefore too complex to be adopted in our Project, since they would require prohibitive computing time (centuries) to design the policies (see Section 7.8). Thereby the *Harvest* criterion should be disaggregated into a set of criteria that can be associated with indicators that are easier to compute. In order to identify them, consider a simple non-dynamic crop model, as for example the FAO model (FAO, 1986 and 1995). It expresses the biomass Y_a of the harvest as a function of the potentially obtainable biomass Y_m (what one would obtain without water limitations) and of the evapotranspiration deficit δ, i.e.

$$Y_a = (1 - k\delta)Y_m$$

The coefficient k, termed *production coefficient*, depends on the crop that is being considered, while the evapotranspiration deficit δ is a function of the *supply deficit* (or *water failure*), namely the sum of the daily differences between the volume of water delivered to the crop and the *reference water demand*. Among the criteria at a hierarchical level lower than *Harvest* we must therefore include the *Supply deficit*. Note that, for the rice crop, which is the primary crop on the right bank of the Ticino, one must look very carefully at the supply deficit in the *submersion* period in which the paddies are flooded (from the end of March to the beginning of May).

The FAO model is very simple and does not take account of the *stress* that a crop is subject to in conditions of water failure. The water that the crop uses comes in part from the ground, from which it is absorbed by its roots, and in part from precipitation which is directly intercepted by its leaves. Soil is a porous substance, made up of a solid matrix and pores, which are filled with air, water and solutes. At the end of sufficiently abundant precipitation events or irrigation cycles, all the pores are filled with water and the ground is said to be *saturated*. Then, the force of gravity causes a part of this water to percolate slowly towards the water table and the amount of water in the pores, termed *soil moisture*, decreases. When the capillary action manages to counterbalance gravity, percolation becomes negligible. At this equilibrium point, only a fraction of the original water is still present in the ground: this amount of water is called the *field capacity* of the ground. Without hydrological inputs the soil moisture continues to diminish through the suction of the plants and direct evaporation. When the soil moisture reaches very low levels, the root apparatus of the plant is no longer able to resist the capillary action of the ground. Absorption ceases, the cells of the plant begin to lose humidity, and the non-woody tissues lose their turgidity and wilt. If the lack of water continues, the plant is destined to die and, even if water is given, its development has been affected. The moisture content of the soil at which this process begins is called *wilting point*. The *plant-available water*, namely the amount of soil water that can be exploited by the plant, is equal to the difference between the field capacity and the wilting point and it is only a fraction of the water that falls on the ground from precipitation or irrigation. However, the amount of water that the plant can actually exploit is only a fraction of the plant-available water and when that amount is lower than the volume of water lost through evapotranspiration, the plant goes into stress.[3]

When evaluating an alternative, it is important to take into account that stress depends on the time distribution of the deficit. A high deficit, equally distributed over a long period, is less dangerous than a lesser deficit concentrated in a short period. The plant, in fact, can develop forms of compensation and specialization in response to water scarcity, by adapting its metabolism. If water scarcity is present from its germination, the plant compensates by developing smaller organs and grows more slowly. On the contrary, if the plant is used to having an abundance of water and all of the sudden there is none, the same deficit hits it much harder than in the previous case.

In conclusion, to estimate the harvest it is necessary to introduce a second criterion: *Crop stress*.

Lastly, let us consider the criterion *Distribution costs*, which takes into account the economic and social costs induced by the regime of releases. In a consortium it is hardly ever possible to serve all the users at the same time, because the water pressure is usually insufficient. Irrigation is done in turns, so that at any time only a subgroup of the users is serviced. To achieve the desired distribution, the network's hydraulic organs (generally floodgates) are put in certain positions and kept there for periods of time whose duration

[3]To avoid this stress the duration of the irrigation turns is defined as a function of the characteristics of the ground.

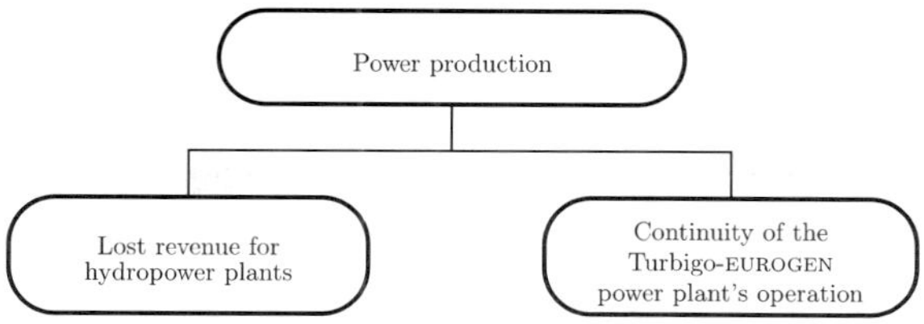

Figure 4.13: The hierarchy for the ENEL *Power* sector.

depends on the water volumes diverted to the consortium. On the basis of the volumes that are foreseen at the beginning of each turn, a distribution plan is established which defines the day and the hour at which each user will be serviced. If the allocated volume varies during the turn, to maintain the planned distribution the position of the floodgates and the plan must be modified. If this happens there is an increase in the number of manoeuvres that must be made, and so the distribution costs increase, the water distribution turns lengthen and inconvenience for the farmers follows.

These issues are examined in detail in Sections 2.2.2.2 and 4.5.7 of the DVD.

4.2.2.3 ENEL Power

The sector criterion is *Power production*, which is influenced by the regime of releases and by the water diverted to the Industrial Canal (Figure 4.13). The regime influences both *hydropower production* and *thermoelectric power generation* (see Sections 2.2.2.3 and 4.5.3 of the DVD). With regard to *hydropower production*, the criterion used by ENEL is the *Lost revenue* induced by water failures with respect to the Abstraction License of the hydropower plants. This criterion is preferable to the total supply deficit, since ENEL pursues profit, which depends not only on the water volume run through the turbines, but also on the price of electricity; and the latter varies according to the month, the day of the week, and the time of day (see the figures in Section 2.2.2.3 of the DVD). With regard to *thermoelectric power generation*, ENEL is interested in guaranteeing the continuity of activity at the Turbigo-EUROGEN power plant, whose cooling towers demand that the Industrial Canal take no less than 25 m^3/s. If this volume is guaranteed, the energy produced by Turbigo-EUROGEN is independent from the volume supplied. For this reason the leaf criterion for this branch is the *Continuity of the Turbigo-*EUROGEN *power plant's operation*. If the volume in the Industrial Canal drops below 25 m^3/s, the plant must be shut off, and this would create serious difficulties for supplying power to the city of Milan. The damage that would follow would be far greater than the consequences that would result from the same deficit for any other downstream user, and so it is rational for the Industrial Canal to be given priority for the first 25 m^3/s of water released from the lake. This norm is actually present in the distribution policy among the canals, as emerged from the interviews with the Stakeholders (Section 6.7.3). The continuity of operation is not as interesting for the hydropower production since, apart from the lost revenue, the effects of shutting off the hydropower plants are negligible. Their combined installation power is only 70 MW and thus much easier to substitute than the 1230 MW produced at Turbigo-EUROGEN.

4.2.2.4 East Sesia Power

The regime of releases influences the volume allotted to the Regina Elena Canal and so also the energy produced in its six small power plants, which are managed by the East Sesia Con-

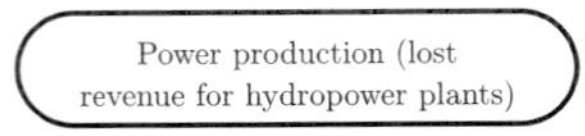

Figure 4.14: The hierarchy for the *East Sesia Power* sector.

sortium. The sector criterion is the same one identified for ENEL: *Power production*. Since the energy produced is for the Consortium, there is no problem of service continuity, and the only evaluation criterion is the *Lost revenue for hydropower plants* (see Sections 2.2.2.4 and 4.5.4 of the DVD). Note that in this case the leaf criterion coincides with the sector criterion (Figure 4.14).

4.2.2.5 Downstream Tourism

Unlike what happens around the lake, along the Ticino there is no real tourist industry, only a few restaurants and *trattorie*, which are frequented more for the rural environment where they are found than for the proximity of the river. Therefore, the sector criterion is not the *Loss of activity due to vacating tourists* as in the upstream areas, but the *Accessibility to tourists* of the area belonging to the two Ticino Parks (Figure 4.15), which are used by the inhabitants of the nearby cities for recreational activities. These activities require access to beaches on the muggy weekends in July and August, to the river for canoe trips, to paths for walking and cycling, ..., which become difficult or impossible when the Ticino is too high (for swimming, canoeing, walking and cycling) or too low (for swimming and canoeing). The two leaf criteria are *Access to the land* and *Access to the water* (see Sections 2.2.2.5 and 4.5.11 of the DVD).

4.2.2.6 Downstream Environment

The sector criterion is *Environmental quality* (Figure 4.16) of the riparian and riverine ecosystems that prosper in the river and in the oxbow lakes in the valley of the Ticino,[4] whose *Overall condition* is influenced by the flow regime of the river. The fish species that populate the river and some of the birds are damaged by rapid variations in the Ticino flow rate during the breeding seasons. Fishes are sensitive to sudden decreases in flows, which expose the eggs that were laid on the river bottom, while birds can suffer from a rapid rise in river level, which endangers the nests built on the banks. Since these variations in flow are produced by the manoeuvres made at the Miorina dam or at the diversion dams of the

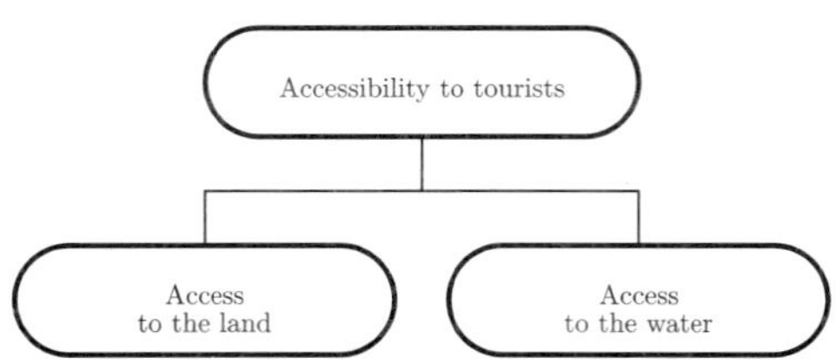

Figure 4.15: The hierarchy for the *Downstream Tourism* sector.

[4]The environmental conditions in the irrigation canals are not taken into consideration, given that the greater part of their route is outside the valley. Their role is to take water to the plateau.

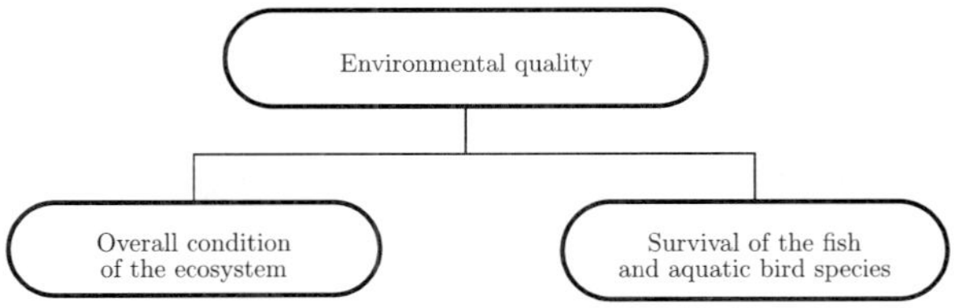

Figure 4.16: The hierarchy for the *Downstream Environment* sector.

canals, we must introduce the leaf criterion *Survival of the fish and aquatic bird species* to take into account these effects (see Sections 2.2.2.6 and 4.5.2 of the DVD).

4.3 Examples of indicators

In the first section of this chapter we defined an indicator as a functional of the trajectories[5] of the hydrological variables (level of the lake, release from the Miorina dam, flow diverted into the canals, ...) that characterize the system. Indicators provide a measure of the *effects* of a given alternative on the system, making it easier for the Stakeholders to define an index, which is a measure of their *satisfaction* with that alternative. Clearly, there must be as many indicators as there are evaluation criteria. In the previous section we defined the latter; now we must associate an indicator to each of them. Defining the indicators is an indispensable and unavoidable operation, but it is long and tedious. For this reason we included it in the DVD (Section 4.5) and not here. Here we will simply reproduce a few definitions, chosen to give the reader an idea of the difficulties that can be encountered in this operation.

Up to this point, we have taken it for granted that the satisfaction associated to a given alternative must be expressed in quantitative terms: this is why we defined an index as a function of the corresponding indicator, which in turn is a functional of the system trajectories. However, since indices must express an Expert's judgement, their values can be directly assigned by the Expert himself, after looking at the trajectories that the different alternatives produce. In this case the procedure from the trajectory to the index is not formalized: it occurs 'in the mind of the Expert'. The judgement of the alternative (and also the index) is expressed in qualitative terms, using an adjective (good, bad, acceptable), or by assigning the alternative a position in a ranking (first, second, third) or by giving it a score (eight, two, five points out of ten). Clearly, when indices are expressed in an *ordinal scale*, arithmetic operations (such as weighted sum) cannot be performed upon them, only comparative judgements.

In the Verbano Project we decided to use only quantitative indices, defined by formalized procedures which require the intermediate step of indicators' definition. The reason for this choice is that, as we will see in Section 8.3, the trajectories that were used to estimate the indicators are very long (9768 days) and the number of alternatives to be evaluated is high (about 150). Therefore, it is almost impossible for an Expert to evaluate the alternatives by directly comparing all the trajectories. It seemed more rational to identify an automatic procedure that could substitute the Expert. To do this, it is necessary to work alongside the Expert, to understand how he would judge an alternative and to translate this way of judging into a formal procedure that constitutes an operational definition of the indicator.

[5]Which are obtained by simulating the dynamics of the system, subject to the alternative under examination, over a sufficiently long evaluation horizon.

To be assured that the procedure so identified is well defined, and that it could therefore be used as a substitute for the Expert in the evaluation of the alternatives, a validation test should be performed. For this, several alternatives are chosen and for each of them the indicator value is computed. At the same time, the Expert is asked to rank the alternatives by directly comparing the trajectories they produce. Then, the ranking provided by the indicator is compared with the one provided by the Expert. If they are different, the procedure identified must be re-examined. Sometimes its identification fails because the Expert judges the trajectory at a glance, on the basis of its shape, and has difficulty in verbalizing how he made his judgements. When this happens, the indicator provides a ranking of the alternatives different from the one of the Expert, because its definition neglects some effects of the phenomenon that the Expert notices when he looks at the trajectory. The Analyst should try again with patience, by analysing this results together with the Expert and trying to understand where the error comes from. Once the test has been passed successfully, the indicator is identified, but if one feels that the Expert's judgement is essentially intuitive, its a good idea to repeat the test, after some time has passed, to make sure that the judgement is coherent and stable over time.

Notation An indicator will be denoted with the letter i, accompanied by a code that is constructed in the following way: some letters from the name of the sector, followed by the letter U or D, which distinguishes its belonging to a sector from the *u*pstream group or the *d*ownstream group and at the end a progressive number within the sector. For example, the code $i_{\text{Env\_D\_1}}$ identifies the first indicator in the *Downstream Environment* sector.

In the mathematical formulation of the functional (i.e. of the procedure) that defines an indicator we will use the following symbols:

- t denotes the generic instant of time (which as we will see in Section 6.3 is a day) in the time horizon over which the indicator is computed. Correspondingly, $[t, t+1)$ is the time step (day). The state variables, such as the level of the lake h_t^{SC}, refer to 8 o'clock in the morning on day t; the flow variables, such as the inflow a_{t+1} and the release r_{t+1}, refer to the volume of water that flows through a given section in the interval $[t, t+1)$.

- H denotes the time horizon over which the indicator is computed. It is called *evaluation horizon*, because it is the time horizon over which the performances of the alternatives are evaluated (Section 8.3). The horizon H includes N years, among which the generic year is indicated with the symbol y. To simplify the notation this last symbol will be used both with an ordinal meaning, to indicate one of the N years (i.e. $y = 1, \dots, N$), and also to denote the corresponding time horizon (i.e. $t \in y$ means that t is a day of the year y). The symbols $|H|$ and $|y|$ indicate the total number of days in the horizon H and in the year y, because $|\cdot|$ denotes the operator that provides the *cardinality* of the set.

- Next to an indicator the symbol ($\downarrow$) indicates that the Stakeholders want to minimize its value, while the symbol ($\uparrow$) indicates the opposite.

4.3.1 ENEL Power

We have seen that the two branches of the hierarchy in Figure 4.13, which are related to hydropower production and thermoelectric power generation respectively, are quite different from each other. In the first only economic effects show up and so the corresponding indicator will be expressed in monetary terms. As for the second, the effects of the closure of the Turbigo-EUROGEN plant are difficult to quantify; what is certain is that the economic impact of the closure would be much greater than any negative effect concerning the hydropower production. Therefore, it is not advisable to express the corresponding indicator in monetary terms, but rather in another unit of measure, such as the number of days that the plant is unable to operate.

Let us consider first the leaf criterion *Lost revenue for hydropower plants*. Associating it with an indicator is a simple task, because the criterion itself suggests the indicator's definition: the monetary value of the total production deficit. More precisely, the verbal definition of the indicator is the following

$i_{\text{EnP\_D\_1}}$ (↓) average annual lost revenue [k€/yr] for the ENEL hydropower plants

which we now must translate into a computation procedure.

The value of $i_{\text{EnP\_D\_1}}$ is computed as the sum of the daily lost revenues over the evaluation horizon, divided by the number N of years. The daily lost revenue is the product of the price of energy and the overall production deficit in the complex of the ENEL hydropower plants. For a given hydropower plant of this complex, each day the production deficit is defined as the difference between the energy that the plant would have produced on that day if a volume that matched the water demand (see Section 2.2.2) had run through the turbines, and the energy that it actually produced with the volume that was provided. More precisely, let us denote with $\overline{Q}_t$ [m^3/s] the flow demanded in day t and with $\underline{Q}$ [m^3/s] the minimum flow that can be run through the turbines of that plant. If on day t the average water flow that is actually run through the turbines is q_t [m^3/s], the *energy E_t produced* [kWh] within the day is

$$E_t = \begin{cases} \psi g \eta \gamma q_t H & \text{if } q_t \geqslant \underline{Q} \\ 0 & \text{if } q_t < \underline{Q} \end{cases} \tag{4.1}$$

where ψ is a dimensional conversion coefficient (whose value is 0.024 hours), g [m/s^2] the gravitational acceleration, η the turbine efficiency, γ [kg/m^3] the water density and H [m] the hydraulic head.

Since the turbine efficiency η depends on the technology used in the power plants, a factor which does not depend upon the regulation of the lake, it is advisable to assume the value 1 for it, so that E_t takes on the meaning of *hydraulic energy provided to the plant*. The *maximum hydraulic energy $E_t^{\max}$* that can be legally provided to the plant is then

$$E_t^{\max} = \psi g \gamma \overline{Q}_t H \tag{4.2}$$

Therefore the production deficit $D_t(q_t)$ is

$$D_t(q_t) = E_t^{\max} - E_t = \psi g \gamma H \left(\overline{Q}_t - q_t\right) \tag{4.3}$$

The price of energy is time-variant and depends on the month, the day and the hour of day being considered. The highest price is paid in the peak hours in winter,[6] the lowest in the evening and night hours in the month of August and during weekends. Since the hydropower plants we are considering are run-of-river plants, and as such they produce over all the 24 hours, the price to be adopted for the energy produced on day t is the average

[6] By 'winter' we mean the months from October to March inclusive. With 'summer' we mean April to September, August excluded.

price θ_t on that day (the values [k€/kWh] of θ_t are specified in Section 4.5.3 of the DVD). The lost revenue on day t is therefore given by $\theta_t D_t(q_t)$.

In conclusion, the indicator $i_{\text{EnP\_D\_1}}$ is expressed by the following functional

$$i_{\text{EnP\_D\_1}} = \frac{1}{N} \sum_{t \in H} (\theta_t \cdot D_t^{\text{Enel}})$$

where D_t^{Enel} is the *overall production deficit for the* ENEL *power plants*, i.e. the sum of the production deficits in the four plants of Porto della Torre, Vizzola, Tornavento and Turbigo Superiore. The water flow through the first is denoted with q_t^{Ptor}, while the other three are fed by the flow (q_t^{IC}) that enters the Industrial Canal, so that the total production deficit is

$$D_t^{\text{Enel}} = D_t^{\text{Ptor}}(q_t^{\text{Ptor}}) + D_t^{\text{Viz}}(q_t^{\text{IC}}) + D_t^{\text{Torn}}(q_t^{\text{IC}}) + D_t^{\text{Turb}}(q_t^{\text{IC}})$$

4.3.2 Upstream Flooding

For this sector we present the indicators that are associated with the two leaf criteria descending from the *Damage to structures* criterion: *Average conditions* and *Worst event* (Figure 4.4).

As in the case of hydropower production, the first thought is to adopt the *economic value* of the damage as an indicator: we would then define the *average annual damage* and the *worst event damage*. The procedure for estimating these indicators is, however, much more complicated than in the previous case, where the average lost revenue could be obtained by adding up the daily lost revenues. Damage to structures is not made up of the sum of daily damage, since this damage is produced by contact with water and does not depend on the duration of the flooding (if it does, it is only in a very modest way). Rather than daily damages, we have to aggregate the damage incurred during each *flood event*. But what is a flood event? Its definition could seem simple: an event occurs when the lake overflows its shores, i.e. when the lake level exceeds the flood threshold. But in which location? The flood threshold varies according to the place we are considering. Furthermore, the lake surface cannot be thought of as horizontal during a flood: if it were, the water would stay still and the flow from the tributaries would not be transferred to the lake's effluent. The motion of the water is produced by the water-level differences created by the very tributaries' flows between their intake points and the lake outlet. The difference might be significant. In Verbano, up to 35 cm of difference in the water-levels between Locarno and Sesto Calende have been recorded. Therefore, the flooding begins at different times in different townships along the banks of the lake. It follows that we must define the damage separately for each township and then aggregate them over the entire lake. This is the reason why in Figure 4.4 we have disaggregated the sector criterion for the individual townships.

Let us then consider one township, Locarno for example. The event begins when the flooding threshold $h_{\text{fl}}^{\text{Loc}}$ is exceeded; the threshold can be defined as the level at which the first damage occurs. But when does the event end? We could respond that it ends when the lake returns below that same threshold. But here a question arises: if the level of the lake rises again to exceed that threshold once more, are we dealing with the same event or a new one? The question is not an idle one, since it is clear that the damage produced by a flood which occurs after another is quite different, depending on how much time has passed between the first and the second event. If several months have passed when the second occurs, the damage from the first will have been repaired and the flooding will take place in a renewed structure. On the contrary, if the second follows the first only a day later, the damage will be greatly inferior, or maybe non-existent, because the structure will not yet have been repaired. We must therefore establish more clearly at which point a flood event is concluded. The Experts suggested that we consider the event to be over after the level of the lake has remained below the flood threshold for at least 7 consecutive days.

Now that the definition of an event has been established, we must estimate the damage that it produces. The damage to a structure depends on the part that was submerged, and so it is a function of the *peak of the flood*. Thus we must find the relationship between damage and level. To do this we need a lot of information:

- the level at which the first damage occurs;
- the positions and the elevations of structures and activities at risk of flooding;
- the economic value and the time required to repair each of these structures;
- the economic value of a working day for each of these activities;

and lastly we need to know

- the lake's surface profile as a function of the lake's level at Sesto Calende and (if necessary) of the tributaries' flows.

With this data we could estimate the overall damage D that is caused by a flood event along the banks of the lake, as a function of the level h^{SC} measured at Sesto Calende when the peak is reached. To do this, for every level value h^{SC} and for each township j, the local damage $D^j(h^{SC})$ must be estimated by aggregating, structure by structure, the value of its flooded portion and the losses due to the interruption of the activities that occurs there. Then the overall damage D is computed by aggregating the local damage $D^j(h^{SC})$ for each township, i.e.

$$D(h^{SC}) = \sum_j D^j(h^{SC})$$

In 1983 Bianchi and Laniado (1983), in a study dedicated specifically to Verbano, adopted this procedure with success and obtained the curves in Figure 4.17, where the curve *Overall* is the function $D(h^{SC})$, while the others are analogous curves for the three regions with Verbano shores.

When the function $D(h^{SC})$ is known, the values of the two indicators *Average conditions* and *Worst event* can easily be computed: given the trajectory of the lake level subject to the alternative under study, we can find out the peak level values for all the flood events contained in that trajectory, and compute the overall damage for each one of them through the function $D(h^{SC})$. In this way we get a sequence of damage values, from which it is easy to compute both the *average annual damage* and the *damage from the worst event*.

This procedure is very precise, but it requires that the function $D(h^{SC})$ be available. The one estimated by Bianchi and Laniado cannot be adopted in our Project, since the socio-economic structure on the lake's hinterland has been significantly modified since 1983. At the same time, it was not possible to conduct another survey with the budget that we were allocated.[7] The lack of data forced us to abandon this procedure and to adopt another definition for the two indicators.

[7] In the study by Bianchi and Laniado the survey took three years of work.

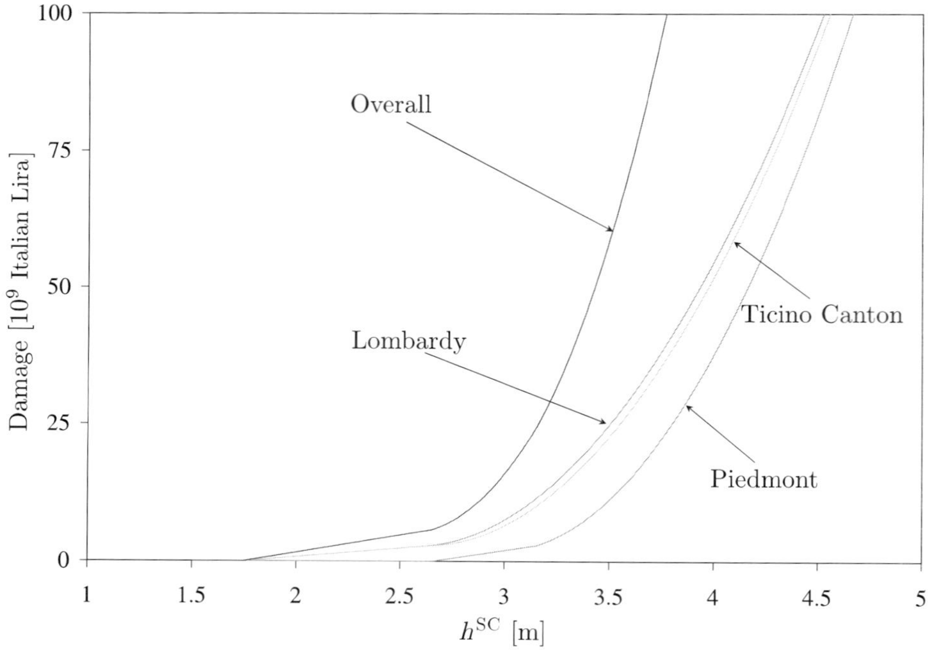

Figure 4.17: The damage produced by a flooding, as a function of the lake level h^{SC} at Sesto Calende, for the overall lake and for the three regions with Verbano shores (Bianchi and Laniado, 1983).

In cases such as this, one should look for an indicator that is not the exact translation of the leaf criterion being considered, but that is a *proxy indicator*, namely a variable in a logical relationship with the leaf criterion and related to some of the effects of the alternative through a functional, objective and potentially quantifiable link (Keeney and Raiffa, 1976). The proxy indicator thus assumes the role of an *indirect ordinal estimator of the leaf criterion* and the degree of satisfaction of the latter can be evaluated through the value assumed by the proxy indicator because, *in the mind of the Expert*, there is a relationship between the proxy indicator and the criterion. Note that since the proxy indicator is an ordinal quantity, the values that it assumes can only be compared with each other, and it will not be possible to conduct arithmetic operations upon them.

To find a proxy indicator, first observe that for each township j the function $D^j(h^{SC})$ can be estimated, even if all the data have not been collected, provided that we find an Expert who knows the physical and economic conditions in the territory very well. For each level h^{SC}, in fact, we can show him/her an image of the flooded surface, superimposed on a map of the structures, and ask him/her to estimate the amount of damage. People working at the city technical office could be good candidates for this task.

However, the estimates of the damage obtained in this way, being values of a proxy indicator, cannot be averaged over time to estimate the average annual damage. To estimate these values we ought to give the Experts an image of the *average flooded surface area*, instead of the sequence of images of the flooded surface area in the different events. Analogously an image of the *flooded surface area in the worst event* is required for the estimate of the *damage from the worst event*. These two images can be obtained from the trajectory of the lake level subject to the alternative under study, by individuating the peak levels h^{SC} for all the flood events and by computing the flooded area S_{fl}^j that corresponds to each of them, through the function $S_{fl}^j(h^{SC})$. The latter function can be easily obtained by means of a *digital elevation model* (DEM) of the lake and surrounding area.

In conclusion, the most convenient solution is to adopt the (average and maximum) flooded area as a proxy indicator for (average and maximum) flooding damage. In the *Evaluation* phase (see Chapter 9) an Expert will associate each proxy indicator with a *value function* in order to translate the value of the surface area into a value of satisfaction. The value function provides an ordinal estimate of the amount of damage (the greater the damage, the smaller the value). Since the Expert must know the territory well to make this translation, it is a good idea to keep the estimates specific to the single townships and not try to obtain the sum of the flooded areas for the entire lake perimeter.

For the operational computation of the proxy indicators, the following functions must be identified:

- the *lake's surface profile* $Q^j(h^{SC})$ that, given the lake level h^{SC} at Sesto Calende, provides the elevation Q^j of the lake in township j;
- the function $S_{fl}^j(Q^j)$ that relates the elevation Q^j of the lake in township j to the area S_{fl}^j of the flooded surface in the same place. It is obtained by means of a DEM.

From the functions $Q^j(h^{SC})$ and $S_{fl}^j(Q^j)$, the function $S_{fl}^j(h^{SC})$ can be obtained: this expresses the flooded area in township j as a function of the level at Sesto Calende. From the trajectory of the lake level produced by the alternative under examination over the time horizon H, the sequence of peak levels $\{h_i^{SC}\}_{i \in I_j}$ is extracted, I_j being the set of flood events i in township j. Each event begins when the level exceeds the flood threshold h_{fl}^j in that township and ends when the level drops and stays below the threshold for at least 7 consecutive days. The following two indicators can then be computed:

$i_{Fl\_U\_1\_j}$ (↓) maximum flooded area [km^2] in township j

$$i_{Fl\_U\_1\_j} = \max_{i \in I_j}\left[S_{fl}^j\left(h_i^{SC}\right)\right] \tag{4.4}$$

$i_{Fl\_U\_2\_j}$ (↓) average annual flooded area [km^2/year] in township j

$$i_{Fl\_U\_2\_j} = \frac{1}{N}\sum_{i \in I_j}\left[S_{fl}^j\left(h_i^{SC}\right)\right] \tag{4.5}$$

One last observation: once we know the extent of the flooded area, it is possible to estimate the damage when we are considering a single flood event (as for the first indicator), but not when many events are to be considered (as for the second), except for when the relationship between area and damage is linear, as would happen when the value of the structures and the economic activities is uniformly distributed in the area being considered. When this is not the case, the average flooded area is not sufficient information to estimate the average damage, since the latter depends on the whole probability distribution of the flooded area, not only on its average value. If, as is usually the case, the value of the structures increases with their elevation on the lake, the damage per unit of area increases with the flooded surface and, as a consequence, the average area provides an underestimate of the damage. Therefore it would be advisable that the Expert keep this distortion in mind, when he is asked to identify the corresponding value function (see Section 9.3).

Note nevertheless that, even if the Expert is not able to do so, the two indicators are still good proxies of the corresponding leaf criteria, since they are *indirect ordinal estimators* of the effect under examination: the flooding damage. In order that an indicator be a proxy of the damage, it is not necessary that there be a linear relationship between its value and the damage; it is sufficient that, for each pair of alternatives that the Expert is asked to compare, his preference between the two is the same if based upon either the proxy indicator or the effect itself. Our indicators satisfy this condition, because it is evident that greater flooded areas produce greater damage and vice versa.

The identification of the function $S^j_{\text{fl}}(h^{\text{SC}})$ and of the value function takes, on average, a couple of days of work for each township and requires the availability of a DEM. Thus, it was not possible to consider all the 44 townships around the lake and in the Project we had to select just two of them, one Italian and the other Swiss, which were picked from those that suffer the highest damage: Verbania and Locarno. Thereby the index j in (4.4)–(4.5) takes on only two values: (*Vb* for Verbania and *Loc* for Locarno).

4.3.3 Downstream Environment

Of this sector we illustrate the search for the indicator associated with the leaf criterion *Overall condition of the ecosystem*, whose formalization is rather difficult.

In order to define the indicator, the first step is to determine the condition that corresponds to the maximum satisfaction of the criterion proposed. To identify this condition, we interview the Expert: "the condition of an ecosystem is best when it is free to evolve in natural regime", he explains. "Therefore the best environmental quality for the River Ticino downstream from the lake would be obtained by suspending the regulation of the lake and the abstraction from the canals".

It is obvious that the indicator could be defined on the basis of the effects that the releases into the Ticino produce on the ecosystem, but it would not be convenient to do so, because, in order to give it a value, we would need a model that could relate the trajectory of the release to the environmental effects that they produce. Identifying such a model would be a very long and costly operation, with uncertain results. It would be preferable to compute the indicator directly from the trajectory of the release, but this is possible only if the Expert is able to assess the conditions of the ecosystem by simply observing that trajectory. To verify this possibility, we provide the Expert with several graphics, such as in Figure 4.18, which, for a given year, show the trajectories of the lake inflow, natural outflow and release in correspondence to different alternatives. The graphics in Figure 4.18 show the alternatives $A0$ and $A4$: the first was introduced in Section 3.7.1 while, for $A4$, it is now sufficient to know that it differs from $A0$ only in the regulation policy. Looking at the figure, the Expert points out that $A0$ produces a flow rate that is equal to the MEF for very long periods, which end only when the inflow to the lake increases significantly. This is sufficient for him to conclude that $A0$ would not produce good conditions. On the contrary, the conditions produced by $A4$ should be good, because the release trajectory it produces is very similar to the one of the natural outflow, even if its average value is lower, due to the abstractions of the canals.

The way in which the Expert makes his judgements seems to be linked to the time that the flow in the Ticino is equal to some specific values, so we think it opportune to show him the river flow *duration curve*: a function which, for each possible flow value, provides the

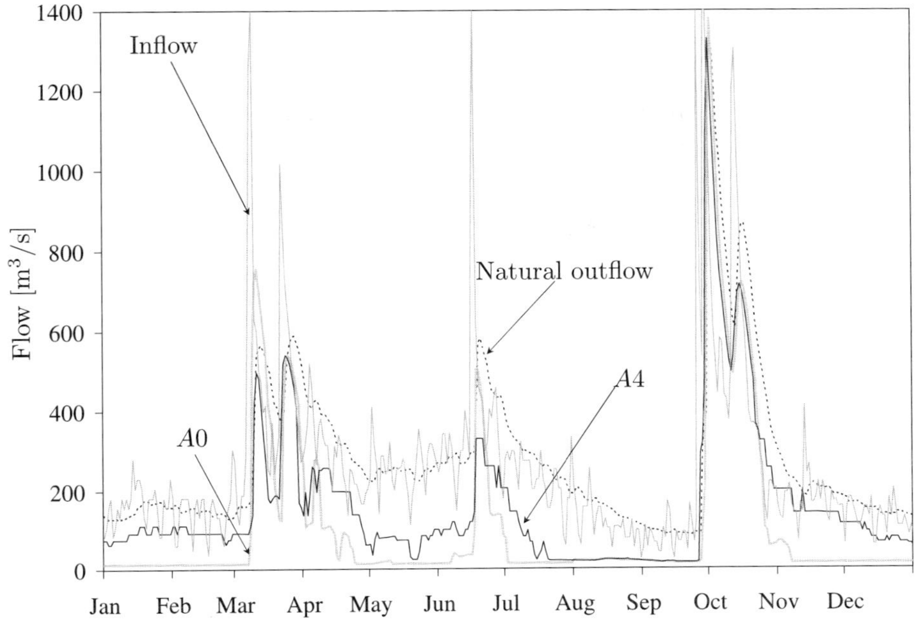

Figure 4.18: Comparison between the trajectory of natural lake outflow in 1991 and the trajectory of the flow that would have been released into the Ticino in the same year with alternatives $A0$ and $A4$.

average annual number of days that the river flow rate exceeds that value. If the Expert could express his judgement based on the duration curve, we would obtain a distinct advantage, since his judgement would be based upon information that considers the entire evaluation horizon, rather than just one or two years: what actually the Expert can intuitively judge looking at a trajectory. In fact, it is very difficult or practically impossible for the Expert to make an accurate comparison if he must consider longer trajectories: the duration curve always has the same length, 365 days, whichever time horizon is used to compute it.

Figure 4.19 shows the duration curves that correspond to the trajectories in Figure 4.18, while Figure 4.20 presents their enlargement in the region of low flow values. The duration curves of $A0$ and $A4$ are almost identical for flow values higher than 300 m^3/s, which occur on average not more that 50 days in a year. Furthermore, while the duration curve of the natural outflow is very smooth, the $A4$ curve shows many steps[8] and the $A0$ curve has a very long horizontal stretch in correspondence to 13 m^3/s (the MEF value), which extends from the 195th to the 365th day. The presence of a horizontal stretch, for a given flow value, reveals that this value occurs on many days of the year: precisely as many days as the difference between the horizontal coordinates of the stretch's starting and ending points. For example, with alternative $A0$, the MEF flow is maintained for 170 days on average each year. On the contrary, the flow values corresponding to very steep stretches have been recorded on rare occasions, or even never, if the duration curve is vertical. A curve that has no steps or horizontal stretches is indicative of a good variability in the river flow.

Observing Figures 4.19 and 4.20, the Expert expresses the same judgements as by analysing the release trajectories; but now his judgements are much more reliable than be-

[8]The presence of these steps is caused by the technique used for the policy design (see Section 7.8).

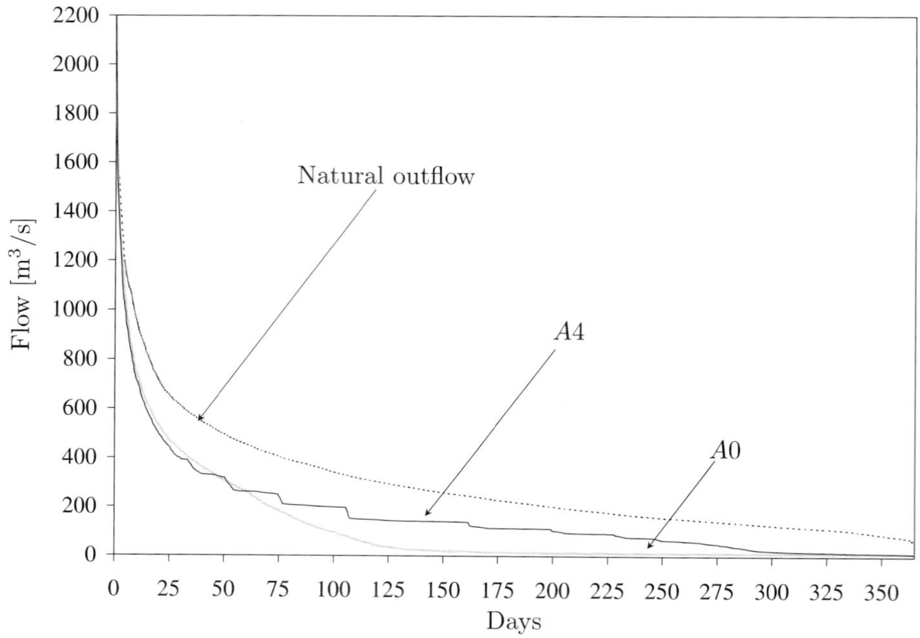

Figure 4.19: Duration curves of natural outflow and of release produced by alternatives $A0$ and $A4$.

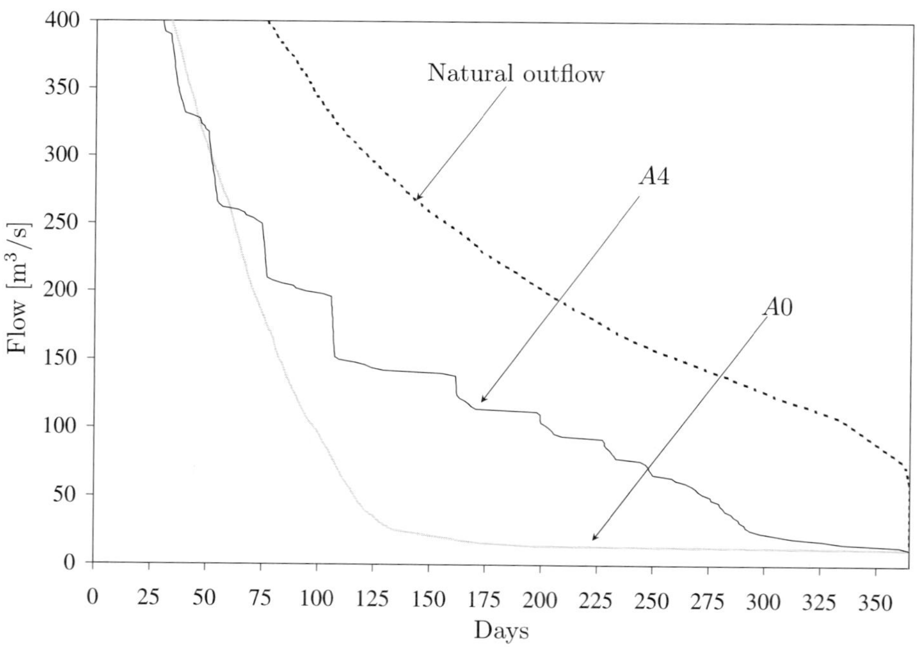

Figure 4.20: An enlargement of the duration curves of Figure 4.19 in the region of low flows.

fore, since they are based not on just one or two years, but on the whole evaluation horizon. However, we have not yet found an automatic procedure to generate such judgements: they were based on the shape of a function and all attempts to derive a set of indicators from that function turn out to be unsuccessful. We must conclude that the duration curve is, without any doubt, meaningful information, but it does not allow us to identify the indicator that we are looking for. The synthesis it produces leads to the loss of a part of the information: it allows us to know the average number of days that a certain flow occurs, but not whether these days are consecutive or disperse.

Until this point we have followed the following procedure: we started by discovering the type of information which the Expert is used to working with, we analysed his way of making a judgement and tried to formalize it, but we were unable to complete this last step. However the attempt was not totally fruitless, for two reasons: first, it helped us to understand that the judgement is connected to the distance between the trajectories of the natural outflow and of the regulated release; second, it produced new information. We showed, in fact, the Expert many alternatives and obtained a judgement for each one, and thus we were able to rank them from his point of view. We can now use these results and look for an indicator that expresses a measure of the distance between the natural outflow and the regulated release, and that produces the same ranking that was given by the Expert for the alternatives he was shown. We tried several indicators and the best was

$i_{\text{Env\_D\_1}}$ (↓) Distance from natural regime [m^3/s]

$$i_{\text{Env\_D\_1}} = \sqrt{\frac{1}{|H|} \sum_{t \in H} \left(r_t^n - q_t^{\text{TIC}}\right)^2} \tag{4.6}$$

where r_t^n is the natural regime outflow and q_t^{TIC} the flow at Panperduto in the regulated regime.

When we validated this indicator by comparing pairs of alternatives, we realized, however, that the ranking provided by the indicator coincided with the one provided by the Expert only when the values of $i_{\text{Env\_D\_1}}$ in the two alternatives were very different. When they were similar, often the Expert preferred the alternative associated with the higher value and when this happened, that alternative was the one with the higher MEF value. Reflecting on this result, we realized that the indicator $i_{\text{Env\_D\_1}}$ does not always express the overall condition of the ecosystem correctly. Consider, for example, an alternative that almost always keeps the release into the Ticino very close to the natural outflow, except in short periods in which the release is very low. The value of $i_{\text{Env\_D\_1}}$ would be low, since over a long horizon only a few days with flows very different from the natural ones is not sufficient to produce a high value for the indicator. From an environmental point of view, however, those few days with very low flow produce negative impacts that are not compensated by the long periods in which the release is similar to the natural outflow. Thus a correct evaluation of the alternatives requires that we also consider the minimum value of the flow (that is the MEF), the frequency with which flows equal to the MEF occur, and the average duration of the periods in which this happens. We concluded that, to assess the effects of an alternative for the leaf criterion *Overall condition of the ecosystem*, the indicator $i_{\text{Env\_D\_1}}$ is not sufficient, and should be accompanied by three others: the MEF value ($i_{\text{Env\_D\_2}}$); the frequency ($i_{\text{Env\_D\_3}}$) with which flows equal to the MEF occur; and the average duration ($i_{\text{Env\_D\_4}}$) of the periods in which this happens. Further tests demonstrated, however, that the values of

$i_{\text{Env\_D\_3}}$ and $i_{\text{Env\_D\_4}}$ are highly correlated with the value of $i_{\text{Env\_D\_1}}$ and so those two indicators are redundant. In conclusion, we associated only two indicators to the leaf criterion *Overall condition of the ecosystem*, but in Section 11.4.2 we will see that the Expert is not able to integrate them into a single index.

4.3.4 Irrigation

To conclude these examples, we present the indicators associated to the two leaf criteria descending from *Harvest* (Figure 4.12). They help us to show how the *Project Goal* directly influences the choice of indicators.

The first of the two leaf criteria is the *Supply deficit*. The indicator that is spontaneously associated to it is the *Average annual deficit* in the whole irrigation district, defined as the sum of the average annual deficit in each of the individual districts served by the canals (Regina Elena, Villoresi and Industrial Canals). To formally define each of these deficits, we must define the demand with respect to which they have to be computed. According to the FAO model (Section 6.9.2 of the DVD), for each day t the *gross demand* of the cth canal is defined as the flow w_t^c that must be allotted to it so that the effective evapotranspiration rate of the crops is equal to the potential one. Once the second is known, the gross demand w_t^c can be computed with the procedure described in Section 6.9.2, where the *Irrigation* model is presented, taking into account the crop plan that was adopted by the district, the rotation of turns between the consortium members, the expected precipitation and the water loss from the canals. As shown in Figures 4.21 and 4.22, both for the Regina Elena Canal and for the Villoresi Canal, the gross demand so computed differs from the demand (W_t^c) declared by the Consortiums (see Figure 2.8), for reasons that vary from canal to canal:

- The declared demand W_t^c can be inferior to the gross demand w_t^c, since the canal cannot convey more than the declared demand. This is what happens in the Regina Elena Canal in the months of June and July; as a consequence, its district suffers a structural deficit in those months.

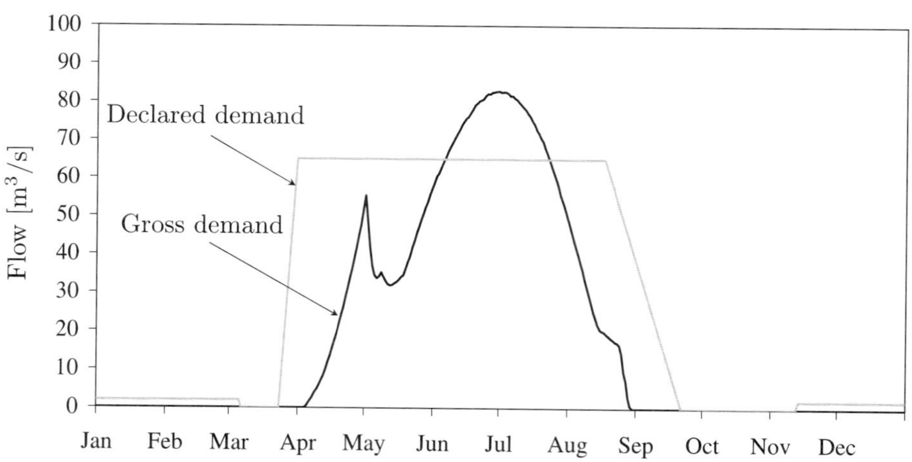

Figure 4.21: Comparison between the gross demand, computed with the FAO model, and the declared demand for the Regina Elena Canal.

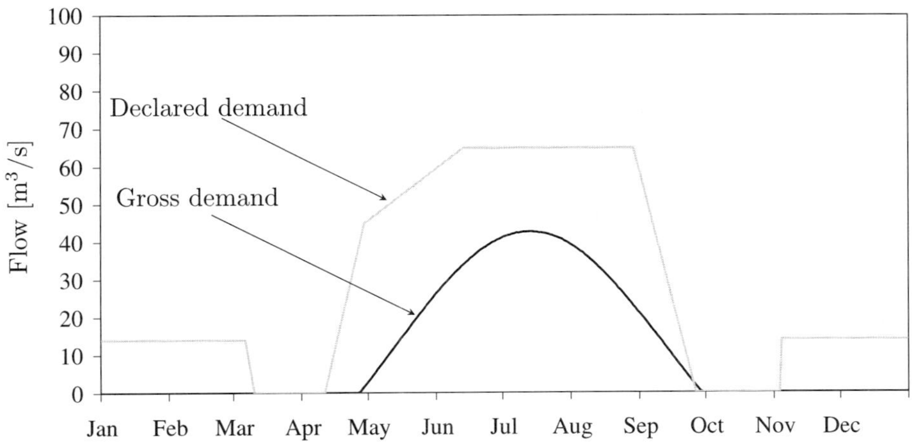

Figure 4.22: Comparison between the gross demand, computed with the FAO model, and the declared demand for the Villoresi Canal.

- The declared demand is related to the Abstraction Licence, which was defined at a time when the irrigated surfaces were much greater than those existing today. This is the case for the Villoresi Canal.
- In the period when the rice paddies are submerged (from the end of March to the beginning of May) the manager of the Regina Elena Canal prefers not to consider the contribution from the expected precipitation (which has been subtracted from the crop demand when computing w_t^c), since if it should be lacking the damage could be quite significant.
- The Industrial Canal presents a different situation, since ENEL gained the right to use the water flows that the Abstraction Licence allocates to the Old Users for irrigation, and which are actually supplied to them after having been run through the turbines. From a legal standpoint, this is a delicate issue: even if at a given moment the gross demand is lower than the flow allotted by the Abstraction Licence (which in this case coincides with the declared demand), ENEL still has the right to divert the latter amount.

The managers of the canals have the legal right to derive a flow equal to the whole declared demand (and no more than that) at any time. Since in the definition of the Project Goal we have said that a review of the declared demand will not be considered, it seems correct to define the deficit not with respect to the gross demand w_t^c, but in relation to the declared demand W_t^c, that from now on we will call simply *demand*. The reader might find this decision surprising; nevertheless it is coherent with the Goal assigned to the Project (Section 2.6). It is thus quite understandable now why the Project Goal must always be accurately defined and publicly declared before the work begins: it constitutes a premise, an axiom (and as such some might consider it debatable), which can condition the results, at times quite significantly.

Nevertheless, it is not credible that the difference between declared demand and gross demand, especially when it is as significant as in the case of the Villoresi Canal (see Figure 4.22), could actually vanish without a trace. This is not the case, in fact: in Section 6.7.3 we will see that, when distributing the release among the canals, if the conditions are critical the East Ticino Consortium makes voluntary reductions to the flow diverted into the Villoresi Canal, the reason for which is certainly the difference between the canal's gross demand and the declared demand (see Figure 4.22).

For the cth irrigation canal among the three (Regina Elena Canal, referred to as *REC*, Villoresi, *VC*, and Industrial, *IC*),[9] the deficit [m^3/s] on day t is defined as

$$d_t^c = \begin{cases} 0 & \text{if } q_t^c \geqslant W_t^c \\ (W_t^c - q_t^c) & \text{otherwise} \end{cases} \tag{4.7}$$

where q_t^c [m^3/s] is the flow diverted into the canal. The *average annual deficit of the canal* [Mm^3/year] is given by

$$D^c = \frac{1}{N}\frac{\Delta}{10^6}\sum_{t\in H} d_t^c \tag{4.8}$$

where Δ is a coefficient of dimensional conversion between seconds and days (86 400 s/day). Then we can define the indicator

$i_{\text{Irr\_D\_1}}$ (↓) average annual volume of total deficit [Mm^3/year]

$$i_{\text{Irr\_D\_1}} = \sum_c D^c$$

Let us now consider the leaf criterion *Crop stress*. The Experts from the two consortiums pointed out that crop losses in a given agricultural year, namely the difference between the potential harvest and the actual harvest, is an increasing function of the maximum *stress* produced during that year. The stress $\digamma_t$ on day t is defined as the average daily deficit in the last δ days, δ being the duration of the irrigation turn. Since this varies from zone to zone, according to the characteristics of soil and crop, we fixed the value of δ at a conventional duration of fourteen days. The maximum stress $\digamma_y$ in year y is given by

$$\digamma_y = \max_{t\in y} \frac{1}{14}\sum_{\tau=t-13}^{t}\left(d_\tau^{\text{VC}} + d_\tau^{\text{REC}} + d_\tau^{\text{IC}}\right) \tag{4.9}$$

It is not true, however, that a very small stress results in a loss of harvest, nor that the loss continues to grow indefinitely as the maximum stress increases: below a certain threshold there is no loss, and the loss cannot increase further once the whole harvest has been lost. Therefore, we should define a measure of the potential damage stemming from the maximum stress. To do this, we adopted a function $f(\cdot)$, whose value varies between 0 and 100, and in which the thresholds were defined by Experts on the basis of their experience (see Figure 4.23). They said that if the maximum stress is inferior to 40 m^3/s, the loss to the harvest is negligible and consequently the damage can be considered equal to zero; then the damage increases linearly until, for stress values equal to 120 m^3/s or over, the damage is at a maximum. In conclusion, we associate the following indicator to the *Crop stress* criterion

[9] Remember (see Section 2.2.2) that the Regina Elena and Industrial Canals have both an irrigation demand and a hydroelectric demand. Clearly only the first should be considered in the computation of the irrigation deficit. The symbol W_t^c refers to that.

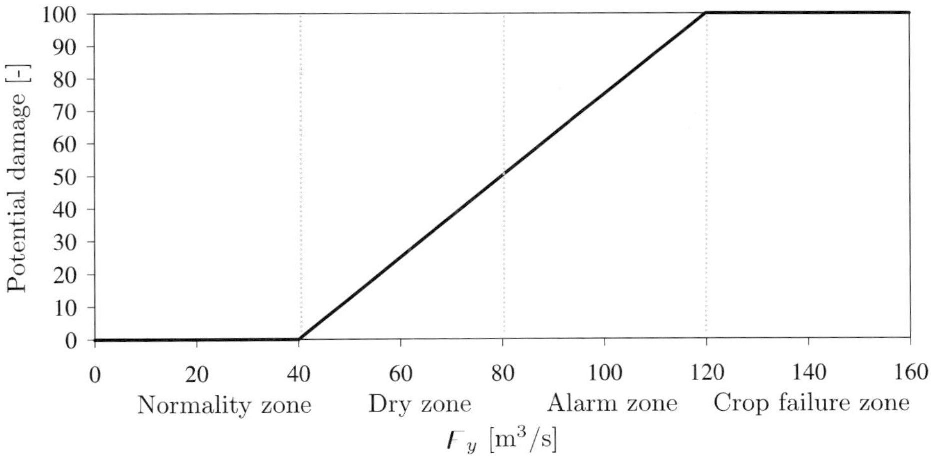

Figure 4.23: The function $f(\cdot)$ that expresses the potential damage stemming from the maximum annual stress F_y.

$i_{Irr\_D\_2}$ (↓) average potential damage (stemming) from stress [–]

$$i_{Irr\_D\_2} = \frac{1}{N}\sum_{y=1}^{N} f(F_y) \tag{4.10}$$

To conclude, we will add an important consideration about the indicators we have just defined. They were introduced in order to help assess the satisfaction of the two criteria *Supply deficit* and *Crop stress*, but they can also be thought of as two proxy indicators for the *Harvest* criterion. In fact, analogously to what we have seen in the previous section for the *Downstream environment* sector, we can avoid the difficulty of computing the value of the *Average biomass produced*, which would be the natural indicator for the *Harvest* criterion (and, at the same time, the output of the crop-system) by considering the *flow diverted* into the canals (i.e an input of the crop-system). The two indicators *Average annual deficit* and *Average annual potential damage from stress* appear to be *indirect ordinal estimators* of the *Average biomass produced*. The Expert will have to identify a relation between the values of the first two and the satisfaction of the farmers for the last, by defining a value function. The only difference with respect to the *Downstream Environment* is that in this case we only need two proxy indicators, rather than four, to reproduce the experience of the Expert and to avoid the use of a dynamic model for the crop-system, which would be too costly and complex.

4.4 Alternatives, indicators and indices

The previous sections should have given the reader a good feel for the concepts that we introduced in a piecemeal way in this chapter. In this section we will re-examine them in an organic way.

Our aim is to associate each alternative with a vector of indices which measure the satisfaction that the alternative produces with respect to each criterion. In other words, we want to use the indices to compare the alternatives. An index must allow us to solve pairwise comparisons: given two alternatives $A1$ and $A2$, it must be possible to identify the best one with respect to criterion C, by comparing the values that its associated index assumes in correspondence to $A1$ and $A2$. An index must thus allow us to rank the alternatives with respect to its criteria, i.e. with respect to a given interest. It must be a function of the alternative and it can be defined either on an *ordinal* scale (*qualitative index*) or an *absolute* scale (*quantitative index*).

Sometimes the index value for an alternative can be *qualitatively* estimated by interviewing the Stakeholders or by referring to an Expert who represents them (Figure 4.24a). This procedure can conceptually always be adopted, but it has two main drawbacks:

- the evaluation might appear to be too subjective, particularly to those that feel disfavoured by the estimate so obtained; therefore it could easily be the subject of contention;
- when there are many alternatives, the task of the Expert is arduous and so tiredness and boredom may impair his judgement as the work proceeds.

For these reasons, with help from the Expert, one should identify a procedure to estimate the value of the index (Figure 4.24b) automatically. The procedure must 'incorporate' 'reproduce' and 'contain' the Expert's experience in an operational and repeatable way, and be identifiable through interviews. We used and will use the term *index* to indicate both the procedure for identifying it and the value it produces.

In our Project, as often happens, the system is a *dynamical system*, whose characteristic quantities vary in time: think for example of the level of the lake, the release, etc., which we accordingly call *variables*. The system's behaviour over a given horizon (the one along which we want to evaluate the effects of the alternative) is described by the trajectories of these variables, which we call *outputs*[10] of the system. An alternative modifies the system and thus also the trajectories of its outputs. We can then subdivide the procedure for estimating the index into two steps:

- The simulation of the behaviour of the system subject to the alternative, to obtain the output trajectories that it produces. Note that for the simulation a mathematical model of the system is required.
- The computation of the index value through a procedure, usually expressed by a functional, which defines the index as a function of the output trajectories.

This possibility is illustrated in Figure 4.24c. The reader will notice that this way of proceeding is the one that we took for granted in the preceding sections, when we used the term index to indicate the procedure shown in Figure 4.24c.

We have tried to follow this way in Section 4.3.3 when discussing the leaf criterion *Overall condition of the ecosystem*, from the *Downstream Environment* sector, but, as we have seen, we were not able to formalize the way in which the Expert carries out the passage from the trajectory (in that case of the release into the Ticino) to the satisfaction of the criterion.

[10]The meaning of the word is the one given by System Theory: output is a variable that depends only on time and on the state of the system. In our case, the system outputs are the level and the release from the lake, and the flows diverted into the canals. Be careful not to not give the term a physical meaning, as "something that comes out from the lake": in this sense the level could not be considered as an output.

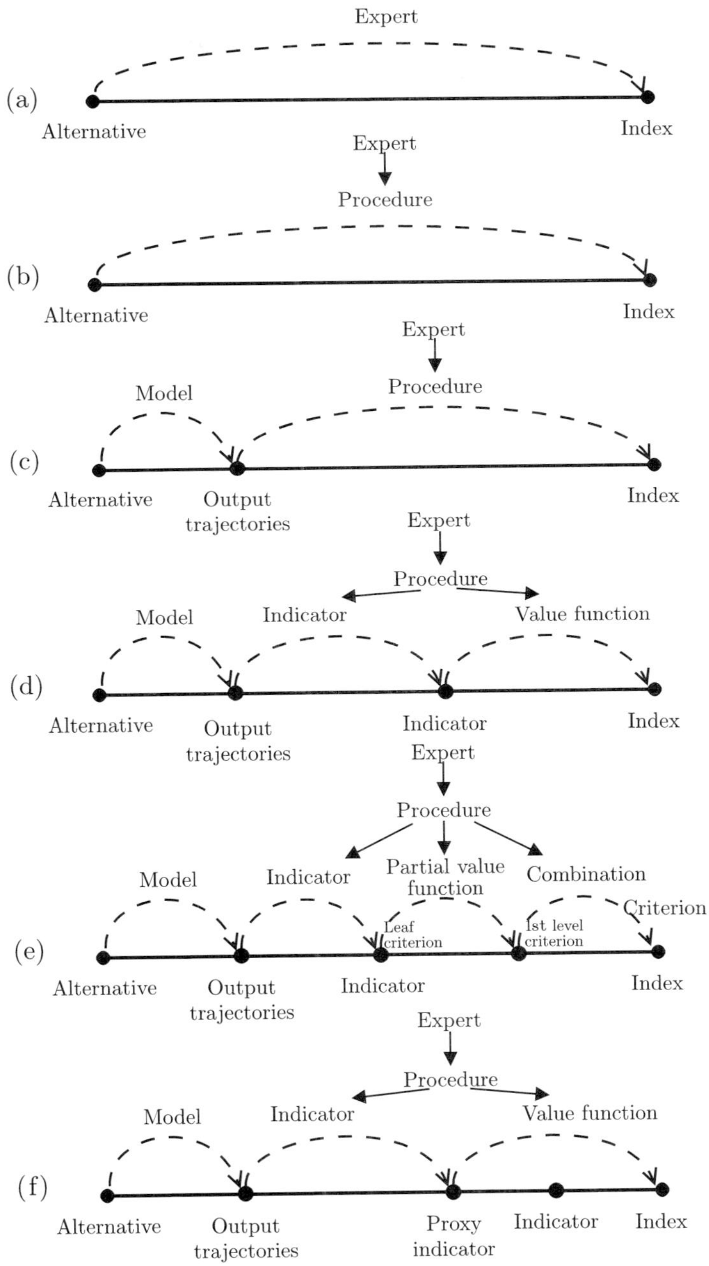

Figure 4.24: Different ways to associate an index to an alternative.

Difficulties of this type often arise and direct identification of the index is rarely possible. To simplify the task, the passage from the output trajectories to the index is broken down into two steps, by introducing an *indicator*, i.e. an 'intermediate' quantity between the trajectories and the index (Figure 4.24d), which can be measured directly or estimated with a procedure,[11] which makes it easier to evaluate the satisfaction of the corresponding criteria. The indicator must be chosen in such a way that, on the one hand, it is easy to identify the functionals with which it is computed once the output trajectories are obtained, and, on the other hand, it is possible for an Expert to pass from the indicators to the index (Beinat, 1997). Therefore, a guideline for choosing the indicators is to make the distance between indicators and index as short as possible, so that the space for the Expert's subjective opinion, and thus the triggers for contention, are minimized. Unfortunately, the complexity of reality, along with limits to knowledge, data availability and computing time, make it impossible, in most cases, to completely eliminate this distance.

The passage from indicator to index is technically covered by the *value function*, which we will formally introduce in Chapter 9. Here we simply anticipate that it is a mathematical relationship that 'reproduces' the Expert's judgement and that it is identified through interviews and experiments.

Sometimes the association of an indicator to a criterion is easy and natural: think, for instance, of what happened for the ENEL *Power* sector (Section 4.3.1). When this is not the case, the criterion may be broken down into a set of *sub-criteria*. If it is possible to associate an indicator to each of these sub-criteria, and then work back from them to the index, the set of criteria may be operationally substituted for the original criterion. One example is how we identified the indicators for the *Irrigation* sector. If we cannot identify the indicators, we may proceed with a new breakdown of the sub-criteria for which the identification failed. With this procedure a *hierarchy of criteria* is individuated, which, in mathematical terms, can be defined as a partially ordered set of criteria. Note that the value function associated to the originally proposed criterion is a combination of the partial value functions associated to the leaf criteria of this hierarchy (Figure 4.24e).

However, when an indicator cannot be easily associated to a criterion (or a sub-criterion), breaking this latter down into other sub-criteria is not the only possibility. In fact, in the *Upstream Flooding* sector we identified the leaf criterion *Damage to structure* and tried to associate to it the indicator *Economic value of the damage*; but when we discovered it was too expensive to estimate its value, instead of breaking the *Damage to structure* criterion down into sub-criteria, we turned to the proxy indicator *Flooded area*. We can reflect on this way of proceeding by looking at Figure 4.24f: it shows that the proxy indicator is none other than the indicator that we are looking for. It is, in fact, the input of the value function, while the original tentative indicator is completely unused; its role was only to help us identify the proxy.

4.4.1 Factors that influence the choice of the indicators

The reader will probably have formed the idea that the choice of the criteria and the definition of the indicators are guided essentially by the Project Goal. Ideally, this is how it should be, but actually the choice of both criteria and indicators is influenced by the conditions imposed upon the use of the indicators in the following phases of the procedure.

[11] Like index, the term *indicator* also is used in practice to denote both the procedure and the estimated value that results from applying it.

In Chapter 1 we saw that a model of the water system is required, in order to simulate the system behaviour both for the design of the alternatives and for their evaluation. In particular, as we explained in the previous section, we are interested in determining the trajectories of the variables that are required to compute the indicators. The choice of the latter thus poses conditions for the structure of the model that can be adopted. But the relation should also be considered in the opposite direction, which means that indicators whose computation requires knowledge of variables that we cannot simulate should not be chosen. We will deal with the choice and identification of the model in Chapter 6.

Other than these general considerations, there are other more specific elements that must be taken into account. In Chapter 3 we saw that designing a different regulation policy is one of the possible actions of the Project. The design of a policy requires that the regulation objectives be included among the indicators, which means that when choosing the indicators the regulation objectives must be taken into account. This is reasonably obvious. Less obvious is the fact that the algorithms used to design regulation policies impose conditions on the form of the indicators that can be used (see Section 7.7 and, for more details Chapter 12 of THEORY). These conditions should also be taken into consideration when the indicators are defined. Sometimes this is easy, but at other times the identification of indicators that respect such conditions is difficult, if not impossible, without altering the corresponding criterion.

In particular, one of those conditions is that the indicator must be a combination of the instantaneous values of a quantity (called *step indicator* or *step cost*) over the whole evaluation horizon. At every time instant, this value depends only on variables relative to that same time: in mathematical terms, the indicator is a functional (e.g. the summation) of the step-indicators, each of which is a function of the system outputs at a given time instant. This condition is the so-called *hypothesis of separability* and an indicator that respects it is a *separable* indicator. In order that this hypothesis hold true, the value of the step indicator at time t must not depend on the value assumed by that same indicator at preceding time instants: in other words, the step indicator cannot be the state of a dynamical system.

When the hypothesis of separability is proven to be true, as for example in the case of the ENEL *Power* sector, the identification of the indicator is simple. When it is not, it is always possible to trace the problem back to a case in which the hypothesis does hold, provided that the state of the system is suitably enlarged. For example, if the instantaneous indicator at time t depends on its own value at time $t-1$, it is sufficient to include a *state transition function* in the model of the system that expresses the dynamics of the indicator. Let us consider a practical example: the *Harvest* criterion for the *Irrigation* sector. In Section 4.2.2 we said that the harvest could be considered as a leaf criterion and the *Average annual biomass produced* would be the associated indicator. To compute such an indicator, maintaining separability, we can describe the harvest as the sum of the daily values of a new variable which is always zero except for the day of the harvest, when it is equal to the biomass of the crop. The biomass is the output of a dynamical system, because the biomass on a given day depends on the state of the crop on that day, which in turn depends on the state of the crop on the previous day and the water that it was given in the last 24 hours. With this description, we would obtain a separable indicator for the harvest, but this description 'has a cost': it requires the use of a crop growth model, besides the model of the lake.

The cost of model identification is never negligible; on the other hand, a simple well-chosen non-dynamical relation is often more precise and reliable than a dynamical model that has been badly calibrated to save money or time. But there is one more reason why the use of a dynamical model for the indicators should be avoided when possible: it implies the enlargement of the state of the system and this has severe consequences for the design of the regulation policy. Such a policy, in fact, should have as arguments not only the state of the physical system, but also the states of all the systems that appear in the description of the indicators. To understand why this happens, consider again the example of the *Harvest* criterion. If the crop is a dynamical system (as, in fact, it is), the release decision cannot be made without considering its state. Think of the extreme case in which the crop is dead: clearly it is useless to irrigate it and the flow which would have been destined for it can, more usefully, be used somewhere else. Knowledge of the state is essential for the correct description of the management problem, but this becomes more complicated and costly to solve. It is the task of the Analyst to guess when an increase in complexity is justified by the advantages that can be obtained from it and when, instead, it is better to simplify the description. The latter possibility is the one we turned to in the case of the harvest criterion, which we decided to represent by means of the two proxy indicators presented in Section 4.3.4. The difficulties that we encountered in that section are linked to the attempt to simplify the description, and avoid modelling the dynamical system. Clearly, by omitting this description we lost accuracy. Also the difficulties that we encountered in the *Upstream Flooding* sector could be interpreted from this point of view.

4.5 The indicators of the Verbano Project

Chapter 5

Defining actions and formulating the Design Problem

EB, FP and RSS

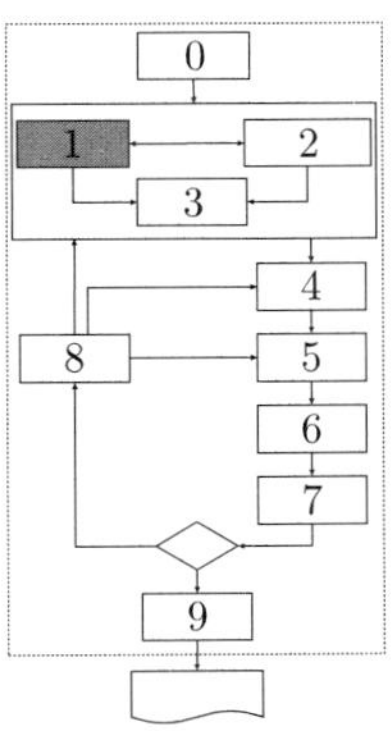

Figure 5.1: The phases of the PIP procedure. Phase 1, which is analysed in this chapter, is highlighted.

In Chapter 3 we illustrated the *intervention* options for the Verbano Project. Now, for each intervention option we must establish the way in which it will be carried out. To reach this target, we will describe each intervention as the result of one or more *actions*, each of which can be univocally specified, from an operational point of view, by defining some parameters and/or functions. Integrated and coordinated packages of actions form the *alternatives* of our Project.

In this chapter we will first deal with the definition of the actions, i.e. we return to Phase 1 of the PIP procedure (see Figure 5.1). Then, we will provide the reader with some notions about the formulation of the Design Problem, through which the 'most interesting' alternatives will be identified. This Problem (whose solution is the task of Phase 4 of the PIP procedure) will be the topic of Chapter 7, but we need to anticipate some issues on the subject, in order to facilitate the understanding of next chapter, which is devoted to the identification of the water-system model (Phase 3): remember that not always can the phases of the PIP be sequentially developed, recursions are often necessary.

5.1 Types of actions

We can subdivide actions into two categories: structural and non-structural.

Structural actions are concerned with physical modifications of the system. This category includes excavating the lake outlet, which would produce a change in the stage–discharge relation.

Non-structural actions are concerned with functional or operational modifications of the system. They can be normative actions, i.e they define legal constraints that must be respected in the system operation, or regulatory actions. Examples of the first type are modifying the regulation range or stating the MEF, since both establish constraints that must be respected in the lake regulation and in the distribution of the release. An example of the second is modifying the regulation policy.

Every possible combination of these actions constitutes an alternative A, which is therefore univocally defined by a 4-tuple (s, r, m, p) of actions, where:

- s is a stage–discharge relation,
- r a regulation range,
- m a value of MEF,
- p a regulation policy.

The set $\mathcal{A}$ of alternatives is thus the following

$$\mathcal{A} = \left\{A\colon\ A = (s, r, m, p) \text{ with } s \in \mathcal{S},\ r \in \mathcal{R},\ m \in \mathcal{M},\ p \in \mathcal{P}(s, r, m)\right\} \tag{5.1}$$

where:

- $\mathcal{S}$ is the set of stage–discharge relations considered,
- $\mathcal{R}$ the set of regulation ranges,
- $\mathcal{M}$ the set of MEF values,
- $\mathcal{P}$ the set of regulation policies associated with the triple (s, r, m).

The set $\mathcal{P}$ depends on the specific triple of actions (s, r, m) being considered because the structural action (s) and the normative actions (r, m) influence both the stage–discharge relation (and thus the set of feasible releases at each value of the lake level) and the constraints that must be satisfied in management. In other words, the triple (s, r, m) defines the space within which the regulation decisions can be taken.

At this point, one may wonder how many alternatives there are in the set $\mathcal{A}$. To answer this question, we must find out how many elements there are in each of the sets $\mathcal{S}$, $\mathcal{R}$, $\mathcal{M}$ and $\mathcal{P}$. We know (Section 3.1) that two structural actions have been proposed; since alternative $A0$ must also be considered, the set $\mathcal{S}$ is made up of three elements and thus is a finite set. The situation is different for the other three sets: for example, defining a regulation range means fixing, for every day of the year, (a) the level above which it is necessary to completely open the dam, and (b) the level at which the release cannot be greater than the inflow. It is evident that there are infinite possibilities for fixing these limits, so $\mathcal{R}$ has an infinite number of elements. The same is true for the sets $\mathcal{M}$ and $\mathcal{P}$. Therefore, $\mathcal{A}$ too contains an infinite number of alternatives. Figure 5.2 shows a simple representation of a portion of the set $\mathcal{A}$ in the space (r, m, p), by means of two parallelepipeds[1] corresponding to two of the three structural actions s (the third, which is similar, has been omitted for brevity's sake).

[1] To simplify the figure we assume that there exists a one-to-one relation between actions and points in the space (r, m, p).

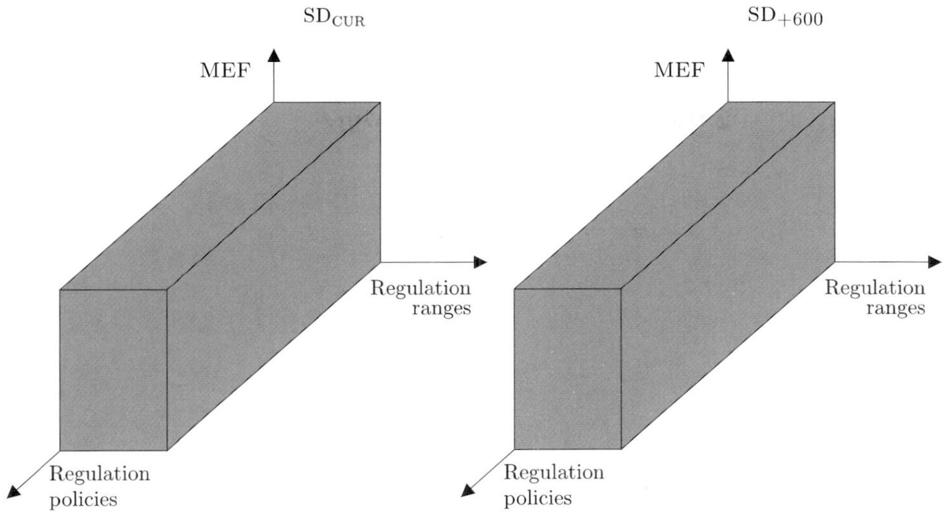

Figure 5.2: A portion of the set $\mathcal{A}$ of alternatives.

The reader should have understood what we have said so far about the sets $\mathcal{S}$, $\mathcal{R}$ and $\mathcal{M}$ (for their formal definition, see Section 5.3), while (s)he may be a little confused about the set $\mathcal{P}$. In Chapter 3 we broadly discussed the stage–discharge relation, the regulation range and the MEF, while until this point we have been very vague about the regulation policies. Thus the moment has come to formalize this concept.

In the next section we define the regulation policy and discuss how it can be designed, under the hypothesis that the triple of actions (s, r, m) has already been fixed; in the following sections we broaden our field of analysis and consider the case when also the triple (s, r, m) can vary.

5.2 Regulation actions

5.2.1 The regulation policy

To introduce the concept of policy, suppose for a moment that it is not our job to discuss the distribution policy among the Ticino and all the canals; in other words, suppose that the only *decision* u_t (also termed *control*) that we have to make each day t, at 8 o'clock in the morning, is the water volume to be released in the following 24 hours, i.e. in the interval $[t, t+1)$. A *regulation policy* p (Maas et al., 1962) is a sequence

$$p = \{m_t(\cdot);\ t = 0, 1, \ldots\} \tag{5.2}$$

(often periodic) of *control laws* $m_t(\cdot)$, each of which provides the control u_t, given the available *information* $\Im_t$ at time t, i.e.

$$u_t = m_t(\Im_t) \tag{5.3}$$

In the simplest case, shown in Figure 5.3, $\Im_t$ is the lake storage s_t and so the regulation is based on the classic scheme of *closed loop control*.

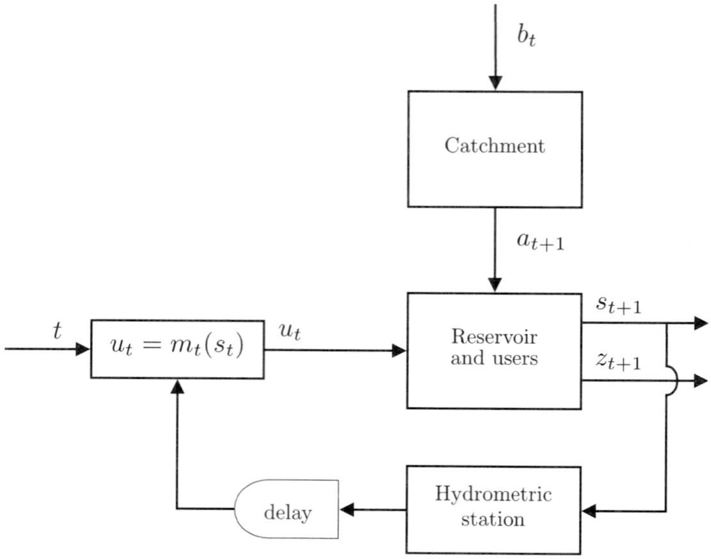

Figure 5.3: A closed loop control scheme; b_t is the precipitation in the catchment, a_{t+1} the inflow to the lake in the interval $[t, t+1)$, u_t the control, s_t the lake storage and z_t the state of the water users.

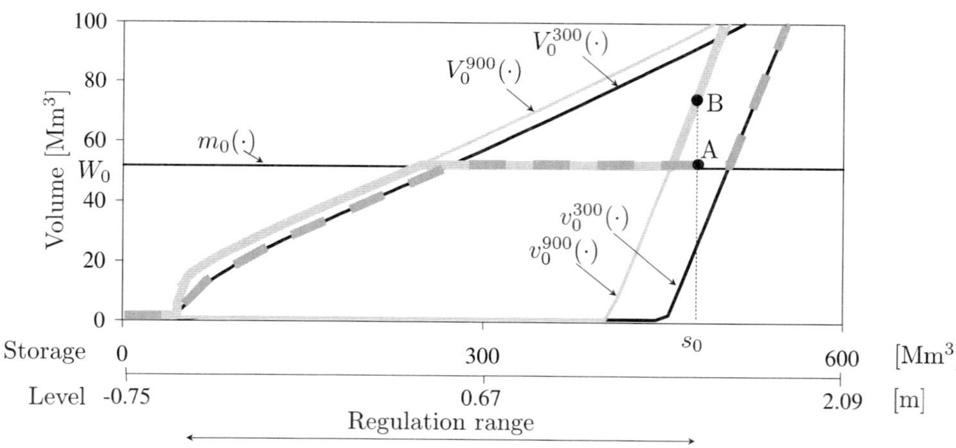

Figure 5.4: The control law $m_0(\cdot)$, which minimizes the irrigators' water deficit, and the maximum (V) and minimum (v) volumes that can be released in correspondence with two different inflow values (black and grey thin solid lines).

Let us also suppose that the regulation objective is to minimize the irrigators' water deficit: in that case, the best choice for the control u_t is to release a volume that is equal to the water demand W_t, whatever the storage of the lake s_t. Therefore, at every time instant, for example at time $t = 0$, the control law in the plane (s_0, u_0) is expressed by a horizontal line that coincides with the value W_0 (52 million m^3 in the example in Figure 5.4). The decision is taken, as we have said, at 8 o'clock in the morning. At that time, given the information $\Im_0$, the control law provides the control u_0, i.e. the volume that *should be released* over the following 24 hours. But the amount that is *actually* released over the next 24 hours will not necessarily be exactly the same: no matter what the decision is, the actual release can neither be more than the maximum volume V_0 that can be released (with the gates completely open), nor less than the minimum volume v_0 that can be released (defined by the regulation range). Clearly both V_0 and v_0 also depend on the inflow that will enter the lake during the day. The inflow is random by nature, and so it will not be known until the following day, i.e. at time $t = 1$. Consequently, also the actual release r_1 (i.e. the volume released from the lake between 8 am on day 0 and 8 am on day 1) will not be known until that time: it is for this reason that we adopt the subscript 1 to denote it (more generally, we indicate the actual release in the interval $[t, t + 1)$ with the symbol r_{t+1}). For example, if the average inflow over the course of the day $[0, 1)$ were 300 m^3/s, the minimum volume that could be released would be given by the function[2] which in Figure 5.4 we indicate with the symbol $v_0^{300}(\cdot)$, and the maximum volume by the function $V_0^{300}(\cdot)$. The bold dashed line thus provides the actual release r_1, as a function of the initial storage s_0. If, instead, the inflow were 900 m^3/s, the minimum and maximum volume released would be given by the functions $v_0^{900}(\cdot)$ and $V_0^{900}(\cdot)$ and the actual release would be described by the bold solid line. Note that if the storage s_0 were, say, 470 million m^3, in the first case the actual release r_1 would be equal to the decision u_0 (which, as we have seen, is equal to the demand W_0: see point A in the figure) while in the second case it would be higher (point B). This is because in the second case the 900 m^3/s inflow would make the lake level exceed the upper extreme of the regulation range, at a certain time during the day, and at that time the dam would have to be completely opened. In this second case, at 8 o'clock on the following day, the storage s_1 would be 476 million m^3, while it would be equal to 444 million m^3 in the first case.[3]

This example helps us to understand why we want to define the regulation mode through a policy, i.e. a sequence of control laws, rather than specifying a simple sequence of controls; or, in other words, why the closed loop control scheme is preferable to the much simpler *open loop* control scheme (Figure 5.5). Suppose the inflow is deterministic, i.e. its value at every future time instant is exactly known. Then, once the control u_0 is fixed, the actual release can be univocally determined at time 0, as well as the storage s_1. This means that at time 0 we are able to define also the control u_1, and the controls for all the following time instants. In other words, if at time $t = 0$ we deterministically knew the sequence $\{\bar{a}_t\}_{t=0,1,\ldots}$ of future inflows, at that time we could establish the sequence of optimal controls $\{\bar{u}_t\}_{t=0,1,\ldots}$ and derive the corresponding sequence of the storage $\{\bar{s}_t\}_{t=0,1,\ldots}$. Now, suppose that we have chosen such a sequence of controls and that we apply the first of them, namely $\bar{u}_0$. However, the inflow is actually not deterministic and so the storage s_1 at time 1 will not necessarily be the same storage that we were expecting. In that case, the Regulator will be in a difficult

[2]The function is computed by integrating the continuity equation of the lake (see (6.25) and (6.26)) over a horizon of 24 hours and by assuming that the instantaneous release is given by the stage–discharge relation and that the instantaneous inflow is always equal to the specified value (300 m^3/s).

[3]The latter value can be easily computed as $470 - 52 + 300 * 0.0864$, where 300 is the inflow rate.

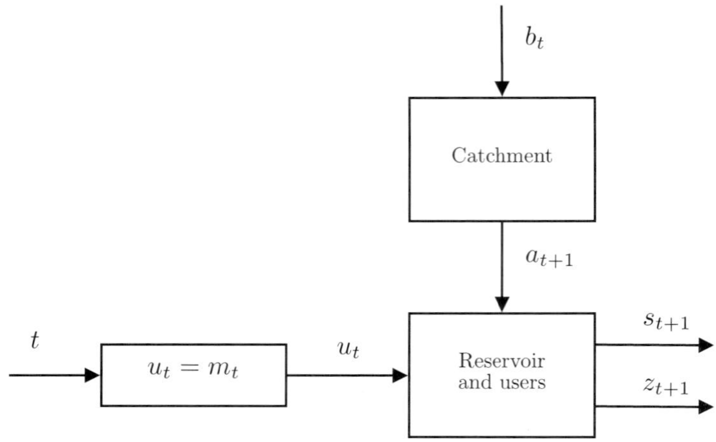

Figure 5.5: An open loop control scheme.

situation: he should, in fact, apply the control $\bar{u}_1$ that we suggested to him, but he will know that that decision was made with reference to a situation different from the one he is observing. So what should he do? The obvious response is to provide him with a decision for every possible situation, i.e. for every possible value of s_1; in other words, give him a control law, and therefore a policy.

In conclusion, the open loop control (i.e. a priori determining the sequence of controls to be applied at each of the future time instants) will be a rational control mode only if the inflow is deterministically known. But, since the inflow is not deterministic, the rational approach is to adopt a policy, i.e. to specify what to do for every situation that can occur.

In the last example we supposed that there was only one regulation objective and so we were able to talk about 'optimal' decision and 'optimal' policy. But what happens when, as in the case of Verbano, one must take into account a plurality of objectives that, as we know, are in conflict? In this case there is no longer one 'optimal' policy, but many different policies can be of interest; each of them produces different values for the different objectives. Let us suppose, for example, that we want to take into account the irrigators' needs, and that we also want to reduce the flooded area around the shores of the lake. Now we might think that the control law should provide for opening the dam completely even before the upper extreme of the regulation range is reached, as shown in Figure 5.6. In this way a compromise is reached between the two objectives. We could, however, think about other policies, in which the compromise that is reached is different. In general, when conflicting objectives must be taken into account, many different policies can be designed, not a single one. Thus a question arises: what kind of criterion should we use, in order to determine which policies must be considered in the decision-making process?

Clearly it makes no sense to consider policies that are not *feasible*, i.e. that do not satisfy all the constraints to the Problem. However, even among the feasible policies, only the *efficient* policies are interesting. A policy p is efficient when no other feasible policy exists that provides a better performance with respect to at least one objective, without worsening at least one of the others. All the non-efficient policies (also called *dominated* policies) can be eliminated from the decision-making process, since no one would ever be interested in them. For each triple (s, r, m), we must thus individuate, from the feasible policies, all

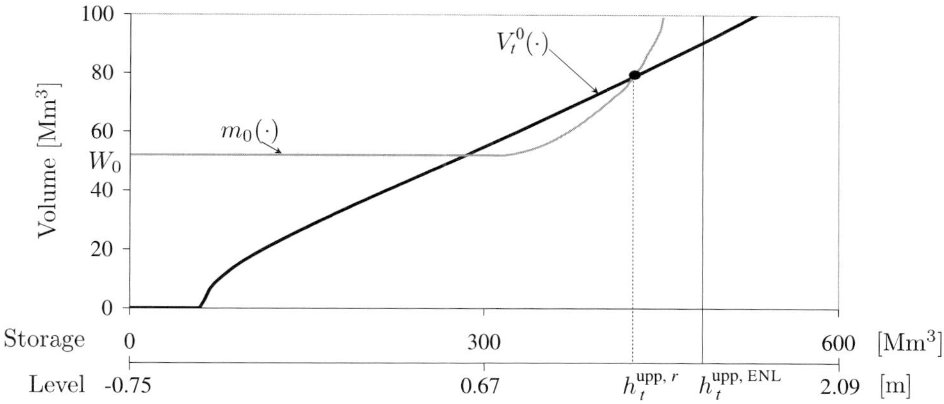

Figure 5.6: The control law $m_0(\cdot)$ that realizes a compromise between minimizing the users' deficit and minimizing the flooded area on the lake's shores.

and only the efficient policies, the set of which is called the *Pareto Frontier*. This set can be identified by formulating and solving an opportune Multi-Objective (*Optimal*) *Control Problem*, which we will define in general terms in Section 5.2.6 and analyse in detail in Chapter 7.

5.2.1.1 Set-valued policies

Even if, from a logical standpoint, the most appropriate approach to the regulation problem seems to be the adoption of a policy (closed loop control scheme), in practice such a method has rarely been applied. The reasons for this failure are many and complex. For now we will illustrate only one of them, which is linked to the definition of the policy itself.

The definition presented in the previous section is the one commonly adopted in *Control Theory*, where the task of the policy is to carry out the automatic control of a given system. But when the policy is used in the management of a natural resource, such as a lake, it is not required to 'automatically' manage the system, but rather to support the management of the (human) Regulator, by suggesting 'good' decisions, among which the Regulator can make his choice. In this perspective, it seems more rational that the policy, instead of proposing a single control $m_t(\Im_t)$, should propose the set $M_t(\Im_t)$ of all the equivalent controls[4] (Aufiero et al., 2001), i.e. of all the controls that provide long term performances that are 'equivalent' with respect to the objectives of the Control Problem of which the policy is a solution (for more detail see Section 5.2.6). From the set $M_t(\Im_t)$, the Regulator can choose the control that, in his opinion, best responds to the current situation. Thus it is advisable to substitute the concept of a *point-valued* policy with that of a *set-valued* policy, defined as a sequence of set-valued control laws $M_t(\Im_t)$, i.e.[5]

$$p = \{M_t(\cdot);\ t = 0, 1, \ldots\} \tag{5.4}$$

[4] For a more precise definition see Section 8.4.

[5] In THEORY a set-valued policy is denoted with the symbol P to distinguish it from a point-valued policy. Since in this text the difference between the two is just briefly mentioned, we decided to use the symbol p to denote both of them, so as not to overload the reader with symbols. The reader who has read or is reading THEORY should keep this in mind.

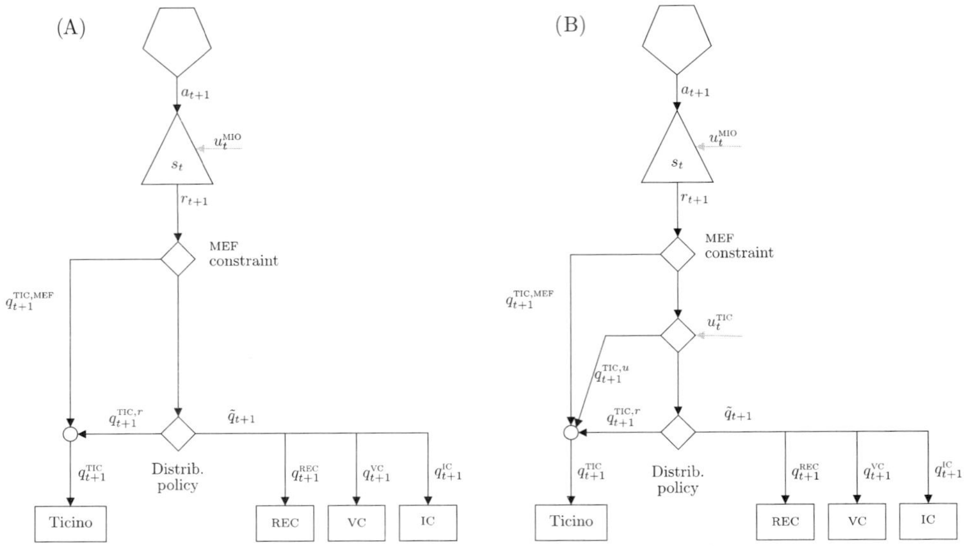

Figure 5.7: Comparison between the current distribution scheme (A) and the scheme that will be used to design regulation policies that take environmental needs into consideration (B).

5.2.2 The control

Managing the waters of Verbano first of all means making a *release decision*, namely choosing the value of the variable that in the previous section we denoted with u_t. From now on we will indicate that decision with the symbol u_t^{MIO}, to highlight that it concerns the volume that the Regulator would like to release at the Miorina dam. In the current management of the lake, this is the only decision made by the Regulator (scheme A in Figure 5.7): in fact, as we observed in Section 3.4, the distribution among the various canals and the Ticino is carried out according to a fixed distribution policy, which will be described in Section 6.7.3.

However, we have already observed that this policy is accepted only by the irrigators and hydropower producers, and not by the Ticino Parks: in the design of the regulation policy we must therefore take their dissatisfaction into account. The Project Goal does not allow us to redefine the distribution policy, but only to decide how the volume r_{t+1} released at Miorina must be divided into two terms: the volume q_{t+1}^{TIC} that is left to the Ticino and the overall volume $\tilde{q}_{t+1}$ diverted into the canals. The first is the sum of the volume $q_{t+1}^{\mathrm{TIC,MEF}}$ guaranteed by the MEF constraint and the possible residual volume $q_{t+1}^{\mathrm{TIC},r}$ after the distribution among the canals. The second ($\tilde{q}_{t+1}$) is subdivided among the Regina Elena, Villoresi and Industrial Canals (volumes q_{t+1}^{REC}, q_{t+1}^{VC} and q_{t+1}^{IC} respectively) on the basis of a distribution policy that cannot be discussed.

With the aim of improving the environmental quality of the River Ticino, a second control variable u_t^{TIC} is introduced, which represents the volume allocated to the Ticino in the interval $[t, t+1)$, in addition to the MEF (scheme B in Figure 5.7). This control variable will be named *environmental allotment decision*. For the same reasons that we illustrated in Section 5.2.1 when speaking of the release decision, the environmental allotment decision cannot always be perfectly realized either. This is precisely the reason why it is advisable to distinguish between the decision u_t^{TIC} and the volume $q_{t+1}^{\mathrm{TIC},u}$ that is actually left to the River Ticino following that decision. The overall volume left to the river will thus be given by

$$q_{t+1}^{\text{TIC}} = q_{t+1}^{\text{TIC,MEF}} + q_{t+1}^{\text{TIC},r} + q_{t+1}^{\text{TIC},u} \tag{5.5}$$

Note that, if u_t^{TIC} is equal to zero for every time t, scheme B turns into scheme A, that is the distribution among the Ticino and all the canals is carried out according to the current distribution policy.

In the design of the regulation policies we thus consider the following control vector

$$\mathbf{u}_t = \begin{bmatrix} u_t^{\text{MIO}} \\ u_t^{\text{TIC}} \end{bmatrix} \tag{5.6}$$

and, at every time t, the policy will be defined by a set-valued control law $M_t(\cdot)$, which provides a set of decision pairs $(u_t^{\text{MIO}}, u_t^{\text{TIC}})$.

5.2.3 The design indicators

To individuate the objectives with respect to which the policy will be designed, in the first place we must define the criteria which we want to take into consideration. In Chapter 4 the criteria that will be used to evaluate the effects of the alternatives were defined. Straightaway we are led to believe that the policy must be designed taking account of all these evaluation criteria. However, this is not the case, because evaluating the alternatives means comparing different tuples (s, r, m, p), while the policy design is concerned only with one of its components (p) and it is done once the other three (s, r, m) have been fixed: thus not all the criteria that are needed in the first case are necessarily required in the second.

To make this clearer, we consider an example, relative to the sectors *Upstream Flooding* and *Downstream Flooding*. Downstream floods are caused both by the flow coming from the Miorina dam and by the backflow from the Po (see Section 4.2.2). The first contribution begins to be significant only when the flow rate is greater than 1000 m^3/s, which, regardless of the stage–discharge relation and the regulation range adopted,[6] can happen only when the lake level is higher than 1.5 m and thus the lake is in free regime. In these conditions flow and level are bijectively related by the stage–discharge relation: this means that a regulation policy that is favourable to the *Upstream Flooding* sector, i.e. a policy that lowers the peak lake level during the floods, will also reduce the peak release into the Ticino in the same conditions. In the design of the regulation policy, which is carried out once the stage–discharge relation has been fixed, it is thus sufficient to consider the criteria associated with the *Upstream Flooding* sector. In the evaluation of the alternatives, however, the criteria from the *Downstream Flooding* sector must also be considered, because the conflict between the two sectors emerges when the alternatives characterized by different stage–discharge relations, which produce different behaviours in free regime conditions, are compared. It is important to underline that the above reasoning is based on the hypothesis that the regulation is carried out only within a regulation range and ceases to be valid whenever this hypothesis is removed, as we will explain in Section 14.2.3.

The idea of using only a subset of the evaluation criteria when designing the policy is not born of purely theoretical reasons, but it also depends on the practical difficulty of solving a Control Problem with a high number of objectives. For reasons that will be clarified further on (Section 5.2.6), the computing time necessary to identify the Pareto Frontier grows exponentially with the number of criteria (and thus of objectives) considered. It is

[6] Among those presented in Chapter 3.

thus necessary that this number be not too high. Therefore, with the Stakeholders' agreement, we must select a subset $\mathcal{C}_d$ of criteria, called *design criteria*, which will be considered in the design of the policy.

Let us first consider from which sectors the set $\mathcal{C}_d$ must be derived. On the basis of the description of the physical system and the existing conflicts, presented in Chapters 2 and 3, it is natural to concentrate our attention on the following sectors:

- *Upstream Flooding*
- *Irrigation*
- ENEL *Power*
- *East Sesia Power*
- *Downstream Flooding*
- *Downstream Environment*

However, we have just explained that the *Downstream Flooding* sector is well represented by the *Upstream Flooding* sector, and so we can consider only the latter when designing the policy. For similar reasons, the *East Sesia Power* sector can be omitted and its interests can be represented by ENEL *Power*: in fact, the periods in which power production is most convenient are the same for both of them, since the energy market is the same, and there cannot be conflicts between them, because we have decided not to modify the distribution policy (Section 5.2.2).

In the formulation of the Control Problem, the objectives should in theory be given by the indices associated to the four sectors *Upstream Flooding*, *Irrigation*, ENEL *Power* and *Downstream Environment*. However, in the initial phases of the decision-making process, the Experts rarely have the information necessary to define such indices. For the definition of the objectives we must thus replace the indices with appropriate indicators which we will accordingly call *design indicators* (see Chapter 18 of THEORY) and which must not be confused with the *evaluation indicators* considered so far.

For each of the above sectors, the design indicator must be constructed on the basis of the evaluation indicators of that sector. To identify it, we must consider both the relative importance attributed to the indicators of the sector under study (and for this we must refer to the judgement of the Experts) and the form of the functionals that define those indicators. In fact, as we will see in Chapter 7, the best algorithm that is available today for the Control Problem resolution poses conditions on the mathematical form of the objectives. If the Experts opt for a design indicator that, with the model adopted, does not satisfy those conditions, there are two alternatives: the first is to modify the model, the second to modify the indicator (we show an example of the first solution in Section 6.9, while the second is detailed in Section 7.2). It will thus be only in Section 7.3 that we will be able to define the design indicators. For now we can only anticipate their names, which are indicated in Table 5.1.

Lastly, we must ascertain that the set $\mathcal{C}_d$ of the design criteria is representative of the set $\mathcal{C}$ of the evaluation criteria. For this it is necessary that:

(1) based on the design criteria in $\mathcal{C}_d$, the efficient regulation policies be designed in correspondence to a number of significant tuples (s, r, m), by solving the corresponding Control Problems (see Chapter 7);

Table 5.1. The design indicators

Sector	Design Indicator	Symbol
Upstream Flooding	Average annual flooded area in Locarno and Verbania	i_{fl}
Irrigation	Average annual potential damage from stress	i_{irr}
ENEL *Power*	Average annual lost revenue for the ENEL hydropower plants	i_{Enel}
Downstream Environment	Distance from natural regime	i_{env}

(2) all the evaluation indicators be computed for each of the policies so determined (see Chapter 8);

(3) for each evaluation indicator, the correlation between the set of its values so determined and the values of each one of the other design indicators be computed.

If an evaluation indicator is negatively correlated or uncorrelated with all the design indicators, and the Experts declare that all the values it assumes are unfavourable or even unacceptable, one must conclude that the corresponding criterion is not represented in the set $\mathcal{C}_d$. Then, $\mathcal{C}_d$ must be enlarged to contain it, and this procedure must be carried out once again. Thus we understand that the choice of the design indicators is one of the (many) sources of recursion in the decision-making process.

To avoid the notation becoming cumbersome, we decided to use the same symbol i to refer to both the evaluation and the design indicators; the reader will easily distinguish them according to the context.

5.2.4 The information system

The performance of a policy depends on the choice of its argument $\Im_t$, i.e. of the information it processes. The scheme presented in Figure 5.3 is the simplest case with the information $\Im_t$ coinciding with the storage s_t of the lake. However, it is possible to enrich this scheme by considering other information, for example the state z_t of the water users (if one decides to represent this part of the system as well) or hydro-meteorological information I_t from the catchment. When the latter information is used, the control scheme is no longer just a closed loop, but a *compensation line* is added with the aim of anticipating the effects of meteorological events (Figure 5.8).

By increasing the dimension of the vector $\Im_t$, i.e. the amount of information that is used, one expects the performance of the policy to improve. At the same time, the costs increase: the management cost increases because at every time t all the components of $\Im_t$ must be measured, and so opportune measurement networks must be established and maintained; the design cost increases because, as we will show in Chapter 7, the computing time for the solution of the Control Problem grows exponentially with the dimension of $\Im_t$. Before introducing a new dimension into $\Im_t$ we must therefore evaluate these two opposing effects quite carefully. The second is relatively easy to evaluate, the first much more difficult. Sometimes, however, it is possible to estimate an upper bound for the maximum improvement that an increase of information could produce. Suppose, for example, that the Control Problem is formulated with the two objectives *average annual flooded area* and *average annual volume of deficit* and that the efficient policies are designed only on the basis of the storage s_t.

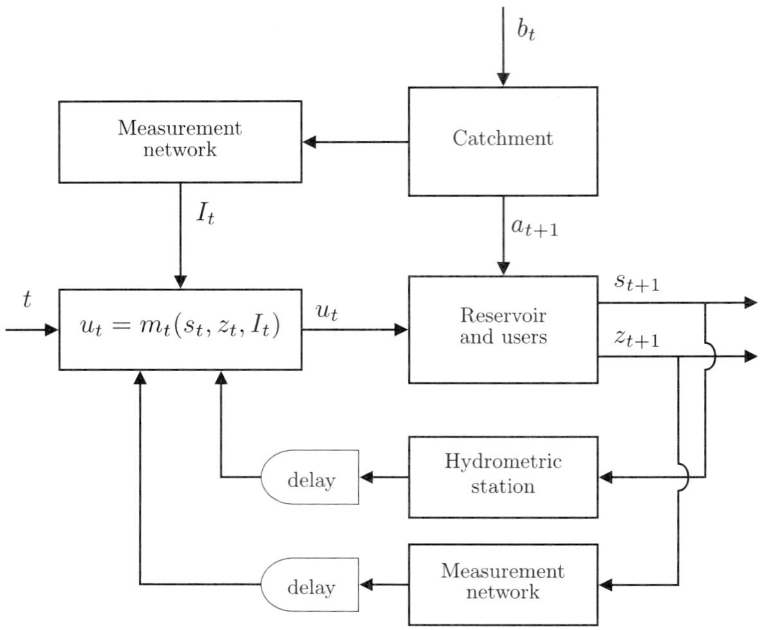

Figure 5.8: A control scheme with a compensation line.

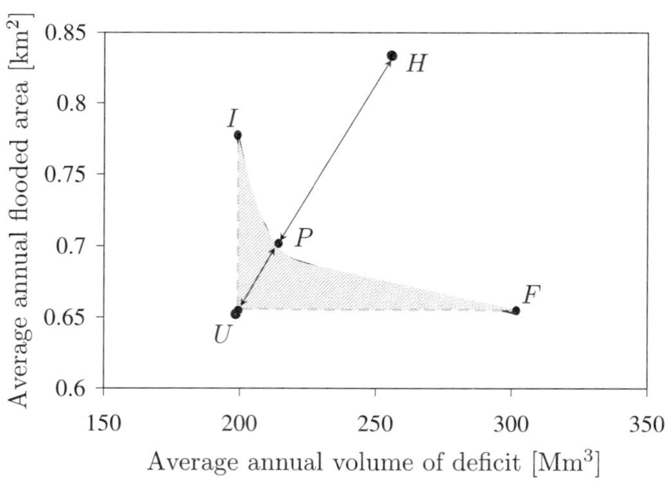

Figure 5.9: The plan of the objectives and the Pareto Frontier of the Control Problem in which the stage–discharge relation and the regulation range are the current ones, and the information $\Im_t$ is simply the storage.

The results are represented in Figure 5.9, where the black line is the Pareto Frontier that is obtained by proceeding in this way. The extreme points of the frontier are generated by the extreme policies, which are designed taking account of only one objective: point I represents the best performance for *Irrigation*, point F the best for *Upstream Flooding*. The *Utopia* point U, whose coordinates are the values of the objectives in correspondence to the two extreme policies, represents the Regulator's dream. Generally this point is unfeasible since in order to obtain it, antithetic regulation policies must be adopted. For example, to realize point I, a volume equal to the irrigation demand should be released whenever possible (we discussed this in Section 5.2.1), while to realize point F, it is necessary always to leave the dam completely open.[7] It is easy to understand that when the Utopia point is feasible, it is the only point on the Pareto Frontier. The current performance H of the system is represented on the same plane. Consider segment HU: the point P at which it intersects the Pareto Frontier is an interesting compromise point. Since it is located on the segment HU, it improves both the objectives; moreover, it improves them in such a way that the ratio of the increments is equal to the ratio of the components of H, measured with respect to U, i.e. the same ratio of the objectives's historical values. In other words, by choosing P, the weights that are implicitly attributed to the two objectives are those that correspond to the importance that past practice attributed to *Irrigation* with respect to *Upstream Flooding*. It follows that the ratio of the lengths $\overline{PU}$ and $\overline{HU}$ is a good index of how much the performances of the system could be further improved, by including other variables besides s_t in the information $\Im_t$. This is true when, as in the example, the coordinates of U do not depend on the information adopted (remember how they were computed). Segment PU represents an upper estimate of the improvement that could still be obtained. In the case of Verbano Project (see again Figure 5.9), by adopting a policy based only on the storage s_t (point P), 75% of the potential improvement HU is obtained; an extension of the information system could improve the performance by as much as another 25%. By comparing this estimate with the cost of the monitoring networks, an Expert is able to evaluate whether it would be advisable to consider more complex information systems or not.

Unfortunately, the previous analysis cannot be correctly applied when the value of even one of the coordinates of U depends on the information $\Im_t$. This occurs, for example, if we consider the *average annual stress* instead of the *average annual volume of deficit*. In this case, in fact, the extreme policy for irrigation is no longer independent from the information system adopted, i.e. it is not known a priori and instead it must be computed by resolving the corresponding Control Problem.[8]

To understand which information $\Im_t$ might be usable in our case, consider Figure 5.10, which shows a scheme of the Verbano system. The exogenous information is the state I_t^{Ca} of the catchment and the state of the Po, i.e. the flow rate I_t^{Po}. The lake is described by its storage s_t and the irrigators by the state z_t^{REC}, z_t^{VC} and z_t^{IC} of the crops. All this information might be usefully included in $\Im_t$. However, in Section 5.2.3 we decided not to consider the

[7]Actually the situation is a little more complicated, as an alert reader may notice, since, by simply leaving the dam open, the point F will not necessarily be generated. In fact, with the dam always open, it is likely that at some time instants water which in the future could be used to reduce the supply deficit would be released, without it being necessary for the reduction of flooding. However, it is true that the value of the coordinates of point F corresponding to *average annual flooded area* can be determined by leaving the dam constantly open. An analogous observation is valid for I, so that one can conclude that what we affirm in the text is correct in relation to the components of U.

[8]It is not true that by releasing a volume equal to the demand at every time instant the minimum water stress will be achieved: the reader can easily realize this with a few counter-examples.

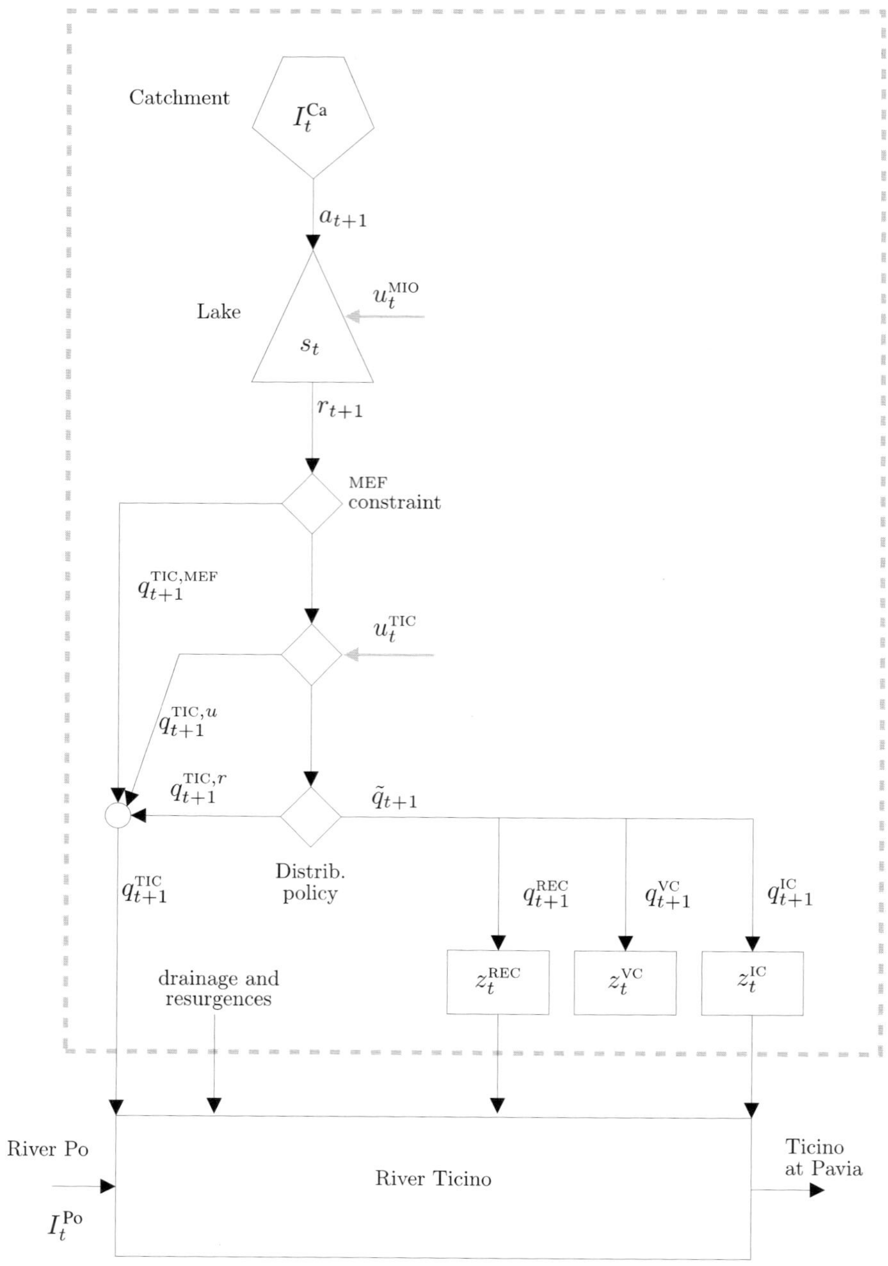

Figure 5.10: The Verbano system. The grey dashed line surrounds the part considered in the policy design.

sector *Downstream Flooding* when designing the policies; it follows that only the part of the system enclosed by the dashed line is considered and so we do not need the information I_t^{Po} about the flow rate of the Po. It remains to be decided whether to use all or only a few of the other four variables. In Chapter 6, once the models of the different components have been identified, we will explain why the choice is made for the minimal solution: consider only the lake storage s_t.

5.2.5 The constraints

Not all the variables that describe the system can vary without restriction; there are, in fact, many constraints which bind them. First of all, the dynamical (or static) relationships that link *inputs* and *outputs* of the different components, in other words the models of such components, which are considered in the next chapter. Other constraints derive from the normative actions: for example, the norm that imposes the passage to the free regime when the lake level rises above the upper extreme h_t^{upp} of the regulation range. Also the MEF is expressed with a constraint. Note that the value of h_t^{upp} depends on the regulation range that is adopted, the outflow in free regime depends on the stage–discharge relation, and the value of the MEF depends on the corresponding norm. Therefore, the structural and normative actions are reflected in the parameters that define the constraints of the Control Problem. This proves what we already guessed in Section 5.1: the Control Problem that defines the policy changes as the triple (s, r, m) changes.

Nevertheless, we have not yet included all the constraints that must be included in the formulation of the Control Problem. In order to individuate the remaining ones, we observe firstly that the reasons that led to the selection of the four design indicators in Table 5.1 are the same ones that suggested the interventions, i.e. the structural and normative actions s, r and m. This would seem to introduce a redundancy, since the same reasons are considered twice: first in the design indicators (and thus in the objectives of the Problem, as we will see later) and once again in the constraints. To solve this apparent 'non-sense', it is advisable to examine the relationship between constraints and objectives more carefully. Understanding this will be useful in the future to establish how to explore the alternatives space.

Adopting a given design indicator means looking for a regulation policy that provides a good performance for the corresponding sector, *in the long-term*. Note this specification: it means that the possibility that the behaviour might be unacceptable, at *some* time instants, cannot be excluded. If we want to exclude this eventuality, it is necessary to impose constraints that guarantee a minimum performance at *every* time instant. For example, we have seen that in the *Downstream Environment* sector the overall condition is evaluated on the basis of the indicator $i_{\text{Env\_D\_1}}$, which measures the distance between the regulated regime and the natural one. In Section 4.3.3 we noted that this indicator, when evaluated over a long horizon, is not very sensitive to sporadic events in which the flow rate in the Ticino produced by the regulation is zero. Since events of this type are extremely damaging to the river ecosystem and they cannot be compensated by a high number of days in which the flow rate is close to the natural one (as the indicator would seem to indicate), it is necessary to introduce the MEF constraint.

A similar situation occurs in the ENEL *Power* sector. In Section 4.2.2 we explained that, if the 25 m^3/s water volume required by the cooling towers of the Turbigo-EUROGEN power plant were not supplied, the plant would have to be put out of service, causing very serious damage, which is incommensurably large with respect to the damage that might be produced by a water failure of 25 m^3/s imposed upon any other downstream user. Therefore, it is rational that the Industrial Canal have the priority over the first 25 m^3/s of release from the lake.[9] Again, this must be expressed through a constraint.

When introducing constraints of this nature, one must be very conscious that they cause a reduction of the set of feasible controls. Such a reduction, in turn, influences the optimization of the objectives: the stricter the constraints, the worse the values of the objectives. A constraint must thus be adopted only after much consideration, when it cannot be avoided.

5.2.6 Formulating the Control Problem: a draft

The time has come to describe the Optimal Control Problem with which the efficient policies are generated for a given triple (s, r, m). We are not able to define it precisely as yet, because we have not defined the model of the system, which we will deal with in the next chapter. Therefore, we must postpone its formal description until Chapter 7. For now, we will simply draft its structure, by assuming, in order to keep the exposition simple, that we are looking for a point-valued policy.

All the indicators introduced in Chapter 4 are functions of the system behaviour over a certain horizon. We call *design horizon* the one over which the design indicators are computed. The behaviour of the system will vary as the policy p varies, but it will depend also on the presence of stochastic and uncertain factors, such as precipitation and prices, that act upon it. It follows that the value assumed by the design indicator i_k (with $k = \text{fl, irr, env, Enel}$) is not deterministically known, given the policy p; instead, it is a stochastic or uncertain variable. Therefore, we must consider a suitable statistic J_k of it, as, for example, the worst value or the expected value, which we will call *objective* (see Section 7.6.1). An objective J_k is thus a deterministic function of the policy, i.e.

$$J_k = J_k(p) \quad k = \text{fl, irr, env, Enel} \tag{5.7}$$

When all the objectives must be minimized,[10] the *Multi-Objective Control Problem* (or *Control Problem* for short) assumes the following form

$$\begin{aligned} &\min_{p \in \Omega} \left[J_{\text{fl}}(p), J_{\text{irr}}(p), J_{\text{env}}(p), J_{\text{Enel}}(p) \right] \\ &\text{subject to any constrains to be specified} \end{aligned} \tag{5.8}$$

where Ω is the set of feasible policies.

The solution to problem (5.8) does not provide a single policy, but the set of all the efficient policies: namely the Pareto Frontier. In order to solve it, it is necessary to trace it back to a family of Single-Objective Control Problems, each of which provides a specific efficient policy as solution. The family of Single-Objective Problems may be derived from problem (5.8) with different methods. For example, when adopting the *weighting method*, the single objective $\bar{J}$ of the new Problem is defined as the weighted sum of the objectives J_k of the original Problem, i.e.

$$\bar{J} = \sum_{k=\text{fl, irr, env, Enel}} \lambda_k J_k \tag{5.9}$$

[9] The reader might object that our reasoning is correct as far as the water users are concerned, but that the environmental damage in the long run can be incommensurably greater that the 'incommensurable' damage produced by one stop of Turbigo-EUROGEN. (S)he would be right if we had not introduced the MEF and a design indicator concerning the *Downstream Environment* sector.

[10] If any of them have to be maximized, it is sufficient to consider their opposites.

where the parameters λ_k are called *weights* and satisfy the following condition

$$\sum_{k=\text{fl},\text{irr},\text{env},\text{Enel}} \lambda_k = 1 \tag{5.10}$$

Problem (5.8) is thus equivalent to the following family of Problems, which are parametric in the vector $\boldsymbol{\lambda} = |\lambda_k|$,

$$\min_{p \in \Omega} \sum_{k=\text{fl},\text{irr},\text{env},\text{Enel}} \lambda_k J_k$$

$$\text{subject to any constraints to be specified} \tag{5.11}$$

In practice, it is impossible to compute the infinite number of policies that make up the Pareto Frontier, because we cannot solve problem (5.11) an infinite number of times, in correspondence to the infinite possible vectors of weights. We thus consider only a finite number of Problems, that is, a finite number of vectors $\boldsymbol{\lambda}$, each of which provides a policy, whose performance corresponds to a point on the Pareto Frontier. Once these points have been computed we can guess the form of the frontier. To facilitate this task, the values to consider for vector $\boldsymbol{\lambda}$ should be carefully chosen.

5.3 Structural and normative actions

Until now we have seen how to identify the most interesting regulation actions, i.e. how to design the efficient policies for a given triple (s, r, m) of structural and normative actions. Now we must define which triples are to be considered, that is, define the sets $\mathcal{S}$, $\mathcal{R}$ and $\mathcal{M}$ introduced in Section 5.1. In theory, all the possible actions must be considered but, to reduce the cost of the design, it is advisable that all the 'uninteresting' actions, namely those that would not be chosen by any of the Stakeholders, be eliminated from the analysis a priori.

5.3.1 Stage–discharge relation

In Section 3.1 we explained that Switzerland proposed two alternative interventions for excavating the lake outlet, which would produce two different stage–discharge relations, named SD_{+300} and SD_{+600}. We denote the two corresponding actions with the same symbols. They both have no realization cost, since the excavation costs are completely covered by the proceeds from selling the gravel that is removed.

From Figure 3.4 we can see that, for any level value, the release produced by the SD_{+600} stage–discharge relation is higher than the one produced by SD_{+300}. It follows that, if a regulation range were not imposed, the set of feasible controls associated with the first stage–discharge relation would include the one associated with the second (observe Figure 5.11 in which the sets of feasible release decisions u^{MIO} are shown). Therefore all the policies designed with the action SD_{+300} could be applied also if the action SD_{+600} were implemented, but not vice versa. Thus we could proceed by considering only the action SD_{+600}, and the task of preventing high flows from producing unacceptable performances for the *Downstream Flooding* sector could be completely entrusted to the regulation policy, provided that we included a suitable indicator for that sector among the design indicators. Then, once the best compromise alternative had been chosen, it would be easy to ascertain

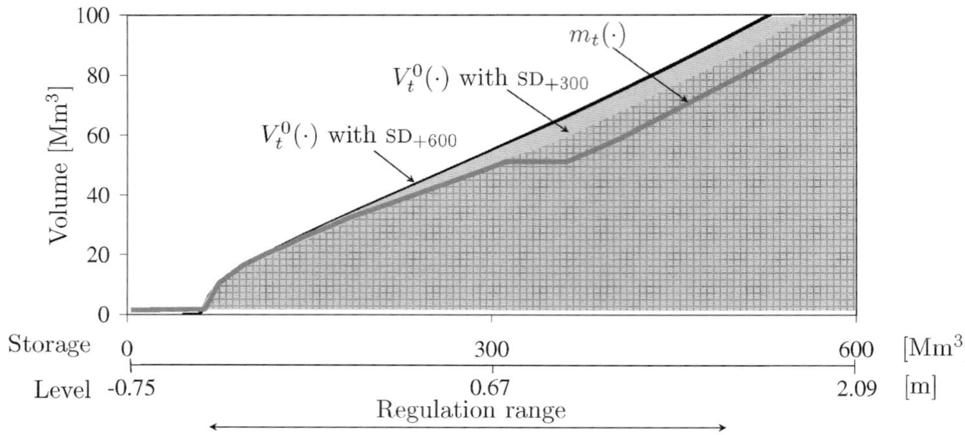

Figure 5.11: The space of feasible release decisions with SD_{+300} (grid area) and with SD_{+600} (grey area) if the regulation range did not exist. The thick grey line represents a possible control law.

whether it was really necessary to realize SD_{+600} or whether SD_{+300} was sufficient. To do this, if the chosen policy was point-valued, it would be enough to verify whether, at every time t, the control law $m_t(\cdot)$ was inferior to the curve $V_t^0(\cdot)$ that describes the maximum possible release with zero inflow for SD_{+300} (see Figure 5.11). If, instead, it were a set-valued policy, at every time t, the check would have to be carried out considering the upper extreme of the set $M_t(\cdot)$.

When instead, as in the Verbano Project, a regulation range is imposed, for levels higher than its upper extreme, the dam must be completely open and so, in these conditions, the outflow is established by the stage–discharge relation rather than the policy. Then, it is no longer true that all the possible effects of SD_{+300} can be produced also with SD_{+600}. The action SD_{+300} must thus be analysed too.

Note, nevertheless, that only the *Downstream Flooding* sector could benefit from a limitation of the outflow. Therefore, if the people from this sector found that an intervention based on the action SD_{+600} was interesting, there would no longer be any reason to analyse the alternatives based on SD_{+300}. In order to limit the design costs, it was thus decided, with the favourable opinion of the representatives of the *Downstream Flooding* sector, not to consider the action SD_{+300}, with a commitment to return and study its effects in the event that none of the alternatives examined were satisfying to them.

Other than the action SD_{+600}, it is obvious that also the possibility of not modifying the lake outlet must be considered, i.e. maintaining the current stage–discharge relation. We denote this action with the symbol SD_{CUR}. In conclusion, we define the following set $\mathcal{S}$ of structural actions:

$$\mathcal{S} = \{\text{SD}_{\text{CUR}}, \text{SD}_{+600}\}$$

5.3.2 Regulation range

The concept of a regulation range was introduced in Section 3.3; now we can formalize it by defining a range r as the sequence of lake level pairs $(h_t^{\text{low},r}, h_t^{\text{upp},r})$ over the year, each of which defines the levels within which the Regulator may freely establish the release at time t. We have already seen that, when the lake level reaches the lower extreme of the regulation range, the Regulator is obliged to release a flow that is no greater than the inflow, so that the level cannot drop any further. Conversely, when the level reaches the upper extreme of the range, the Regulator must open the dam completely, to achieve the maximum outflow.

There are numerous proposals to modify the current range (see Figure 3.7 and Section 3.3), but none of them provides for shifting the lower extreme from the current value of −0.50 m (with respect to the zero level at Sesto Calende gauging station). Therefore that value will remain the same in all the alternatives considered. We indicate this extreme with the symbol h^{low}, without the subscript t, because it is time-invariant. The proposals are relative to shifting the upper extreme h_t^{upp}. Some suggest that it be lowered in some periods of the year, with respect to the current value $h_t^{\text{upp,CUR}}$, with the purpose of mitigating upstream flooding. But the structural action SD$_{+600}$ is also aimed at this objective: so any reduction in h_t^{upp} would be redundant. Other than redundant, it would also be counterproductive, because the change to the range is seen today as a counterproposal to the Swiss request to excavate the lake outlet: its aim is thus to increase the active storage for regulation. We therefore decided not to consider ranges r characterized by values of the upper extreme $h_t^{\text{upp},r}$ lower than the current one. Finally, none of the proposals considers raising the upper extreme over +1.50 m above zero level at Sesto Calende (in the following we will refer to that extreme with the symbol $h^{\text{upp,ENL}}$, where the acronym ENL stays for 'enlarged'). The set $\mathcal{R}$ of interesting ranges is thus made up of all the ranges r that, on every day t, have the lower extreme $h^{\text{low},r}$ equal to h^{low} and the upper extreme $h_t^{\text{upp},r}$ between the current upper extreme $h_t^{\text{upp,CUR}}$ and $h^{\text{upp,ENL}}$, i.e.

$$\mathcal{R} = \left\{r\colon\ h^{\text{low},r} = -0.50 \text{ m};\ h_t^{\text{upp,CUR}} \leqslant h_t^{\text{upp},r} \leqslant h^{\text{upp,ENL}}\right\}$$

The space within which the regulation ranges can be defined is therefore the grey area in Figure 5.12.

We use the symbol RANGE$_{\text{CUR}}$ to refer to the current range and RANGE$_{\text{ENL}}$ for the range whose upper extreme is always equal to $h^{\text{upp,ENL}}$. All the ranges in the set $\mathcal{R}$ contain RANGE$_{\text{CUR}}$ and they are all contained in RANGE$_{\text{ENL}}$. We can thus consider these two ranges as extreme cases, and it is not difficult to understand that these are the only actions to be studied, because it is useless to study the infinite number of ranges included between them. Observe, in fact, that the regulation range was introduced in the Regulation Licence of 1940, to safeguard the *Upstream Flooding* sector: the Stakeholders from this sector were worried that the Regulator might fill the lake too much, to favour the downstream users. Since at that time they did not have another way of imposing a certain regulation rule, the only possibility was to decouple the problem of upstream flooding from the downstream users' problems: when the lake level falls within the range, the Regulator acts, at his own discretion, in favour of the water users; when the level falls outside the range, he is only concerned about the shoreline inhabitants. Today we have a new tool: the regulation policy. As we have explained, the policy can be designed by taking into account all the interests in play. Then, the interests of all Stakeholders, and not just those of the shoreline inhabitants, can be guaranteed by requiring that the Regulator follow the selected policy. Nothing else is needed: the regulation range is superfluous. Nevertheless, psychological reasons require that it not be abolished, but just modified. To understand why, observe that, once a policy is established, the regulation range can be deduced from it. If it is a point-valued policy,

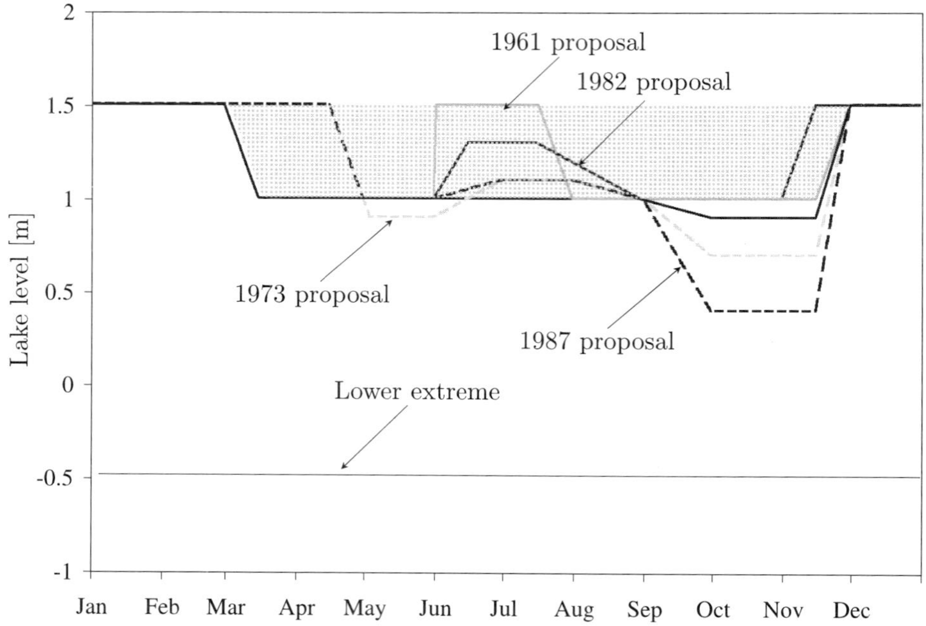

Figure 5.12: The space (grey area) within which the upper extreme of the regulation range can vary, according to the possible normative actions under study, and the proposals put forward in previous studies.

in fact, the upper extreme of the range, on day t, will be the level $h_t^{\text{upp},r}$ at which the release suggested by the control law $m_t(\cdot)$ is equal to the outflow in free regime, computed in correspondence to a zero inflow (see Figure 5.13). If, instead, it is a set-valued policy, the previous computation must be made considering the lower extreme of the set $M_t(\cdot)$ in place of $m_t(\cdot)$. It follows that we can design all the policies by assuming RANGE$_{\text{ENL}}$: if the weight

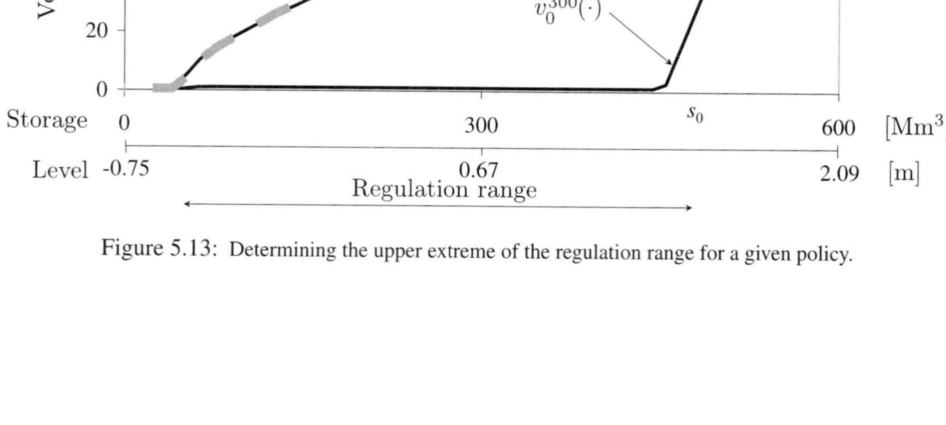

Figure 5.13: Determining the upper extreme of the regulation range for a given policy.

attributed to the *Upstream Flooding* sector is sufficiently high, the control law will impose the maximum possible release well before the level $h^{\text{sup,ENL}}$ is reached, and this is equivalent to considering a time-variant range contained within $\text{RANGE}_{\text{ENL}}$. Considering only the latter is thus completely sufficient.

As well as $\text{RANGE}_{\text{ENL}}$, we clearly must consider also the possibility of modifying nothing, i.e. of keeping the range as it is. In conclusion, we will consider the following set $\mathcal{R}$ of ranges

$$\mathcal{R} = \{\text{RANGE}_{\text{CUR}}/\text{RANGE}_{\text{ENL}}\}$$

5.3.3 The MEF

As we explained in Section 3.4, the positions of the Stakeholders regarding the MEF are very different. The proposal from ENEL to realize a *time-invariant* MEF, equal to 13 m^3/s, is not appreciated by the Ticino Parks, who would like a more severe constraint. However, since this value represents an improvement with respect to the current value of 3 m^3/s, it will be the minimum value to be considered. The proposal from the Parks suggests values between 24 and 27 m^3/s: therefore we have decided to consider time-invariant MEF of 20, 25 and 30 m^3/s as well as that of 13 m^3/s.

This discretization allows us to explore the range of proposals that were presented by the Stakeholders. But why limit the analysis to time-invariant MEF? Considering *time-variant* MEF, for example MEF whose value follows the trajectory of natural outflows, could be interesting both from the environmental quality standpoint and from the users' standpoint. As for the first, in fact, it would bring the release into the Ticino closer to the natural outflow, and it would avoid the flow in the river being unvaried for long periods, a condition that was judged very negative by the Expert (see Section 4.3.3). As for the users, a time-variant MEF would require a smaller sacrifice of the lake' storage capacity, given that the highest values of the MEF would be imposed just in those periods when the inflows are high. It thus seems reasonable to consider time-variant MEF as well.

In order for it to be applicable in practice, the MEF must be a constraint that is defined a priori: it cannot be defined on line, i.e. day by day, for example as a fraction of the flow that would be registered that day in natural conditions (i.e. if the lake were not regulated). It must be periodic, with a period of one year, and linked to a statistic of the natural outflow process: the most attractive one is the median[11] $r_t^{n,\text{median}}$.

We will compare time-invariant and time-variant MEF values that guarantee the same annual water volume V_{MEF} in the Ticino. A time-invariant MEF that imposes a daily release not lower than $\bar{q}^{\text{MEF}}$ guarantees an annual volume V_{MEF} equal to $\bar{q}^{\text{MEF}} \cdot 86\,400 \cdot 365$, where 86 400 are the seconds in a day and 365 the days in a year. We define the corresponding time-variant MEF so that the annual volume V_{MEF} is the same and the trajectory of the flow q_t^{MEF} that it imposes daily has the same form as the trajectory of the median $r_t^{n,\text{median}}$. Moreover, in order for the existence of the river ecosystems not to be compromised, it is advisable that the flow never fall below a minimum value, which it is logical to assume equal to 13 m^3/s. Therefore, for every time instant t, the flow q_t^{MEF} is given by the following expression

$$q_t^{\text{MEF}} = 13 + r_t^{n,\text{median}} \cdot \frac{V_{\text{MEF}} - V_{13}}{V_{\text{median}}} \tag{5.12}$$

[11] The annual trajectory of the median, estimated over the years 1974–1998, was adopted.

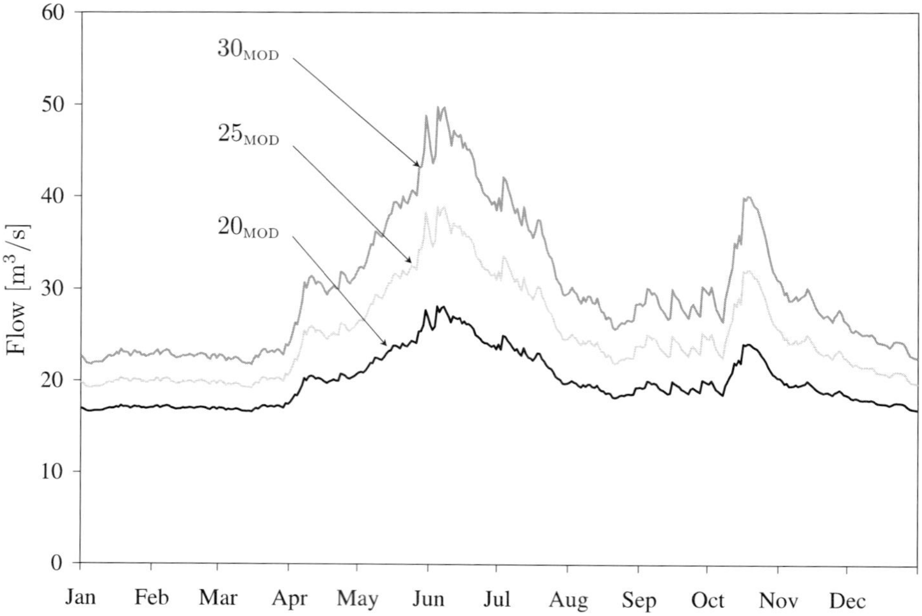

Figure 5.14: Trajectories of the time-variant MEF.

where V_{13} denotes the annual volume that corresponds to the time-invariant MEF of 13 m^3/s and V_{median} is the annual volume associated to the median $r_t^{n,median}$. Using equation (5.12) it is possible to associate a time-variant MEF to each of the three selected time-invariant MEFs. A MEF will be denoted with the flow value that it guarantees (instantly or in the average), followed by the subscript CON (constant) when time-invariant, or MOD (modulated) when time-variant. In Figure 5.14 the trajectories of the three modulated MEF are shown.

In conclusion, the set $\mathcal{M}$ of MEF values is defined as follows

$$\mathcal{M} = \{13_{CON}, 20_{CON}, 25_{CON}, 30_{CON}, 20_{MOD}, 25_{MOD}, 30_{MOD}\}$$

5.4 Solving the Design Problem

Once the sets $\mathcal{S}$, $\mathcal{R}$ and $\mathcal{M}$ have been defined as described in the previous section, the set $\mathcal{A}$ of the alternatives, given by (5.1), has the structure shown in Figure 5.15: it is composed of two disjoint sets, each one associated to a structural action. These, in turn, are made up of two disjoint subsets, each one associated to a regulation range. Finally, each subset is composed of seven segments, each of which is associated to a MEF value. The points of such segments represent the infinite possible policies. So, at least from a theoretical point of view, the number of alternatives is still infinite. However, we have already observed that, in practice, it is possible to design only a finite number of policies. Thus we will not design the whole set $\mathcal{A}$, but only a subset $\bar{\mathcal{A}}$. The question arises whether that subset is a 'good' representation of the entire set $\mathcal{A}$. In order to individuate a criterion with which to judge such 'goodness', remember that our aim is to provide the Stakeholders with the information they need to identify a set of *reasonable* (or *compromise*) *alternatives* (Figure 5.16), i.e. a subset

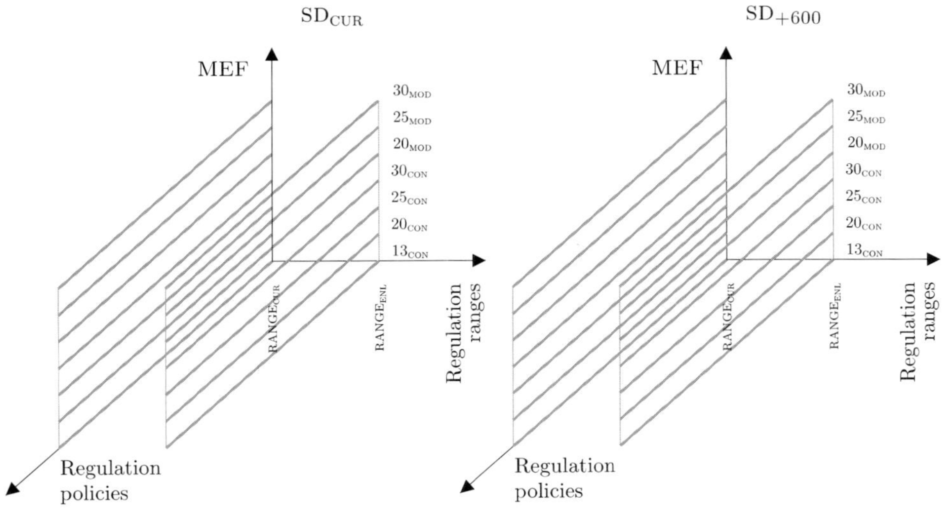

Figure 5.15: The set $\mathcal{A}$ of the alternatives after the discretization of the sets S, R and M of actions.

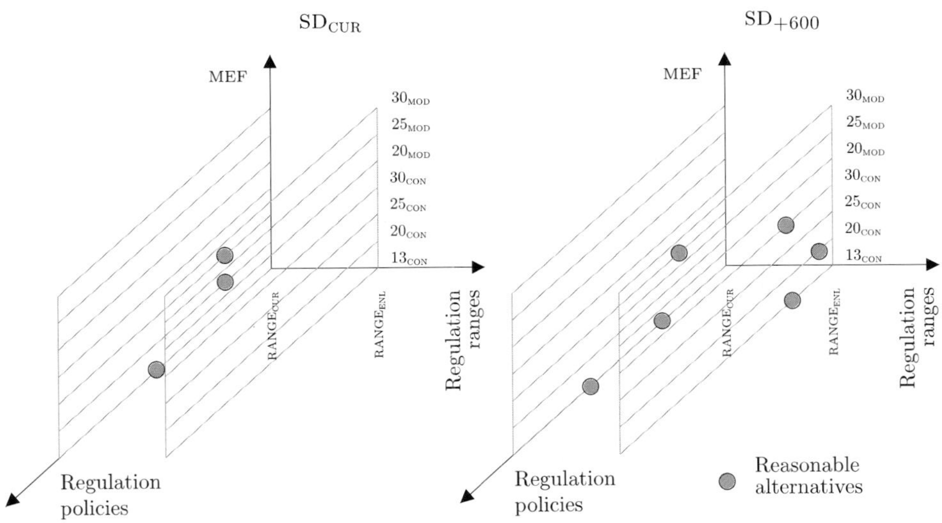

Figure 5.16: The reasonable alternatives.

of $\mathcal{A}$ composed of the alternatives that obtain a significantly wide agreement. Only these alternatives will be proposed to the final Decision Makers, who will choose the *best compromise alternative* among them, or at least on the basis of them (Phase 9 of PIP procedure). The set $\bar{\mathcal{A}}$ can thus be considered a good representation of $\mathcal{A}$ when the Stakeholders are convinced that there cannot be alternatives in $\mathcal{A}$ that could get an agreement significantly wider than those produced by the reasonable alternatives found in $\bar{\mathcal{A}}$.

To identify $\bar{\mathcal{A}}$, we first of all observe that, as we explained in Section 5.2.1, the alternatives defined by non-efficient regulation policies are not interesting. Choosing them

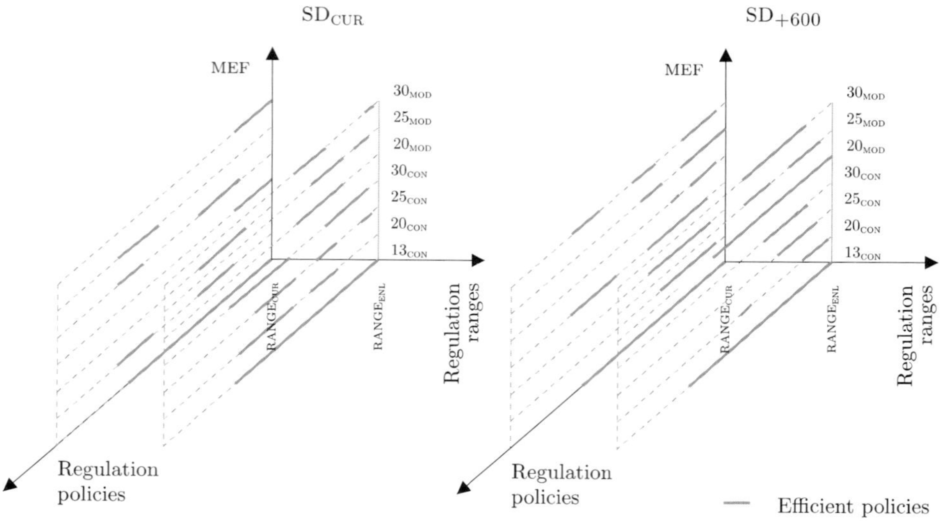

Figure 5.17: The set of alternatives that correspond to efficient policies.

would be irrational, because for each of them there is at least one efficient alternative that, for the same performances for some Stakeholders, offers better performances for others. Therefore, it is not necessary to consider the whole set $\mathcal{P}$ of possible policies; but only the subset that corresponds to the efficient policies is interesting. Of the set $\mathcal{A}$, it is thus necessary to consider only the part represented in grey in Figure 5.17.[12] This is precisely the reason why a Multi-Objective Control Problem of the form (5.8) must be formulated and solved for every triple (s, r, m) of structural and normative actions. By using the weighting method (see Section 5.2.6), the Multi-Objective Problem is transformed into a family of Single-Objective Problems, each one of the form (5.11) and thus parametric in $\boldsymbol{\lambda}$. The solution of the Single-Objective Problem for a given triple (s, r, m) produces one of the infinite points in the corresponding segment (for example, the thick segment associated to the triple (SD$_{+600}$/RANGE$_{\text{ENL}}$/30$_{\text{MOD}}$) in Figure 5.17). In practice, we will solve only a finite number of such Problems, defined by a selection of possible values of $\boldsymbol{\lambda}$. These values must be chosen in such a way that it is possible to reconstruct the form of the Pareto Frontier (see Figure 5.18). In this way, the points shown in Figure 5.19 can be generated. Thus it will not be possible to consider the axis of policies as a continuum, just as it is not possible for the other two axes. We indicate the set of alternatives that we *could* design with $\check{\mathcal{A}}$: it is given by

$$\check{\mathcal{A}} = \left\{A\colon\ A = (s, r, m, p),\ \text{with}\ s \in \mathcal{S},\ r \in \mathcal{R},\ m \in \mathcal{M},\ p \in \check{\mathcal{P}}(s, r, m)\right\} \quad (5.13)$$

where $\check{\mathcal{P}}(s, r, m)$ is the subset of the efficient regulation policies that we chose to consider for each triple (s, r, m) in $\mathcal{S} \times \mathcal{R} \times \mathcal{M}$.

Let us denote with $|\check{\mathcal{P}}|$ the number of policies in $\check{\mathcal{P}}(s, r, m)$ (which for simplicity's sake we assume to be equal for every triple (s, r, m)) and with $|\mathcal{S}|$, $|\mathcal{R}|$ and $|\mathcal{M}|$ the cardinality

[12] Also the number of efficient policies is infinite, but it is an infinite of a lower-order than the total number of policies. This is why, in Figure 5.17, the subset of efficient policies is represented with a segment which is contained within the one that represents all the policies.

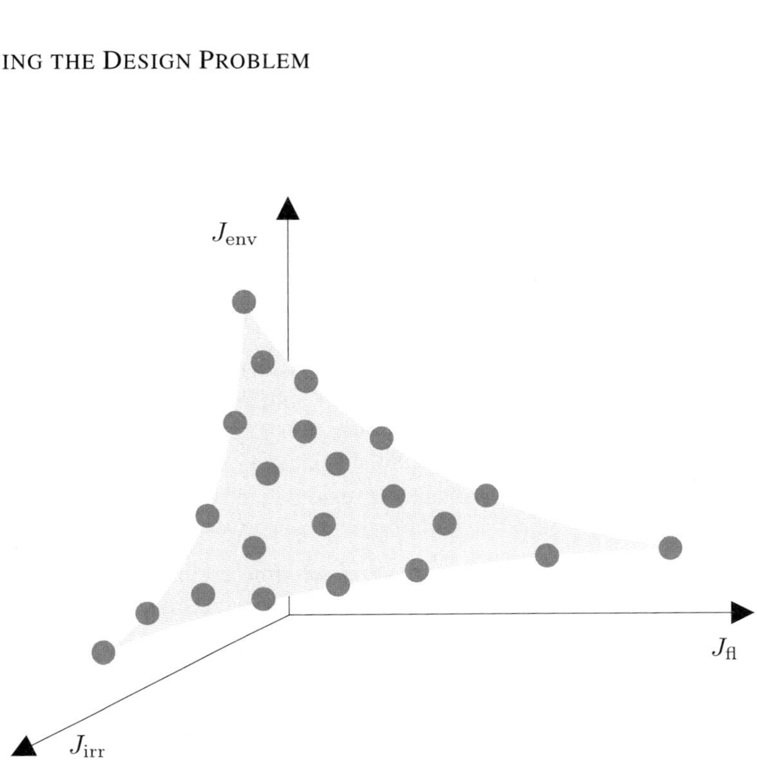

Figure 5.18: The Pareto Frontier in the objectives space for a given triple (s, r, m).

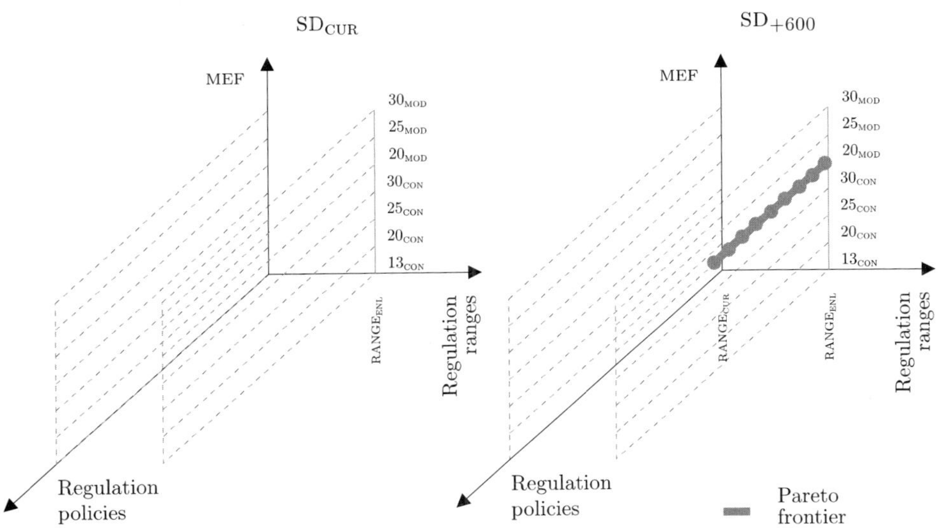

Figure 5.19: The efficient alternatives of Figure 5.18, plotted in the alternatives space.

(i.e. the number of elements) of the sets $\mathcal{S}$, $\mathcal{R}$ and $\mathcal{M}$, equal to 2, 2 and 7 respectively. The total number $|\breve{\mathcal{A}}|$ of alternatives to be designed is thus given by

$$|\breve{\mathcal{A}}| = |\mathcal{S}| \cdot |\mathcal{R}| \cdot |\mathcal{M}| \cdot |\mathcal{P}|$$

The complexity of the Project depends on the number $|\breve{\mathcal{A}}|$: the computing time necessary to design the policies grows linearly with $|\breve{\mathcal{A}}|$, while the complexity of the interaction with the Experts and the Stakeholders increases according to $|\breve{\mathcal{A}}|^2$, since they must compare all the possible pairs of alternatives in order to identify the reasonable alternatives. The number $|\breve{\mathcal{A}}|$ is very large: even if we decided to design only 20 policies for each Control Problem, we would have to design and evaluate 560 alternatives $(20 \cdot 2 \cdot 2 \cdot 7)$ and ask the Stakeholders to make 313 600 comparisons (560^2). Clearly, this task cannot be carried out. For this reason, the set $\bar{\mathcal{A}}$ of the alternatives that we will actually examine must necessarily be a subset of $\breve{\mathcal{A}}$.

There are three questions that we should pose to ourselves in order to individuate $\bar{\mathcal{A}}$:

1. Is it necessary to take all the combinations of structural and normative actions into consideration, or could some of them be eliminated a priori?
2. Is it necessary to design the entire set $\bar{\mathcal{A}}$ before the evaluation and comparison of the alternatives?
3. Are there combinations of actions that should be examined before others?

Let us begin with the first question. Since the modification of the regulation range is an Italian counterproposal to the excavation of the lake outlet, it does not make much sense to consider the combination (SD_{CUR}/$\text{RANGE}_{\text{ENL}}$), that is modifying the range without excavating the lake outlet, because the Swiss are very likely to refuse the alternatives that contain this combination, even if only for purely psychological reasons. The set of acceptable alternatives is thus the one shown in Figure 5.20: it contains 21 combinations of structural

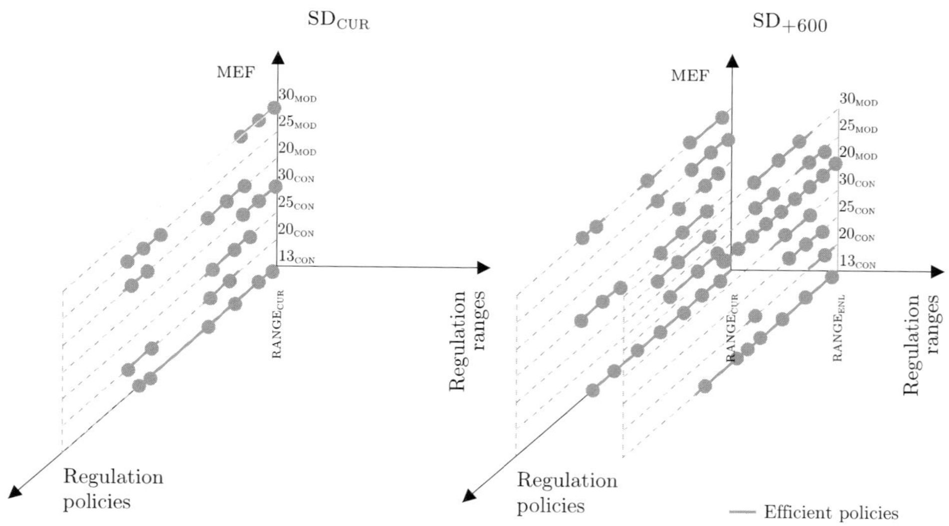

Figure 5.20: The set $\breve{\mathcal{A}}$ when only acceptable combinations of structural and normative actions are considered.

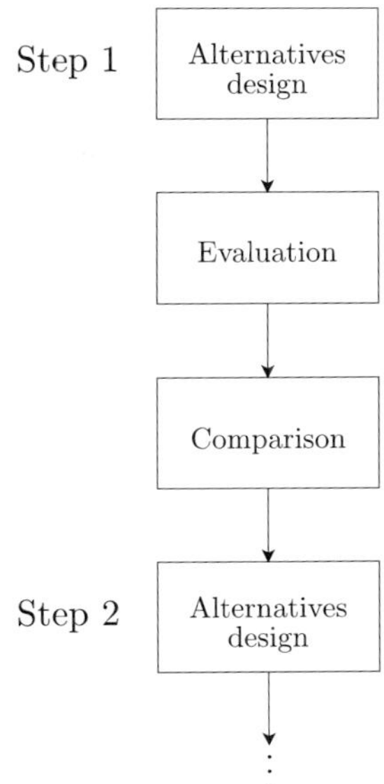

Figure 5.21: A recursive procedure for designing the alternatives.

and normative actions (instead of $28 = 2 \cdot 2 \cdot 7$) and thus requires that the same number of Multi-Objective Problems be formulated and solved.

However, the number of alternatives is still too high: we should design and evaluate 420 ($20 \cdot 21$) alternatives and ask the Stakeholders to make 176 400 (420^2) comparisons. To further reduce these numbers, let us consider the second question. The answer to this question is negative: in other words we can define a recursive procedure that, at each step (say the ith), allows us to design a smaller set $\bar{\mathcal{A}}_i$ of alternatives, which are then evaluated and compared with the 'best' alternatives identified in the previous steps $(i-1)$. Then, a new set of 'best' alternatives is obtained, which provides a basis for deciding which alternatives should be designed in the next step (see Figure 5.21). The set $\bar{\mathcal{A}}$ is the union of all the sets $\bar{\mathcal{A}}_i$ produced in this way. In this perspective, the third question appears to be very important. So let us look first for its answer, since this could suggest what the first step of the recursive procedure should be. Then we will define the following steps of the procedure, and only at the end will we demonstrate that it allows for the identification of all the reasonable alternatives.

Choosing a stage–discharge relation or a regulation range or a value of MEF that is different from the current situation means modifying the constraints of the Control Problem and thus favouring certain objectives and disfavouring others (see Table 5.2). Note that when we say that by passing from $\text{RANGE}_{\text{CUR}}$ to $\text{RANGE}_{\text{ENL}}$ an improvement is expected in the performances with respect to certain objectives (for example, the one representing the *Irrigation* sector), we do not mean that all the alternatives with $\text{RANGE}_{\text{ENL}}$ are better than those that can be obtained with $\text{RANGE}_{\text{CUR}}$. This, in fact, is certainly not true, because the performances of the single alternatives also depend on the regulation policy. What we mean is that with $\text{RANGE}_{\text{ENL}}$ we expect to find alternatives that are better than the best alternative that can be obtained with $\text{RANGE}_{\text{CUR}}$. The policies that characterize these better alternatives, if they exist, could be identified by assigning a weight much higher than the others to the

Table 5.2. Expected effects on the performances of several sectors when changing the structural and normative actions (⇑ improves, ⇓ worsens)

Action			Effects on Stakeholders' satisfaction
change	from	to	
Stage–discharge relation	SD_{CUR}	SD_{+600}	*Upstream Flooding* ⇑ *Downstream Flooding* ⇓
Regulation range	$RANGE_{CUR}$	$RANGE_{ENL}$	*Irrigation* and *Power* ⇑ *Upstream Flooding* ⇓
MEF	13_{CON}	30_{CON}	*Downstream Environment* ⇑ *Irrigation* and *Power* ⇓

objective of the favoured sector. Similarly, by saying that a sector is disfavoured by passing from a given combination of actions to another one, we mean that the new combination can lead to alternatives whose performances are worse for that sector than those obtained with the worst alternative of the first combination.[13]

In Section 5.2.5 we show that the constraints guarantee that in the short-term critical situations are avoided. We also highlight the fact that constraints must be avoided in the formulation of the Problem whenever they express interests that could be formalized through objectives. It is thus advisable that the combination of actions which require the softer constraints are examined first, so that the corresponding policies can take value in the largest possible subset of the decision space. In our case, the combination ($SD_{+600}/RANGE_{ENL}/13_{CON}$) must be considered first: all the other combinations reduce the space for the release decision, as shown in the example in Figure 5.22.

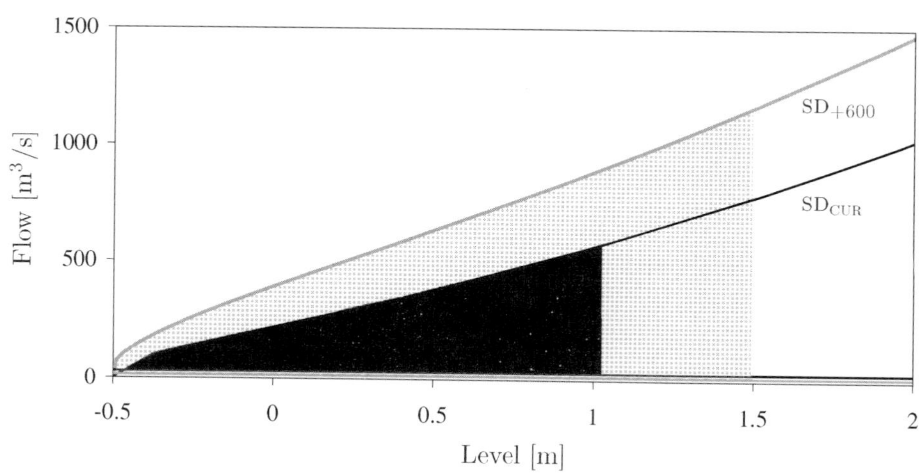

Figure 5.22: The set of possible release decisions on any day between the 16th March and the 31th October, with the combination ($SD_{CUR}/RANGE_{CUR}/30_{CON}$) (black); and with the combination ($SD_{+600}/RANGE_{ENL}/13_{CON}$) (grey).

[13]Note the fact that, because of the way the actions are defined in the Verbano Project, a sector cannot be both favoured and disfavoured at the same time; the previous definitions could lose their meaning in other contexts.

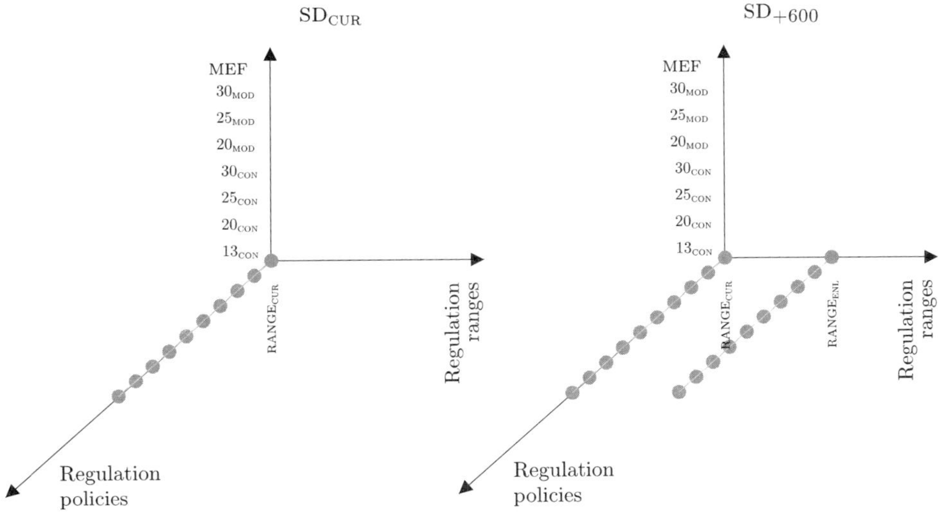

Figure 5.23: The set $\bar{\mathcal{A}}_1$ of the alternatives to be designed in the first step.

By observing that figure, however, it can be noticed that the two combinations of structural and normative actions are differentiated not only by the space for the release decisions, but also by the free regime conditions: in such conditions, in fact, SD_{+600} produces a greater release than SD_{CUR}. Italy might not accept this increase in outflow, for fear that it would accrue to the downstream flooding, and pose a veto on this stage–discharge relation. One thus understands that the constraints can take on another function: to guarantee the social acceptability of the alternatives being considered. For this reason the combination (SD_{CUR}/$\text{RANGE}_{\text{CUR}}$/$13_{\text{CON}}$) must be considered as well. On the other hand, Switzerland might consider a modification in the regulation range unacceptable, even if the lake outlet were excavated. The third combination that must be analysed is thus (SD_{+600}/$\text{RANGE}_{\text{CUR}}$/$13_{\text{CON}}$).

These are the three combinations that must be considered first.

5.4.1 The first step

In the first step of the recursive procedure shown in Figure 5.21, we consider only the Control Problems that are characterized by the following combinations of structural and normative actions

$$\begin{aligned} &\text{Problem 1:} \quad (\text{SD}_{+600}/\text{RANGE}_{\text{ENL}}/13_{\text{CON}}) \\ &\text{Problem 2:} \quad (\text{SD}_{\text{CUR}}/\text{RANGE}_{\text{CUR}}/13_{\text{CON}}) \\ &\text{Problem 3:} \quad (\text{SD}_{+600}/\text{RANGE}_{\text{CUR}}/13_{\text{CON}}) \end{aligned} \tag{5.14}$$

Thus, the alternatives that we must design are those shown in Figure 5.23.

In order to solve each of these Problems, we must design a set of policies that produce, in the objectives space, a sufficiently dense grid along the Pareto Frontier. The operation is not simple since we cannot choose the policies to be designed by directly considering the Pareto Frontier; we can only act in the weights space by choosing the vector $\boldsymbol{\lambda}$ that

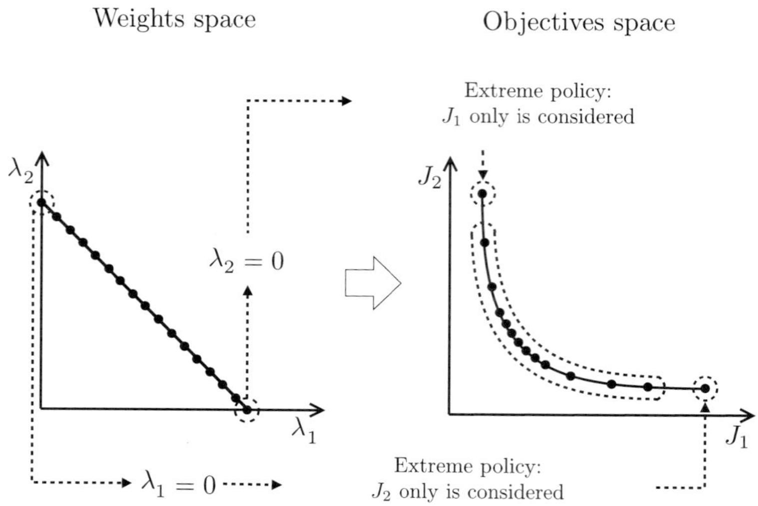

Figure 5.24: An uniform grid in the weights space (left panel) and the corresponding policies in the objectives space (right panel).

appears in problem (5.11). The most obvious thing is to start by fixing a grid of constant density in the weights space, as shown in Figure 5.24 (where, for simplicity's sake, we considered only two objectives). Once the corresponding problems (5.11) have been solved, the performances of the policies that have been designed could be more concentrated in certain zones of the objectives space and less in others (as it is in Figure 5.24). This derives from the fact that the curvature of the Pareto Frontier in general is not constant. We must therefore eliminate some values of $\boldsymbol{\lambda}$ and/or add others, so that the policies cover the frontier in a reasonably uniform way. We will say then that the Problem has been solved.

Once the three problems (5.14) have been solved, the first set of alternatives will be available. These alternatives are first shown to the Experts, so that their effects can be evaluated (Chapter 9), and then to the Stakeholders, so that they can be compared (Chapter 10). It might be that none of the alternatives that emerge from the comparison (reasonable alternatives) achieve a wide agreement, but that each one gets the agreement of a few Stakeholders and the opposition of many others. In that case we should not be discouraged, since at that point we will have produced only a part of all the possible alternatives, and precisely the opposition that emerged is the key to structuring the second step of the *Design Alternatives* phase.

5.4.2 The second step

In the second step our aim is no longer to explore the whole space of the alternatives, but to enlarge the agreement we have found so far. Thereby, for each of the reasonable alternatives identified in the first comparison step, we look in its neighbourhood for alternatives that obtain the agreement of those Stakeholders that were opposed to it, without losing the agreement of those that were satisfied with it.

We clarify the concept with the support of Figure 5.25, in which the alternatives designed in the first step are represented by grey points, the reasonable alternatives found at

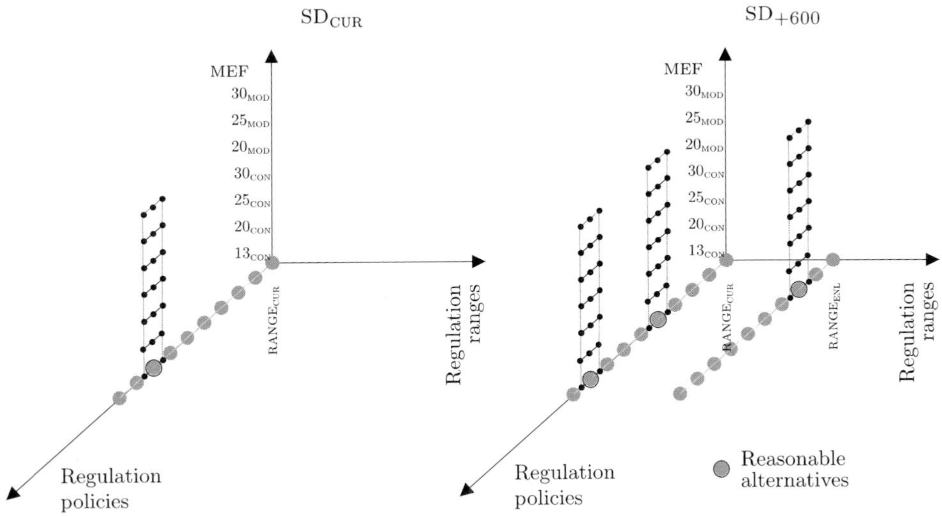

Figure 5.25: The set $\bar{\mathcal{A}}$ of the alternatives that have been designed at the end of the second step: those designed in the first step ($\bar{\mathcal{A}}_1$) are individuated by grey points, those designed in the second ($\bar{\mathcal{A}}_2$) by black points. The alternatives in a circle are the reasonable alternatives identified in the first comparison step.

the end of the first comparison step are marked with a circle, and the alternatives designed in the second step are represented by black points. To understand how the latter are found, let us concentrate on one of the reasonable alternatives, for example on $A1$, characterized by the combination (SD$_{+600}$/RANGE$_{\mathrm{ENL}}$/13$_{\mathrm{CON}}$). To simplify the description, let us suppose that it was designed by means of a Control Problem with three objectives only. Figure 5.26 shows its position in the objectives space.[14] We denote the set of sectors that are supporting its implementation with $\mathcal{S}^+$ (we assume that in the example these are the sectors associated with the objectives J_1 and J_3) and with $\mathcal{S}^-$ we denote the set of sectors that are opposing it (in the example, the sector associated to J_2). Among the alternatives characterized by the same combination of structural and normative actions, let $A2$ be the alternative closest to $A1$ whose performances for the sectors in $\mathcal{S}^-$ are considered to be acceptable (even if they might not be supporting it yet). Alternative $A2$ does not satisfy all the sectors in $\mathcal{S}^+$; otherwise it would have been chosen as the reasonable alternative in the place of $A1$. However, there might be alternatives that have not yet been designed between $A1$ and $A2$, for which the agreement is wider than $\mathcal{S}^+$. If they do exist, they must be characterized by better performances than those of $A1$ for at least some of the sectors in $\mathcal{S}^-$ and by intermediate performances between those of $A2$ and $A1$ for the sectors in $\mathcal{S}^+$: which means, in the example in Figure 5.26, that they must be found in the white triangle. However, these conditions are not sufficient to guarantee a wider agreement, because we do not know how the Stakeholders of the sectors in $\mathcal{S}^+$ will evaluate alternatives with performances that are inferior to $A1$'s: we must ascertain this through comparison.

[14]The comparison of the alternatives is carried out by the Stakeholders in the space of the indices, while the Analyst usually works in the space of the objectives. The performances of the alternatives in the two spaces are clearly linked.

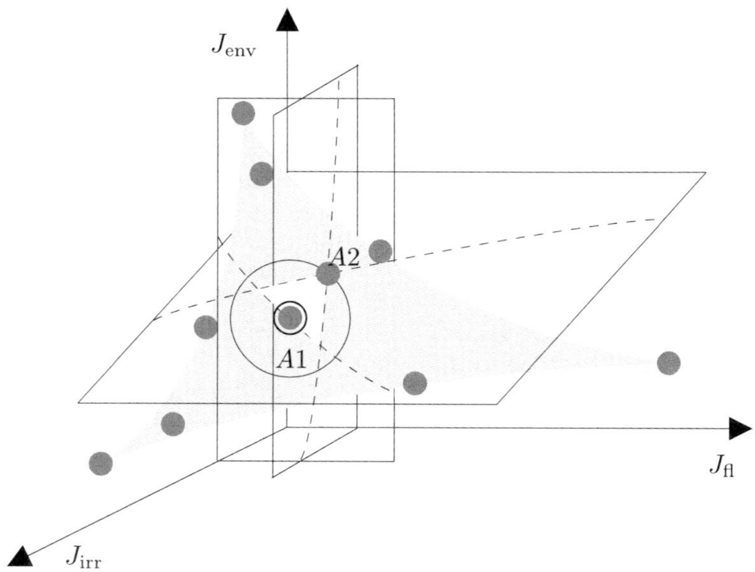

Figure 5.26: The neighborhood (white) of a reasonable alternative $A1$ identified in the first comparison step, where new reasonable alternatives can be searched for in the second step.

The agreement may be further broadened also by considering combinations of structural and normative actions that were not considered in the first step, i.e. alternatives characterized by values of MEF greater than 13 m^3/s.

To understand if and when the increase of the MEF might allow us to broaden the agreement, observe Figure 5.27. It shows the cross sections of the Pareto Frontiers in the plane of the objectives relative to the *Downstream Environment* and *Irrigation* sectors, for different values of MEF and the same pair (s, r). When the weight attributed to the environmental

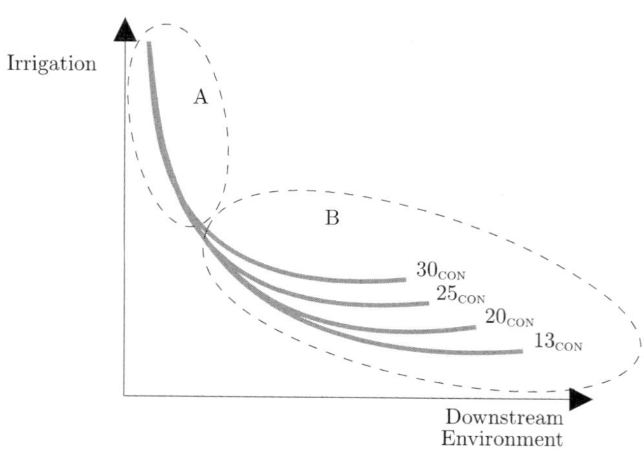

Figure 5.27: A cross section of the Pareto Frontier in the plane of the objectives relative to the *Downstream Environment* and *Irrigation* sectors, for different values of MEF.

objective is high enough (with respect to the one attributed to the irrigation objective), the resulting policies will almost always guarantee a flow in the Ticino greater than 30 m^3/s, so that the corresponding alternatives will have similar performances even if the MEF varies (zone A in Figure 5.27). Therefore, if the comparison that is carried out in the first step produced reasonable alternatives in zone A, in the second step it would not be necessary, nor useful, to consider actions in which the MEF is greater than 13 m^3/s.[15] Conversely, when the importance given to the environmental objective diminishes, the value of the MEF influences the performances of the alternatives that can be designed (zone B). Think, for example, about an extreme case in which only the irrigation objective is considered: then the flow in the Ticino is equal to the MEF each day, except when the lake is in free regime. It will thus be in such cases that values of MEF greater than 13 m^3/s should be explored.[16]

In conclusion, in the second step we must design the alternatives that in Figure 5.25 are represented by black points. They form the set $\bar{\mathcal{A}}_2$ that, together with the reasonable alternatives that were determined in the first step, is presented for the second comparison step.

If, at the end of the second comparison step, an alternative has been found that obtains the agreement of all the Stakeholders, that is a strong candidate to be the best compromise alternative, given that no one is opposed to its implementation. Otherwise, for each reasonable alternative one can try to identify new actions that reduce the negative effects in the sectors that are unsatisfied, without modifying the performances in the others: these actions are called *mitigation measures* (see Phase 8 of PIP procedure). Since they aim at mitigating the negative effects produced by a specific alternative, they cannot be proposed a priori. This is the reason why we differentiate them from the structural and normative actions. We explain how they can be identified in Chapter 14. Once mitigated, the reasonable alternatives obtained at the end of the second step are the reasonable alternatives that are to be considered in the *Final Decision* phase.

5.4.3 Why this procedure is valid

To demonstrate that the above procedure is able to individuate all the reasonable alternatives, we must prove that, for each of them, at least one reasonable alternative can be found among the ones designed in the first step of the procedure. To do this, we observe first of all that, if the reasonable alternative is associated to the MEF 13_{CON} action, it must have been identified in the first step. On the contrary, when it is not associated to the MEF 13_{CON}, it is an alternative such as $A5$ in Figure 5.28. From its position, we deduce that the *Irrigation* sector is not willing to accept values of its objective greater than $\bar{J}_{\text{irr}}$, i.e. the value associated

[15] Note that we can estimate whether we are in zone A even without solving Problems with different values of MEF: we just ascertain that the frequency with which the flow in the Ticino is equal to 13 m^3/s is low.

[16] By observing Figure 5.27 the reader may note, perhaps with surprise, that for a given value of the irrigation objective, the environmental objective worsens as the MEF increases. This apparent *non-sense* is explained by the fact that an increase in MEF from 13_{CON} to 30_{CON} causes not only an increase in the river flow in those days in which it would have been less than 30 m^3/s, but also a reduction in the remaining days of the year, because by hypothesis the irrigation objective must remain constant and the water volumes must respect the mass balance. Furthermore, since the increase in the MEF reduces the decision space, a simultaneous improvement of the *Downstream Environment* objective cannot be achieved. It follows that its value either remains the same (in which case zone B does not exist) or worsens, as shown in Figure 5.27. However, the worsening of the objective is not necessarily displeasing a priori, because, as we explained in Section 4.3.3, the interests of *Downstream Environment* are not completely expressed by one objective (the indicator $i_{\text{Env\_D\_1}}$), but by the pair ($i_{\text{Env\_D\_1}}$, MEF). In other words the MEF must be considered as a 'second environmental objective' so that the frontier that is obtained with action 30_{CON} cannot be considered as dominated by the one obtained with 13_{CON}.

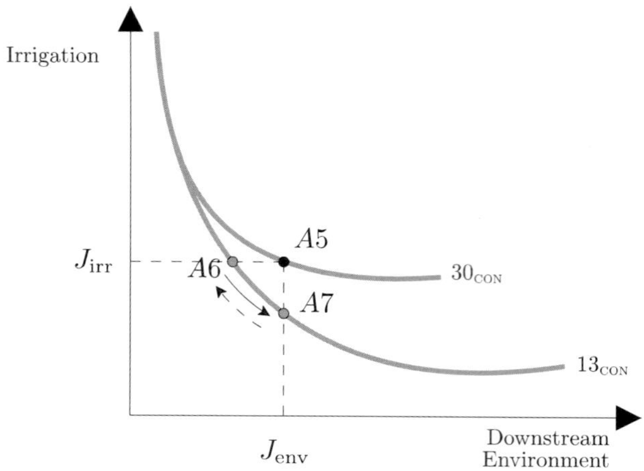

Figure 5.28: A cross section of the Pareto Frontier for different values of the MEF.

to $A5$. Therefore, in the first step, they might have accepted the compromise of alternative $A6$, but they were attracted by alternatives with better performances.[17] In other words, they have tried to move along the frontier associated to the MEF 13_{CON}, in the direction of the solid arrow. Instead, the *Downstream Environment* sector has not necessarily accepted the value $\bar{J}_{env}$ and thus the compromise of alternative $A7$ in the first step, because even if the value of the objective is equal to the one of $A5$, the value of MEF is smaller. Anyway, it is certain that at the first step they tried to move along the frontier associated to the MEF 13_{CON} according to the direction of the dashed arrow. It follows that between alternatives $A6$ and $A7$ there must be at least one reasonable alternative that has been identified in the first step. In the second step, when values of MEF greater than 13_{CON} are considered, the Stakeholders moved from this reasonable alternative towards the reasonable alternative $A5$.

The procedure that we identified proves that the answer to the second question on page 130 is negative: it is not necessary to design all the alternatives before the comparison.

[17]This means smaller values of the objective, since this is to be minimized.

Chapter 6

Identifying the Model

EB, AC, FC, RSS and EW

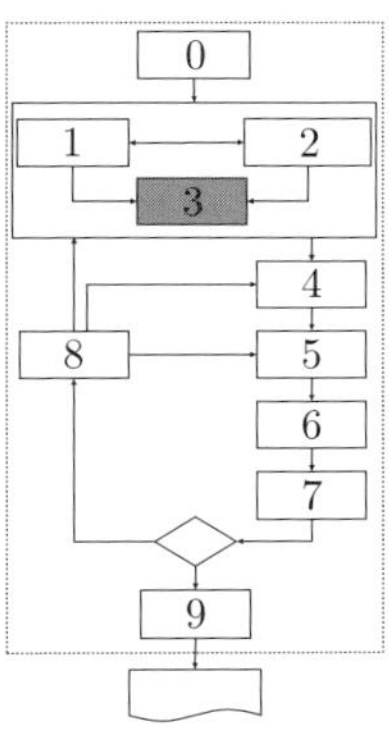

Figure 6.1: The phases of the PIP procedure. Phase 3, which is analysed in this chapter, is highlighted.

In the previous chapter we explained that, given a triple of structural and normative actions, the efficient regulation policies are designed by solving an (Optimal) Control Problem. We do not yet have all the necessary components, so we were only able to provide a first draft of the formulation of that Problem, but we have shown its essence: each policy is designed by minimizing the weighted sum of the objectives, for a given vector of weights. The policy is constituted by a sequence of control laws (Section 5.2.1), one for each time t, which provide pairs of control decisions $(u_t^{\text{MIO}}, u_t^{\text{TIC}})$, as a function of the lake storage present at that time instant. The precise definition of the Problem thus requires that we specify the relation that exists between the control and the hydrological quantities upon which the value of the objectives depends. This relation is expressed through a model of the water system, whose identification is the subject of this chapter (Phase 3 of the PIP procedure in Figure 6.1).

Once the alternatives are identified, i.e. the design of the regulation policies will be completed (Phase 4), it will be necessary to estimate their effects (Phase 5) on all the indicators that we chose in Phase 2 (Chapter 4). To do this, a model of the system must be available that can compute not only the variables that appear in the design indicators (and thus in the objectives), but also all those appearing in the evaluation indicators. For example, a model of the propagation of the flood waves along the River Ticino is necessary to compute the river level at Pavia, which in turn is necessary for evaluating the effects of the floods in that city.

6.1 The structure of the model

The identification of the model begins from the analysis of the system structure, represented through a scheme such as the one in Figure 6.2. It represents the system as an aggregate

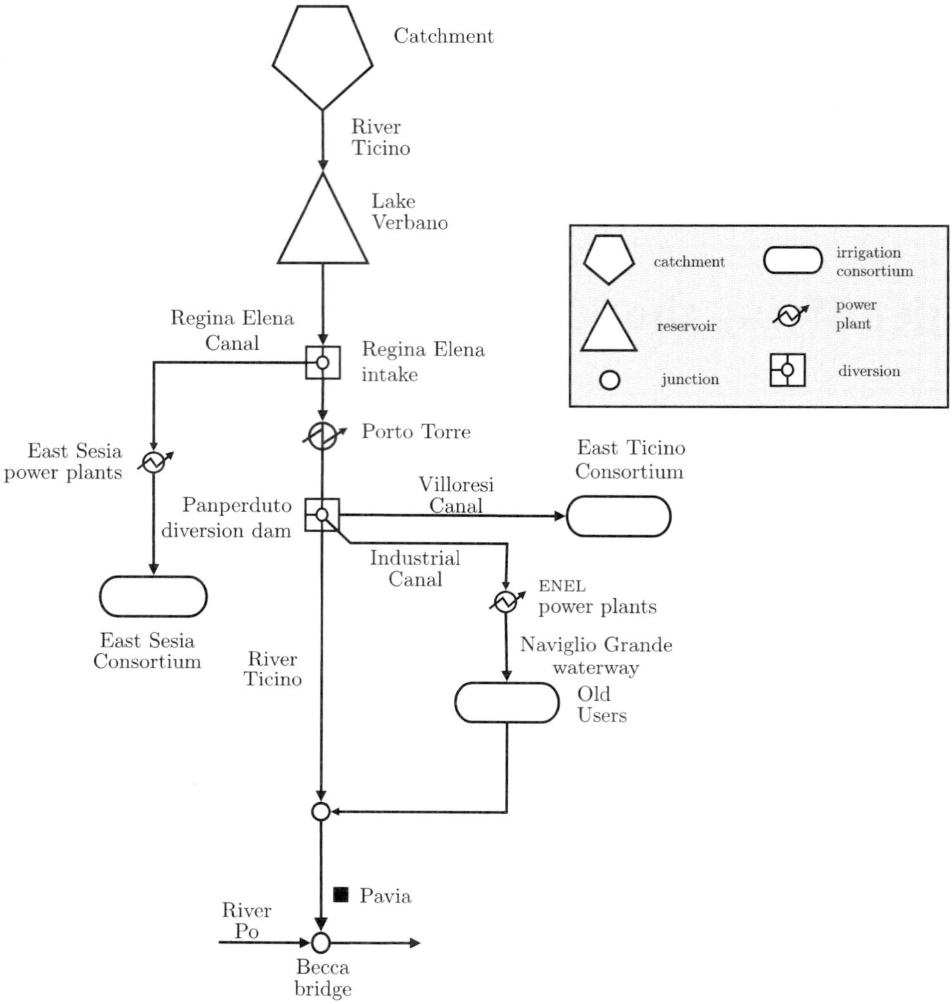

Figure 6.2: The physical scheme of the Verbano water system.

of simpler subsystems, termed *components*, each of which carries out a single operation. For example the lake stores the inflow, while the diversion dam at Panperduto distributes the flow among several water users and the Ticino. This representation is helpful because the model can be more easily constructed by first identifying a model for each component and then interconnecting such models, according to the topology of the system. It is not always advisable, however, to follow this path. The aim of the model is to reproduce the relationships that exist between the control variables and the hydrological variables which appear in the definition of objectives and indicators. Thus it is often more useful that the model be based on the relationships between the *logical components* of the system, which represent the functions performed in the system, and which do not always correspond to physical components. In the following these logical components will be also called *units*. The model will thus be constructed by aggregating the models of the individual units.

Even though we explained it in other words, this distinction was anticipated in Section 5.2.2, where Figure 5.7 shows that when decision[1] u_t^{TIC} is included among the control variables, the distribution of the flow downstream from the Miorina dam must be represented with a sequence of three logical nodes (the *distribution nodes* for MEF, for environmental purposes and for distribution among water users), which do not correspond to the two existing diversion dams (at the Regina Elena intake and at Panperduto). The logical scheme of the model is therefore the one shown in Figure 6.3. By observing the figure, one may notice another point in which the scheme of the model differs from the scheme of the physical system: the connection of the unit which represents the power plant at Porto della Torre. In the physical scheme this plant is situated downstream from the intake of the Regina Elena Canal and upstream of the Panperduto diversion dam, where the distribution among the rest of the water users takes place. In the logical scheme this is impossible, because the distribution among the water users is decided at the same time for all of them. The unit that represents Porto della Torre is thus fed by those variables whose sum corresponds to the flow provided to the power plant and it has no output, because the variables that represent the water flowing out from the plant are already included in other branches of the diagram, as inputs to the units that use it. Note that the logical components can, if necessary, violate the mass balance (as is the case of node 3 in Figure 6.3); while in the physical scheme of the system this is forbidden. Moreover, in the physical scheme, arcs represent physical fluxes (water flows) while, in the logical one, they represent information fluxes (signals). Consider, for example, node 5 in Figure 6.3: it does not represent a distribution of the flow, but the same signal (value of the flow downstream from node 4) being sent to nodes 6 and 7.

For simplicity's sake, in the following, logical nodes that represent a distribution of the flow will also be termed *diversion* (*node*), and the adjective *logical* will be omitted; the reader must remember that, as we have just said, this does not imply that a corresponding diversion dam exists.

6.2 The models of the single units

The next step in the creation of the model of the water system is the identification of a model for each of the units that appear in the logical scheme in Figure 6.3.

We explained in Section 3.5 that in normal conditions the Regulator assumes a release decision once a day and this is why in Section 5.2.1 we defined the policy as a sequence of control laws, one for each day t. Since the Regulator decides in discrete time instants, a discrete-time description of the individual units should be adopted.[2] A continuous-time representation would be unnecessarily onerous, because of the computing time and the large amount of data that it would require.

For each unit, we must identify the vector $\mathbf{u}$ of the *input* variables (the exogenous variables that influence its behaviour), the vector $\mathbf{y}$ of the *output* variables (the variables that we want to observe) and define the relationship that links them. The simplest case is the one in which the output $\mathbf{y}_t$ at time t depends only on the input $\mathbf{u}_t$ at the same time instant. In this case we say that the unit is *non-dynamical*[3] and its model will thus be defined as a

[1] The increase in volume left to the Ticino in the interval $[t, t+1)$, with respect to the volume that is left to it according to the current distribution policy, with the aim of improving its environmental quality.

[2] However, this does not mean that the modelling time step must necessarily be equal to the decision time step: it could be a submultiple of it.

[3] Not to be confused with *stationary*, which means that its characteristics do not change over time.

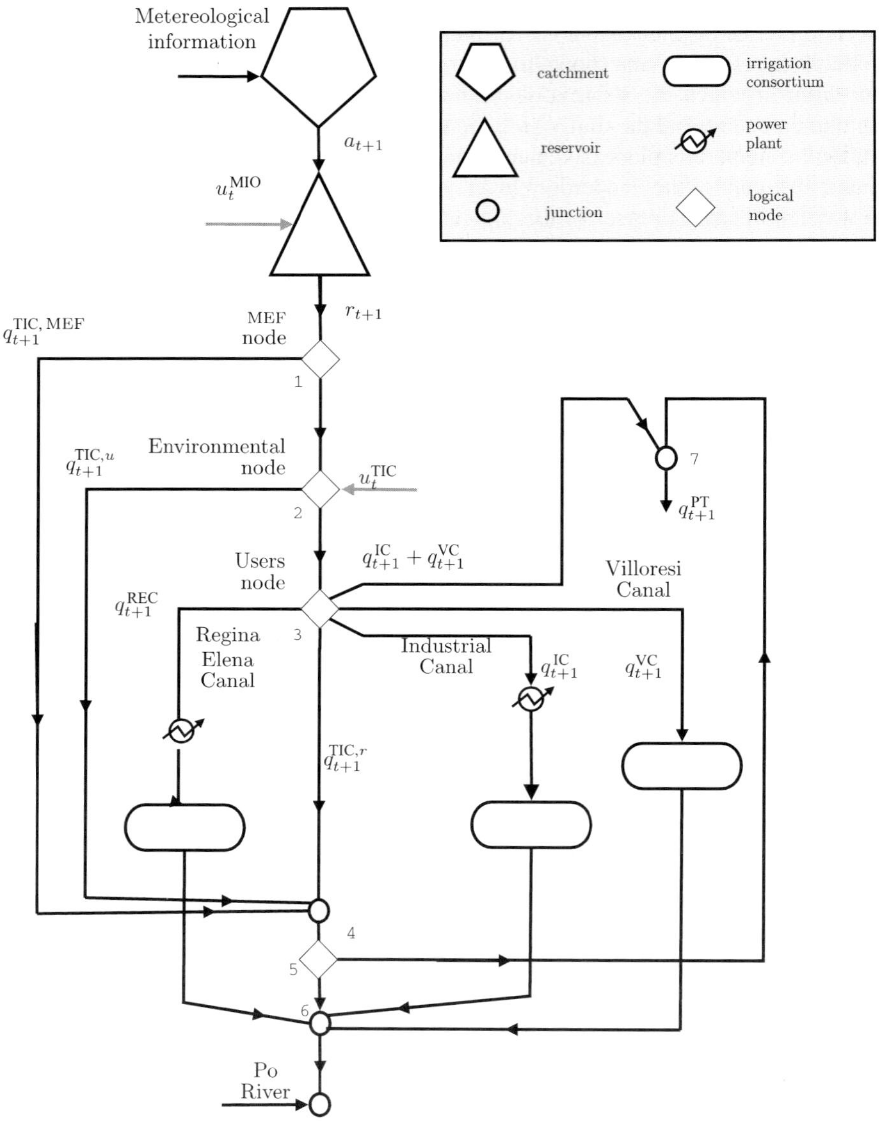

Figure 6.3: The logical scheme of the Verbano water system.

relationship of the following type

$$\mathbf{y}_t = h_t(\mathbf{u}_t, \boldsymbol{\theta}_t) \tag{6.1}$$

where $\boldsymbol{\theta}_t$ is a vector of parameters and the relationship $h_t(\cdot)$ generally depends on time.

However, it is possible that the output at time t does not depend only on the value of the input in the same time instant, but also on the input values in the previous instants, i.e. it depends on past history. In this case we say that the *unit* is *dynamical*. For example, the level of the lake at time t (output) does not only depend on today's inflow (input), but also on yesterday's inflow and on all the previous days', until infinity. Clearly yesterday's inflow will influence the current level more than the day before yesterday's, and similarly for the preceding days. However, to describe the dynamics of the unit it is not necessary to take account of all the infinite past input values, since these values modified the condition of the unit, i.e. its *state*, and it is this state that at time t influences the output. Returning to the lake example, the current level depends only on the storage (state) and it is this value that is recursively modified by the inflow. With $\mathbf{x}_t$ we indicate the vector of all the variables that must be known at time t in order to compute both the output at time t (the current level in the example) and all the future outputs once the future inputs are known. This vector is the state of the unit. The unit model will then be composed of two parts: one similar to (6.1), which links the state and (sometimes) the input to the output, which is called *output transformation*, and another that describes the evolution of the state – given the present state and the input – which is called *state transition function*. The model of the unit will thus have the following form

$$\mathbf{x}_{t+1} = f_t(\mathbf{x}_t, \mathbf{u}_t, \boldsymbol{\vartheta}_t) \tag{6.2a}$$

$$\mathbf{y}_t = h_t(\mathbf{x}_t, \mathbf{u}_t, \boldsymbol{\vartheta}_t) \tag{6.2b}$$

where $\boldsymbol{\vartheta}_t$ is a vector of parameters.

From what has been said the reader might conclude that the distinction between a dynamical unit and a non-dynamical one is sharp. Actually, this is not the case. Consider a unit that describes a stretch of river, or a canal, with no distributed contribution along its path, either positive (inflow) or negative (leakage). With this hypothesis, the input variable is the volume q_t^i that flows into the upstream section in the time interval $[t, t+1)$, while the output variable will be the volume q_t^o that flows out during the same interval. Under the further hypothesis that the stretch of river under study does not have any *detention capacity*, the output will be equal to the input after a time τ (termed *translation time*) has passed, i.e.

$$q_t^o = q_{t-\tau}^i \tag{6.3}$$

Note that this relation affirms that the river stretch is a dynamical system, whose state is given by $q_{t-1}^i, q_{t-2}^i, \ldots, q_{t-\tau}^i$, namely by all the volumes that have entered the stretch before time t and that have not yet exited. The output q_t^o is equal to the last component $(q_{t-\tau}^i)$ of the state; the input is the inflow volume q_t^i in the time step $[t, t+1)$. Nevertheless, if the translation time τ is considerably shorter than the time step Δ used in the model, equation (6.3) can be approximated by the following relation

$$q_t^o = q_t^i \tag{6.4}$$

which describes the river stretch as a non-dynamical system. Thus, the same unit can be seen as dynamical or non-dynamical depending on the way it is described (for example, depending on the choice of the modelling time step).

As we will show in Section 7.8, the computing time required to design a regulation policy grows exponentially with the number of state variables in the water system model. It follows that, whenever possible, river stretches and canals should be described with non-dynamical equations, such as (6.4), instead of dynamical representations, such as (6.3), even if by so doing the model is less accurate.

What we have said for the river stretches can be generalized to all the units that appear in the scheme in Figure 6.3. For each unit we must assess how important it really is that it be described with a dynamical relation, such as (6.2), or whether it is possible to adopt a non-dynamical description, such as (6.1). We will discuss this in Sections 6.5–6.7. However, before examining the models of the individual units in detail we would like to discuss two general points: the duration Δ of the *modelling (time) step* and the notation that we have adopted.

6.3 The time step

We have already talked about the time step in Section 3.5; at that point, however, the reference was to the step with which the Regulator takes decisions, i.e. to the *decision (time) step*. We observed that the decision step is not actually constant, because in critical situations, when for example there is a high risk of flooding, the Regulator takes decisions with a step that is much shorter than in normal conditions. Nevertheless, considering that we are studying regulation at the planning level (Chapter 1), the decision step can be assumed as constant and equal to a day without any significant loss of accuracy.

The choice of decision step clearly influences the choice of modelling step, because the latter must be equal to the first or a submultiple of it. Since the decision step is equal to a day, the modelling step will be

$$\Delta = \frac{1}{k} \quad \text{with } k \text{ integer number}$$

When choosing the value of k, one must consider:

- the sampling time of the hydrological measurements used for model *calibration* and *validation*;
- that the computing time needed to solve the Control Problems and to evaluate the alternatives increases linearly with k;
- that the discrete-time description of the system should not produce an excessive loss of information (remember that in reality the system is time continuous).

Almost all the time series in our possession have a daily time step and this forces us to assume $k = 1$. This assumption also has the advantage of reducing the computing time to a minimum. It remains to be verified that this does not bring about an excessive loss of information. The sufficient condition that this not occur is that the modelling time step Δ be shorter (by about one order of magnitude) than the smallest *time constant* of all the units' models (Shannon's *Sampling Theorem*, Shannon, 1949). The time constant is a measure of the speed at which the output of an asymptotically stable system converges to the *equilibrium output*, when the input is kept constant. It can be properly defined only for linear time-invariant systems, i.e. for systems in which equation (6.2) is of the form

$$\mathbf{x}_{t+1} = A\mathbf{x}_t + B\mathbf{u}_t \tag{6.5a}$$

$$\mathbf{y}_t = C\mathbf{x}_t + D\mathbf{u}_t \tag{6.5b}$$

where A, B, C and D are matrices. If $\lambda^{\max}$ is the dominant *eigenvalue* of matrix A, i.e. the eigenvalue of largest magnitude, the time constant T_c of the system is

$$T_c = \frac{1}{\mathrm{Re}(\lambda^{\max})} \tag{6.6}$$

If the system is non-linear, we can refer to the time constant of the linear model obtained by *linearizing* the original one around a significant point, for example a point representing average operating conditions.

In conclusion, by adopting the value of a day for Δ there is no significant loss of information if, for each unit, the order of magnitude of T_c is at least ten days. If one of the units does not satisfy this condition, the accuracy of the model is still acceptable if the value of T_c is greater than several days.

6.4 The notation

In general, not all the input variables are under our direct control. This is why we introduce the distinction between inputs whose value is established by us, which are called *controls*, and inputs that assume a value independently of our decisions, which we call *disturbances*. Among the disturbances that act on the units, some are *random variables*, for example the meteorological conditions or the inflow to the lake; thus we call them *random disturbances*. Before a random disturbance occurs, we can only provide a statistical or set-membership description for it (namely, its *probability distribution* or the set of its possible outcomes), depending on whether it is a stochastic or uncertain disturbance (see Section 5.7 of THEORY); the exact value of the disturbance will be known only after it has occurred. For example, the inflow to the lake in the time interval $[t, t+1)$ will be known only at the end of that interval, i.e. at time $t+1$; before then, at time t for example, we can describe the inflow with a probability distribution. In other words, before time $t+1$ the inflow in the interval $[t, t+1)$ is a random variable; from the time instant $t+1$ onwards it is a *deterministic variable*. To help us remember when this occurs, we decided to describe this kind of variables with a symbol whose subscript indicates the instant at which the random disturbance becomes a deterministic disturbance. For example the inflow to the lake in the time interval $[t, t+1)$ is denoted with the symbol a_{t+1}.

It is important to note that the presence of a random disturbance propagates uncertainty. In other words, all the variables that depend on it are random, until it is realized. For example, the level h_{t+1} of the lake tomorrow depends on the inflow a_{t+1} that will occur over the course of the present day (t) and so h_{t+1} is a random variable until the time instant $t+1$, as indicated by its subscript, on the basis of the convention described above. This convention is not applied to the disturbances alone, but also to all the system variables, and so it can be enunciated in general terms by affirming that *the subscript of a variable denotes the time instant at which it assumes a deterministic value*. We have already used this convention in Section 5.2.1, when denoting the release with r_{t+1}.

Instead, when a function is considered, the subscript denotes the initial time instant of the interval over which it is applied, even if, at that instant, some of its arguments may not yet have a value that is deterministically known and consequently its value is random. This is why it is possible that a function such as $f_t(\cdot)$ has variables with the subscript $t+1$ among its arguments.

Note that this convention leads us to write equations (6.1) and (6.2), which describe the models of the units, in a way that might seem a little strange at first glance, but which proves to be very useful in practice. Suppose, for example, that some of the input variables of equation (6.2) are controls (we still indicate them with $\mathbf{u}_t$), while others are deterministic ($\mathbf{w}_t$) and random (ε_{t+1}) disturbances. Equation (6.2a) thus assumes the following form

$$\mathbf{x}_{t+1} = f_t(\mathbf{x}_t, \mathbf{u}_t, \mathbf{w}_t, \varepsilon_{t+1}, \boldsymbol{\vartheta}_t) \tag{6.7a}$$

Equation (6.2b) stays unchanged if $h_t(\cdot)$ does not depend on ε_{t+1}, while it takes the following form

$$\mathbf{y}_{t+1} = h_t(\mathbf{x}_t, \mathbf{u}_t, \mathbf{w}_t, \varepsilon_{t+1}, \boldsymbol{\theta}_t) \tag{6.7b}$$

if it does, since then the output relative to the interval $[t, t+1)$ is no longer deterministically known at time t and thus it is denoted with $\mathbf{y}_{t+1}$.

There is still one convention left to define: the dimension adopted to represent the water flows, such as the release from Miorina. The simplest way to express the daily decision u_t^{MIO} is to indicate the volume that the Regulator would like to release from Miorina in the following twenty-four hours. It is, in fact, this volume, together with the inflow volume, that influences the current lake storage and transforms it into the storage of the following day. Nevertheless, expressing the release in m^3/d seems strange; we are more used to thinking in terms of m^3/s, i.e. in terms of average daily flow rate. The latter can be obtained, once the release u_t^{MIO} $[\text{m}^3/\text{d}]$ is known, with the following relation

$$\frac{u_t^{\text{MIO}}}{86\,400} \tag{6.8}$$

where 86 400 is the number of seconds in a day. To avoid a useless proliferation of symbols, we use the symbol u_t^{MIO}, just as any other symbol relative to a water flow, to indicate both the daily volume and the average daily flow rate. The reader should not get confused, as long as (s)he keeps in mind that *when we speak about water flow (rates), it is understood that we are dealing with average daily flow rates*. In the figures the values will always be expressed in m^3/s, while in the text we will speak of daily volume whenever it may facilitate understanding.

6.5 The catchment

Defining a model of the catchment means providing a mathematical representation of the inflow process. More precisely, for reasons that will be illustrated in Section 6.6.2, we need a model of the *effective* (or *net*) *inflow* a_{t+1}, which is defined as the difference between the total volume a_{t+1}^{tot} that flows into the lake in the time interval $[t, t+1)$ and the volume that evaporates in the same time interval.

It is essential to bear in mind that:

- The model will not be used to forecast inflows, but to design regulation policies, which requires, as we explained in Section 5.2.6, that we solve an appropriate Stochastic (Optimal) Control Problem. In order that the computing time for its solution be acceptable, we must use a simple model such as a *stochastic empirical model*, which does not try to explain the formation of inflow with a detailed representation of the physical phenomena involved, but simply describes the inflow process as a realization of an opportune *stochastic process*, i.e. a sequence of stochastic variables.

- The algorithm that we will adopt to solve the (Stochastic, Optimal) Control Problem (Section 7.7.1) requires that the model must not contain either deterministic or stochastic disturbances, except for a particular type of random disturbances: *white noises*. As we will see in the next paragraphs, this means in practice that, at every time t, the inflow a_{t+1} in the following time interval must be computed only on the basis of the inflows registered in the previous time intervals.

Before describing the procedure that was used for identifying the catchment model, we briefly present some properties of stochastic processes and the most common classes of stochastic empirical models. A more complete description is provided in Appendices A2 and A6 of THEORY.

6.5.1 Properties of stochastic processes

A stochastic process is *stationary* when all the *joint probability densities*

$$\phi_t(y_t, y_{t+1}, \dots, y_{t+m}) \quad \forall(t, m)$$

of its variables do not depend on time. Moreover, a stationary process is said to be *Gaussian* when every variable y_t has a *Gaussian* (or *normal*) *distribution* $N(\mu, \sigma^2)$. Finally, it is named *standard Gaussian* when $\mu = 0$ and $\sigma^2 = 1$. In the following we will see that the inflow process can be traced back to this last case.

Note that the only available information is the time series of effective inflows $\{a_t\}_{t \in H}$ recorded by the Ticino Consortium over the horizon H (1st January 1974 – 31st December 1998).[4] This means that we know only a finite sample of a realization of the inflow process,[5] rather than the entire process (the set of all the possible realizations). We must therefore suppose that the process is *ergodic*, so that we can use statistics of the time series (such as its temporal average) as estimates of the corresponding statistics of the process (in the example, the expected value of the process).

6.5.2 Classes of stochastic models

The simplest classes of empirical models for stationary and Gaussian stochastic processes are:

- *Autoregressive models* AR(p): the value y_{t+1} that the variable y assumes in the interval $[t, t+1)$ is linked to the values assumed by the same variable in p previous time instants, i.e.

$$y_{t+1} = \alpha_1 y_t + \alpha_2 y_{t-1} + \dots + \alpha_n y_{t-p+1} + \varepsilon_{t+1} \tag{6.9}$$

  where ε_t is a random disturbance produced by a white stochastic process,[6] with a distribution $N(0, \sigma_\varepsilon^2)$.

- *Moving average models* MA(q): the value y_{t+1} is linked to the last q values of the disturbance

$$y_{t+1} = \varepsilon_{t+1} + \gamma_1 \varepsilon_t + \gamma_2 \varepsilon_{t-1} + \dots + \gamma_q \varepsilon_{t-q+1} \tag{6.10}$$

[4] The reasons behind this choice were explained in Section 2.1.1.

[5] The entire realization would be given by $\{a_t\}_{t=-\infty}^{t=+\infty}$.

[6] Which means that the value of the disturbance at time t is uncorrelated with the values of the disturbance in all the previous instants.

- *Autoregressive moving average models* ARMA(p, q): both AR and MA parts are present

$$y_{t+1} = \alpha_1 y_t + \alpha_2 y_{t-1} + \cdots + \alpha_p y_{t-p+1} + \\ + \varepsilon_{t+1} + \gamma_1 \varepsilon_t + \gamma_2 \varepsilon_{t-1} + \cdots + \gamma_q \varepsilon_{t-q+1} \tag{6.11}$$

  This last class encompasses the two preceding ones: we see immediately that the class of ARMA$(p, 0)$ models coincides with the AR(p) class and that the class of ARMA$(0, q)$ models coincides with the MA(q) class.

The characteristics of these processes are completely defined by the (normalized) *autocorrelation function* $\varrho(\tau)$, whose plot is called *correlogram*.

6.5.3 Analysing stationarity and standardization

We have seen that a stochastic process is stationary when all the (infinite) joint probability distributions of its variables do not depend on time. The description of the inflow process for Verbano, provided in Section 2.1.1, clearly shows that we cannot consider it to be stationary. In fact, even if after 1973 (the year in which the last hydropower reservoir built in the catchment came into service) one can reasonably say that the inflow presents the same statistical characteristics from one year to the next, we cannot assume that they do not vary over the course of the year. For example, Figure 2.3 shows that the median of the inflows does change over the course of the year. It follows that the joint probability distributions of the process also change periodically. Therefore, the inflow process is not stationary, but *cyclostationary*, i.e.

$$\phi_t(a_t, a_{t+1}, \ldots, a_{t+m}) = \phi_{t+T}(a_{t+T}, a_{t+T+1}, \ldots, a_{t+T+m}) \quad \forall m$$

where $T = 365$ days. A model of the classes presented in the previous section can thus be adopted only once the time series $\{a_t\}_{t \in H}$ of the recorded inflows has been transformed into a time series that can be produced by a stationary process.

To do this, first of all we ask ourselves how to identify the probability distribution of the inflow on a given day. Consider a particular day of the year, referred to as τ, a number[7] between 0 and 364. Let H_τ be the set of days t of H such that $t_{\text{MOD}\,T}$ is equal[8] to τ. For example, if τ corresponds to 31st of January, H_τ will contain all values t corresponding to 31st January that are present in the horizon H. We can now extract the sample $\{a_t\}_{t \in H_\tau}$ and use one of the traditional methods of statistical inference (see Appendix A2 of THEORY) to identify the probability distribution of the inflow on day τ. For example, if we hypothesize that the inflow has a normal distribution, from the sample $\{a_t\}_{t \in H_\tau}$ we must first estimate the expected value μ_τ and standard deviation σ_τ, and then verify whether the hypothesis is acceptable with a suitable statistical test (for example by adopting the Kolmogorov–Smirnov test, as shown in Figure 6.4).

Once the distribution has been identified, the data from the sample $\{a_t\}_{t \in H_\tau}$ are *normalized*, i.e. transformed into a sample $\{\tilde{a}_t\}_{t \in H_\tau}$ which could be extracted from a normal (or quasi-normal) distribution, using a transformation that varies according to the distribution that was originally identified.[9] The data obtained in this way are then *standardized*,

[7] In this simple presentation, we will not take into account the existence of the 29th of February in leap years.

[8] The script $t_{\text{MOD}\,T}$ denotes the residue of the division of t for T.

[9] If the hypothesis of the normal distribution was accepted, the transformation is simply given by the identity

$$\tilde{a}_t = a_t$$

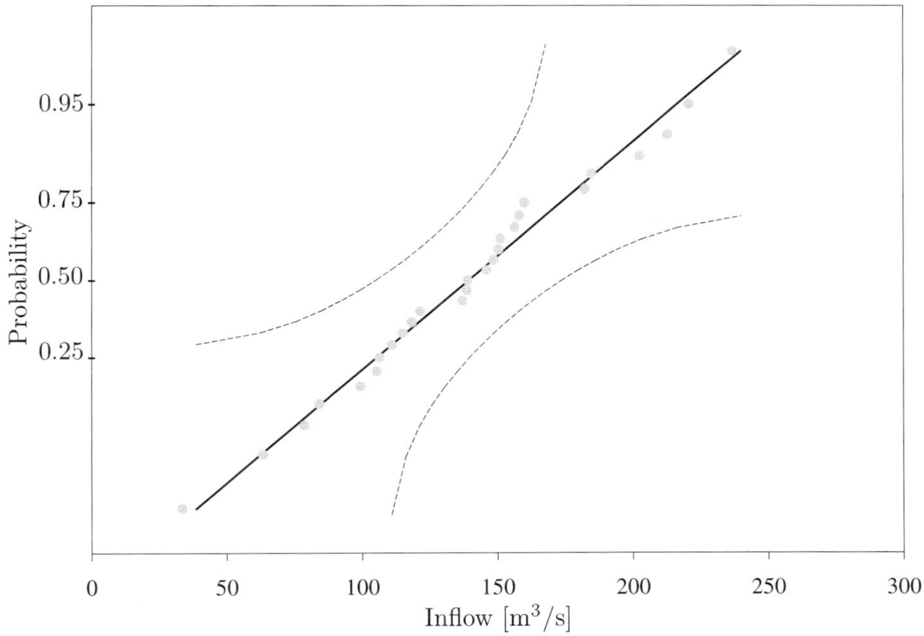

Figure 6.4: Sample cumulative distribution function of the inflow data on 31st January from the different years in horizon H. The confidence bounds (5%), computed with the Kolmogorov–Smirnov test, show that the hypothesis of normal distribution is acceptable.

i.e. transformed into a sample $\{y_t\}_{t \in H_\tau}$ which could be extracted from a standard normal distribution $N(0, 1)$. This is obtained by applying the transformation

$$y_t = \frac{\tilde{a}_t - \tilde{\mu}_{t\text{MOD}\,T}}{\tilde{\sigma}_{t\text{MOD}\,T}} \tag{6.12}$$

where $\tilde{\mu}_{t\text{MOD}\,T}$ and $\tilde{\sigma}_{t\text{MOD}\,T}$ are the expected value and standard deviation estimated with the sample $\{\tilde{a}_t\}_{t \in H_\tau}$.

By repeating the previous operations for each day τ of the year, the original time series $\{a_t\}_{t \in H}$ is transformed into a new time series $\{y_t\}_{t \in H}$ that we can consider to be produced by a stationary and Gaussian process y_t. Then we can try to describe such a process with a model that belongs to one of the classes introduced in the previous subsection. Note that the operations that we have carried out so far define a bijective transformation that links a_t to y_t, for every possible value of τ. It follows that any value of y produced by the model can be easily[10] transformed back into an inflow value a.

In general, the transformation depends on the distribution that was defined: for example, if we start from a log-normal distribution, the transformation is given by

$$\tilde{a}_t = \ln(a_t)$$

[10]For example, if the inflow has a normal distribution, the formula for transforming y into a is

$$a_t = \sigma_{t\text{MOD}\,T}\, y_t + \mu_{t\text{MOD}\,T} \tag{6.13}$$

6.5.4 Criteria for choosing the model

The data set is first of all subdivided into two time series (H^1 and H^2) that are as homogeneous as possible, in the sense that they contain the same typology of events[11]; in other words, dry spells and floods must appear in both the series with similar frequency. By subdividing the data set of inflows to Verbano into a time series from 1st January 1974 to 31st December 1988 and another one from 1st January 1989 to 31st December 1998, this condition is satisfied. One of the two series (the first in the Verbano Project) is used for model calibration, and the other for validation.

To *calibrate* a model means to estimate the value of its parameters that make it explain the observed data 'at best', that is according to a specified criterion. One of the most common is the *least squares criterion*, which selects the parametrization that minimizes the sum of squared deviations[12] ε_t, defined as the differences between the observed value ($\bar{y}_t$) and the value ($\hat{y}_t$) computed by the model, i.e.

$$\varepsilon_t = \bar{y}_t - \hat{y}_t$$

Several algorithms, called *estimators*, exist for parameter estimation based on this criterion (see Appendix A4 of THEORY).

To *validate* a model means to ascertain that the calibrated model is capable of 'explaining' data in a period that is different from the one used for calibration. An index that is often used for quantifying how well the model 'explains' the observed data is the *Coefficient of Determination* (or *Explained Variance*), defined as

$$R^2 = 1 - \frac{\sigma_\varepsilon^2}{\sigma_y^2} \tag{6.15}$$

where σ_y^2 is the variance of process $\{y_t\}$ and σ_ε^2 is the variance of the model residual. The definition is given with reference to the process y_t, because it is of this process that we want to construct a stochastic model. Nevertheless, given that ultimately we are interested in a description of the inflow series $\{a_t\}$, it is more meaningful to provide the Coefficient of Determination with respect to this variable. In the following we will thus use the index

$$R^2 = 1 - \frac{\sigma_e^2}{\sigma_a^2} \tag{6.16}$$

where σ_a^2 is the variance of the inflow time series and σ_e^2 the variance of the residuals with respect to it. By construction, the value[13] of R^2 is always less than or equal to 1. The better the model 'explains' the data, the smaller is the variance σ_e^2 of the residual and thus the higher R^2.

---

i.e. the inverse of (6.12); while if a has a log-normal distribution, the inverse transformation is

$$a_t = e^{\sigma_{t \text{MOD} T} y_t + \mu_{t \text{MOD} T}} \tag{6.14}$$

[11]For more details, see the example presented in Section 6.9.1.3 of the DVD.

[12]They are also called *residuals* or errors.

[13]Note that R^2 can also be negative. When this happens, the simple model

$$a_t = \mu_t$$

explains the data better than the model being considered.

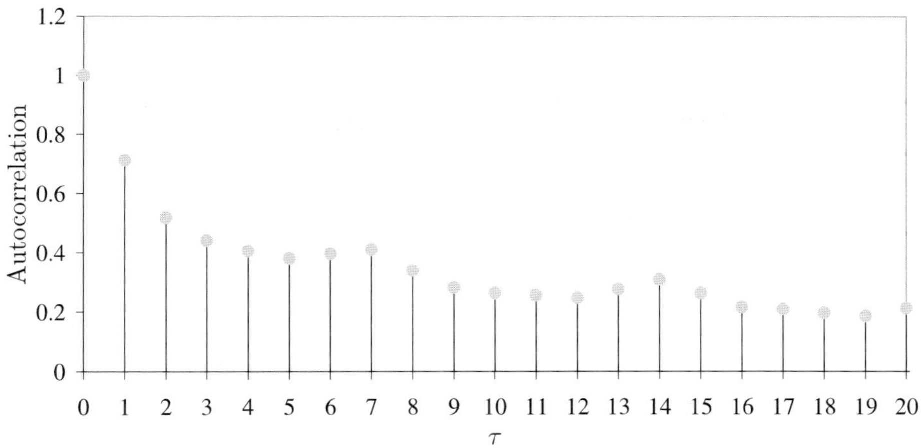

Figure 6.5: Sample correlogram $r(\cdot)$ of the inflow time series for Verbano.

A 'good' model should provide a high value of R^2 both in calibration and in validation: obtaining high values of R^2 over the calibration data set is not sufficient to affirm that the model is good, since the model might reproduce the data with which it was calibrated very well, but not be able to reproduce other series. This occurs, for example, when the ratio between the number of data used in calibration and the number of parameters to estimate is not high enough: in this case the model is said to be *overparametrized*.

The value of R^2 is not, however, the only index to be considered in the choice of model. Consider, for example, a model in the AR(p) class: by increasing the order p of the model, the value of R^2 clearly increases both in calibration and in validation (at least until problems due to overparametrization do not rise). However, as p increases, the number of state variables[14] of the model also increases, and with them the computing time required for the solution of the Control Problem (see Section 7.8). Therefore, if we identify a model with $n+1$ state variables that provides a value of R^2 better than a model with n state variables, before adopting it we must ask ourselves if the best performance that it offers compensates the greater computing burden required to design a policy. Only if the response is positive will the new model be adopted.

6.5.5 The inflow process model

As we have said, the characteristics of a stationary, Gaussian, stochastic process are completely defined by its autocorrelation function $\varrho(\cdot)$. In practice, $\varrho(\cdot)$ is approximated by its sample estimate $r(\cdot)$, which is computed over the time series $\{y_t\}_{t\in H^1}$ that was derived from the Verbano inflow time series. The correlogram that was obtained, i.e. the plot of $r(\cdot)$, is shown in Figure 6.5. Observe that its trend is substantially decreasing, but it presents local maxima for $\tau = 7$ and $\tau = 14$. This is a strong indication that the inflow process is not

[14]From the definition of state variable, it follows that the state variables of a AR(p) model are the last p measures of the flow rate. Similarly, in an ARMA(p, q) model, the state is given by the last p measures of the flow rate and the last q estimates of the residual. Nevertheless, in this second case it can be proved that there exists a state-space model (*internal description*) that uses only n state variables, n being the maximum between p and q (Peterka, 1986) (see also Appendix A6 of THEORY).

completely natural, but rather it is influenced by a perturbation which is due to human activity and which is periodic with a period of one week. To better describe the inflow process, it would be opportune to individuate this perturbation and eliminate it from the time series, given that, because of its anthropic origin, it is likely to have a simpler deterministic description.

In Section 2.1.1 we explained that inflows to Verbano are strongly influenced by the presence of Alpine reservoirs for hydropower production. In particular Figure 2.6 highlights that one of the effects of this presence is a weekly fluctuation in the flow rate of the inflow. Since the periodicity is the same as the one observed in the autocorrelation function, it is more than likely that the perturbation's origin is the regulation of the hydropower reservoirs. To eliminate the effect of the regulation from the time series $\{a_t\}$ (and thus from $\{y_t\}$), the storage time series for each reservoir should be available; then, for each reservoir, we could add (or subtract) the water volume stored (or released) to the net daily inflow that has been measured, taking into account the travel time from the reservoir to the lake. Acquiring the storage time series for all the reservoirs proved to be impossible, however, so we had to give up trying to remove the component with anthropic origins from the inflow process.

Based on the data set $\{y_t\}_{t \in H^1}$, we have calibrated 7 models of the AR(p) class, with $p = 1, \dots, 7$, and two models of the ARMA(p, q) class, with $p = 1$ and $q = 1, 2$. Their performances are reported in Table 6.1, in which, for comparison, the performances of AR(0) model are also shown. Note that the Coefficient of Determination on the calibration data set increases by 350% when passing from AR(0) to AR(1) while, when passing from the latter to more complex models, the maximum increase that can be obtained is only 2%, which is not enough to justify the enormous increase in computing time that would occur if they were adopted. However, the increase in computing time is very significant even when passing from AR(0) to AR(1): the total time required to generate all the alternatives passes from about 200 days to around 6000 days. Even the adoption of an AR(1) model is thus impossible in practice. This is why we decided to design all the alternatives using the simple AR(0) model, and use an AR(1) model only to increase the performances of the best compromise alternative, once it had been identified. The computing time required for a single alternative design is acceptable: from 15 hours to 15 days depending on whether it provides for the environmental allotment decision or not.

In the Verbano Project we thus adopt the following model

$$a_{t+1} = \sigma_{t_{\text{MOD}\,T}} \cdot \varepsilon_{t+1} + \mu_{t_{\text{MOD}\,T}} \tag{6.17a}$$

$$\varepsilon_{t+1} \sim N(0, 1) \tag{6.17b}$$

6.5.6 Use of precipitation data

On page 147 we anticipated that the algorithm that we adopt to solve the Control Problem requires that in the model of the system (and so also in that of the catchment) no exogenous variables (i.e. deterministic disturbances) be used, such as precipitation data for example. It is important to ask ourselves what the cost of this limitation is, in terms of Coefficient of Determination, to evaluate whether it would be preferable to use another algorithm.

The simplest class of stochastic, empirical models with exogenous variables is the ARX($p, r; k$) class, defined by the following expression

$$y_{t+1} = \alpha_1 y_t + \alpha_2 y_{t-1} + \cdots + \alpha_p y_{t-p+1} + \varepsilon_{t+1} + \beta_1 w_{t+1-k} + \cdots + \beta_r w_{t+1-k-r+1} \tag{6.18}$$

Table 6.1. Performances of different inflow models over the calibration data set H^1 = (1st January 1974 – 31st December 1988), and over the validation data set H^2 = (1st January 1989 – 31st December 1998)

Model	Coefficient of Determination	
	Calibration	Validation
AR(0)	0.170	0.288
AR(1)	0.594	0.595
AR(2)	0.593	0.594
AR(3)	0.600	0.601
AR(4)	0.606	0.608
AR(5)	0.607	0.604
AR(6)	0.600	0.613
AR(7)	0.596	0.601
ARMA(1, 1)	0.593	0.593
ARMA(1, 2)	0.601	0.600

Table 6.2. Performances of different inflow models with exogenous variables, over the calibration data set (25th September 1992 – 31st December 1995) and the validation data set (1st January 1996 – 19th October 1998)

Model	Coefficient of Determination	
	Calibration	Validation
ARX(1, 1, 1)	0.859	0.717
ARX(1, 1, 2)	0.621	0.504
ARX(1, 2, 1)	0.858	0.713
ARX(1, 2, 2)	0.619	0.506
ARX(1, 3, 1)	0.857	0.715
ARX(1, 3, 2)	0.617	0.498

where w_{t+1} is the exogenous deterministic variable and k is the delay with which that variable influences y_{t+1}. Since the available time series of precipitation is shorter than the inflow time series, we had to use a calibration and validation data set[15] shorter than those used in the previous section. The performances of several different order ARX models, reported in Table 6.2, allow us to draw the following conclusions:

- using the measure w_t of precipitation on the previous day proves to be very advantageous for estimating y_{t+1}: for example, the Coefficient of Determination increases by 45% when passing from AR(1) to ARX(1, 1; 1);

- it is of no advantage to consider the measure w_{t-1} of precipitation two days early, as emerges from the performances of the ARX(1, 1, 2) model and from the fact that ARX(1, 2, 1) and ARX(1, 3, 1) have the same performances as ARX(1, 1, 1).

In conclusion, if it were possible to use a deterministic input, we could obtain an improvement in the model's performance. However, to solve the Control Problems, we would

[15] Calibration = (25th September 1992 – 31st December 1995); Validation = (1st January 1996 – 19th October 1998).

Table 6.3. Transfer time of the waves

Transfer time	
Tributary	Hours
Toce	6
Tresa	7
Maggia	9
Ticino	11

have to substitute the SAA algorithm (Section 7.7.1) with a *model-free* algorithm, such as *Q-learning planning* (see Section 13.4 of THEORY). By doing so the computing times would increase by about 100 times and so we have to do without this improvement.

6.6 The lake

The dynamics of the storage s_t of the lake is described by the following mass balance equation

$$s_{t+1} = s_t + a^{tot}_{t+1} - r_{t+1} - S(s_t)e_{t+1} \tag{6.19}$$

where a^{tot}_{t+1} and r_{t+1} are the inflowing volume (*inflow*) and the released volume (*release*) respectively, in the time interval $[t, t+1)$; e_{t+1} is the evaporated volume per unit of surface area in the same time interval and $S(\cdot)$ is the function that provides the surface area of the lake as a function of the *storage*. The lake is thus described by a dynamic model, in which the state is the storage s_t (for more detail see Section 5.1 of THEORY).

Equation (6.19) is valid only if the size of the lake is such that the transfer times of the inflows from the input sections of the different tributaries to the output section of the lake are negligible, with respect to the duration of the time step Δ. Table 6.3 reports the estimates (Rinaldi, 1980) of those times for Verbano's four main tributaries. They are all clearly less than a day, and so equation (6.19) can be considered valid.

Now we will examine the terms that appear in equation (6.19) in detail.

6.6.1 The lake storage

Every morning at 8 o'clock the Ticino Consortium measures the lake level h^{SC}_t at the Sesto Calende hydrometric station, whose 'zero' is set at the elevation of 193.016 m above sea level, according to the Italian *ellipsoid*.[16] The storage s_t is then estimated by assuming that the lake is a *cylinder* and thus with the following relation

$$s_t = S \cdot h^{SC}_t + s_{inf} \tag{6.20}$$

where S is the average area of the lake surface (211.5 km^2) and s_{inf} is a constant. The latter establishes the storage value at zero in correspondence with the level -0.75 m, thus reducing the numerical values being considered for s_t. Clearly the value of s_{inf} can be arbitrarily chosen without influencing the results, since they depend only on variations of the storage.

[16] It is important to underline the fact that the heights are defined according to the *Italian ellipsoid*, since Italy and Switzerland use different ones.

The hypothesis of cylindrical storage is valid only if the following conditions hold:

1. The surface of the lake water can only shift vertically, remaining parallel to itself,[17] in other words its form is time-invariant up to a vertical shift and, as such, independent of the values of both the level h_t^{SC} and the inflow a_{t+1}^{tot}.

2. The area S of the surface is independent from h_t^{SC}.

Before verifying these two conditions (we will do so in Sections 6.6.3 and 6.6.4) it is important to show how equation (6.19) is simplified when equation (6.20) is verified and the practical consequences that follow.

6.6.2 Inflow and evaporation

As we have said, the term a_{t+1}^{tot} that appears in equation (6.19) represents the total inflow to the lake in the interval $[t, t+1)$. The inflow rate is difficult, if not impossible, to measure, especially during precipitation events when not only is the inflow due to the tributaries, but also to the distributed runoff along the lake shores. Observe, however, that if the hypothesis of cylindrical storage is valid, equation (6.19) can be reduced to the following expression

$$s_{t+1} = s_t + a_{t+1} - r_{t+1} \tag{6.21}$$

where the term a_{t+1}, which replaces the expression $a_{t+1}^{tot} - Se_{t+1}$, is the *effective inflow* (or *net inflow*), i.e. the inflow net of evaporation. Equation (6.21) can be used to compute a_{t+1}, once the other three terms that appear in it are known. Since these three terms can all be measured (the release r_{t+1} can be determined from the position of the Miorina dam's gates, while the storage is obtained from the measures of level h_t^{SC} through equation (6.20), as we explained in Section 6.6.1), equation (6.21) is adopted by the Ticino Consortium to estimate the inflow volume between 8 o'clock on one day and 8 o'clock on the following day. Bear in mind that, from now on, *we will use the effective inflow values, even if, for brevity's sake, we will simply call them 'inflow values'*. Since the inflows are estimated using equation (6.21), the accuracy of the estimate depends directly on the error that is made in assuming storage to be cylindrical and on the precision of the level measurement.

6.6.3 The lake surface profile

Now we will deal with validating the conditions that allow us to consider the lake storage cylindrical (Section 6.6.1). The first condition requires that the form of the lake surface can be assumed to be time-invariant, up to a vertical translation. To verify this we must identify the expression that relates the water level h_t^{SC} at Sesto Calende with the level at any other point on the lake surface, and ascertain whether in such an expression h_t^{SC} figures as an additive term. Given the long narrow shape of the lake (Figure 6.6) it is reasonable to assume that, in a given instant, the level (above zero at Sesto Calende) is the same at all the points which are at the same distance d from Sesto Calende. Therefore, if the hypothesis that we must verify is valid, between h_t^{SC} and the level h_t^l at a location l, situated at distance d, the following relation exists

$$h_t^l = h_t^{SC} + h(d) \tag{6.22}$$

[17] The lake surface is never horizontal, as intuition would lead us to believe: if it were, the inflows from the tributaries could not move towards the emissary. The surface is inclined towards the lake mouth and, for the same reason, more inclined the greater is the inflow.

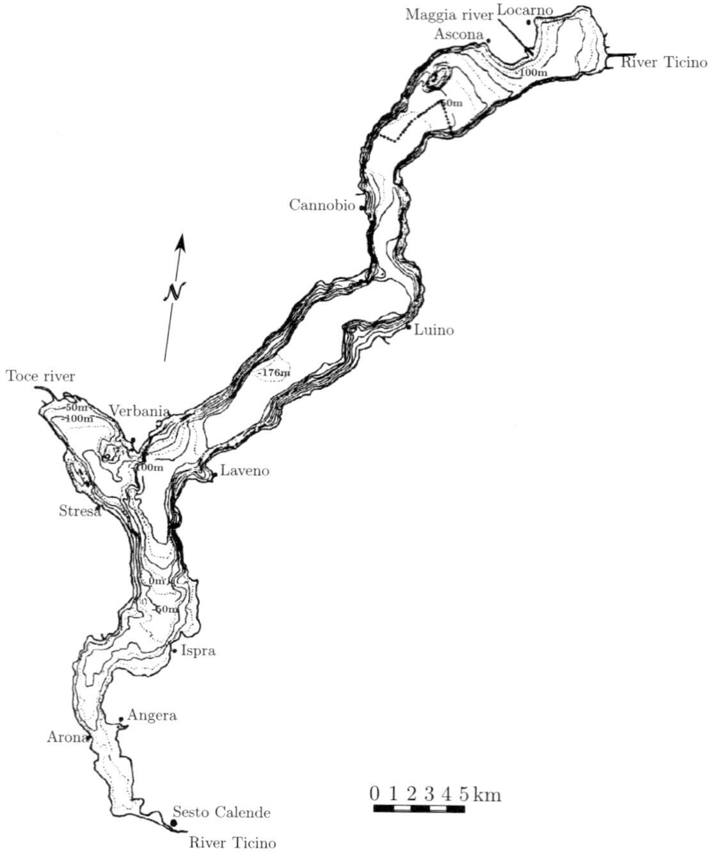

Figure 6.6: The bathymetry of the lake.

which means that the difference $(h_t^l - h_t^{SC})$ between the two levels is independent of the level h_t^{SC}.

In Figure 6.7 the differences in level between Locarno (h_t^{Loc}) and Sesto Calende (h_t^{SC}) are shown, as the latter (h_t^{SC}) varies, according to the estimation carried out by Fantoli (1897), by Citrini (1973) and by the IST[18] (Istituto Scienze della Terra[19]). The three estimates are essentially in agreement and show that the difference $h_t^{Loc} - h_t^{SC}$ is not constant but grows with the level at Sesto Calende. Therefore, it would seem that the shape of the lake surface could not be considered time-invariant.

Nevertheless, in a recent work, Maione and Mignosa (1995) show that the lake surface profile is piecewise linear, i.e. it can be considered linear in the lake's stretch between Sesto Calende and a section situated 3 km upstream, and in the stretch between the latter and Locarno. See for example Figure 6.8, which shows the lake's surface profile when h_t^{SC} is equal to 2 m. This is explained by the fact that in the last three kilometers the lake' section

[18] Private communication.

[19] www.ist.supsi.ch, last visited on September 2006.

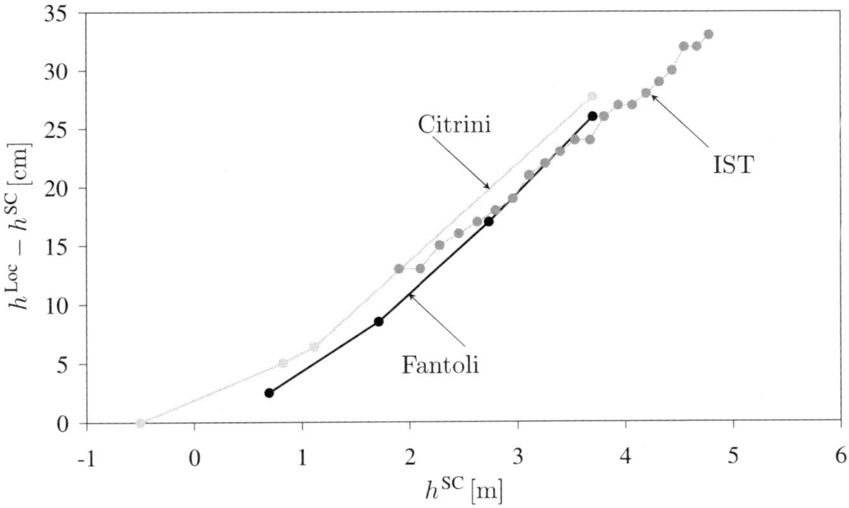

Figure 6.7: Difference between the lake level h_t^{Loc} at Locarno and h_t^{SC} at Sesto Calende (re-elaboration of the data in Fantoli (1897), Citrini (1973) and IST (Istituto Scienze della Terra)).

is much narrower than in the further upstream part of the lake, an thus in that stretch the lake is almost a river and the level is influenced by Miorina's *drawdown curve*. Considering this profile and using the time series of measures $(h_t^{\mathrm{SC}}, h_t^{3\mathrm{km}}, h_t^{\mathrm{Loc}})$[20] provided by IST, we identified the following relation between h_t^{SC} and h_t^l

$$h_t^l = \begin{cases} (1.0645 \cdot h_t^{\mathrm{SC}} - 0.0301) \cdot \frac{d}{3} & 0 \leqslant d < 3 \\ (0.0072 \cdot h_t^{\mathrm{SC}} + 0.0141) \cdot \frac{d}{57} + 1.06404 \cdot h_t^{\mathrm{SC}} - 0.031 & 3 \leqslant d \leqslant 60 \end{cases} \tag{6.23}$$

from which the level differences shown in Figure 6.9 where obtained. Note that the trends of the differences $(h_t^{\mathrm{Loc}} - h_t^{\mathrm{SC}})$ and $(h_t^{3\mathrm{km}} - h_t^{\mathrm{SC}})$ are similar to those in Figure 6.7. Nonetheless, the fact that these trends are parallel, together with the fact that the trend of the difference $(h_t^{\mathrm{Loc}} - h_t^{3\mathrm{km}})$ is almost horizontal, proves that in practice as the level increases the lake

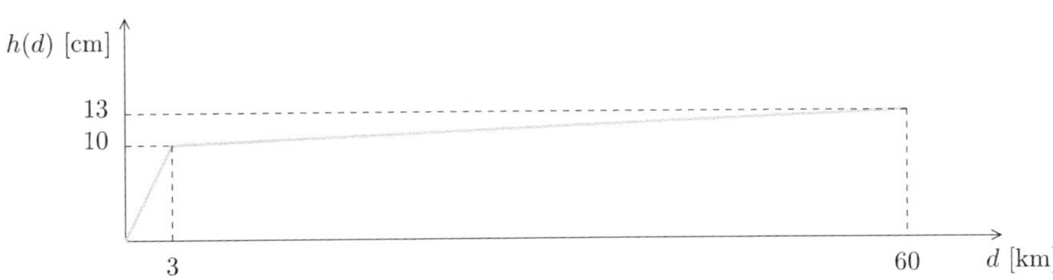

Figure 6.8: Lake level $h(d)$ with respect to Sesto Calende water level as a function of the distance d from Sesto Calende, when the release at Miorina dam is equal to 1000 m^3/s and the Sesto Calende water level h_t^{SC} is equal to 2 m (from Maione and Mignosa, 1995).

[20]The symbol $h_t^{3\mathrm{km}}$ denotes the water level in the section located 3 km upstream of Sesto Calende.

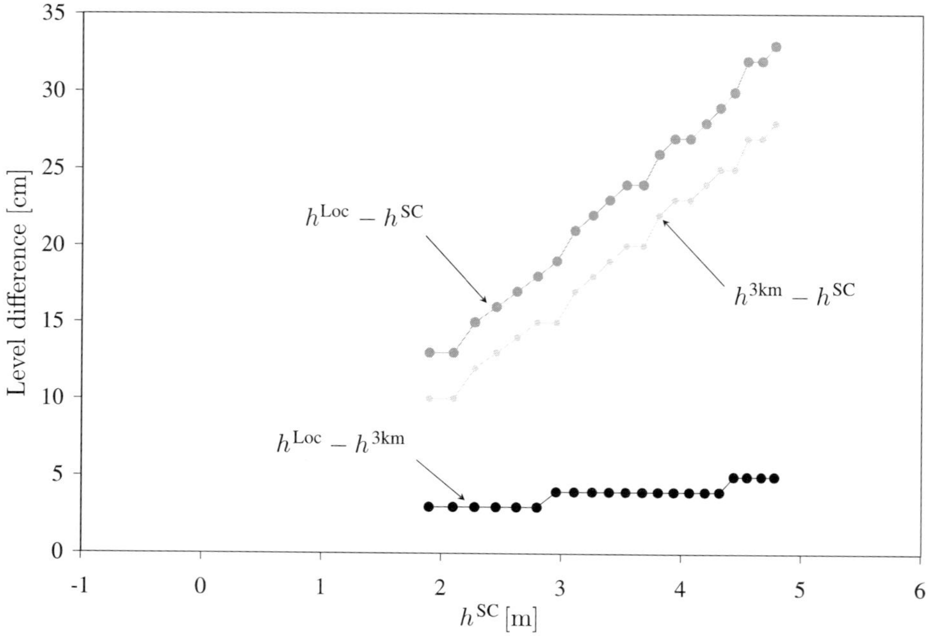

Figure 6.9: Level differences as a function of the lake level h_t^{SC} at Sesto Calende. The level h_t^{3km} refers to the section located 3 km upstream of Sesto Calende; the level h_t^{Loc} refers to Locarno (data from IST).

surface profile changes only in the three kilometers that precede the mouth of the lake. It is thus only in this stretch that the form of the lake surface cannot be supposed to be time-invariant, while this hypothesis, and thus equation (6.22), are acceptable for the rest of the lake. By using equation (6.22) in place of equation (6.23), the error in the estimation of the inflow is less than 1% (this is the estimation error obtained in correspondence to the greatest daily variation of the lake level ever recorded, namely 1.52 m during the 1993 flood). The slightness of this error is explained by observing that the area of the lake is 211.5 km^2, while the area of the last lake stretch is only 0.62 km^2. Note that with this approximation, the error in the estimation of the level is much more significant (7% in the same event).

We can thus conclude that:

1. The hypothesis that the form of the lake surface is time-invariant is acceptable for computing the storage, so we will estimate its value with equation (6.20) and describe its dynamic with equation (6.21).

2. We will use equation (6.23) to estimate the lake level in a section at distance d from Sesto Calende. In particular, we will use it to estimate the lake levels at Locarno and Verbania, which are necessary for computing the indicators of the *Upstream Flooding* sector (see Section 4.5.5 of the DVD). The distance d of the two cities is 60 and 29 km respectively.

Table 6.4. Water-level/surface-area relation for Verbano

h_t^{SC} [m]	S [km^2]
0.0	208.00
1.0	212.00
2.0	216.97
3.0	221.82
4.0	229.17
5.0	234.35
6.0	240.61
7.0	246.32
7.5	249.82
8.0	253.32

6.6.4 The expansion of the lake area

We must now check the second condition: the area S of the lake surface is independent of the level h_t^{SC}. Figure 6.6 shows that the shores of the lake are very steep, except in the Fondotoce zone, near to Verbania. This favours the validity of the hypothesis.

Consider the lake's (water-level/surface-area) pairs determined by Fantoli (1897) and shown in Table 6.4. They show that the area increases with the level by about 2.8% in the regulation range and by about 10% in the first four metres, i.e. for the level variations produced by large floods. The first difference can be considered negligible, while the second cannot. It must be noticed, however, that the percentage error assumes those values only when the inflow rate is actually measured, while it is much smaller when the inflow rate is estimated with equation (6.21). In this case, in fact, the two errors, the one made in estimating the inflow rate with equation (6.21), the other made in computing the future storage with the same equation, tend to cancel each other out. Their mutual cancellation is almost perfect when the alternative under examination adopts SD$_{\text{CUR}}$, i.e. the same stage–discharge relation with which the inflows were estimated, while it is a little larger when SD$_{+600}$ is adopted. To avoid this error, we should give up the hypothesis of cylindricity, but then the inflows should be estimated with equation (6.19). For this we would need not only measures of lake level and release, but also measures of daily evaporation (e_{t+1}). The latter, however, are not among the measures currently taken. Thus it would be necessary to replace the measure with an estimate, but by doing so we would run the risk of errors that are greater than those we are trying to avoid.

We can thus conclude that both the conditions that allow us to consider the storage as a cylinder are valid and so we can describe the dynamics of the lake with equation (6.21).

6.6.5 The release

Since Verbano is a regulated lake, the release r_{t+1} in the time interval $[t, t+1)$ depends on the release decision u_t^{MIO}, taken at 8 o'clock on day t. The release, however, is not always equal to the decision, because it depends on the inflow trajectory over the course of the day as well. In fact, the release must be feasible, i.e. it must be between the *minimum* (v_t) and the *maximum* (V_t) *volume* that can be feasibly released through the day. Let us illustrate how these volumes can be computed.

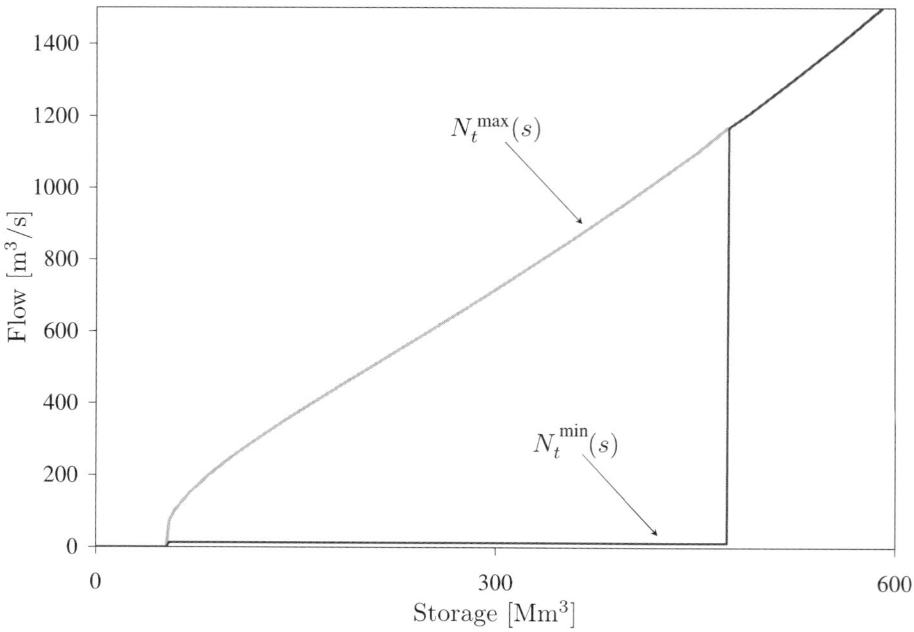

Figure 6.10: The maximum and minimum instantaneous storage–discharge relations $N_t^{\max}(\cdot)$ and $N_t^{\min}(\cdot)$ associated to the combination ($\text{SD}_{+600}/\text{RANGE}_{\text{ENL}}/13_{\text{CON}}$).

When the lake level is non-inferior to the lower extreme h_t^{low} of the regulation range, the maximum instantaneous release must be zero; otherwise it is obtained by leaving the Miorina dam completely open. The relation between storage and maximum instantaneous release is termed *maximum instantaneous storage–discharge relation* and denoted by $N_t^{\max}(\cdot)$. Notice that this function depends on the combination of structural and normative actions that is adopted.

As for the minimum instantaneous release, it is equal to the MEF when the level is within the regulation range. Otherwise, if the level exceeds the upper extreme, the minimum release is equal to the maximum instantaneous release, because the dam must be completely opened; if, instead, the level is below the lower extreme it must be zero. The relation between storage and minimum instantaneous release is termed *minimum instantaneous storage–discharge relation* and denoted by $N_t^{\min}(\cdot)$. Again, it depends on the adopted combination of structural and normative actions. For example, in Figure 6.10 the maximum and minimum instantaneous storage–discharge relations $N_t^{\max}(\cdot)$ and $N_t^{\min}(\cdot)$ associated to the combination ($\text{SD}_{+600}/\text{RANGE}_{\text{ENL}}/13_{\text{CON}}$) are shown. In Section 5.3.2 we assumed that for all the regulation ranges the lower extreme is time invariant, while the upper extreme can vary periodically. It follows that the maximum instantaneous storage–discharge relation $N_t^{\max}(\cdot)$ is time-invariant, while the minimum one is periodic. Its periodicity is also due to the periodicity of the MEF.

Given the maximum and minimum instantaneous storage–discharge relations and a pair (s_t, a_{t+1}), the maximum and minimum volumes that can be released in the interval $[t, t+1)$ can be estimated only if the trajectory of the instantaneous inflow $q^a_{[t,t+1)}(\cdot)$ is known for every time instant ξ in that interval. Unfortunately, we do not have this information, since we only know the volume a_{t+1} of the overall inflow in the interval. Thus, the trajectory of the inflow over the course of the day can only be hypothesized. The most simple hypothesis is to assume that it is constant, i.e.

$$q^a_{[t,t+1)}(\xi) = \frac{a_{t+1}}{\Delta} \quad \forall \xi \in [t, t+1) \tag{6.24}$$

with Δ being the duration of the day (86 400 s).

With equation (6.24) we can now compute the minimum volume v_t that can be released in the interval $[t, t+1)$ as a function of the initial storage s_t and of the overall inflow a_{t+1}. To do this, first we must compute the trajectory $s_{[t,t+1)}(\cdot)$ of the storage over the course of the day, when the instantaneous release is $N_t^{\min}(\cdot)$. This is obtained by integrating the differential equation

$$\frac{ds}{d\xi} = \frac{a_{t+1}}{\Delta} - N_t^{\min}(s) \tag{6.25}$$

(instantaneous mass balance) over the interval $[t, t+1)$ with the initial condition s_t. Once the trajectory of the storage has been obtained, the volume v_t is given by the following integral

$$v_t(s_t, a_{t+1}) = \int_t^{t+1} N_t^{\min}\big(s_{[t,t+1)}(\xi)\big)\, d\xi \tag{6.26}$$

Observe that the function $v_t(\cdot)$ is periodic with time, since $N_t^{\min}(s)$ is such.

The maximum volume V_t that can be released is obtained with a similar procedure, provided that $N_t^{\min}(\cdot)$ is replaced by $N_t^{\max}(\cdot)$ in equations (6.25) and (6.26).

Once the functions $v_t(\cdot)$ and $V_t(\cdot)$ are known, we can express the release r_{t+1} as a function of the release decision u_t^{MIO} assumed at time t, the storage s_t at the beginning of the day and the inflow volume a_{t+1}, i.e.

$$r_{t+1} = R_t\big(u_t^{\text{MIO}}, s_t, a_{t+1}\big) = \begin{cases} v_t(s_t, a_{t+1}) & \text{if } u_t^{\text{MIO}} < v_t(s_t, a_{t+1}) \\ V_t(s_t, a_{t+1}) & \text{if } u_t^{\text{MIO}} > V_t(s_t, a_{t+1}) \\ u_t^{\text{MIO}} & \text{otherwise} \end{cases} \tag{6.27}$$

The function $R_t(\cdot)$ is called *release function*. It is based on the assumption that the Regulator operates the regulation gates in such a way that the release r_{t+1} coincides with the release decision u_t^{MIO}, if possible, and, if not, takes the closest feasible value. This behaviour is known in the literature as *normal behaviour*. We do not consider other hypotheses of behaviour.

From the periodicity of $v_t(\cdot)$ the periodicity of the release function $R_t(\cdot)$ clearly follows

$$R_t(\cdot) = R_{t+kT}(\cdot) \quad t = 0, 1, \ldots; \; k = 1, 2, \ldots \tag{6.28}$$

Figure 6.11 shows a section of the function $R_t(u_t^{\text{MIO}}, s_t, a_{t+1})$ when the value of a_{t+1} is fixed. The reader should note that on the ordinate axis of this graph the average daily flow rate is reported, which, as we have said, is a way to express the volume that is released over the day, while in Figure 6.10 the instantaneous flow rate appears.

We can use $R_t(\cdot)$ to rewrite equation (6.21) by highlighting the dependence of the release on the inflow and the control

$$s_{t+1} = s_t + a_{t+1} - R_t\big(u_t^{\text{MIO}}, s_t, a_{t+1}\big) \tag{6.29}$$

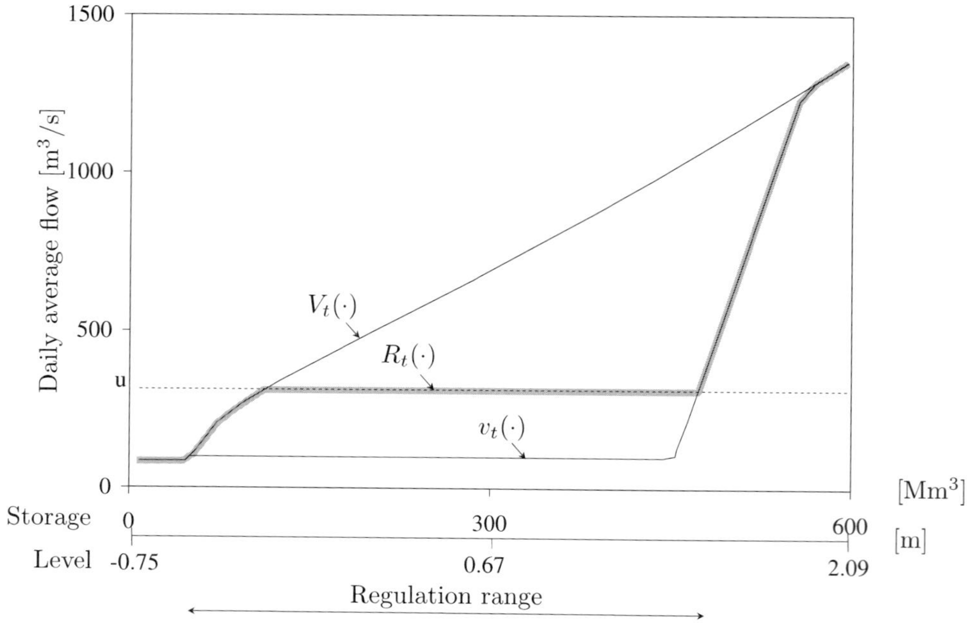

Figure 6.11: A section of the release function $R_t(\cdot)$ (grey line) when a_{t+1} is equal to 300 m^3/s.

6.6.6 The release decision

Each morning the Regulator has to establish the volume u_t^{MIO} to be released from the lake over the following 24 hours. The task of the policy, once it is determined, will be to suggests the 'best' values to the Regulator each day, but in order to design the policy it is necessary to specify the set $U_t^{\text{MIO}}(s_t)$ of values from which the 'best' can be selected. The lower extreme of this set will certainly be equal to zero, because clearly water cannot be transferred from the River Ticino to the lake. The upper extreme could be set at infinity because, even if the decision were to release a physically impossible volume of water, equation (6.27) would produce a feasible release. Nevertheless, by doing so, in the solution of the Control Problem we would have to evaluate and compare an infinite number of possibilities at every time instant (even if the decision were, as it will be, *discretized*). Therefore $U_t^{\text{MIO}}(s_t)$ must have an upper bound. This could be the maximum volume $V_t(s_t, a_{t+1})$ that can be released on day t, if it were not a random variable because of the uncertainty in the inflow. Nevertheless, in Section 7.8.3 we will see that, for the practical resolution of the Control Problem, we must assume that the probability distribution of a_{t+1} is limited, i.e. that there exists a maximum inflow $\bar{a}_{t+1}$ over which the probability density is equal to zero. Then we can assume the maximum volume $V_t(s_t, \bar{a}_{t+1})$ that can be released on day t if the maximum inflow $\bar{a}_{t+1}$ occurs, given the initial condition of lake storage s_t at time t, as an upper bound of $U_t^{\text{MIO}}(s_t)$. Since there is no reason to exclude decisions between these two extremes, we can conclude that

$$U_t^{\text{MIO}}(s_t) = \left\{u_t^{\text{MIO}}\colon\ 0 \leqslant u_t^{\text{MIO}} \leqslant V_t(s_t, \bar{a}_{t+1})\right\} \tag{6.30}$$

The four equations (6.29), (6.27), (6.20) and (6.30) provide a complete description of the lake and thus constitute the model we were looking for. Note that the dependency on the

Table 6.5. Estimates of the time constant [days] according to two different linearization methods

Relation	Tangent at $h^{SC} = 0.5$ m	Best interpolator in $[-0.5, +1.5]$
SD_{CUR}	6.9	5.0
SD_{+600}	6.5	4.5

alternative under study is expressed by the function $R_t(\cdot)$, since this function depends on $v_t(\cdot)$ and $V_t(\cdot)$, which in turn depend on the actions considered in the alternative.

6.6.7 The time constant of the lake

As explained in Section 6.3, once the model of the lake has been identified we must ascertain whether it satisfies the *Sampling Theorem*, i.e. we must verify that the modelling time step Δ, which in our case is equal to one day, is shorter than the smallest time constant of the system, by about one order of magnitude. We must thus compute the time constant of the lake in free regime, since it is in such conditions that fastest decreases in the lake level can occur and thus the choice of the time step is more critical.

In free regime conditions, the dynamics of the storage is described by an equation similar to (6.29), in which $R_t(\cdot)$ is replaced by $N_t^{\max}(\cdot)$. This equation is non-linear, because the instantaneous storage–discharge relation $N_t^{\max}(\cdot)$ is non-linear. Therefore, in order to define the time constant, we must first linearize the model (see Section 6.3). This is equivalent to linearizing the storage–discharge relation, i.e. approximating it with a function of the following form

$$\tilde{N}(s) = \alpha s + \beta \tag{6.31}$$

The time constant of the model is thus

$$T_c = \frac{1}{\alpha} \tag{6.32}$$

The linearization of the storage–discharge relation can be carried out in two different ways. In the first, the storage–discharge relation is approximated by its tangent evaluated at the level value that is most frequent. In the second, the straight line is determined that best approximates it in a given interval of levels. We have explored both of the possibilities, by adopting the level 0.5 m for the first method, and the interval $[-0.5, +1.5]$ for the second, which corresponds to the enlarged regulation range. The corresponding values of the lake's time constant are reported in Table 6.5. As the reader can see, these values vary according to the linearization method that is adopted and the stage–discharge relation being considered, but in all cases the condition imposed by the Sampling Theorem is sufficiently well satisfied. Therefore, adopting a time step of one day does not cause a significant loss of information.

6.7 The diversion nodes

As we have already observed, the three diversion nodes that appear in the logical scheme of the model (see Figure 6.3) do not correspond to diversion dams in the physical system, but together reproduce the decision process with which the release r_{t+1} is distributed among the users and the Ticino. Each diversion node is described by an algebraic relation which links its outflows to its inflows. We will examine the three nodes in their logical sequence.

6.7.1 The MEF diversion node

The volume that enters this diversion node is the release r_{t+1} from the Miorina dam and its aim is to meet the MEF constraint. The constraint requires that a water volume q_t^{MEF} be left to the Ticino, which is possible only when the release r_{t+1} is not less than q_t^{MEF}. Otherwise, the best that can be done is to let the entire release r_{t+1} into the river. In other words, the volume $q_{t+1}^{\text{TIC,MEF}}$ that is left to the Ticino in the interval $[t, t+1)$ in compliance with the MEF is defined by the following expression

$$q_{t+1}^{\text{TIC,MEF}} = \min\{q_t^{\text{MEF}}, r_{t+1}\} \tag{6.33}$$

Consequently the remaining volume for distribution among the users is $r_{t+1} - q_{t+1}^{\text{TIC,MEF}}$.

6.7.2 The environmental diversion node

We explained in Section 5.2.2 that some alternatives might include the possibility for the Regulator to leave to the Ticino not only the volume $q_{t+1}^{\text{TIC,MEF}}$ imposed by the MEF, but also an additional volume u_t^{TIC}, with the aim of improving the quality of the river environment. The environmental allotment decision could only be implemented when the release is sufficient to guarantee a water volume q^{Turb} for the Industrial Canal, which is necessary for cooling the Turbigo-EUROGEN power plant. In fact, as we explained on page 83, the damage that would occur if the running of the plant were interrupted would be extremely high.

Given that $r_{t+1} - q_{t+1}^{\text{TIC,MEF}} - q^{\text{Turb}}$ is the residual volume available after having complied with the MEF and satisfying the Turbigo-EUROGEN water demand, the volume $q_{t+1}^{\text{TIC},u}$ delivered to the Ticino following the decision u_t^{TIC} is given by

$$q_{t+1}^{\text{TIC},u} = \max\{\min[u_t^{\text{TIC}}, r_{t+1} - q_{t+1}^{\text{TIC,MEF}} - q^{\text{Turb}}], 0\} \tag{6.34}$$

Consequently the residual volume r_{t+1}^r that can be distributed among the other water users is

$$r_{t+1}^r = r_{t+1} - q_{t+1}^{\text{TIC,MEF}} - q_{t+1}^{\text{TIC},u} \tag{6.35}$$

6.7.2.1 The environmental decision

The set $U_t^{\text{TIC}}(s_t)$ of values, within which the decision u_t^{TIC} can be chosen, clearly cannot contain negative values or values that are higher than the release decision u_t^{MIO}; therefore

$$U_t^{\text{TIC}}(s_t) = \{u_t^{\text{TIC}}: 0 \leqslant u_t^{\text{TIC}} \leqslant u_t^{\text{MIO}}\} \tag{6.36}$$

The set U_t^{TIC} depends on the storage s_t since its upper extreme u_t^{MIO} (release decision) depends on s_t through the control law.

6.7.3 The users' distribution node

The diversion node for the distribution of water volumes among the various users is described by the *distribution policy*, to which we have already referred in Section 3.4.

6.7.3.1 The distribution policy

A policy for distributing the release among the water users and the River Ticino has never been codified in a written form. The distribution is decided by the Regulation Committee of the Ticino Consortium, which is composed of its director and the representatives from the Irrigation Consortia and ENEL, who come together whenever the existing or expected conditions require. Interviews with the members of the Committee, and the analysis of the time series of daily allotment made in the years 1981–1994, allowed us to identify a quantitative policy that, given the time series of releases from the Miorina dam, reproduces the time series of the flows to the canals and the Ticino very well. The procedure we used to identify this policy and the description of how it was validated are presented in Section 6.7.3 of the DVD; here we simply describe its form.

At each time instant t, the policy provides the *allotment vector*

$$\left| q_{t+1}^{\mathrm{REC}}, q_{t+1}^{\mathrm{VC}}, q_{t+1}^{\mathrm{IC}}, q_{t+1}^{\mathrm{TIC},r} \right|$$

as a function of the release by means of one of the following five *allotment rules*, which are selected based on the value of the volume r_{t+1}^{r} to be distributed:

1. Total satisfaction

The overall water demand W_t of the irrigators and hydropower producers is

$$W_t = W_t^{\mathrm{REC}} + W_t^{\mathrm{VC}} + W_t^{\mathrm{IC}} \tag{6.37}$$

where W_t^{REC}, W_t^{VC} and W_t^{IC} are the total demands of the users that are served by the Regina Elena, Villoresi and Industrial Canals respectively. When

$$r_{t+1}^{r} \geqslant W_t \tag{6.38}$$

all the water users can be satisfied without difficulty. The water is distributed by allocating to each user the amount that meets the demand, and any residual volume $q_{t+1}^{\mathrm{TIC},r}$ is left to the Ticino. The allotment rule is therefore given by

$$q_{t+1}^{\mathrm{REC}} = W_t^{\mathrm{REC}} \tag{6.39a}$$

$$q_{t+1}^{\mathrm{VC}} = W_t^{\mathrm{VC}} \tag{6.39b}$$

$$q_{t+1}^{\mathrm{IC}} = W_t^{\mathrm{IC}} \tag{6.39c}$$

$$q_{t+1}^{\mathrm{TIC},r} = r_{t+1}^{r} - W_t \tag{6.39d}$$

2. Hydropower reduction

When the volume r_{t+1}^{r} is inferior to the overall demand W_t the users cannot be completely satisfied. Therefore it is necessary to reduce the supply to several or all of them. ENEL and the East Sesia Consortium have agreed to be the first to reduce the flow diverted into their canals, up to a maximum reduction that is equal to the difference between the overall demand and the irrigation demand. They agree to reduce the flow because the Abstraction Licences for hydropower use were considered, by tacit agreement, subordinate to irrigation ones (see Section 7.6.3), even before the advent of the Galli Law (1994), which explicitly states this subordination. The iterative procedure with which the supply is reduced and subdivided between the two users (which we will now illustrate) is based on the different 'seniority' of their rights.

The maximum acceptable reductions for each of the two canals are

$$q_t^{\text{maxred,IC}} = W_t^{\text{IC}} - W_t^{\text{IC,irr}} \tag{6.40a}$$
$$q_t^{\text{maxred,REC}} = W_t^{\text{REC}} - W_t^{\text{REC,irr}} \tag{6.40b}$$

where $W_t^{\text{REC,irr}}$ and $W_t^{\text{IC,irr}}$ are the irrigation demands of the East Sesia Consortium and the Old Users, which are directly or indirectly satisfied by the Industrial Canal. We denote with $q_t^{\text{maxred,hyd}}$ the sum of such supply reductions.

If

$$W_t - q_t^{\text{maxred,hyd}} \leqslant r_{t+1}^r < W_t \tag{6.41}$$

the allotment rule is the following

$$q_{t+1}^{\text{REC}} = W_t^{\text{REC}} - q_{t+1}^{\text{red,REC}} \tag{6.42a}$$
$$q_{t+1}^{\text{VC}} = W_t^{\text{VC}} \tag{6.42b}$$
$$q_{t+1}^{\text{IC}} = W_t^{\text{IC}} - q_{t+1}^{\text{red,IC}} \tag{6.42c}$$
$$q_{t+1}^{\text{TIC},r} = 0 \tag{6.42d}$$

where the amount of the reductions $q_{t+1}^{\text{red,REC}}$ and $q_{t+1}^{\text{red,IC}}$ is established with the following iterative procedure. Firstly, reduce the abstraction of the Regina Elena Canal, up to a maximum which corresponds to an average daily flow of 25 m^3/s. If r_{t+1}^r is greater than the overall demand reduced by that volume (i.e. if $r_{t+1}^r > W_t - 25 \cdot 86\,400$), a reduction $q_{t+1}^{\text{red,REC}}$ equal to $W_t - r_{t+1}^r$ is assigned to the Regina Elena Canal. Otherwise, the abstraction of the Industrial Canal must be reduced too, with the constraint that such reduction does not exceed an average daily flow of 18.5 m^3/s. If, after these steps, the total reduced demand is still higher than the amount of water available, the flow provided to the Regina Elena Canal is further diminished, up to the maximum reduction $q_t^{\text{maxred,REC}}$. If this is still not sufficient, the abstraction of the Industrial Canal is also diminished to the maximum reduction value $q_t^{\text{maxred,IC}}$.

3. Irrigation reductions

When the hydropower reduction is not sufficient to achieve equilibrium between demand and supply, it is also necessary to reassess the distribution of the irrigation volume. In these conditions the East Ticino Consortium, which has the best ratio between Abstraction Licence and current demand, agrees to reduce the abstraction up to a maximum value $q_t^{\text{maxred,VC}}$. Therefore, when

$$W_t - q_t^{\text{maxred,hyd}} - q_t^{\text{maxred,VC}} \leqslant r_{t+1}^r < W_t - q_t^{\text{maxred,hyd}} \tag{6.43}$$

the allotment rule is the following

$$q_{t+1}^{\text{REC}} = W_t^{\text{REC}} - q_t^{\text{maxred,REC}} \tag{6.44a}$$
$$q_{t+1}^{\text{VC}} = r_{t+1}^r - q_{t+1}^{\text{REC}} - q_{t+1}^{\text{IC}} \tag{6.44b}$$
$$q_{t+1}^{\text{IC}} = W_t^{\text{IC}} - q_t^{\text{maxred,IC}} \tag{6.44c}$$
$$q_{t+1}^{\text{TIC},r} = 0 \tag{6.44d}$$

4. The conflict

Now we will examine the case in which the volume to be distributed is so low that the previous reductions are not sufficient to reach an equilibrium between water demand and supply. Since Verbano was regulated, this condition has rarely occurred (the last time was in September 1995). This is why the members of the Regulation Committee are not able to define a precise procedure, as they did for the previous cases. They say that they come to an agreement on a case-by-case basis, taking into account the particular conditions at that moment, "*through the application of an unwritten code of behaviour … in relation to the trend of the crisis*" (Maione and Mignosa, 1995). Nevertheless, we must identify an allotment rule for this case as well, and to do this we used the following two assumptions:

- the minimum flow required by the Turbigo-EUROGEN power plant is guaranteed for the Industrial Canal, whenever it is physically possible;
- the remaining available volume is distributed among the canals by adopting three distribution coefficients $\alpha_t^{\text{REC}}, \alpha_t^{\text{VC}}, \alpha_t^{\text{IC}}$ (such that $\alpha_t^{\text{REC}}+\alpha_t^{\text{VC}}+\alpha_t^{\text{IC}} \leqslant 1$) for each t, whose value varies in time, depending on the state of development of the crops in the irrigation districts.

We estimated the values of these parameters using the time series of daily allotment in the years 1981–1994 and validated them with an indirect cross-examination (Section 6.7.3 of the DVD).

With these assumptions, when

$$q^{\text{Turb}} \leqslant r_{t+1}^r < W_t - q_t^{\text{maxred,hyd}} - q_t^{\text{maxred,VC}} \tag{6.45}$$

the allotment rule is the following

$$q_{t+1}^{\text{REC}} = \alpha_t^{\text{REC}}\left(r_{t+1}^r - q^{\text{Turb}}\right) \tag{6.46a}$$

$$q_{t+1}^{\text{VC}} = \alpha_t^{\text{VC}}\left(r_{t+1}^r - q^{\text{Turb}}\right) \tag{6.46b}$$

$$q_{t+1}^{\text{IC}} = \alpha_t^{\text{IC}}\left(r_{t+1}^r - q^{\text{Turb}}\right) + q^{\text{Turb}} \tag{6.46c}$$

$$q_{t+1}^{\text{TIC},r} = r_{t+1}^r - \left(q_{t+1}^{\text{REC}} + q_{t+1}^{\text{VC}} + q_{t+1}^{\text{IC}}\right) \tag{6.46d}$$

Equations (6.46) do not yet provide a correct description of the allotment rule, however, because they do not take into account the fact that the Villoresi Canal has a minimum operative flow $q_t^{\text{ope,VC}}$, which is imposed by its hydraulic functioning, and below which it is not possible to feed all the inlets of the canal. Consequently, the Villoresi Canal will actually receive what was established by equation (6.46b) only when that volume is greater than $q_t^{\text{ope,VC}}$. Otherwise the abstraction of the Villoresi Canal is limited to the volume $q_t^{\text{min,VC}}$, which guarantees the survival of the fish that live in the canal, and the remaining volume is left to the Regina Elena Canal. By doing so the East Ticino Consortium, which gets water from the Villoresi Canal, acquires a credit from the East Sesia Consortium, which is fed by the Regina Elena Canal. The credit will be repaid next time a conflict occurs: the East Sesia Consortium will reduce its abstraction, temporarily relinquishing a part of its rights to the East Ticino Consortium. The debt is cancelled also when the lake reaches the upper extreme of the regulation range. As we have said, the case of conflict has very rarely occurred and even more rarely has it been necessary to turn to this practice of 'credit repayment'. For this reason we decided to take it into account in a simplified way, by substituting equations (6.46a) and (6.46b) with the following

$$q_{t+1}^{\mathrm{VC}} = \begin{cases} \alpha_t^{\mathrm{VC}}(r_{t+1}^r - q^{\mathrm{Turb}}) & \text{if } \alpha_t^{\mathrm{VC}}(r_{t+1}^r - q^{\mathrm{Turb}}) \geqslant q_t^{\mathrm{ope,VC}} \\ q_t^{\mathrm{min,VC}} & \text{otherwise} \end{cases} \tag{6.47a}$$

$$q_{t+1}^{\mathrm{REC}} = \alpha_t^{\mathrm{REC}}\left(r_{t+1}^r - q^{\mathrm{Turb}}\right) + q_{t+1}^{\mathrm{VC,REC}} \tag{6.47b}$$

where $q_{t+1}^{\mathrm{VC,REC}}$ is the volume transferred by the East Ticino Consortium to the East Sesia Consortium, defined by the following relation

$$q_{t+1}^{\mathrm{VC,REC}} = \begin{cases} 0 & \text{if } \alpha_t^{\mathrm{VC}}(r_{t+1}^r - q^{\mathrm{Turb}}) \geqslant q_t^{\mathrm{ope,VC}} \\ \alpha_t^{\mathrm{VC}}(r_{t+1}^r - q^{\mathrm{Turb}}) - q_t^{\mathrm{min,VC}} & \text{otherwise} \end{cases} \tag{6.47c}$$

5. The extreme case

To complete the definition of the distribution policy, we must finally establish what happens when equation (6.45) is not satisfied, i.e. when

$$r_{t+1}^r < q^{\mathrm{Turb}} \tag{6.48}$$

In practice, not only has this situation never occurred in the past, but it has never occurred with all the alternatives that we have examined. Nevertheless, the distribution policy must be complete, which means it must account for all the possibilities, and so we must specify what would happen if this did occur. Given that none of the users could be satisfied, we assumed that the whole volume r_{t+1}^r would be left to the Ticino.

6.7.3.2 The mass balance in the distribution nodes

In the description of the distribution policy we saw that the volume r_{t+1}^r is distributed among the Regina Elena, Villoresi, and Industrial Canals and the River Ticino. The allotment rules satisfy the mass balance equation

$$r_{t+1}^r = q_{t+1}^{\mathrm{REC}} + q_{t+1}^{\mathrm{VC}} + q_{t+1}^{\mathrm{IC}} + q_{t+1}^{\mathrm{TIC},r} \tag{6.49}$$

However, this does not appear in the logical scheme in Figure 6.3 (see the five arcs coming out from the user's distribution node 3), for the reasons illustrated on page 141.

6.7.3.3 Use of the distribution policy in the alternatives design

In the distribution policy several parameters appear: some are reference parameters (the water demand and the maximum acceptable reduction for each of the water users), others express the decision of the Committee (the distribution coefficients). The estimation of the latter was carried out by using the historical values of the former (Section 6.7.3 of the DVD). However, when the distribution policy is used for designing and evaluating the alternatives, it is not opportune to assign the historical values to the reference parameters; it is preferable to adopt those values that are expected to occur in the future, according to the *design* and *evaluation scenarios* (see Sections 7.5 and 8.3). By doing so, we are assuming that the approach with which the distribution was carried out in the past (expressed by the five cases described above and by the values of the distribution coefficient) will be maintained in the future, while the reference parameters are updated for the conditions that the Regulation Committee is likely to deal with.

6.8 Confluence points

In the model there are two confluence points. The first reconstructs the volume q_{t+1}^{TIC} released into the Ticino

$$q_{t+1}^{\mathrm{TIC}} = q_{t+1}^{\mathrm{TIC,MEF}} + q_{t+1}^{\mathrm{TIC},u} + q_{t+1}^{\mathrm{TIC},r} \tag{6.50}$$

where, as we have said, $q_{t+1}^{\mathrm{TIC,MEF}}$ is the volume provided in compliance with the MEF, $q_{t+1}^{\mathrm{TIC},u}$ is the volume that, according to the regulation choice, is added to the first to improve the river quality, and $q_{t+1}^{\mathrm{TIC},r}$ is the residual volume after the distribution among the canals.

The second confluence point reconstructs the volume q_{t+1}^{PT} provided for the Porto della Torre power plant

$$q_{t+1}^{\mathrm{PT}} = q_{t+1}^{\mathrm{IC}} + q_{t+1}^{\mathrm{VC}} + q_{t+1}^{\mathrm{TIC}} \tag{6.51}$$

From this equation, by recalling equations (6.35), (6.49) and (6.50), the following can be derived

$$q_{t+1}^{\mathrm{IC}} + q_{t+1}^{\mathrm{VC}} + q_{t+1}^{\mathrm{TIC}} = r_{t+1} - q_{t+1}^{\mathrm{REC}} \tag{6.52}$$

which reflects, as it should, the position of the power plant, which is situated between the intake of the Regina Elena Canal and the Panperduto diversion dam (Figure 6.2).

6.9 The sectors and the indicators

In Section 4.4.1 we explained that the aim of the model is to provide the variables that are required for computing the indicators. The latter were already defined in Chapter 4, but we would like to revisit them here in the light of what we have understood so far.

An *indicator* i is a functional of the trajectories of the state, of the control and of the disturbance that acts on the system, over a time period $t = 0, \ldots, h$, which is called *horizon* (either *evaluation horizon* or *design horizon* according to the case), i.e.

$$i = i(\mathbf{x}_0, \mathbf{x}_1, \ldots, \mathbf{x}_h; \mathbf{u}_0, \mathbf{u}_1, \ldots, \mathbf{u}_{h-1}; \varepsilon_1, \varepsilon_2, \ldots, \varepsilon_h)$$

Most of the functionals $i(\cdot)$ defined in Section 4.5 of the DVD are *separable*, that is they can be expressed as the combination (through a suitable operator Ψ, for example the summation) of h functions $g_t(\mathbf{x}_t, \mathbf{u}_t, \varepsilon_{t+1})$, called *step-indicator*, each of which expresses the cost[21] associated to the transition from $\mathbf{x}_t$ to $\mathbf{x}_{t+1}$ and whose arguments are only variable relative to the time interval $[t, t+1)$; and of a function $g_h(\cdot)$ whose argument is the final state $\mathbf{x}_h$, called *penalty on the final state* (this last function is sometimes missing), i.e.

$$i = \Psi\left\{g_t(\mathbf{x}_t, \mathbf{u}_t, \varepsilon_{t+1}),\ t = 0, \ldots, h-1;\ g_h(\mathbf{x}_h)\right\} \tag{6.53}$$

In the next chapter we will explain that, when the design indicators have this form, a powerful algorithm can be used to solve the Design Problem. The step-indicator g_t is a function $g_t(\mathbf{x}_t, \mathbf{u}_t, \varepsilon_{t+1})$ of the state and the control at time t, and also of the disturbance ε_{t+1}. As a consequence of the dependence on the disturbance, the cost g_t cannot be deterministically computed at the moment when the decision is taken: it must be considered as a stochastic

[21] The word 'cost' is used here as a synonym of 'measure' and one should not necessarily attribute to it the meaning of expense, or damage: in fact 'negative costs', i.e. 'benefits', are also possible.

variable. Examples of indicators of the form (6.53) are the indicators from the ENEL *Power*, *Upstream Flooding* and *Downstream Environment* sectors, defined in Section 4.3. It may be surprising that some of the step-indicators that appear in the cited indicators are expressed as functions of variables that do not belong to the triple $(\mathbf{x}_t, \mathbf{u}_t, \varepsilon_{t+1})$. An example of this is the indicator (4.5) which estimates the average annual flooded area [km^2/year] in the township j. Also in these cases, however, the arguments of the step-indicator can always be traced back to variables of the triple $(\mathbf{x}_t, \mathbf{u}_t, \varepsilon_{t+1})$, e.g., in the cited example, the level h^{SC} is computed, through equation (6.20), as a function of the lake storage s_t, which is a component of $\mathbf{x}_t$.

The step-indicator $g_t(\cdot)$ can thus be thought of as a model of the evaluation criterion to which indicator $i(\cdot)$ is associated: in fact $g_t(\cdot)$ has the same form as equation (6.7). Similarly, when designing the alternatives, the step-indicators associated to the five design indicators, which we selected in the previous chapter (see Table 5.1), can be seen as models of the five sectors considered in the Control Problem.

The functional $i(\cdot)$ cannot always be spontaneously defined in a separable form. Consider, for example, indicator (4.10) which estimates the average potential damage from stress and which we rewrite below for the reader's convenience

$$i_{\text{Irr\_D\_2}} = \frac{1}{N}\sum_{y=1}^{N} f(\digamma_y) \tag{6.54}$$

It is expressed as a function of the maximum stress $\digamma_y$ for each year y, which in turn is given by

$$\digamma_y = \max_{t\in y} \frac{1}{14} \sum_{\tau=t-13}^{t} \left(d_\tau^{VC} + d_\tau^{REC} + d_\tau^{IC}\right) \tag{6.55}$$

The indicator $i_{\text{Irr\_D\_2}}$ is not separable. This is not because in equation (6.54) the index of the summation is the year y and not the day t, since it is easy to re-write equation (6.54) in the following form

$$i_{\text{Irr\_V\_2}} = \frac{1}{N}\sum_{t=0}^{h} g_t(\digamma_t) \tag{6.56}$$

where the summation operates on days, provided that the step-indicator $g_t(\digamma_t)$ is defined as

$$g_t(\digamma_t) = \begin{cases} 0 & \text{if } t \neq \text{end of the year} \\ f(\digamma_a) & \text{otherwise} \end{cases} \tag{6.57}$$

The non-separability comes from the fact that the maximum stress $\digamma_y$ is a function of all the deficits that occurred over the course of the year and so it does not depend only on the variables relative to time t, as required by equation (6.53).

A non-separable indicator can always be made separable, provided that the state of the system is suitably enlarged. In our example, it is sufficient to define the maximum stress as

$$\digamma_{t+1} = \begin{cases} 0 & \text{if } t = \text{end of the year} \\ \max\left\{\frac{1}{14}\sum_{\tau=0}^{13} z_t^\tau, \digamma_t\right\} & \text{otherwise} \end{cases} \tag{6.58a}$$

where

$$z_t^0 = d_t^{\mathrm{VC}} + d_t^{\mathrm{REC}} + d_t^{\mathrm{IC}} \tag{6.58b}$$

$$z_t^\tau = z_{t-1}^{\tau-1} \quad \text{for } \tau = 1, \ldots, 13 \tag{6.58c}$$

and equation (6.56) turns into a separable function of the following state vector

$$\mathbf{x}_t = \left|s_t, F_t, z_t^0, \ldots, z_t^{13}\right|'$$

In fact, the three deficits that appear in equation (6.58b) can be expressed as a function of the triple $(s_t, \mathbf{u}_t, \varepsilon_{t+1})$ by means of equation (6.27) and of the equations that describe the three diversion nodes.

Once again, note that the step-indicator $g_t(\cdot)$, defined by equation (6.57), can be thought of as the output transformation of a dynamic model of the form (6.7), whose state transition function is given by equation (6.58). The model describes the *Irrigation* sector, and the vector $|F_t, z_t^0, \ldots, z_t^{13}|$ is precisely the state of the crops to which we referred in Section 5.2.4.

As we have noted, all the indicators can be seen as (static or dynamic) models of the interests (i.e. the criteria) to which they refer; and the sectors, which group such interests, can be interpreted as units of the global model of the water system. What we have seen here and the description provided in Chapter 4 are sufficient to allow the reader to reconstruct the models of all the sectors, with the exception of one, the *Downstream Flooding* sector, which we deal with in the next section.

6.9.1 The model of Downstream Flooding

The indicators associated to the criteria of the *Downstream Flooding* sector are defined as functions of the Ticino level h_t^{Pv} at Pavia (see Section 4.5.6 of the DVD), but still we do not have an expression that allows us to determine this level as a function of the release r_t at the Miorina dam.

To identify such an expression, we must first of all study the phenomenon. In flood conditions, the Miorina dam is completely opened and the lake is in free regime. Very often floods occur because of heavy precipitation, and if it rains on the plains as well, as is often the case, the farmers stop irrigating. In these conditions the flow rate diverted to the canals is reduced, to avoid flooding along their course (in fact, canals are designed with ever-decreasing sections since, in normal conditions, the flow rate decreases as the distance from the intake increases and the water is abstracted for irrigation; but if the abstraction is suspended, flooding may occur). It follows that the fraction of the release that is left to the Ticino at Panperduto is much higher than in normal conditions. As the flood wave proceeds downstream, its peak is reduced due to the fact that, while proceeding, the wave must fill the river channel in which the flow is usually low before the arrival of the flood. Finally, the flood wave reaches Pavia, where the level of the river depends not only on the flow that comes from upstream, but also on the backflow effect from the River Po. The confluence between the Ticino and the Po, at the Becca Bridge, is only 8 km downstream from the Covered Bridge (Ponte Coperto) at Pavia (where the Ticino level is measured) and the difference of level between the two sections is only about 1.50 m in normal conditions. Therefore, if in these conditions the flow of the Po increases, its waters can flow up the river channel of the Ticino, to Pavia and even further upstream; the greater the Po flow, the further upstream it goes. If at the same time the flow in the Ticino also increased, its level would have to be much higher in order for its flow to run out into the Po. This is the phenomenon that we termed *backflow effect*.

From this description we understand that the process of propagation of the flow wave is very complex and that its detailed description requires a lot of information (measures of precipitation on the plains, manoeuvres effected in the canals, ...). We must therefore be content with a simplified description. It seems reasonable to relate the level at Pavia to the release at Miorina, without taking into account the abstractions and the lateral inflow downstream from Panperduto. These approximations cause two errors which are of opposite signs and should roughly cancel each other out. We must, however, explicitly consider the state of the River Po.

Usually the propagation of flood waves is described using partial differential equations, i.e. using models that are continuous in time and space, such as the classical *de Saint Venant* equations, or one of their approximations (see for example Abbott, 1979 and Graf, 1998). The use of these models, even if they are suitably discretized, requires hydraulic data – levels and/or flow rates – measured with a very short time step, which in our case should not be longer than 2 or 3 hours. Instead, the estimates produced by the model of the Verbano system refer to average daily flow and the available time series of levels at Pavia are also composed of daily measures, so that both of them cannot be used to feed and calibrate models of this type. We must therefore abandon the classic hydraulic approach and look for a different tool. The most suitable one seemed to us to be the *neural network*, which can be thought of as a non-linear regressor. Here we provide a brief description of it and the way it was calibrated, and we recommend that interested readers refer to Section 6.9.1 of the DVD and Appendix A8 of THEORY for greater detail.

Our aim is to estimate the sequence of Ticino level h_t^{Pv} at Pavia, given a sequence of release r_t at Miorina and a sequence of flow q_t^{Po} in the Po. The neural network, whose nature is described below, allows us to do this, as long as we have some sequences of these three quantities with which the network can be 'trained', i.e. with which its parameters can be estimated. The steps to be carried out are the following:

- analyse the available time series of the three quantities h_t^{Pv}, r_t and q_t^{Po};
- choose a structure for the network;
- train it;
- validate it.

6.9.1.1 Analysing the available time series

The confluence between the Ticino and the Po is at the Becca Bridge (see Figure 6.12). Consequently, we should use a value q_t^{Po} of the flow of the Po in that section, but this should be derived from a measure of level taken a bit further upstream, where the effect of backflow of the Ticino in the Po can be considered non-influential. Unfortunately, the first hydrometric station on the Po upstream of the Becca Bridge is located upstream of the confluence with the Sesia River (point A in Figure 6.12), and so the flow data that it provides have no significance for our aims, because they do not contain the contributions from significant tributaries: the Sesia River itself and the Tanaro River. We are thus obliged to use a hydrometric station downstream from the confluence. The closest is the station at Piacenza (point B), which has the defect that the flow rate registered at that point also

Figure 6.12: Hydrometric stations on the River Po.

depends on the flow from the Ticino. In this way a sort of vicious circle is generated, which we are forced to accept, as we have no other solutions.

The time series available are thus the following:

- level $h_t^{\mathrm{Pv,his}}$ of the River Ticino at the Covered Bridge hydrometric station (Pavia), measured at 12:00 noon on day t, in the years 1943–1998;
- average daily release r_{t+1}^{his} at Miorina between 8:00 a.m. on day t and 8:00 a.m. on day $t+1$, in the years 1943–1998;
- average daily flow $q_{t+1}^{\mathrm{Po,his}}$ of the River Po at Piacenza hydrometric station, between 12:00 midnight on day t and 12:00 midnight on day $t+1$, in the years 1926–1997.

With this data we want to train a neural network that provides the level of the Ticino at Pavia at 12 noon, given the values of the other variables at a certain time instant, which we still have to specify. The network will be used to estimate the sequence of levels of the Ticino at Pavia, in correspondence to each alternative. In order that the estimate be reliable, the time series used in the training must reflect a stationary situation, i.e. they must not contain *trends*. If this condition is not satisfied, the origin of the trend must be identified and the trend must be removed.

Figure 6.13 shows pairs of values $(r_{t+1}^{\mathrm{his}}, h_t^{\mathrm{Pv,his}})$ that were measured during the flood events (including their initial and final phases) registered from 1943 to 1998. Different shades of grey are used to specify the decade in which the event falls (lighter for more recent events and darker for more distant ones). Notice that, in general, the level increases as the release increases, but also that, in correspondence to the same release, the level can vary in a range of about 2.5 m. This dispersion is due in part to the backflow effect caused by the Po and in part to the *loop rating* nature of the stage–discharge relation in dynamic conditions.[22] Neither of these phenomena depend on the year being considered; nonetheless, the points in Figure 6.13 corresponding to more recent years are in general lower on the graph than those corresponding to more distant years. Therefore, there must be a third

[22]During a flood event, the relation between level and flow in a given section is not a one-to-one relation but it varies according to whether the flow is increasing or decreasing.

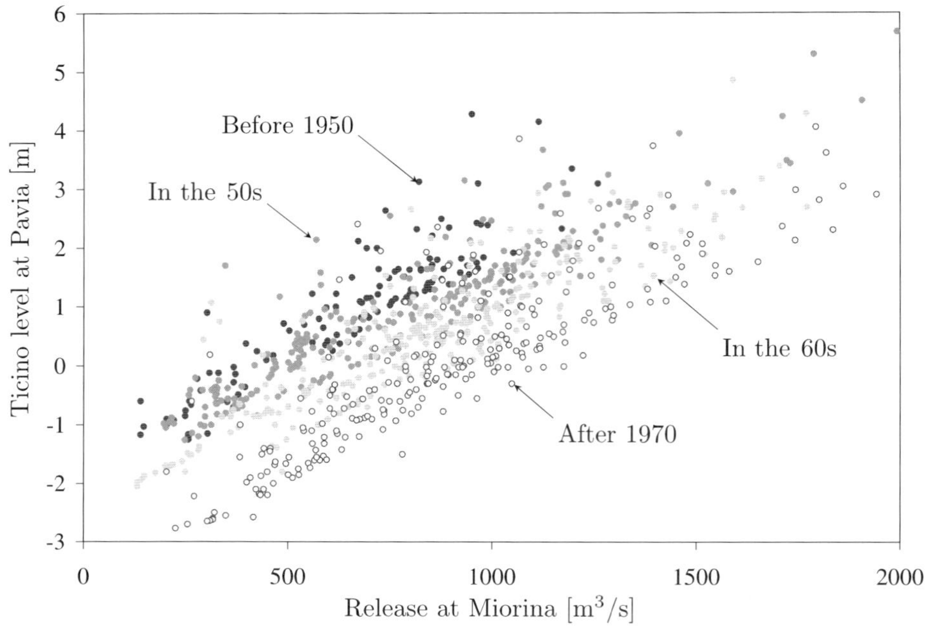

Figure 6.13: Ticino level $h_t^{\mathrm{Pv,his}}$ at the Covered Bridge (Pavia) vs release at Miorina.

cause of the dispersion, whose nature is non-stationary and which must be identified and removed if possible.

The analysis of the data from the periodical measurements carried out in the years 1957–1972 by the Hydrographic Office to update the stage–discharge relation at the Covered Bridge (see Section 6.9.1 of the DVD) revealed that the observed behaviour is produced by the erosion of the river channel, probably induced by an analogous phenomenon occurring along the Po. Because of this erosion, the level $h_t^{\mathrm{Pv,his}}$ depends on the elevation Q_t^{bot} of the bottom of the river channel at the moment of the measurement. To remove the *trend* from the original time series of level $h_t^{\mathrm{Pv,his}}$, the latter must be transformed into the level h_t^{Pv} that would have been measured if the elevation of the bottom had always stayed at a reference value ($Q_{\mathrm{ref}}^{\mathrm{bot}}$), which we will assume to be equal to the elevation of the river bed in 1997, the year in which the last measure was taken. The level h_t^{Pv} can thus be computed as

$$h_t^{\mathrm{Pv}} = h_t^{\mathrm{Pv,his}} - \Theta_t \tag{6.59}$$

where Θ_t is the cumulative lowering of the bottom between the year in which day t falls and 1997. The trend of the lowering is shown in Figure 6.14, along with the measures used for its estimation.

The neural network will be trained and validated using the series h_t^{Pv}, and so the level estimates that it will provide will refer to the elevation of the bottom in 1997, which we will assume remains unvaried for the entire *evaluation horizon* (Section 8.2).

The available data series has been subdivided in two parts: the first is used for the training of the network and the second for its validation.

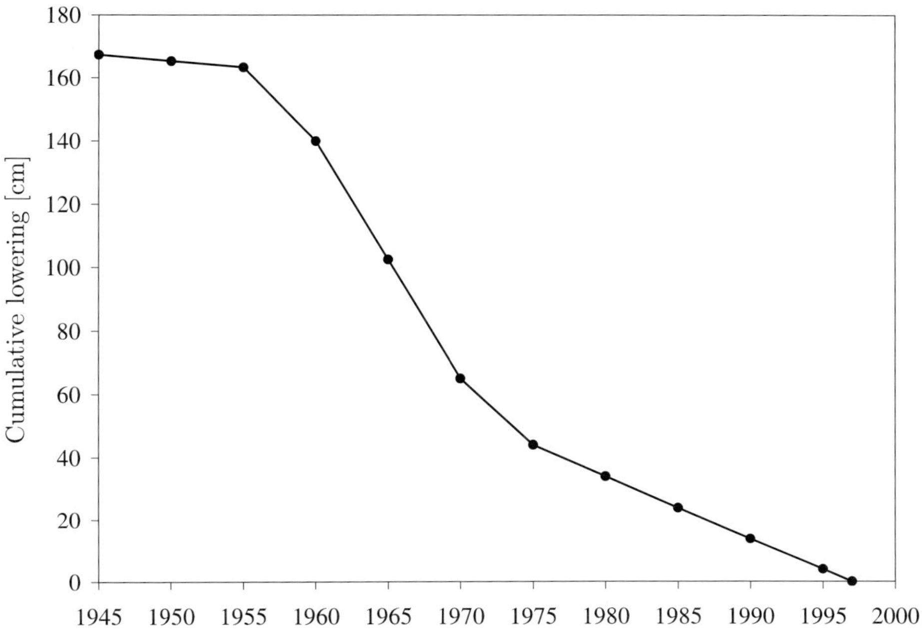

Figure 6.14: The cumulative lowering Θ_t of the bottom of the River Ticino channel (continuous line) and the measures used for its estimation (points).

6.9.1.2 The architecture of the neural network

The architecture of the neural network is shown in Figure 6.15. It is composed of two levels: a hidden layer, constituted of n sigmoid neurons, and an output layer, composed of one linear neuron. The activation functions associated to the neurons in the first layer are hyperbolic tangents, while in the neuron belonging to the second layer the activation function is linear. The relation between the vector $\mathbf{I}_t$ of the input variables and the output, i.e. the level h_t^{Pv} of the Ticino at Pavia, is thus the following

$$h_t^{\mathrm{Pv}} = \boldsymbol{\Gamma}^2\big[\mathbf{tanh}\big(\boldsymbol{\Gamma}^1\mathbf{I}_t + \boldsymbol{\beta}^1\big)\big] + \beta^2 \tag{6.60}$$

where $\boldsymbol{\Gamma}^1[n, m]$ and $\boldsymbol{\Gamma}^2[n, 1]$ are the matrices of the weights of the connections among the neurons, m the number of input variables, $\boldsymbol{\beta}^1[n]$ the vector of the thresholds associated to the neurons in the hidden layer, $\boldsymbol{\beta}^2$ the threshold of the neuron in the output layer and **tanh** a vectorial operator that transforms every component of its vector argument through the hyperbolic tangent function.

6.9.1.3 Inputs and outputs of the neural network

The output of the neural network is the level h_t^{Pv} of the Ticino at Pavia, at 12 noon. The input vector $\mathbf{I}_t$ should contain the average daily release at Miorina and the flow rate of the Po. Both of them, if necessary, can be considered at more than one time instant, to take account of the dynamical nature of the detention process. However, we cannot a priori select the time instant (or instants) at which each input must be considered. The level observed at Pavia at 12 noon on day t, in fact, is influenced by the flow at Miorina and at Piacenza at previous

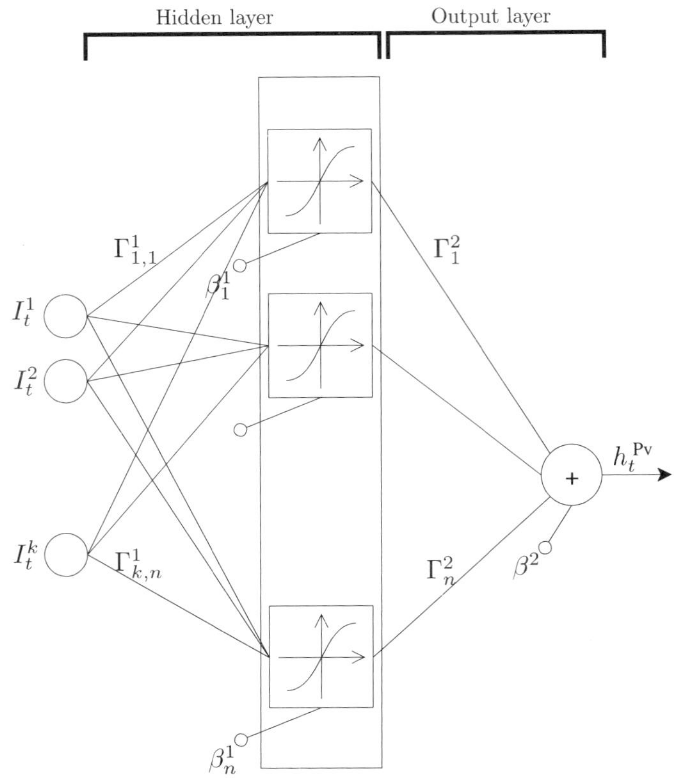

Figure 6.15: The architecture of the neural network used to represent the propagation of the flood wave from Miorina to Pavia.

time instants: the latter must be suitably chosen in order to take into account the travel times of the waves (descending and ascending respectively) along the Ticino. We only have rough estimates of those times, but even if they were more precise, we would not know how to use them, since we do not have continuous measures of the flow at Miorina and at Piacenza, but only the average daily value. Another complication is due to the nature of the available data: as shown in Figure 6.16, while r_t is the average flow in the time interval between 8:00 a.m. on day $t-1$ and 8:00 a.m. on day t, q^{Po}_{t+1} is the average between 12:00 midnight on day t and 12:00 midnight on day $t+1$.

We must therefore define the vector $\mathbf{I}_t$ of the inputs by trial and error, to determine both the time delays and the number of time instants at which to consider each variable.[23] We experimented with the following vectors: $(r_t, q^{\text{Po}}_{t+1})$, $(r_{t-1}, r_t, q^{\text{Po}}_{t+1})$, $(r_t, q^{\text{Po}}_t, q^{\text{Po}}_{t+1})$, $(r_{t-1}, r_t, q^{\text{Po}}_t, q^{\text{Po}}_{t+1})$. Clearly, the results do not only depend on the choice of $\mathbf{I}_t$, but also on the number n of sigmoid neurons considered; thus also the effect of this value was explored by trial and error. The goodness of the different combinations is judged by observing

[23] Note that when a neural network is used to simulate the propagation of the flood wave (as in this case), it is possible to include future values of the flow, i.e. relative to time instants successive to t, among the inputs, and assign deterministic values to them. These values are known because what we are simulating is relative to the past. If, instead, the network was to be used as a predictor, these inputs could not be considered.

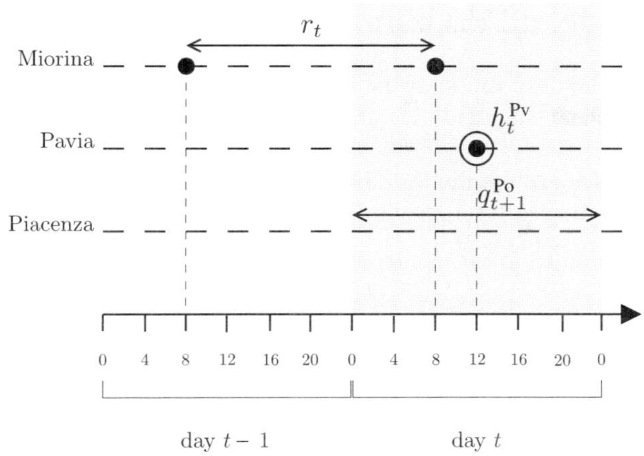

Figure 6.16: The time instant in which the level h_t^{Pv} at Pavia is measured, and the time intervals over which the release r_t at Miorina and the flow q_{t+1}^{Po} at Piacenza are averaged.

the performances obtained with respect to different indicators, in both training and validation. These indicators must consider different aspects, not only the precision with which the levels are computed, but also how many times a flood simulated by the model actually occurred in reality, and how many times the simulation was mistaken.

6.9.1.4 The configuration of the neural network

The network that provided the best results in terms of standard deviation between simulated and measured peaks has the vector $\mathbf{I}_t = (r_t, q_t^{\mathrm{Po}}, q_{t+1}^{\mathrm{Po}})$ as an input and uses 5 neurons in the hidden layer. Its performances are reported in Table 6.6.

6.9.2 The model of Irrigation

Table 6.6. The performances of the adopted neural network in training and in validation

	Training 46 floods	Validation 33 floods
Standard deviation of the peak error [m]	0.22	0.31
Correlation between the peaks	0.98	0.97
Maximum peak error [m]	0.53	0.70
Minimum peak error [m]	0.00	0.05
Correctly simulated flooding [%]	80.00	82.00
Not simulated flooding [%]	20.00	18.00
Simulated flooding that did not occur [%]	3.00	13.00

6.10 The model of Verbano water system

We have completed the description of the models of the single units, and we would like to summarize what has been said and give an overall view of the model of the Verbano water system.

The input variables are the random disturbance ε_{t+1}, which drives the inflow process through equation (6.17), and the control $\mathbf{u}_t$, constituted by the pair of decisions $(u_t^{\mathrm{MIO}}, u_t^{\mathrm{TIC}})$. The state of the system is the vector

$$\mathbf{x}_t = |s_t, \mathcal{F}_t, z_t^0, \ldots, z_t^{13}|' \tag{6.61}$$

where s_t is the storage of the lake, $\mathcal{F}_t$ is the maximum stress that the crops are subjected to until time t and $z_t^0, \ldots, z_t^{13}$ are the total supply deficits in the irrigation districts in the last 14 days. The state transition function is given by the pair (6.29), (6.58). The output transformation is given by the inversion of equation (6.20), by (6.22), (6.27), and by the equations that describe the diversion nodes.

If one wants to design policies that consider downstream flooding explicitly (by adding a suitable objective in the formulation of the Control Problem), the model can no longer be limited to describing the units in the grey dashed box in Figure 5.10 but should also describe the propagation of the flood wave from Miorina to Pavia along the Ticino. Consequently, the state vector $\mathbf{x}_t$ should be extended to include the state variables of the unit that describes the River Ticino.

Even if we do not consider the downstream flooding, the number of state variables is very high and the computing time for the policy design is impracticable (hundreds of years). Fortunately for us, in Section 7.3.2 we will show that, by adopting an approximate expression for indicator $i_{\mathrm{Irr\_D\_3}}$ (*average potential damage from stress*), it is possible to exclude the variables $\mathcal{F}_t, z_t^0, \ldots, z_t^{13}$ from the state, so that it is reduced to the lake storage alone. We will adopt this approximation; consequently the equations of the model, which constitute the constraints of the Control Problem, are the following

$$a_{t+1} = \sigma_{t_{\mathrm{MOD}\,T}} \cdot \varepsilon_{t+1} + \mu_{t_{\mathrm{MOD}\,T}} \tag{6.62a}$$

$$\varepsilon_{t+1} \sim N(0, 1) \tag{6.62b}$$

$$s_{t+1} = s_t + a_{t+1} - r_{t+1} \tag{6.62c}$$

$$r_{t+1} = \begin{cases} v_t(s_t, a_{t+1}) & \text{if } u_t^{\mathrm{MIO}} < v_t(s_t, a_{t+1}) \\ V_t(s_t, a_{t+1}) & \text{if } u_t^{\mathrm{MIO}} > V_t(s_t, a_{t+1}) \\ u_t^{\mathrm{MIO}} & \text{otherwise} \end{cases} \tag{6.62d}$$

$$h_t^{\mathrm{SC}} = \frac{s_t}{S} - s_{\mathrm{inf}} \tag{6.62e}$$

$$h_t^l = h_t^{\mathrm{SC}} + h(d) \tag{6.62f}$$

where $v_t(\cdot)$ and $V_t(\cdot)$ are defined by equation (6.26)

$$q_{t+1}^{\mathrm{TIC,MEF}} = \min\{q_t^{\mathrm{MEF}}, r_{t+1}\} \tag{6.62g}$$

$$q_{t+1}^{\mathrm{TIC},u} = \max\{\min[u_t^{\mathrm{TIC}}, r_{t+1} - q_{t+1}^{\mathrm{TIC,MEF}} - q^{\mathrm{Turb}}], 0\} \tag{6.62h}$$

$$\mathbf{u}_t = \begin{bmatrix} u_t^{\mathrm{MIO}} \\ u_t^{\mathrm{TIC}} \end{bmatrix} \tag{6.62i}$$

$$r^r_{t+1} = r_{t+1} - q^{\text{TIC,MEF}}_{t+1} - q^{\text{TIC},u}_{t+1} \tag{6.62j}$$

$$q^{\text{REC}}_{t+1} = \begin{cases} W^{\text{REC}}_t & \text{if } r^A_t \leqslant r^r_{t+1} \\ W^{\text{REC}}_t - q^{\text{red,REC}}_{t+1} & \text{if } r^B_t \leqslant r^r_{t+1} < r^A_t \\ W^{\text{REC}}_t - q^{\text{maxred,REC}}_{t+1} & \text{if } r^C_t \leqslant r^r_t < r^B_t \\ \alpha^{\text{REC}}_t (r^r_{t+1} - q^{\text{Turb}}) + q^{\text{VC,REC}}_{t+1} & \text{if } r^D_t \leqslant r^r_t < r^C_t \\ 0 & \text{if } r^r_{t+1} < r^D_t \end{cases} \tag{6.62k}$$

$$q^{\text{VC}}_{t+1} = \begin{cases} W^{\text{VC}}_t & \text{if } r^A_t \leqslant r^r_{t+1} \\ W^{\text{VC}}_t & \text{if } r^B_t \leqslant r^r_{t+1} < r^A_t \\ r^r_{t+1} - q^{\text{REC}}_{t+1} - q^{\text{IC}}_{t+1} & \text{if } r^C_t \leqslant r^r_t < r^B_t \\ \max\{\alpha^{\text{VC}}_t (r^r_{t+1} - q^{\text{Turb}}), q^{\text{min,VC}}_t\} & \text{if } r^D_t \leqslant r^r_t < r^C_t \\ 0 & \text{if } r^r_{t+1} < r^D_t \end{cases} \tag{6.62l}$$

$$q^{\text{IC}}_{t+1} = \begin{cases} W^{\text{IC}}_t & \text{if } r^A_t \leqslant r^r_{t+1} \\ W^{\text{IC}}_t - q^{\text{red,IC}}_{t+1} & \text{if } r^B_t \leqslant r^r_{t+1} < r^A_t \\ W^{\text{IC}}_t - q^{\text{maxred,IC}}_{t+1} & \text{if } r^C_t \leqslant r^r_t < r^B_t \\ \alpha^{\text{IC}}_t (r^r_{t+1} - q^{\text{Turb}}) + q^{\text{Turb}} & \text{if } r^D_t \leqslant r^r_t < r^C_t \\ 0 & \text{if } r^r_{t+1} < r^D_t \end{cases} \tag{6.62m}$$

$$q^{\text{VC,REC}}_{t+1} = \begin{cases} 0 & \text{if } \alpha^{\text{VC}}_t (r^r_{t+1} - q^{\text{Turb}}) \geqslant q^{\text{ope,VC}}_t \\ \alpha^{\text{VC}}_t (\tilde{r}_t - q^{\text{Turb}}) - q^{\text{min,VC}}_t & \text{otherwise} \end{cases} \tag{6.62n}$$

in which $q^{\text{red,REC}}_{t+1}$ and $q^{\text{red,IC}}_{t+1}$ are defined by the recursive procedure described in point 2 of Section 6.7.3

$$r^A_t = W_t \tag{6.62o}$$

$$r^B_t = W_t - q^{\text{maxred,hyd}}_t \tag{6.62p}$$

$$r^C_t = W_t - q^{\text{maxred,hyd}}_t - q^{\text{maxred,VC}}_t \tag{6.62q}$$

$$r^D_t = q^{\text{Turb}} \tag{6.62r}$$

$$q^{\text{TIC}}_{t+1} = q^{\text{TIC,MEF}}_{t+1} + q^{\text{TIC},u}_{t+1} + \left(r^r_{t+1} - q^{\text{REC}}_{t+1} - q^{\text{VC}}_{t+1} - q^{\text{IC}}_{t+1}\right) \tag{6.62s}$$

$$q^{\text{PT}}_{t+1} = q^{\text{IC}}_{t+1} + q^{\text{VC}}_{t+1} + q^{\text{TIC}}_{t+1} \tag{6.62t}$$

with

$$\varepsilon_t \in S_{\varepsilon_t} = (-\infty, +\infty) \tag{6.62u}$$

$$\mathbf{u}_t \in U_t(s_t) = \left\{ \left(u^{\text{MIO}}_t, u^{\text{TIC}}_t\right) : \begin{array}{c} 0 \leqslant u^{\text{MIO}}_t \leqslant V_t(s_t, \bar{a}_{t+1}) \\ 0 \leqslant u^{\text{TIC}}_t \leqslant u^{\text{MIO}}_t \end{array} \right\} \tag{6.62v}$$

$$s_t \in S_{s_t} = [0, +\infty) \tag{6.62w}$$

where $V_t(s_t, \bar{a}_{t+1})$ is the maximum volume that can be released on day t if the maximum inflow $\bar{a}_{t+1}$ occurs.

Chapter 7

The Control Problem

EB, AC, RSS and EW

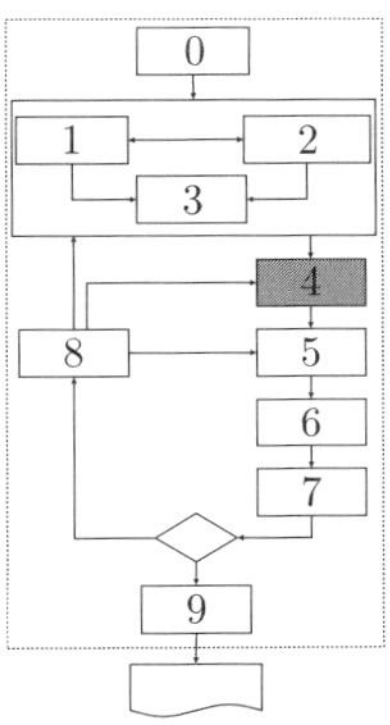

Figure 7.1: The phases of the PIP procedure. Phase 4, which is analysed in this chapter, is highlighted.

In the previous chapter we identified the model of the water system and thus, for each combination of structural and normative actions, we know the form of the constraints of the (Optimal) Control Problem (5.8) that was tentatively formulated in Section 5.2.6. To completely define it, we must still identify the design indicators, from which we will obtain the Problem objectives. Once this is done, we will finally have all the elements that are needed to design the regulation policies and thus complete Phase 4 of the PIP procedure, namely *Designing Alternatives* (Figure 7.1).

7.1 The design time horizon

At the end of the previous chapter we decided to use model (6.62) in the Control Problem. Its input is the pair $(\mathbf{u}_t, \varepsilon_{t+1})$, but when the system is subject to the regulation policy[1] p and thus the control $\mathbf{u}_t$ is defined at every time instant by the control law $m_t(s_t)$, the input is reduced to the disturbance ε_{t+1} alone.

Consider now a *realization* of the disturbance over a time horizon of h days, in other words a sequence of values[2]

$$\boldsymbol{\varepsilon} = |\varepsilon_1, \varepsilon_2, \ldots, \varepsilon_h| \tag{7.1}$$

from the sequences that can be generated by model AR(0) (Section 6.5). Given the initial state of the system, namely the lake storage s_0 at the beginning of the horizon, both the state trajectory $|s_0, s_1, \ldots, s_h|$ and the control trajectory $|\mathbf{u}_0, \mathbf{u}_1, \ldots, \mathbf{u}_{h-1}|$ are univocally defined along the whole horizon.[3] These trajectories can be computed using the policy and

[1] To simplify our description, from now on we will assume that the policy is a point-valued policy.

[2] Note that, considering a time horizon from 0 to h, from (6.62) it follows that the first value of the disturbance to be considered is ε_1 and the last is ε_h.

[3] The sequence of controls terminates with $\mathbf{u}_{h-1}$ and not with $\mathbf{u}_h$, because it is $\mathbf{u}_{h-1}$ which influences s_h – the last state of the horizon being considered – while $\mathbf{u}_h$ influences s_{h+1}.

the state transition equation recursively, namely by recursively solving the following system of equations

$$s_{t+1} = f_t(s_t, \mathbf{u}_t, \varepsilon_{t+1}) \qquad t = 0, 1, \ldots, h-1 \tag{7.2a}$$

$$\mathbf{u}_t = m_t(s_t) \in U_t(s_t) \qquad t = 0, 1, \ldots, h-1 \tag{7.2b}$$

$$|\varepsilon_1, \varepsilon_2, \ldots, \varepsilon_h| \tag{7.2c}$$

$$p = \{m_t(\cdot);\ t = 0, 1, \ldots, h-1\} \tag{7.2d}$$

The behaviour of the system is described by a vector $\mathcal{F}$ which is composed of the following trajectories

$$\mathcal{F} = |s_0, s_1, \ldots, s_h, \mathbf{u}_0, \mathbf{u}_1, \ldots, \mathbf{u}_{h-1}, \varepsilon_1, \varepsilon_2, \ldots, \varepsilon_h| \tag{7.3}$$

Once $\mathcal{F}$ is known, the trajectories of all the variables that appear in the indicators defined in Chapter 4 are also known. In fact, these trajectories are contained in $\mathcal{F}$ or can be derived from it, using the equations that describe the models of the sectors (Section 6.9). Each indicator can thus be seen as a function of the following type

$$i = i(\mathcal{F}) \tag{7.4}$$

In the same section we also mentioned the fact that when the indicator is separable, i.e. when it can be expressed in the form

$$i = \sum_{t=0}^{h-1} g_t(s_t, \mathbf{u}_t, \varepsilon_{t+1}) + g_h(s_h)$$

there is a powerful algorithm that allows us to solve the Control Problem.

The first step that must be taken in order to identify the design indicators is thus the choice of the value of h, more precisely the definition of the design horizon. Pay attention to the fact that the design horizon is the horizon over which the policies are designed, but it is not necessarily the same horizon over which the effects of the alternatives are estimated (evaluation horizon, see Section 8.3).

The simplest case consists in considering a finite horizon. In this case the form of the design indicators is the following

$$i = \sum_{t=0}^{h-1} g_t(s_t, \mathbf{u}_t, \varepsilon_{t+1}) + g_h(s_h) \tag{7.5}$$

where the function $g_h(\cdot)$ is called *penalty* (*on the final state*) and it expresses the cost that is paid[4] if at time h the system is in state s_h. If we assume a finite horizon and let $g_h(\cdot)$ equal zero, it means that we are not interested in what happens after time h: from that moment it is as if the system no longer existed for us. With this assumption it is very probable that the policy, when approaching the end of the horizon, would exploit the entire resource, for example by discharging the entire storage of the lake. Therefore, the choice of a finite horizon is only acceptable if either the system being considered really has a finite life (which is not true for Verbano) or a penalty $g_h(\cdot)$ is specified, to account for our expectations for

[4]Remember that the word 'cost' is used as a synonym of 'measure' and it can have either a negative or positive meaning (a 'negative cost' is a 'benefit').

the period that extends past h into infinity. This last specification is, however, generally anything but easy, and runs the risk of being very arbitrary.

When the interest in the system extends into an infinite future, as in the Verbano Project, it is more logical to consider an infinite horizon, i.e. to assume that h tends towards infinity. Over such a horizon, however, the value of i also tends to infinity, whichever the system trajectory, and so the indicator loses its meaning. To solve this difficulty there are two alternative routes. The first is to discount the future costs by means of a *discount factor* γ, so that the indicator assumes the form

$$i = \lim_{t\to\infty} \sum_{\tau=0}^{t} \left(\frac{1}{1+\gamma}\right)^{\tau} g_\tau(s_\tau, \mathbf{u}_\tau, \varepsilon_{\tau+1}) \tag{7.6}$$

which is termed *total discounted cost* (*over an infinite horizon*) or TDC. By assuming it, one states that the costs in the near future are more important than the costs in the distant future.

The second route consists in considering the *average cost* (*over an infinite horizon*), i.e.

$$i = \lim_{t\to\infty} \frac{1}{t+1} \sum_{\tau=0}^{t} g_\tau(s_\tau, \mathbf{u}_\tau, \varepsilon_{\tau+1}) \tag{7.7}$$

which does not discriminate between the costs in the near future and those in the distant future. The choice between (7.6) and (7.7) depends on the attitude of the Experts to the future costs and on the type of quantities considered in the definition of the indicators. In general, the discount factor is introduced when dealing with monetary quantities, but not when dealing with physical quantities, as is our case. Therefore *we adopted the average cost over an infinite horizon for all the design indicators.*

7.2 How a design indicator is identified

Given what we said in the previous section, the design indicators assume the form (7.7). To identify each of them, it would seem necessary to identify an infinite number of step-cost functions. The workload can be reduced, however, by assuming that the step-cost is a periodic function, of period T equal to one year, i.e.

$$g_t(\cdot) = g_{t+kT}(\cdot) \quad t = 0, 1, \ldots; \; k = 1, 2, \ldots \tag{7.8}$$

Then, for each indicator, we must identify only T functions. The hypothesis is founded on the fact that all the functions that define the model of the water system can be considered to be periodical, with a period equal to T.

In Section 5.2.3 we decided to consider four design indicators relative to the sectors: *Upstream Flooding*, *Irrigation*, ENEL *Power* and *Downstream Environment*. For each of these sectors, in Chapter 4 we defined a hierarchy of evaluation criteria and the indicators that are useful to quantify their satisfaction. With this information, we must now derive a single design indicator for each sector that is a measure of the overall satisfaction of the sector and that is expressed by a separable functional. We must define a procedure to identify it and this procedure will assume different forms according to the answers to the following questions:

1. Are all the indicators positively correlated with each other, in correspondence to the triple ($\text{SD}_{\text{CUR}}/\text{RANGE}_{\text{CUR}}/13_{\text{CON}}$) of structural and normative actions?

2. Are all the indicators expressed by separable functionals?

Note that the existence of a positive correlation among the indicators must be ascertained for any given triple (s, f, d), since the design of the policies is carried out separately for each possible triple (Section 5.2.3). The procedure starts by checking whether positive correlation holds for the triple that describes the current normative and structural situation. Once the design indicators have been defined, it must be verified that, even in correspondence to the other triples, each of the indicators that has been identified permits us to generate policies that respond to the expectations of the Expert who proposed it (Section 7.4).

Sometimes the existence of a correlation cannot simply be deducted from the definition of the indicators. Then, it is necessary to compute the values that they assume in correspondence to different policies and verify whether positive correlation exists among these values. But the design of the policies requires that an objective be associated to each indicator (Sections 5.2.6 and 7.6.1) and that the Control Problems thereby defined be solved. The objectives can be directly formulated when all the indicators are separable functionals; otherwise successive approximations must be used. Therefore the procedure must branch off in two cases according to the response to the second question.

7.2.1 Case I: separable indicators

The procedure is articulated in a number of steps, interconnected by tests, according to the block diagram in Figure 7.2.

- **Step 1 –** Since all the indicators i_k (with $k = 1, \ldots, n$) of the sector under study are expressed by separable functionals, they are all potentially useable as design indicators. Given the triple of structural and normative actions ($\text{SD}_{\text{CUR}}/\text{RANGE}_{\text{CUR}}/13_{\text{CON}}$), it is thus possible to define n Control Problems, each one with four objectives. The latter are obtained by applying statistics (see Section 5.2.6) to the indicator i_k and to the indicators i_{c_j} $(j = 1, 2, 3)$ of the other three sectors.[5] The resolution of each of these Problems allows us to determine:
  - a number o of *compromise policies* $p_{k,i}^{\text{comp}}$, each of which is designed by minimizing the weighted sum of the objectives in correspondence to the vector of weights $\boldsymbol{\lambda}^i = |\lambda_1^i, \ldots, \lambda_4^i|$, belonging to a set $\{\boldsymbol{\lambda}^1, \ldots, \boldsymbol{\lambda}^o\}$ of o vectors proposed by the Analyst;
  - the policy p_k^{ext} which minimizes[6] the single objective associated to the indicator i_k, which we will call *extreme policy*, i.e. the policy designed by solving the Multi-Objective Problem with weights $\boldsymbol{\lambda} = |1, 0, 0, 0|$.

  When all n Problems have been solved, the following set $\mathcal{A}_{\text{test}}$ of alternatives is obtained

$$\mathcal{A}_{\text{test}} = \left\{A\colon\ A = (\text{SD}_{\text{CUR}}/\text{RANGE}_{\text{CUR}}/13_{\text{CON}},\ p) \text{ with } p \in \mathcal{P}_{\text{test}}\right\} \tag{7.9}$$

  where

$$\mathcal{P}_{\text{test}} = \left\{p_1^{\text{ext}}, \ldots, p_n^{\text{ext}}, p_{1,1}^{\text{comp}}, \ldots, p_{n,o}^{\text{comp}}\right\} \tag{7.10}$$

  Finally the values assumed by the indicators $i_1, \ldots, i_n$ are computed in correspondence to all the alternatives of set $\mathcal{A}_{\text{test}}$. This is done by means of the procedure for estimating the effects that will be described in Chapter 8.

[5] If the three of them have already been defined, otherwise we must proceed in a recursive way.

[6] Bear in mind that we are assuming that the indicators represent costs and this is why they must be minimized. If, instead, they represent benefits, it is sufficient to change their sign to trace back to our case.

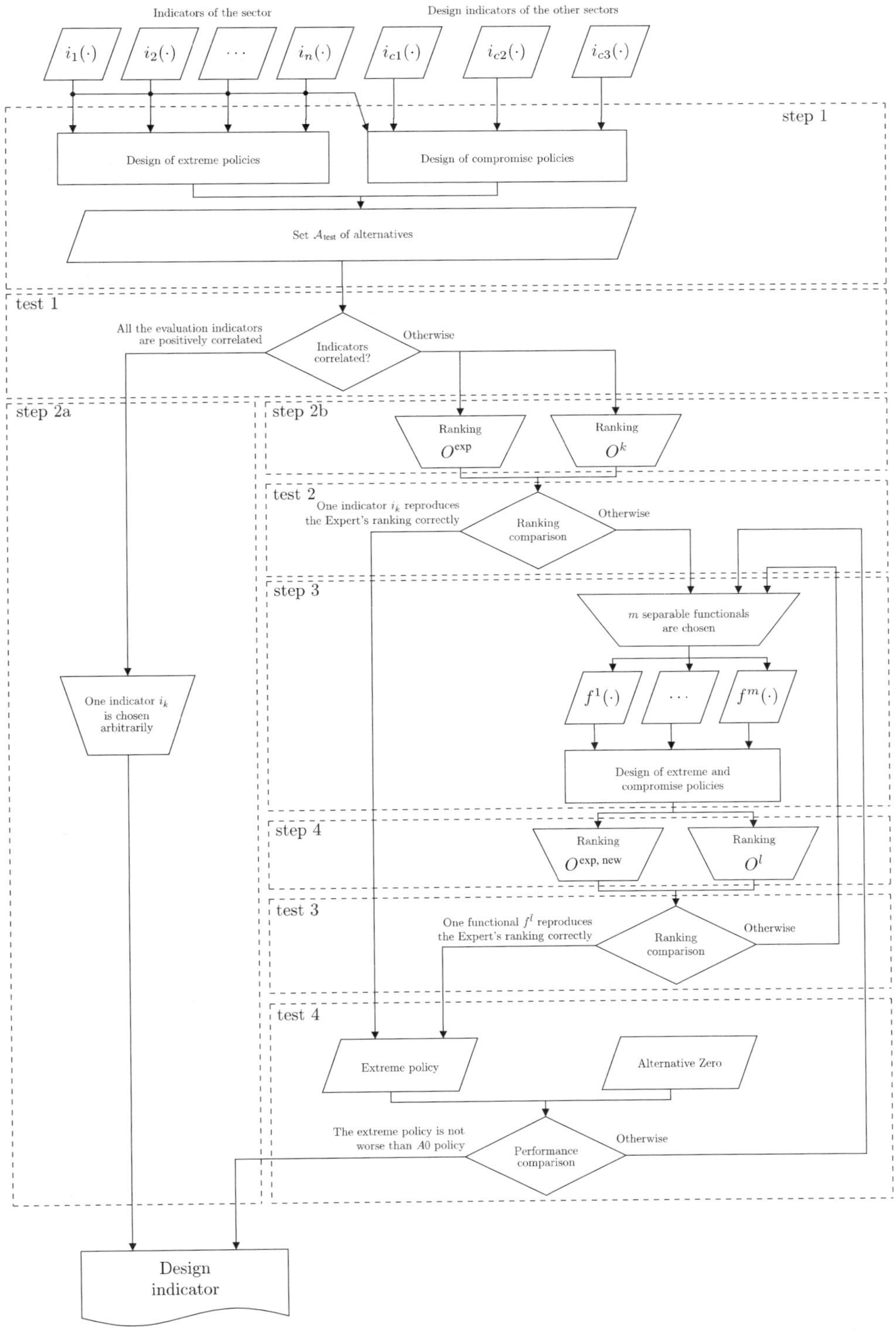

Figure 7.2: Procedure for the definition of the design indicator for a sector whose indicators are all separable.

- **Test 1 –** A number of n-ple of indicator values is now available, with which it is possible to ascertain whether the indicators are all positively correlated, or whether at least one of them is uncorrelated or negatively correlated with the others. In the first case one proceeds with Step 2a, in the second with Step 2b.

- **Step 2a –** If the indicators $i_1, \ldots, i_n$ are all positively correlated, it means that, for the design of the policies, adopting one is completely equivalent to adopting any of the others. It follows that the policies $p_1^{\text{ext}}, \ldots, p_n^{\text{ext}}$ of set (7.10) are all equal, as are all the policies $p_{k,i}^{\text{comp}}$ that provide the same performance for the indicators i_{c_j} ($j = 1, 2, 3$). In other words, the policies do not depend on k. We can thus arbitrarily choose any of the indicators as the design indicator. In the following we will denote it with the symbol $i_{\bar{k}}$. From a practical standpoint, however, indicators defined by continuous functionals, if they exist, are preferable, because their use avoids numerical difficulties that could arise in the solution of the Control Problem (as we will see in Section 7.3.1, where the choice of the design indicator for the *Upstream Flooding* sector is presented). The procedure is terminated.

- **Step 2b –** If one or more indicators are not positively correlated with the others, the policies of set (7.10) change with k, even if they do not necessarily all have to be different. Once the differences among the policies have been highlighted (both in their performance and behaviour), it is necessary to interact with the Expert to understand which of indicators $i_1, \ldots, i_n$ provides the policies that he thinks are the most interesting. To do this, the Expert is asked to make a pairwise comparison of all the alternatives in the set $\mathcal{A}_{\text{test}}$, based on both the indicator values $i_1, \ldots, i_n$ and on the trajectories of the variables that interest him (for example, the lake level for the *Upstream Flooding* sector). For each pair the Expert must say which indicator he prefers and by how much (a weak, significant, strong or very strong preference). In this way a *Pairwise comparison matrix* is obtained from which it is possible to extract a ranking of the alternatives, which we will denote with O^{exp}. The alternatives can then be ordered by decreasing values of each of the indicators $i_1, \ldots, i_n$, to obtain n rankings O^k.

- **Test 2 –** If, among the rankings O^k, a ranking $O^{\bar{k}}$ exists that, in the Expert's opinion, is very similar to vector O^{exp}, then it means that the indicator $i_{\bar{k}}$, which produced $O^{\bar{k}}$, is a candidate to be the design indicator for that sector and the procedure continues with Test 4. If this is not the case, go to Step 3.

  Note that the same ranking $O^{\bar{k}}$ could be produced by different indicators. This happens when these indicators are positively correlated. In this case, as was said in Step 2a, it is possible to choose any one of them.

- **Step 3 –** If, according to the Expert, none of the rankings O^k seems sufficiently similar to O^{exp}, it means that none of the indicators i_k can be assumed as the design indicator. Then, a weighted sum of all the indicators can be used or we can introduce new indicators, i.e. define separable functionals of the trajectories of the system variables, which allow us to reproduce the ranking O^{exp}. Let $f^l(\cdot)$ (with $l = 1, \ldots, m$) be such functionals. For one of them to be a candidate for the role of design indicator, it is not sufficient that it reproduce the ranking O^{exp}, but it is also necessary that it generate alternatives that the Expert considers to be interesting. To make sure of this, we must design new alternatives with each one of these functionals and thus enlarge the set $\mathcal{A}_{\text{test}}$.

This procedure is founded on the observation that, while the indicators $i_1, \ldots, i_n$ must provide the Expert with indications about the effects that an alternative will produce and thus must have a physical meaning, the aim of a design indicator is to generate policies that the Expert considers to be interesting. For this purpose it is not a priori necessary that the indicator have a precise physical meaning.

- **Step 4 –** The Expert is asked to rank the enlarged set $\mathcal{A}_{\text{test}}$ in order to obtain the ranking $O^{\text{exp,new}}$. The m rankings O^l produced by the functionals $f^1, \ldots, f^m$ are computed.

- **Test 3 –** If, among the rankings O^l, a ranking $O^{\bar{l}}$ exists that, in the Expert's opinion, is very similar to ranking $O^{\text{exp,new}}$, the functional $f^{\bar{l}}$ that produced it is a candidate to be the design indicator and the procedure goes on with Test 4. If this is not the case, return to Step 3 to identify a new set of functionals.

- **Test 4 –** Let I be either the indicator $i_{\bar{k}}$ (selected with Test 2) or the functional $f^{\bar{l}}$ (selected with Test 3). To verify whether I actually is the design indicator that we are looking for, it is necessary to compare the effects of the extreme policy $\bar{p}_{\mathsf{I}}^{\text{est}}$ that it generates with those of alternative $A0$. Only if the Expert prefers the first, or at least is indifferent between the two, can the choice of I be considered correct. If, in fact, the alternative $A0$ was preferred, I would not be a good design indicator because it would not be able to define the best policy for the sector under study.

  This check is not necessary if $i_{\bar{k}}$ was selected in Step 2a, i.e. if the indicators are all positively correlated: since the evaluation indicators were validated (Section 4.3), they must generate an extreme policy that gives a performance not worse than $A0$.

7.2.2 Case II: non-separable indicators

The procedure that must be used when some of the indicators are not expressed by separable functionals is described in Figure 7.3. We have seen that non-separable indicators cannot be utilized to define the objectives of a Control Problem, because there is no algorithm to solve it. Therefore, for each non-separable indicator, it is necessary to identify a separable functional that can replace it (Step 0).[7] Note that, while the evaluation indicator must quantify the physical effects relevant to the criterion to which it is associated, by providing values that the Expert is able to interpret, this cannot be expected of the functional that replaces it.[8] Thus the only condition is that, by using the functional in the formulation of the Control Problem, policies will be designed that tend to satisfy the corresponding criterion. The next steps of the procedure, which is essentially identical[9] to Case I, allow us to verify this.

---

[7] For simplicity's sake, Figure 7.3 describes the case in which all the indicators are non-separable.

[8] If, in fact, there is a separable functional that satisfies this condition, it would be both the evaluation and the design indicator and the difficulty would no longer exist.

[9] Except for the fact that after Step 2a, Test 4 must be applied to make sure that the extreme policy, designed with the approximate indicator, is preferred to the policy in alternative $A0$, namely that the adopted approximation is acceptable.

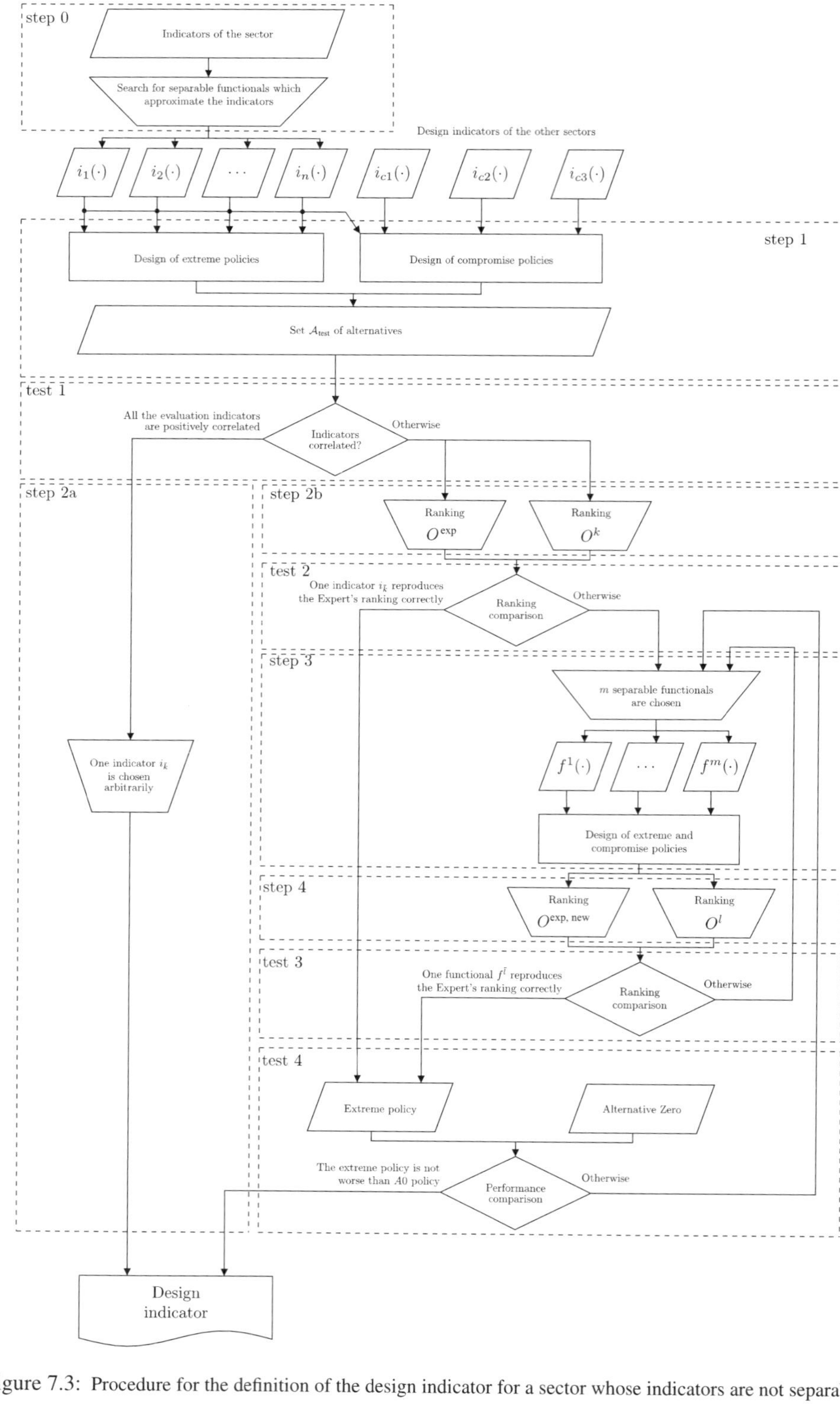

Figure 7.3: Procedure for the definition of the design indicator for a sector whose indicators are not separable.

7.3 The design indicators for Verbano

We will now show how we chose the four design indicators for the case of Verbano. Remember that, to avoid the notation becoming too complex, we use the same symbol i to refer to both the evaluation and the design indicators; the reader will easily distinguish one from the other according to the context.

Before proceeding, it must be noticed that in Section 4.3 the (evaluation) indicators were defined over a finite horizon H, because the evaluation horizon is finite. However, in Section 7.1 we decided to use the average cost over an infinite horizon for the policy design: therefore the definitions of all the indicators must be reformulated on the basis of equation (7.7).

7.3.1 Upstream Flooding sector

As described in Section 4.2.1, at the first level of the hierarchy of criteria for the *Upstream Flooding* sector is a list of townships (actually, we will examine only two: Locarno and Verbania). For each township, three leaf criteria have been defined: *Damage to structure*, *Damage to economic activities* and *Damage caused by interruptions to traffic*. Both the leaf criteria and the indicators with which they are quantified are the same for all the townships, except for the values of a few parameters (the flooding threshold, for example). To identify one design indicator for the entire sector, it is thus sufficient to apply the procedure described in Section 7.2 to a single township.

Let us take Locarno, for example. The damage produced by flooding depends on the level h_t^{Loc} of the lake in the city, which can be expressed as a function of the lake level h_t^{SC} at Sesto Calende, by means of equation (6.23), where d is equal to 60 km. In turn, h_t^{SC} is linked to the storage s_t through equation (6.20), and so the three indicators for this township can be reformulated as follows

- *Average annual flooded area* [km^2/year] *in Locarno*

$$i_{\text{Fl\_U\_2\_Loc}} = \lim_{t\to\infty} \frac{365}{t+1} \sum_{\tau=0}^{t} S_{\text{fl}}^{\text{Loc}}(s_\tau) \tag{7.11}$$

  where the function[10] $S_{\text{fl}}^{\text{Loc}}(s_t)$ provides the flooded surface area at Locarno as a function of the storage s_t;

- *Average annual number* [d/year] *of flooding days in Locarno*

$$i_{\text{Fl\_U\_4\_Loc}} = \lim_{t\to\infty} \frac{365}{t+1} \sum_{\tau=0}^{t} \mu_\tau^{\text{Loc}} \tag{7.12}$$

  with

$$\mu_\tau^{\text{Loc}} = \begin{cases} 1 & \text{if } Q_\tau^{\text{Loc}} \geqslant Q_{\text{fl}}^{\text{Loc}} \\ 0 & \text{if } Q_\tau^{\text{Loc}} < Q_{\text{fl}}^{\text{Loc}} \end{cases} \tag{7.13}$$

[10]The function is easily derived by expressing h_t^{SC} as a function of s_t in $S_{\text{fl}}^{\text{Loc}}(h_t^{\text{SC}})$ (see equation (4.34) in Section 4.5.5 of the DVD).

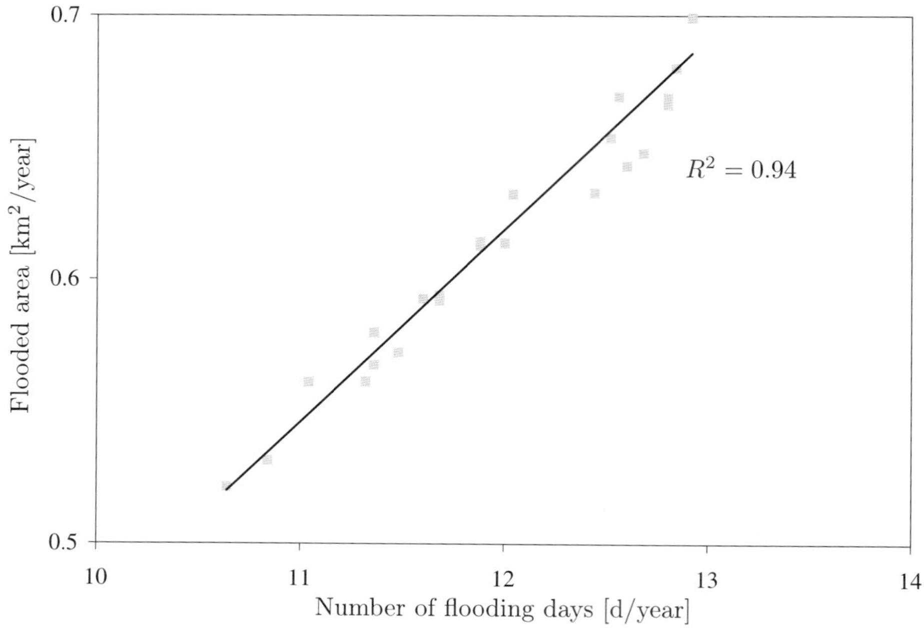

Figure 7.4: The correlation between the average annual flooded area and the average number of flooding days in the township of Locarno.

where Q_τ^{Loc} is the elevation of the lake level at Locarno on day t (which is obtained by transforming h_t^{SC} with (4.30) from Section 4.5.5 of the DVD) and $Q_{\text{fl}}^{\text{Loc}}$ is the elevation at which the first flooding damage occurs in Locarno;

- *Average annual number of days* [d/year] *of interrupted traffic in Locarno*

$$i_{\text{Fl\_U\_6\_Loc}} = \lim_{t\to\infty} \frac{365}{t+1} \sum_{\tau=0}^{t} \beta_\tau^{\text{Loc}} \tag{7.14}$$

with

$$\beta_\tau^{\text{Loc}} = \begin{cases} 1 & \text{if } Q_\tau^{\text{Loc}} \geqslant Q_{\text{traf}}^{\text{Loc}} \\ 0 & \text{if } Q_\tau^{\text{Loc}} < Q_{\text{traf}}^{\text{Loc}} \end{cases} \tag{7.15}$$

where $Q_{\text{traf}}^{\text{Loc}}$ is the elevation at which the first street in Locarno is flooded.

All three of the indicators are separable, so the procedure in Figure 7.2 is applied. Now we can design a number of alternatives and, for each of them, compute the value that the above indicators assume (Step 1). Figure 7.4 shows the scatter diagram of the values that have been obtained for the two indicators *Average annual flooded area* and *Average annual number of flooding days*. Since they are positioned around a straight line, we can infer that the two indicators are strongly correlated. Similarly, we can verify that the third is correlated with the first two. The response to Test 1 is therefore positive and one proceeds with Step 2a. We can adopt either of these indicators as the design indicator: the choice

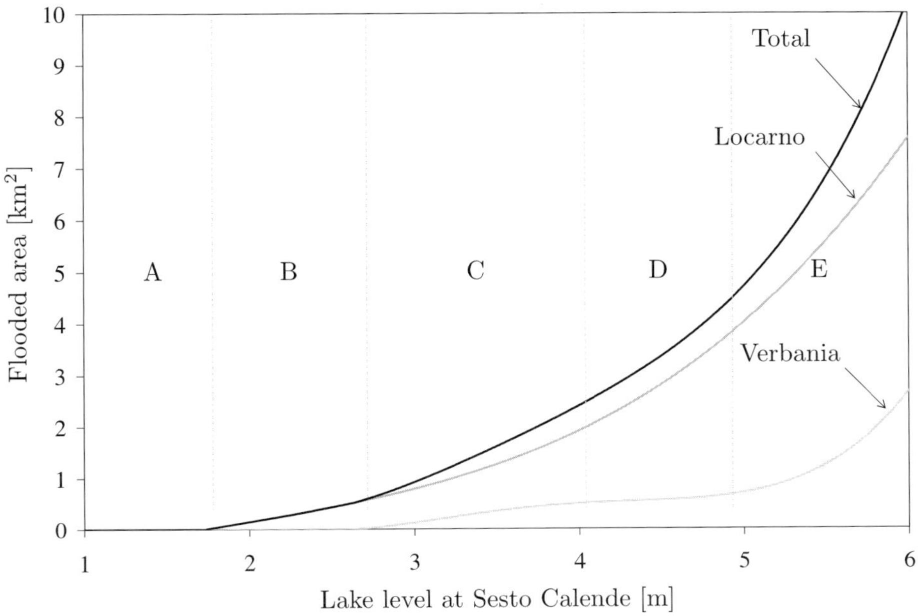

Figure 7.5: Flooded area as a function of the lake level at Sesto Calende. In the computation we took account of the fact that the lake's surface is not horizontal.

makes no difference for the policy design. However, we preferred to choose the indicator *Average annual flooded area*, because the function $S_{\rm fl}^{\rm Loc}(s_t)$ that defines it is continuous, unlike the functions that characterize the other two. To comprehend the reason for this preference, note that if two different policies produce two slightly different values of the level on the same day τ, such that one is just slightly less than the flooding level $Q_{\rm fl}^{\rm Loc}$, and the other exceeds it by a fraction, the indicator *flooding days* would amplify the difference, since the function $\mu_{\tau}^{\rm Loc}$ would assign one more day of flooding to the second policy, and not to the first. On the other hand, the indicator *flooded area* would show that the flooded area is insignificant with the second policy as well.

Once the indicator *Average annual flooded area* has been chosen, we have to specify how the values relevant to the two townships should be aggregated to form the design indicator. The Experts proposed that we give them equal weight. Consequently, the design indicator $i_{\rm fl}$ for the *Upstream Flooding* sector has the following form

$$i_{\rm fl} = \lim_{t\to\infty} \frac{365}{t+1} \sum_{\tau=0}^{t} \left[S_{\rm fl}^{\rm Loc}(s_t) + S_{\rm fl}^{\rm Vb}(s_t) \right] = \lim_{t\to\infty} \frac{365}{t+1} \sum_{\tau=0}^{t} S_{\rm fl}^{\rm Tot}(s_t) \qquad (7.16)$$

The functions $S_{\rm fl}^{\rm Loc}(s_t)$, $S_{\rm al}^{\rm Vb}(s_t)$ and $S_{\rm al}^{\rm Tot}(s_t)$ are shown in Figure 7.5. Their convexity shows that the damage increases more than linearly with the level at Sesto Calende.

The two cities were chosen not only for their economic importance in the lake region, but also because Locarno lies at the lowest point along the bank, while the topography of Verbania is characterized by the presence of a step in the terrain (which is present also in many other townships).

Figure 7.5 shows that the range of lake levels can be subdivided into five sets, which correspond to the following cases:

- *case A*: there is no flooding and so the damage is zero;
- *case B*: damage occurs only at Locarno;
- *case C*: flooding also affects Verbania. The flooded area increases more rapidly with the level than in the previous case;
- *case D*: the flooded area remains almost constant at Verbania with the increase in level, because of the step in altitude, while it increases at Locarno;
- *case E*: the step at Verbania is exceeded and the flooded area increases in both cities as the lake level increases.

7.3.2 Irrigation sector

The hierarchy defined for the *Irrigation* sector considers the *Distribution costs* caused by variability in the supply on the one hand and, on the other, the *Harvest*, to which in turn are associated the leaf-criteria *Supply deficit* and *Crop stress* (see Section 4.2.2). The Expert believes that the distribution costs are negligible with respect to the value of the harvest and suggests that we consider only the second criterion to construct the design indicator.

Over an infinite horizon the indicator of *Average annual volume of total deficit* assumes the following form

$$i_{\text{Irr\_D\_1}} = \lim_{t\to\infty} \frac{365}{t+1} \sum_{\tau=0}^{t} \left(d_{\tau+1}^{\text{VC}} + d_{\tau+1}^{\text{REC}} + d_{\tau+1}^{\text{IC}}\right) \tag{7.17}$$

where d_τ^{VC}, d_τ^{REC} and d_τ^{IC} are the daily irrigation deficits for the Villoresi, Regina Elena and Industrial Canals. The indicator is separable, whatever the state of the system, because the deficits depend only on the control variable at time t. Instead, the indicator $i_{\text{Irr\_D\_2}}$ that measures the *Average annual potential damage (stemming) from stress*, i.e.

$$i_{\text{Irr\_D\_2}} = \frac{1}{N} \sum_{a=1}^{N} f\left(\max_{t\in a} \frac{1}{14} \sum_{\tau=t-13}^{t} \left(d_\tau^{\text{VC}} + d_\tau^{\text{REC}} + d_\tau^{\text{IC}}\right)\right) \tag{7.18}$$

is separable only if the state is the vector[11]

$$\mathbf{x}_t = \left| s_t, F_t, z_t^0, \ldots, z_t^{13} \right| \tag{7.19}$$

In Section 6.10 we said that the sixteen components of $\mathbf{x}_t$ are too many and make it impossible to solve an Optimal Control Problem based on such a state with the computers available today. On the other hand, when the state is constituted by the storage value s_t only, as in the model (6.62) that we chose, the indicator $i_{\text{Irr\_D\_2}}$ is not separable and we have to apply the procedure illustrated in Figure 7.3.

The first step (Step 0) requires identifying a functional that is separable and that approximates $i_{\text{Irr\_D\_2}}$ sufficiently well. To identify this functional, consider equation (7.18), which shows that the potential damage from stress is linked to a sequence of high deficit

[11] See Section 6.9.

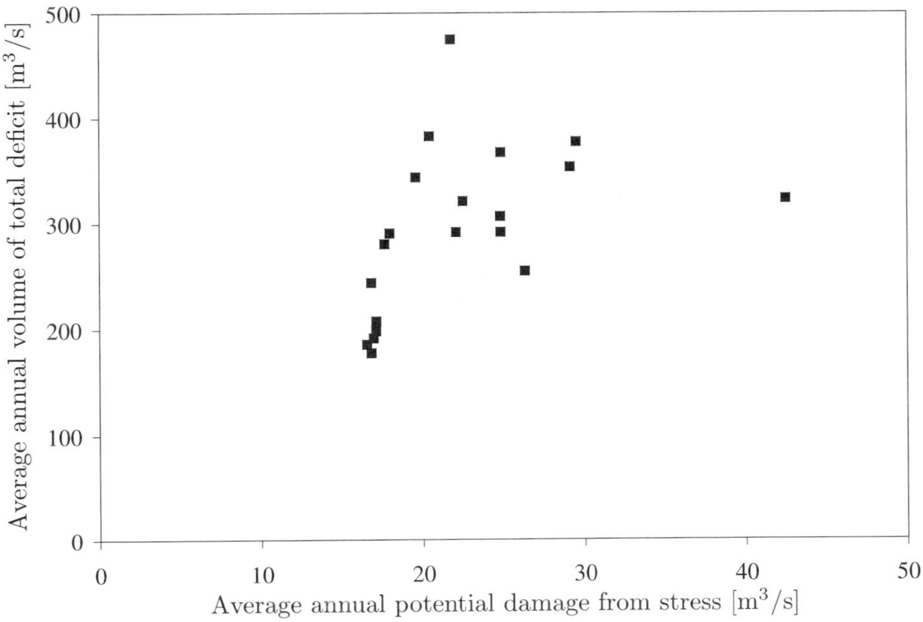

Figure 7.6: Average annual volume of total deficit ($i_{\text{Irr\_D\_1}}$) vs average annual potential damage from stress ($i_{\text{Irr\_D\_2}}$).

values. So, we must identify a separable functional that emphasizes their presence, like, for example, the following

$$\lim_{t\to\infty} \frac{365}{t+1} \sum_{\tau=0}^{t} \left(d_{\tau+1}^{\text{VC}} + d_{\tau+1}^{\text{REC}} + d_{\tau+1}^{\text{IC}}\right)^{16} \tag{7.20}$$

In fact, to avoid the occurrence of a very high deficit on a given day, it is necessary to reduce the supply on the preceding days, and the adoption of indicator (7.20) as design indicator should provide policies that 'spread' deficits over time and thus reduce the potential damage from stress, exactly as $i_{\text{Irr\_D\_2}}$ requires. To make sure that this intuition corresponds to reality, the only way is through experimental verification. For this, we generate a set A_{test} of policies (Step 1) and, for each one of them, we compute the values assumed by the evaluation indicators $i_{\text{Irr\_D\_1}}$ and $i_{\text{Irr\_D\_2}}$. In Figure 7.6 the pairs that are obtained in this way are shown. Since the two indicators are uncorrelated, we must proceed with Step 2b of the procedure.

To understand why these two indicators are not correlated, we can compare the behaviour of the two extreme policies, i.e. policy p_{def}, which minimizes the average annual deficit, and policy p_{stress}, which minimizes the functional (7.20). It is evident that, in principle, they differ only in the release decision u_t^{MIO}, because, given the nature of their objectives, they both suggest that u_t^{TIC} should be kept at zero. As an example, Figure 7.7 compares the two release trajectories that would have occurred during the summer drought of 1989, if the Regulator had always chosen the decision closest to the irrigation demand

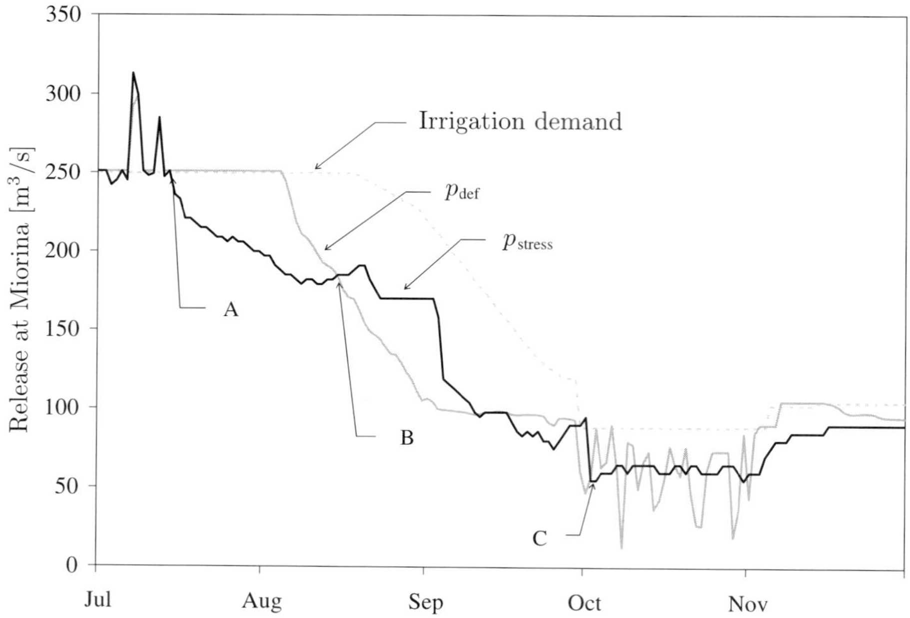

Figure 7.7: The releases produced by the policies p_{def} and p_{stress} in the summer drought of 1989.

(increased by the MEF) from the set $M_t^*(s_t)$ of controls suggested by the two extreme policies.[12]

The policy p_{stress} results in releases which are inferior to the irrigation demand starting from point A, while the policy p_{def} results in releases that equal the demand for longer. The reason for this difference is that p_{stress} is designed based on indicator (7.20), whose step-cost increases more than linearly with the deficit. The result is that it is better to save resources, even if that may provoke or accentuate immediate deficits, in the hope of avoiding larger future deficits. The policy p_{stress} is thus characterized by *risk aversion*. On the other hand, indicator (7.17), with which policy p_{def} is designed, increases linearly with the deficit. It follows that the water demand is satisfied for as long as possible, because 1 m^3 of deficit today is worth as much as 1 m^3 of deficit tomorrow, so there is no reason to produce deficits as long as it is possible to avoid them. From point B onwards, the effective saving from the policy p_{stress} yields its benefits: the deficits from p_{stress} are much less than those from p_{def}. After point C, the release with p_{def} becomes very variable, because the lower extreme of the regulation range has been reached and the release must be equal to the inflow, and the release is set equal to the inflow, thus reflecting its variability. The same does not occur with p_{stress}: in fact, the smaller the lake storage, the greater the tendency to save, and this reduces the width of the fluctuation by delaying the moment at which the lower extreme of the regulation range is reached.

Once the Expert had examined the different behaviours of the two policies, he declared that he preferred p_{stress} and wanted to adopt a functional with the form (7.20) as a design indicator. The choice was confirmed by the fact that the ranking of the alternatives in A_{test} that he proposed agrees with the one produced by using the functional (7.20) (Test 2).

[12]The meaning of this sentence will be completely clear to the reader after reading Section 8.4.

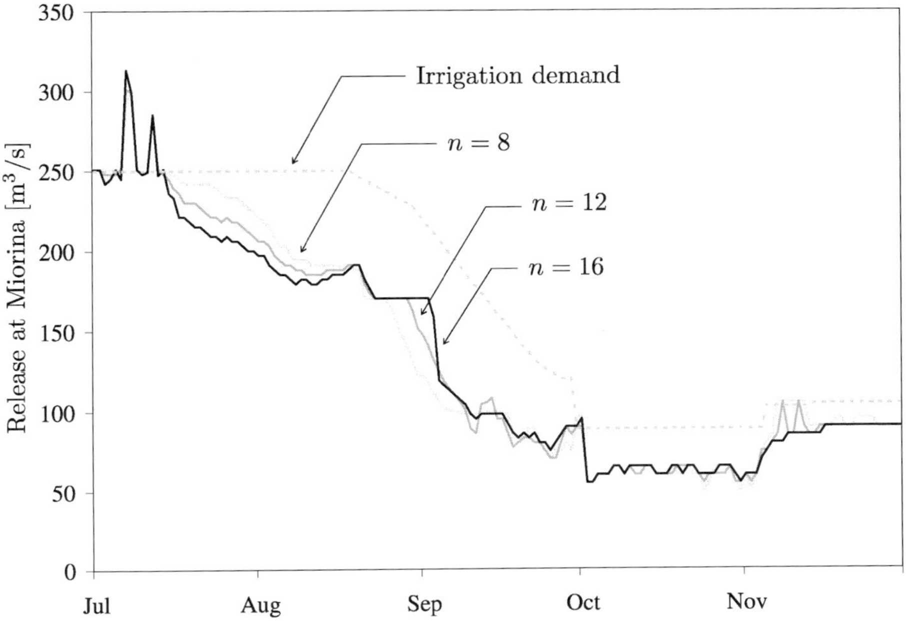

Figure 7.8: The releases produced in the summer drought of 1989 by p_{stress}-type policies designed with three different values of the exponent n in equation (7.21).

We must now verify that the functional (7.20) is the one that best approximates indicator $i_{\text{Irr\_D\_2}}$. To achieve this, note that the exponent 16 that appears in (7.20) was chosen at random. From the above analysis, we know that the functional that the Expert prefers must have the following form

$$\lim_{t\to\infty} \frac{365}{t+1} \sum_{\tau=0}^{t} \left(d_{\tau+1}^{\text{VC}} + d_{\tau+1}^{\text{REC}} + d_{\tau+1}^{\text{IC}}\right)^n \tag{7.21}$$

with $n > 1$. Now we must identify the value of n that best approximates the Expert's judgement. Note that the higher the value of n, the greater the costs attributed to large water deficits and the sooner water savings begin. The exponent n is therefore a 'measure' of the Expert's aversion to risk. To estimate its value, the Expert is shown a number of release trajectories, each generated by a p_{stress}-type policy corresponding to a different value of n. For example, in Figure 7.8 the comparison of the trajectories corresponding to exponents 8, 12 and 16 is shown. The Expert chooses the value 12. With this value, the procedure is repeated until Test 4, which gives a positive result. We conclude that the design indicator i_{irr} we will use for the *Irrigation* sector is the following

$$i_{\text{irr}} = \lim_{t\to\infty} \frac{365}{t+1} \sum_{\tau=0}^{t} \left(d_{\tau+1}^{\text{VC}} + d_{\tau+1}^{\text{REC}} + d_{\tau+1}^{\text{IC}}\right)^{12} \tag{7.22}$$

7.3.3 ENEL Power sector

Following the observations made in Section 4.3.1, the model of the system includes a constraint on the minimum volume diverted into the Industrial Canal, in order to guarantee the *continuity of the Turbigo-*EUROGEN *power plant's operation* as much as possible. Therefore the design indicator for the ENEL *Power* sector should take into account the criterion *Lost revenue for hydropower plants* alone. The indicator $i_{\text{EnP\_D\_1}}$ that is associated to it is defined by a separable functional and can thus be directly used in the Control Problem. The design indicator i_{Enel} is thus the following

$$i_{\text{Enel}} = \lim_{t\to\infty} \frac{365}{t+1} \sum_{\tau=0}^{t} \left(\theta_\tau D_{\tau+1}^{\text{Enel}}\right) \tag{7.23}$$

where $D_{\tau+1}^{\text{Enel}}$ is the *overall production deficit for* ENEL *plants* and θ_τ the *average price of energy* in the interval $[\tau, \tau+1)$, whose values are reported in Section 4.5.3 of the DVD.

7.3.4 Downstream Environment sector

In Section 4.3.3 we concluded that for the *Downstream Environment* sector two indicators should be considered: $i_{\text{Env\_D\_1}}$, which measures the distance of the regulated regime from the natural regime, and $i_{\text{Env\_D\_2}}$, which expresses the MEF value considered. The latter appears in the Control Problem as a constraint, through the function of minimum release $v_t(s_t, a_{t+1})$, defined by (6.26), which is part of the model of the lake. For this reason $i_{\text{Env\_D\_2}}$ should not appear among the design indicators. Only $i_{\text{Env\_D\_1}}$ remains, whose step-cost is

$$g_t^{\text{env}} = \left(q_{t+1}^{\text{TIC}} - r_{t+1}^{n}\right)^2 \tag{7.24}$$

where q_t^{TIC} is the actual flow in the River Ticino on day t and r_t^n is the flow that would occur in natural conditions (i.e. if the lake were not regulated) on the same day. In the function $g_t(\cdot)$ only variables relative to the time period $[t, t+1)$ appear, and so indicator $i_{\text{Env\_D\_1}}$ seems to be a separable functional. Unfortunately, the flow r_{t+1}^n is not a parameter and it does not depend on the state s_t of the regulated lake, but on the state s_t^n that would have occurred in natural regime conditions, and this state is not included in model (6.62). Thus, in order for the indicator to be separable, the state of the model would have to be enlarged. To avoid this, we must approximate it with another indicator and apply the procedure in Figure 7.3.

One possible approximation (Step 0) is obtained by replacing the natural flow r_{t+1}^n on day t with a statistic from a sample of natural flow estimates on the same day. By doing so, we implicitly assume that the process is cyclostationary. In Figure 7.9 two different statistics are compared, the average value and the median. Note that their trajectories are almost coincident in the winter months, while they draw away from each other significantly in April–June and in September–November, when the average value is always higher than the median. The difference is due to the occurrence of flood events, which are rare enough not to influence the median, but which produce water volumes such that the average is increased. Since the statistic that we are looking for must constitute a reference point for the conditions in the regulated regime, the Expert preferred the median, which we denote with $r_t^{n,\text{median}}$. By completing all the steps of the procedure we confirmed the validity of this choice. The design indicator i_{env} for the *Downstream Environment* sector is thus the

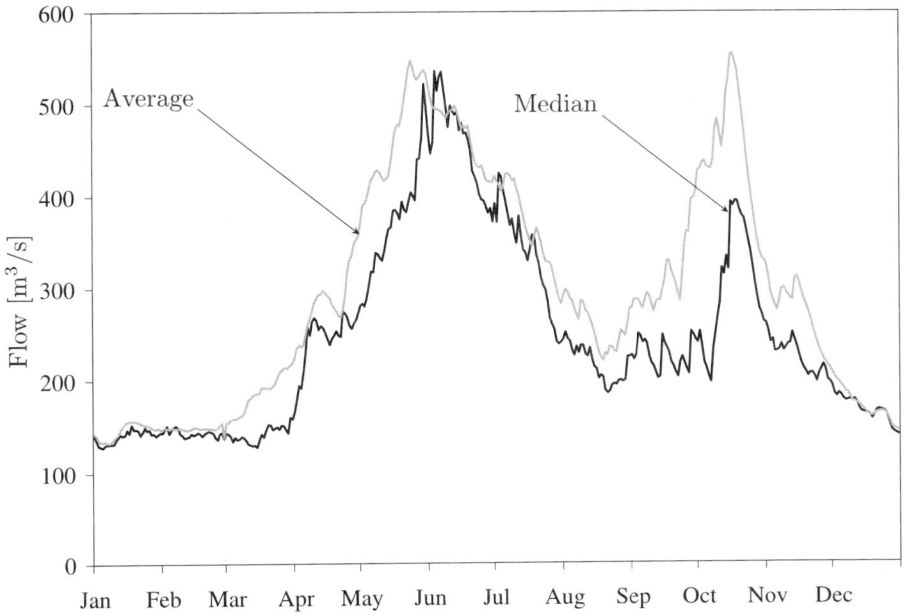

Figure 7.9: Trajectories of the average and median natural flow in the River Ticino (years 1974–1998).

following

$$i_{\text{env}} = \lim_{t \to \infty} \frac{365}{t+1} \sum_{\tau=0}^{t} \left(q_{\tau+1}^{\text{TIC}} - r_{\tau}^{n,\text{median}}\right)^2 \tag{7.25}$$

7.4 Verifying the design indicators

In the previous section we presented the four design indicators that were selected. To arrive at that result, we had to design several different regulation policies. In doing this we adopted the three structural and normative actions that characterize $A0$, namely ($\text{SD}_{\text{CUR}}/\text{RANGE}_{\text{CUR}}/13_{\text{CON}}$). We must now make sure that the design indicators provide policies that the Experts consider to be interesting also for the other triples of actions that we decided to analyse (see Section 5.4). For this purpose, for each of them, we designed a number of regulation policies and estimated their effects. Since they were judged positively by the four Experts, we can consider the set of design indicators to be established.

7.5 The design scenario

Several parameters appear both in the definition of the step-cost functions and in the equations of the model:

- the trajectories of the demands for irrigation;
- the trajectories of the demands for hydropower production;

- the volume q^{Turb} that must be guaranteed at the Turbigo-EUROGEN power plant;
- the trajectory of the daily average energy price θ_t;
- the trajectory of the median natural flow $r_t^{n,\text{median}}$ in the Ticino.

The value of all these parameters must be given. As we anticipated in Section 2.2.2, we adopted:

- the trajectories shown in Figure 7.10 for irrigation and hydropower demands;
- the value 25 m^3/s for the volume q^{Turb};
- the daily average energy prices derived from the values in Figures 2.36, 2.37 of Section 2.2.2.3 of the DVD;
- the trajectory shown in Figure 7.9 for the median natural flow $r_t^{n,\text{median}}$ in the Ticino, which was estimated over the horizon 1974–1998.

These elements define the scenario with respect to which the policies are designed, which is accordingly called *design scenario*.

In Section 8.3 we will present the scenario (called *evaluation scenario*) that will be used to estimate the effects of the alternatives. It contains the design scenario, but is defined by a greater number of elements because the *design indicators* are only a subset of the *evaluation indicators*.

7.6 Formulating the Control Problem

The draft of the (Optimal) Control Problem provided in Section 5.2.6 was incomplete for three reasons:

(1) the constraints to the Problem were not specified;

(2) the control indicators had not been identified;

(3) the objectives were not defined;

(4) the possible existence of a preference structure among the objectives had not been considered.

The constraints have been defined in Chapter 6, the design indicators in Section 7.3. We will now deal with the last two missing elements, to get to a complete definition of the Problem.

7.6.1 The objectives

The system that we are operating on is not deterministic: it contains both stochastic and uncertain variables. A *variable* y_t is said to be *stochastic* when its value is not known, but its probability distribution Φ_t is. A *variable* is *uncertain* when the set of values Ξ_t that it can assume is known, but the probability for each of them is not (for more details see Section 5.7 of THEORY).

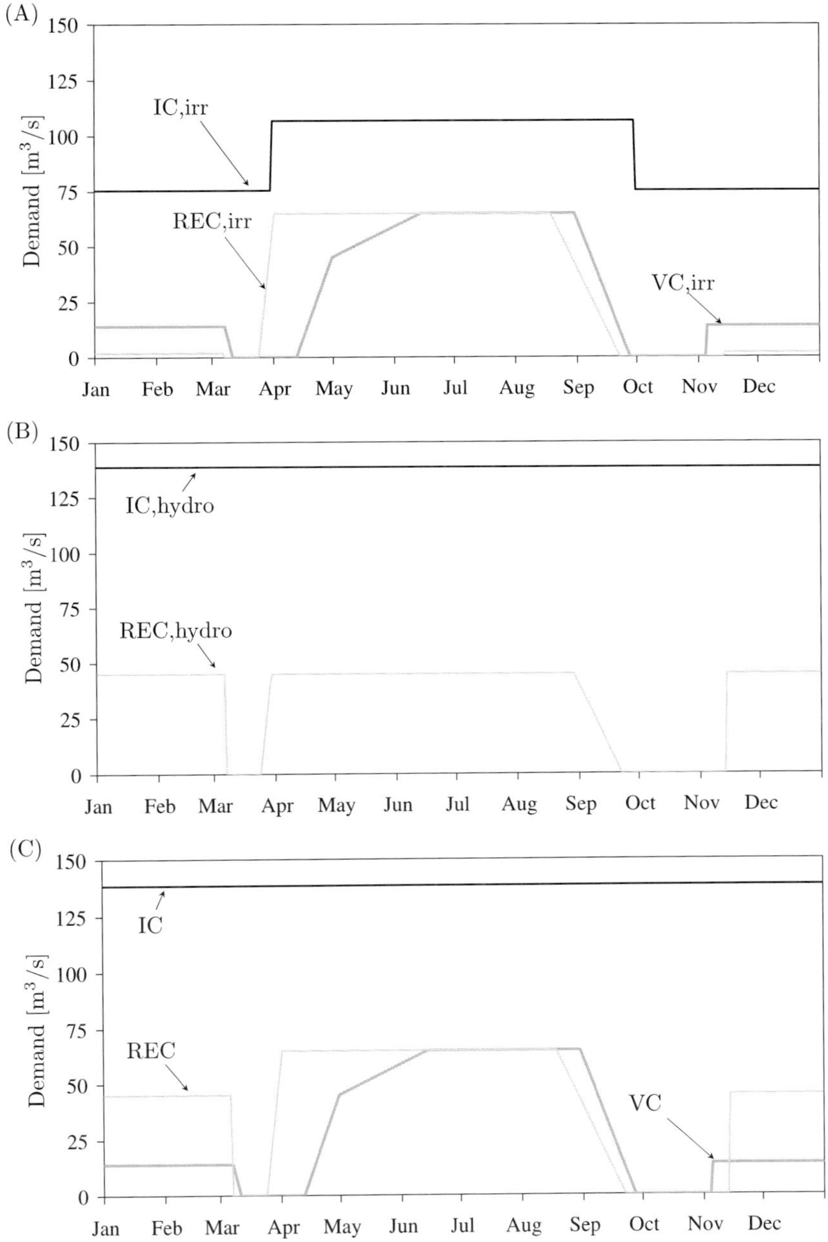

Figure 7.10: Irrigation demands (irr) of Regina Elena (REC), Villoresi (VC) and Industrial (IC) Canals (panel A); hydropower demand (hydro) (panel B) and total demand of each canal (panel C) used in the policy design.

The presence of stochastic (uncertain) elements entails that, even when the policy p has been defined, the values of the design indicators i_k (with $k =$ fl, irr, env, Enel) are stochastic (uncertain), so that they cannot be used for ranking the alternatives. We can, however, obtain a ranking if we consider a statistic J_k for each indicator (see Chapter 9 of THEORY), i.e.

$$J_k = J_k\big(i_k(p)\big) = J_k(p) \quad k = \text{fl, irr, env, Enel} \tag{7.26}$$

In fact, the statistic filters the stochasticity (uncertainty) of the indicator and thus is a deterministic function of the policy. These statistics are the *objectives* that we intuitively introduced in Section 5.2.6. Once a combination of objectives has been chosen, we can look for the policy that minimizes its value, as required by (5.11).

To understand how to obtain the objectives from the design indicators, we deal first with uncertainty and then with stochasticity.

7.6.1.1 Uncertainty

Let us assume for a moment that the inflow is a deterministic variable and that its trajectory over the entire design horizon is known. If the regulation policy is multi-valued, the system is described by the following equations

$$s_{t+1} = f_t(s_t, \mathbf{u}_t, \varepsilon_{t+1}) \qquad t = 0, 1, \ldots, h-1 \tag{7.27a}$$

$$\mathbf{u}_t \in M_t(s_t) \subseteq U_t(s_t) \qquad t = 0, 1, \ldots, h-1 \tag{7.27b}$$

$$|\varepsilon_1, \varepsilon_2, \ldots, \varepsilon_h| \tag{7.27c}$$

$$p = \big\{M_t(\cdot);\ t = 0, 1, \ldots, h-1\big\} \tag{7.27d}$$

Unlike model (7.2), this model does not univocally define the trajectory $\mathcal{F}$ of the system, because we do not know the value assumed by $\mathbf{u}_t$ in every instant, only the set to which it belongs. The choice of a particular value within this set is made by the Regulator and so it is unknown when designing the policies. We should therefore deal with the control as an uncertain variable.

It follows that also the value of the design indicators $i_k(\mathcal{F})$ ($k =$ fl, irr, env, Enel) is not univocally defined, but is an uncertain variable. To measure the performance of policy p, it is thus necessary to consider a statistic of $i_k(\mathcal{F})$. Since the probability distribution of $i_k(\mathcal{F})$ is unknown, the most rational choice is to consider the worst case, namely the maximum value that the indicator can assume. By doing so we adopt a risk aversion attitude, which in the literature is also known as the *Wald criterion* (see Section 9.1.2 of THEORY). Note that considering the worst case is equivalent to considering the performance that the policy guarantees in every case, namely its *certain performance*

$$J_k(p) = \max_{\{\mathbf{u}_t \in M_t(s_t)\}_{t=0,1,\ldots}} i_k(\mathcal{F}) \quad k = \text{fl, irr, env, Enel} \tag{7.28}$$

7.6.1.2 Stochasticity

Let us look at how to deal with stochasticity, which in the Verbano Project is caused by the inflow process. In Section 6.10 we decided to describe it with a cyclostationary AR(0) model, which we estimated in Section 6.5. With this model, for every day t the probability distribution of the disturbance ε_{t+1} is $N(0, 1)$ (see equation (6.17)). In system (7.27) equation (7.27c) must thus be replaced by the following equation

$$\varepsilon_{t+1} \sim N(0, 1) \quad t = 0, 1, \ldots \tag{7.29}$$

so that the statistic $J_k(p)$ in (7.28) turns out to be a stochastic variable. Therefore it will be necessary to consider a statistic of it with respect to the inflow. We could use the Wald criterion again, but, since we know the probability distribution, we can also use a different criterion. We decided for the *Laplace criterion*, which uses the *expected value*.

In conclusion the correct and complete definition of the objective J_k is

$$J_k(p) = \mathop{E}_{\varepsilon_1, \varepsilon_2, \ldots} \left[\max_{\{\mathbf{u}_t \in M_t(s_t)\}_{t=0,1,\ldots}} i_k(\mathcal{F}) \right] \quad k = \text{fl, irr, env, Enel} \tag{7.30}$$

7.6.2 Preferences among the objectives

Now we will consider the preference structure that exists among the objectives. Until this point, we have assumed that all the objectives have the same priority. However, the regulations contained in Law No. 36, of 5th January 1994 (the so-called *Galli Law* from the name of its proponent) make it obligatory to put the hydropower objective in a position that is subordinate to the others. This law establishes, in fact, that "*the use of water for human consumption takes priority over other uses …* [which] *are allowed when there are sufficient resources*" (art. 1, para. I). "*In periods of drought and in cases of water scarcity, during which the abstraction regulations in force are followed, the priority of agricultural use must be guaranteed, after human consumption*" (art. 28, para. IV). And "*With respect for the principle of this law and for the national energy plan … the Inter-Ministerial Committee for Economic Planning is assigned the task of disciplining the use of stored water for hydropower purposes, in order to respond to the emergency water situation*" (art. 30, para. IV).

This hierarchy among the objectives structures the Control Problem and does not allow us simply to make a linear combination of all the objectives, as in (5.9). So what should we do? We will see it in the next section.

7.6.3 The Lexicographic Control Problem

To take account of the preferences among the objectives that are established by legislation, once the triple of structural and normative actions (s, r, m) has been fixed, a first level Control Problem must be formulated, whose objectives are those relative to the *Upstream Flooding*, *Irrigation* and *Downstream Environment* sectors. It has the following form

$$\min_p \left[J_{\text{fl}}(p), J_{\text{irr}}(p), J_{\text{env}}(p) \right] \tag{7.31a}$$

subject to

$$s_{t+1} = f_t(s_t, \mathbf{u}_t, \varepsilon_{t+1}) \qquad t = 0, 1, \ldots \tag{7.31b}$$

$$\mathbf{u}_t \in M_t(s_t) \subseteq U_t(s_t) \qquad t = 0, 1, \ldots \tag{7.31c}$$

$$\varepsilon_{t+1} \sim N(0, 1) \qquad t = 0, 1, \ldots \tag{7.31d}$$

$$p = \{M_t(\cdot);\ t = 0, 1, \ldots\} \tag{7.31e}$$

As we explained in Section 5.2.6, to solve this problem it is necessary to attribute a weight to each of the three objectives J_{fl}, J_{irr} and J_{env}. We denoted the vector of such weights with $\boldsymbol{\lambda}$, and so we can now define the following family of Single-Objective (Optimal) Control Problems, parametric in $\boldsymbol{\lambda}$

$$\min_p \lambda_{\text{fl}} J_{\text{fl}}(p) + \lambda_{\text{irr}} J_{\text{irr}}(p) + \lambda_{\text{env}} J_{\text{env}}(p) \tag{7.32a}$$

subject to the constraints

$$s_{t+1} = f_t(s_t, \mathbf{u}_t, \varepsilon_{t+1}) \qquad t = 0, 1, \ldots \tag{7.32b}$$

$$\mathbf{u}_t \in M_t(s_t) \subseteq U_t(s_t) \qquad t = 0, 1, \ldots \tag{7.32c}$$

$$\varepsilon_{t+1} \sim N(0, 1) \qquad t = 0, 1, \ldots \tag{7.32d}$$

$$p = \{M_t(\cdot);\ t = 0, 1, \ldots\} \tag{7.32e}$$

Once a value of $\boldsymbol{\lambda}$ has been chosen,[13] the solution of this Problem provides an *efficient policy* p^*, namely the sequence of control laws $M_t^*(\cdot)$. As we explained in Section 5.2.1, p^* is a set-valued policy, which means that for every time t and for every value s_t of the state of the system, it provides a set of optimal controls $M_t^*(s_t)$. Since the number of efficient controls can be more than one, it is possible to choose the one (or better, the ones) that minimize the economic loss caused by lost hydropower production. For this purpose, in cascade to problem (7.32), we formulate the following second-level (Optimal) Control Problem

$$\min_p J_{\text{Enel}}(p) \tag{7.33a}$$

subject to

$$s_{t+1} = f_t(s_t, \mathbf{u}_t, \varepsilon_{t+1}) \qquad t = 0, 1, \ldots \tag{7.33b}$$

$$\mathbf{u}_t \in M_t(s_t) \subseteq M_t^*(s_t) \qquad t = 0, 1, \ldots \tag{7.33c}$$

$$\varepsilon_{t+1} \sim N(0, 1) \qquad t = 0, 1, \ldots \tag{7.33d}$$

$$p = \{M_t(\cdot);\ t = 0, 1, \ldots\} \tag{7.33e}$$

Note that, since the controls must be chosen within the set $M_t^*(s_t)$ that is solution of problem (7.32), in equation (7.33c) the set of the feasible controls is no longer $U_t(s_t)$, as in (7.32c), but $M_t^*(s_t)$. Notice also that problem (7.33) is a Single-Objective Control Problem. Its solution is a policy p_i, namely a sequence of control laws $M_t^{i,*}(\cdot)$, which, together with the triple (s, r, m), defines an efficient alternative $A_i = (s, r, m, p_i)$ of the Pareto Frontier. Policy p^* served only as an intermediate element for the computation of p_i, and therefore is abandoned.

We will call problems (7.31) and (7.32) *Primary Multi-Objective Problem* and *Primary (Single-Objective) Problem* respectively. Problem (7.33) is called *Secondary Problem* and the pair (7.31)–(7.33) *Lexicographic Problem*. This latter constitutes the complete formulation of the (Optimal) Control Problem that we sketched out in Section 5.2.6.

Figure 7.12 synthesizes the procedure with which an alternative is identified. By varying the vector of weights $\boldsymbol{\lambda}$, the solution of problem (7.32) describes the Pareto Frontier (see Figure 7.11), i.e. the set of efficient policies associated with the triple (s, r, m) being considered. As this triple varies, the constraints of problem (7.32)–(7.33) also vary, and so different Pareto Frontiers can be obtained. Therefore (7.32)–(7.33) is the instrument that potentially allows us to obtain the whole set $\breve{\mathcal{A}}$ of efficient alternatives, defined by (5.13).

[13] Remember that its component must satisfy the following condition

$$\sum_{k=\text{fl},\text{irr},\text{env}} \lambda_k = 1$$

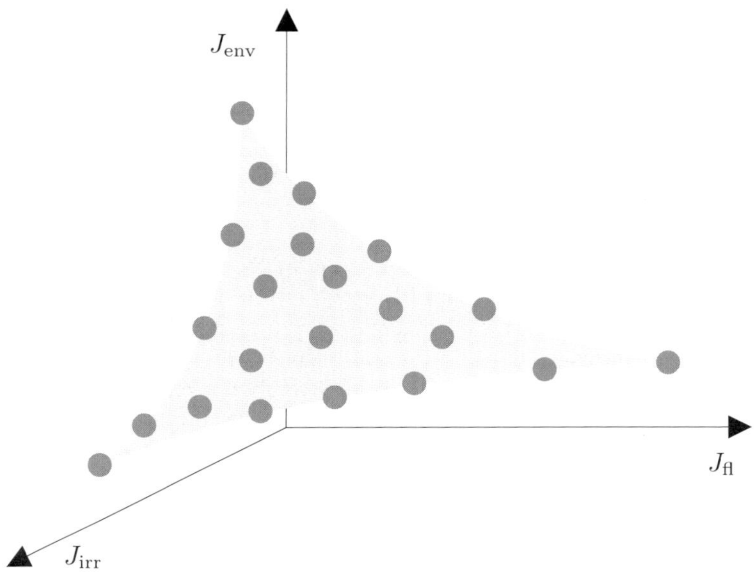

Figure 7.11: A qualitative representation of the Pareto Frontier in the objective space (J_{fl}, J_{irr}, J_{env}). Each point represent the performance of an alternative that solves problem (7.32) for a particular value of $\boldsymbol{\lambda}$.

7.7 Solving the Control Problem: Stochastic Dynamic Programming

It remains now to define the algorithm that allows us to solve the Single-Objective Control Problems (7.32) and (7.33).

The description of the algorithm for a case in which a set-valued policy is sought might be difficult to understand for someone who had not seen algorithms of this type before. To avoid this difficulty, for now, we will limit the description to the case of a point-valued policy. At the end of the section we will present an extension that deals with the algorithm for a set-valued policy.

The Control Problem that we must solve can be regarded as a problem with several stages, in each of which, given the state s_t, the decision $\mathbf{u}_t$ must be made. Each decision $\mathbf{u}_t$ not only has an immediate cost, but it also influences the context in which the following decisions will be made, and therefore their effects, through the state s_{t+1} that it produces at time $t+1$. In making each decision, we must thus weigh up the immediate cost $g_t(s_t, \mathbf{u}_t, \varepsilon_{t+1})$, connected to the transition from s_t to s_{t+1}, and the 'desirability' of the new state s_{t+1} that it leads to. To define a measure of this desirability, the idea is to consider the cost that would be obtained if, starting from s_{t+1}, only optimal decisions were adopted in all the successive transitions, i.e. the decisions that minimize the overall cost from time $t+1$ onwards. Such a minimum overall cost is denoted by the symbol $H^*_{t+1}(s_{t+1})$ and is called *optimal cost-to-go*. From now forward, the asterisk $(*)$ in superscript denotes that the cost is minimum.

If the optimal cost-to-go were known for all the states s_{t+1}, then at time t, for each state s_t, the optimal decision $m^*_t(s_t)$ could be identified by minimizing the expected value

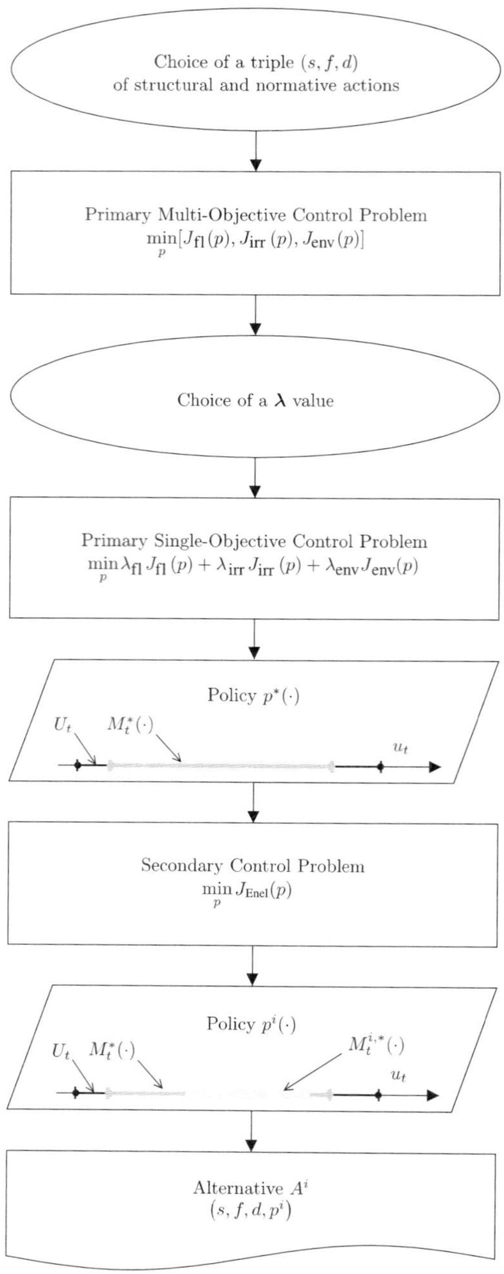

Figure 7.12: A scheme of the procedure for the regulation policy design.

of the sum of the immediate cost g_t and the optimal cost-to-go H^*_{t+1}, with respect to $\mathbf{u}_t$, i.e.

$$m^*_t(s_t) = \arg\min_{\mathbf{u}_t} \underset{\varepsilon_{t+1}\sim N(0,1)}{E} \left[g_t(s_t, \mathbf{u}_t, \varepsilon_{t+1}) + H^*_{t+1}(s_{t+1})\right] \tag{7.34}$$

Consequently, the optimal cost-to-go $H^*_t(s_t)$ associated to state s_t could be computed with the following expression

$$H^*_t(s_t) = \min_{\mathbf{u}_t} \underset{\varepsilon_{t+1}\sim N(0,1)}{E} \left[g_t(s_t, \mathbf{u}_t, \varepsilon_{t+1}) + H^*_{t+1}(s_{t+1})\right] \tag{7.35}$$

This equation, called *Bellman equation*, provides the optimal cost-to-go at time t, given the one at time $t+1$. We can thus think about using (7.35) in a recursive way, proceeding backwards, from the final instant until the initial one. This idea is the basis of *Stochastic Dynamic Programming* (SDP) (Bertsekas, 1976), an algorithm that permits us to compute the optimal costs-to-go for every time t. The optimal policy is completely defined by (7.34) for every time t, once the function $H^*_\cdot(\cdot)$ is known. In turn, this function, which is called *Bellman function*, is univocally defined by the knowledge of the costs-to-go relative to each time t, since these constitute its different time slices.

The reader who is interested in more details is invited to read Chapter 12 of THEORY. Here we just want to underline once again that, due to its recursive nature, this method for computing the optimal cost-to-go (and thus determining the optimal policy) can be applied only if the objectives are expressed by separable cost functionals. This is why the design indicators must be separable as well.

7.7.1 The algorithm

Having understood the concept that is the basis of Stochastic Dynamic Programming, we can present the algorithm to use when the average cost over an infinite horizon is considered, as we decided to do in this Project. This algorithm is known by the name of *Successive Approximations Algorithm* (SAA) (White, 1963) (see Section 12.1.4 of THEORY).

Since the optimal cost-to-go $H^*_t(s_t)$ over an infinite horizon, in the absence of a discount factor, becomes infinite (in mathematical terms, it 'diverges'), it is necessary to substitute it with an indicator that still expresses the 'desirability' of state s_t, but that does not diverge. The simplest solution is to consider, for every state s_t, the difference $h^*_t(s_t)$ between its optimal cost-to-go $H^*_t(s_t)$ and the optimal cost-to-go $H^*_t(\tilde{s}_t)$ of a reference state $\tilde{s}_t$, which has been arbitrarily chosen. This difference is finite even when the time horizon is infinite. By replacing the optimal cost-to-go $H^*_t(s_t)$ in (7.34) with such difference $h^*_t(s_t)$, the same decision is provided.[14] This is how the SAA algorithm works.

More precisely, the SAA algorithm is defined by the following procedure, in which j denotes the iteration being considered, and $\mathcal{S}_{s_t}$ the discrete set to which the state at time t must belong, which is defined by (6.62w).

Algorithm 1 (*SAA*):

Step 0 (*Initialization*): *Set*

$$h^0_t(s_t) = \bar{H}(s_t) \quad \forall s_t \in \mathcal{S}_{s_t};\ t = 0, \dots, T-1 \tag{7.36}$$

[14]In fact, the optimal decision $m^*_t(s_t)$ is not influenced by the constant value $H^*_{t+1}(\tilde{s}_{t+1})$ subtracted from $H^*_{t+1}(s_{t+1})$.

with $\bar{H}(\cdot)$ being an arbitrary function, for example identically zero. Then choose an arbitrary state $\bar{s}_t \in \mathcal{S}_{s_t}$, for $t = 0, \ldots, T-1$. Finally, set $i = 1$.

Step 1: *For $t = T-1, \ldots, 0$ compute the T functions $H_t^i(\cdot)$ defined by the following recursive equation*

$$H_t^i(s_t) = \min_{\mathbf{u}_t} \mathop{E}_{\varepsilon_{t+1}} \left[g_t(s_t, \mathbf{u}_t, \varepsilon_{t+1}) + H_{t+1}^i(s_{t+1}) \right] \quad \forall s_t \in \mathcal{S}_{s_t} \tag{7.37a}$$

subject to

$$s_{t+1} = f_t(s_t, \mathbf{u}_t, \varepsilon_{t+1}) \tag{7.37b}$$

$$\mathbf{u}_t \in U_t(s_t) \tag{7.37c}$$

$$\varepsilon_{t+1} \backsim N(0, 1) \tag{7.37d}$$

where

$$H_T^i(s) = h_0^{i-1}(s) \quad \forall s \in \mathcal{S}_{s0} \tag{7.37e}$$

Step 2 (*Termination test*): *Let*

$$h_t^i(s_t) = H_t^i(s_t) - H_t^i(\bar{s}_t) \quad \forall s_t \in \mathcal{S}_{s_t};\ t = 0, \ldots, T-1 \tag{7.38}$$

If

$$\begin{aligned} &\left| h_t^i(s_t) - h_t^{i-1}(s_t) \right| < \alpha \quad && \forall s_t \in \mathcal{S}_{s_t};\ t = 0, \ldots, T-1 \\ &h_0^j(s) - h_0^{j-1}(s) < \alpha && \forall s \in \mathcal{S}_{s0} \end{aligned} \tag{7.39}$$

with α being a given small and preassigned number, the algorithm terminates and the last T functions $H_t^i(\cdot)$ computed constitute a good approximation of the Bellman function $H^*(\cdot)$, which proves to be periodic of period T and so it is univocally defined by T time slices $H_t^*(\cdot)$, $t = 0, \ldots, T-1$. Thus one assumes

$$H_t^*(s_t) = H_t^i(s_t) \quad \forall s_t \in \mathcal{S}_{s_t};\ t = 0, \ldots, T-1$$

Otherwise increase i by 1 and return to Step 1.

The optimal policy p^* that we are looking for is given by the sequence of control laws $m_t^*(\cdot)$

$$p^* = \left\{ m_t^*(\cdot);\ t = 0, 1, \ldots \right\} \tag{7.40}$$

which are periodic, with period one year, and such that the value $m_t^*(s_t)$ that the tth of them assumes in correspondence to the pair (t, s_t) equals the control $\mathbf{u}_t$ that minimizes (7.37), i.e.

$$\begin{aligned} m_t^*(s_t) = \arg\min_{\mathbf{u}_t} \mathop{E}_{\varepsilon_{t+1} \sim N(0,1)} &\left[g_t(s_t, \mathbf{u}_t, \varepsilon_{t+1}) + H_{t+1}^*(s_{t+1}) \right] \\ &\text{subject to the constraints (7.37b)–(7.37e)} \end{aligned} \tag{7.41a}$$

7.7.2 Extension to set-valued policies

Equation (7.41a) is not necessarily minimized by just one value of $\mathbf{u}_t$: often it is minimized by several values of $\mathbf{u}_t$, which we collect in the following set

$$M_t^*(s_t) = \left\{\mathbf{u}_t \in U_t(s_t) \colon \quad \underset{\varepsilon_{t+1} \sim N(0,1)}{E} \left[g_t(s_t, \mathbf{u}_t, \varepsilon_{t+1}) + H_{t+1}^*(s_{t+1})\right] \text{ is minimal}\right\} \quad (7.42)$$

This set, being a function of s_t, constitutes a set-valued control law. The sequence of these sets is the optimal set-valued policy that we are looking for. For more details see Section 12.3 of THEORY).

7.8 Discretization of the system and computing time

In Section 6.10 we saw that state, control and disturbance belong respectively to the sets $\mathcal{S}_{s_t}$, $\mathcal{S}_{\mathbf{u}_t}$ and $\mathcal{S}_{\varepsilon_t}$. These sets are uncountable and infinite, while the solution of equation (7.37a), which must be carried out for every $s_t \in \mathcal{S}_{s_t}$, requires that $\mathcal{S}_{s_t}$ be countable and finite. The analytical solution, in fact, is impracticable and thus equation (7.37a) must be solved via numerical computation. Similarly, minimizing with respect to the control and computing the expected value also requires that the sets $\mathcal{S}_{\mathbf{u}_t}$ and $\mathcal{S}_{\varepsilon_t}$ be discretized as well. In conclusion, the model should be *discretized*. This operation is not at all trivial and is indeed rather delicate, since it can appreciably influence the results.

For each variable we must decide how many classes of discretization to consider and how to choose them. These two aspects are clearly strictly linked. The greater the number of classes, the better the sets $\mathcal{S}_{s_t}$, $\mathcal{S}_{\mathbf{u}_t}$ and $\mathcal{S}_{\varepsilon_t}$ are approximated, but the computing time required for the solution of the Problem and the memory space allocated to store the results increase. This is a fundamental point that we have mentioned several times. To completely understand the implications, consider a generic system in which the state has dimension n, the control dimension m and the disturbance dimension v. Suppose that each state variable is discretized in 100 classes. Then, the Bellman function would be computed on a grid of 100^n points. In correspondence to each of them, a minimization must be carried out with respect to the control $\mathbf{u}_t$. This is done in the $\mathbb{R}^m$ space. Finally, for each value of $\mathbf{u}_t$, the expected value of the cost-to-go must be computed, which means performing a numerical integration in the space $\mathbb{R}^v$. The computations are therefore time-consuming and it follows that the SAA algorithm can be applied only if dimensions n, m and v are low and if sets $\mathcal{S}_{s_t}$, $\mathcal{S}_{\mathbf{u}_t}$ and $\mathcal{S}_{\varepsilon_t}$ are discretized into a number of classes which is not too large.

When the sets $\mathcal{S}_{s_t}$, $\mathcal{S}_{\mathbf{u}_t}$ and $\mathcal{S}_{\varepsilon_t}$ vary in time, the discretization procedure is applied separately for every time t.

7.8.1 The classes of state

Given the set $\mathcal{S}_{s_t}$, a set $\tilde{\mathcal{S}}_{s_t}$ of the following form, contained in $\mathcal{S}_{s_t}$, must be defined

$$\tilde{\mathcal{S}}_{s_t} = \{s_t \colon \underline{s}_t \leqslant s_t \leqslant \bar{s}_t\} \quad (7.43)$$

where $\underline{s}_t$ and $\bar{s}_t$ are the lower and upper values of the storage s_t to be considered. They must be chosen in such a way that the probability that a value $s_t \notin \tilde{\mathcal{S}}_{s_t}$ may occur is negligible. The procedure for identifying them is explained in the last section of this chapter. In the

Verbano Project it provided the following values

$$\underline{s}_t = 42.3\ \mathrm{Mm}^3 \qquad \bar{s}_t = 1645.47\ \mathrm{Mm}^3 \tag{7.44}$$

which correspond to the following levels

$$\underline{h}_t^{\mathrm{SC}} = -0.55\ \mathrm{m} \qquad \bar{h}_t^{\mathrm{SC}} = 7.03\ \mathrm{m} \tag{7.45}$$

Then we must subdivide the interval $[\underline{s}_t, \bar{s}_t]$ into N adjacent intervals. It is not opportune to adopt intervals with equal width: it is better to use smaller ones for those storage values that occur more frequently, and wider ones for the others. In practice, this means adopting a dense and regular discretization grid in correspondence to the storage values within the regulation range or a little higher, while for the other values the discretization intervals can be progressively wider as the storage increases. Let us denote the midpoint of each interval with $\hat{s}^i$: then the set of possible state values for the discretized system is the following

$$\hat{\mathcal{S}}_{s_t} = \left\{\hat{s}_t^1, \hat{s}_t^2, \ldots, \hat{s}_t^N\right\} \tag{7.46}$$

Set $\hat{\mathcal{S}}_{s_t}$ for Verbano is presented in Section 7.8.1 of the DVD.

In conclusion, in the equations of the SAA algorithm we can replace set $\mathcal{S}_{s_t}$ with set $\hat{\mathcal{S}}_{s_t}$.

7.8.2 The classes of control

In the same way, the set of controls for the discretized system can be identified. It is a rectangular set $\tilde{\mathcal{S}}_{\mathbf{u}_t} \subseteq \mathcal{S}_{\mathbf{u}_t}$

$$\tilde{\mathcal{S}}_{\mathbf{u}_t} = \left\{\left(u_t^{\mathrm{MIO}}, u_t^{\mathrm{TIC}}\right) \colon \underline{u}_t^{\mathrm{MIO}} \leqslant u_t^{\mathrm{MIO}} \leqslant \bar{u}_t^{\mathrm{MIO}} \text{ and } \underline{u}_t^{\mathrm{TIC}} \leqslant u_t^{\mathrm{TIC}} \leqslant \bar{u}_t^{\mathrm{TIC}}\right\} \tag{7.47}$$

The definition of the interval $[\underline{u}_t^{\mathrm{MIO}}, \bar{u}_t^{\mathrm{MIO}}]$ is very simple. The lower value is clearly 0. As for the upper value, note that the regulator can decide the release only when the lake is within the regulation range. Therefore the maximum possible value $u_t^{\mathrm{MIO,max}}$ corresponds to the water volume that can be released when the Miorina dam is completely open and the lake level is equal to the upper extreme of the regulation range, i.e.

$$u_t^{\mathrm{MIO,max}} = N_t^{\max}\left(s_t^{\mathrm{upp}}\right) \cdot \Delta \tag{7.48}$$

where $N_t^{\max}(\cdot)$ is the maximum instantaneous storage–discharge relation, s_t^{upp} the storage that corresponds to the upper extreme h_t^{upp} of the regulation range and Δ the number of seconds in each day (86 400). The value $u_t^{\mathrm{MIO,max}}$ thus depends on the pair (stage–discharge relation/regulation range) that is considered. For example, for the pair (SD$_{+600}$/RANGE$_{\mathrm{ENL}}$), we get

$$N_t^{\max}\left(s_t^{\mathrm{upp}}\right) = 1167\ \mathrm{m}^3/\mathrm{s} \qquad u_t^{\mathrm{MIO,max}} = 100.8\ \mathrm{Mm}^3 \tag{7.49}$$

The definition of the interval $[\underline{u}_t^{\mathrm{TIC}}, \bar{u}_t^{\mathrm{TIC}}]$ immediately follows from the definition of $[\underline{u}_t^{\mathrm{MIO}}, \bar{u}_t^{\mathrm{MIO}}]$, given that between u_t^{MIO} and u_t^{TIC} the following relation holds

$$\begin{cases} u_t^{\mathrm{TIC}} \leqslant u_t^{\mathrm{MIO}} - q_t^{\mathrm{MEF}} - q^{\mathrm{Turb}} & \text{if } u_t^{\mathrm{MIO}} - q_t^{\mathrm{MEF}} - q^{\mathrm{Turb}} > 0 \\ u_t^{\mathrm{TIC}} = 0 & \text{otherwise} \end{cases} \tag{7.50}$$

It follows that

$$\underline{u}_t^{\mathrm{TIC}} = \underline{u}_t^{\mathrm{MIO}} \qquad \bar{u}_t^{\mathrm{TIC}} = \bar{u}_t^{\mathrm{MIO}} - q_t^{\mathrm{MEF}} - q^{\mathrm{Turb}} \tag{7.51}$$

Once the set $\tilde{\mathcal{S}}_{\mathbf{u}_t}$ is defined, we must proceed with its partition and get to the definition of the set

$$\hat{\mathcal{S}}_{\mathbf{u}_t} = \{\hat{\mathbf{u}}_t^1, \hat{\mathbf{u}}_t^2, \ldots, \hat{\mathbf{u}}_t^M\} \tag{7.52}$$

The choice of the classes depends on the water demand pattern that appears in the definition of the design indicators (see Section 7.3) and thus in the objectives. To understand why, let us suppose that we must determine the extreme policy that minimizes the irrigation objective, i.e. that we must solve the following Control Problem

$$\min_p J_{\mathrm{irr}}(p) \tag{7.53}$$

subject to the constraints (6.62). As we said in Section 7.3.2, equation (7.22) defines the design indicator for the *Irrigation* sector on the basis of the following step cost

$$\left[(W_t^{\mathrm{VC}} - q_{t+1}^{\mathrm{VC}})^+ + (W_t^{\mathrm{REC,irr}} - q_{t+1}^{\mathrm{REC}})^+ + (W_t^{\mathrm{IC,irr}} - q_{t+1}^{\mathrm{IC}})^+ \right]^{12} \tag{7.54}$$

On a given day, the overall irrigation demand for the three canals is $W_t = W_t^{\mathrm{VC}} + W_t^{\mathrm{REC}} + W_t^{\mathrm{IC}}$. For example, on day 200 the overall demand W_{200} is equal to 236.85 m^3/s. The discretization of the control should be such that there exist control values that allow for the volume r_{t+1}^r (that is the volume to be divided among the three canals given by (6.35)) to be equal to W_t. Otherwise, to make the step-cost equal zero, a volume greater than W_t should be released and thereby the lake storage would be reduced more than necessary and the probability of producing a deficit in the following days would be increased. To contain this waste, the optimal policy would suggest releasing a volume greater or smaller than W_t on alternate days. This would produce oscillations that would negatively influence the value of the indicator $i_{\mathrm{Irr\_D\_3}}$, which measures the variability of the supply. Note that such an effect would not be induced by the alternative under study; it would be due to a bad choice in the discretization of the control. Among the vectors of controls of the set (7.52) it is thus necessary to include those for which the u_t^{MIO} component satisfies the following relation

$$u_t^{\mathrm{MIO}} = W_t + q_t^{\mathrm{MEF}} + u_t^{\mathrm{TIC}} \tag{7.55}$$

A second observation must be made. The only way to contain the computing time is to reduce the number of classes. Often it is not possible to substantially act upon u_t^{MIO}, but it is easier to act upon u_t^{TIC}. In fact, once the weights attributed to the three objectives J_{fl}, J_{irr} and J_{env} in the formulation of the Problem have been chosen, it is possible to deduce the values for u_t^{TIC} that will be more probably suggested by the efficient policy. For example, when zero weight is attributed to the environmental objective, it is obvious that u_t^{TIC} will always be zero. In Problems of this type it is therefore opportune to replace the set $\hat{\mathcal{S}}_{\mathbf{u}_t}$ with its subset composed of pairs characterized by $u_t^{\mathrm{TIC}} = 0$. On the contrary, consider the case when unitary weight is attributed to the environmental objective and zero to the others. Since the optimal condition for the environment is to get as close as possible to the natural situation, in which no abstraction occurs, the controls suggested by the optimal policy will be such as to satisfy the relation

$$u_t^{\mathrm{TIC}} = u_t^{\mathrm{MIO}} - q_t^{\mathrm{MEF}} - q_t^{\mathrm{Turb}} \tag{7.56}$$

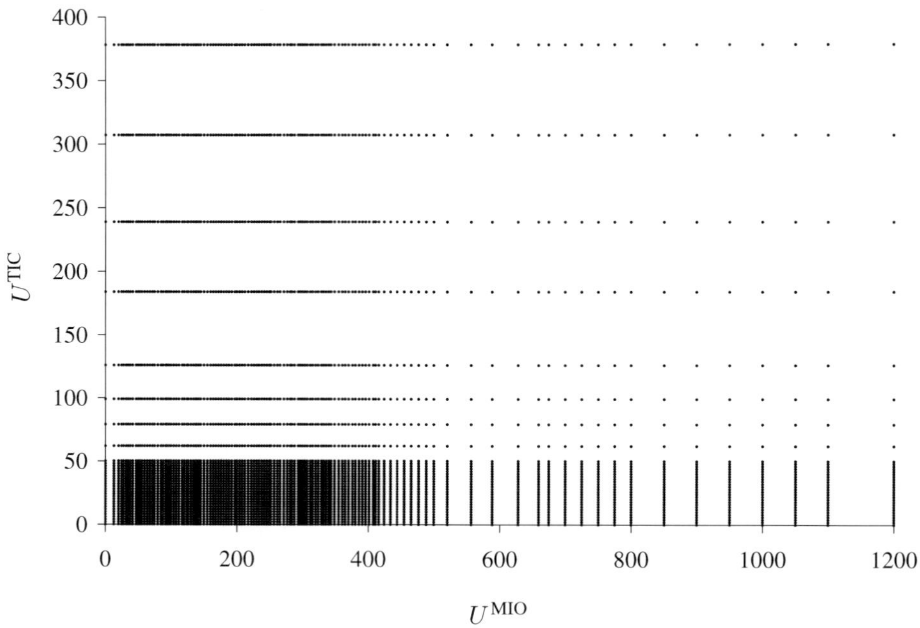

Figure 7.13: The set of controls that are used for the policy design, when the weight attributed to the environmental objective is lower than the weights attributed to the other objectives.

In this case we should consider a subset of $\hat{\mathcal{S}}_{\mathbf{u}_t}$ that contains only the pairs that satisfy equation (7.56). Finally, when all three of the objectives are considered, and the weight attributed to J_{env} is not too high, one expects that the efficient policy will not suggest very high values for u_t^{TIC}. Then it is advisable that $\hat{\mathcal{S}}_{\mathbf{u}_t}$ assume the form shown in Figure 7.13 (the corresponding values are reported in Section 7.8.2 of the DVD). Afterwards, once the policy has been designed, we will have to make sure that it was not arbitrarily constrained by the choice of set $\hat{\mathcal{S}}_{\mathbf{u}_t}$. To do this, it is sufficient to verify that the highest class of values for u_t^{TIC} was never used.

Once $\hat{\mathcal{S}}_{\mathbf{u}_t}$ has been defined, we substitute $U_t(s_t)$ with $U_t(s_t) \cap \hat{\mathcal{S}}_{\mathbf{u}_t}$ in equation (7.37c) of the SAA algorithm.

7.8.3 The classes of disturbance

Similarly, given the set $\mathcal{S}_{\varepsilon_t}$, a set $\tilde{\mathcal{S}}_{\varepsilon_t} \subseteq \mathcal{S}_{\varepsilon_t}$ must be defined

$$\tilde{\mathcal{S}}_{\varepsilon_t} = \{\varepsilon_t \colon \underline{\varepsilon_t} \leqslant \varepsilon_t \leqslant \bar{\varepsilon}_t\} \tag{7.57}$$

in such a way that the probability that $\varepsilon_t \notin \tilde{\mathcal{S}}_{\varepsilon_t}$ may occur is negligible. Since the disturbance ε_t has normal distribution $N(0, 1)$ (see Section 6.5), it is sufficient to set $\underline{\varepsilon_t} = -\vartheta\sigma_{t_{\text{MOD}\,T}}$ and $\bar{\varepsilon}_t = \vartheta\sigma_{t_{\text{MOD}\,T}}$, where ϑ is chosen to be positive and sufficiently large. For example, if $\vartheta = 4$, the probability that $\varepsilon_t \notin \tilde{\mathcal{S}}_{\varepsilon_t}$ is equal to $3 \cdot 10^{-5}$.

Then a partition $(\tilde{\mathcal{S}}^1_{\varepsilon_t}, \tilde{\mathcal{S}}^2_{\varepsilon_t}, \ldots, \tilde{\mathcal{S}}^v_{\varepsilon_t})$ of $\tilde{\mathcal{S}}_{\varepsilon_t}$ is defined, for example by subdividing $\tilde{\mathcal{S}}_{\varepsilon_t}$ into v intervals of equal width. The set of the midpoints of such intervals is the set

$$\hat{\mathcal{S}}_{\varepsilon_t} = \{\hat{\varepsilon}^1_t, \hat{\varepsilon}^2_t, \ldots, \hat{\varepsilon}^v_t\} \tag{7.58}$$

of the discretized values of ε_t. The probability $\hat{\phi}_t(\hat{\varepsilon}^i)$ associated to $\hat{\varepsilon}^i$ is computed with the following formula

$$\hat{\phi}_t(\hat{\varepsilon}^i_t) = \int\limits_{\tilde{\mathcal{S}}^i_{\varepsilon_t}} \phi_t(\eta)\, d\eta \tag{7.59}$$

where ϕ_t is the *probability density function* for a $N(0, 1)$ variable.

Since ε_t has become a discrete variable, the expected value that appears in equation (7.37a) transforms into a weighted sum over all the elements of $\hat{\mathcal{S}}_{\varepsilon_{t+1}}$ and equation (7.37a) assumes the following form

$$H^j_t(s_t) = \min_{\mathbf{u}_t} \sum_{\hat{\varepsilon}_{t+1} \in \hat{\mathcal{S}}_{\varepsilon_{t+1}}} \left[g_t(s_t, \mathbf{u}_t, \hat{\varepsilon}_{t+1}) + H^j_{t+1}(s_{t+1})\right]\hat{\phi}_t(\hat{\varepsilon}_{t+1}) \quad \forall s_t \in \hat{\mathcal{S}}_{s_t} \tag{7.60}$$

The classes of disturbance considered in the Verbano Project are reported in Section 7.8.3 of the DVD.

7.8.4 Computing the storage bounds

The minimum and maximum storage values $\underline{s}_t$ and $\bar{s}_t$, namely the lower and upper bounds of $\tilde{\mathcal{S}}_{s_t}$, must be chosen in such a way that the probability that a value $s_t \notin \tilde{\mathcal{S}}_{s_t}$ may occur is negligible. The lower bound $\underline{s}_t$ is defined by the following condition

$$\begin{aligned}
\underline{s}_t &= \arg\max_{s_t} s_t \\
s_{t+1} &= s_t + \underline{a}_{t+1} - R_t\left(\bar{u}^{\mathrm{MIO}}_t, s_t, \underline{a}_{t+1}\right) \geqslant s_t \\
\underline{a}_{t+1} &= \sigma_{t_{\mathrm{MOD}\,T}} \cdot \underline{\varepsilon}_{t+1} + \mu_{t_{\mathrm{MOD}\,T}}
\end{aligned}$$

It defines $\underline{s}_t$ as the largest storage that is not lower than the storage obtained at time $t+1$ by starting from that storage with the maximum release and the minimum inflow. Similarly, the upper bound $\bar{s}_t$ is defined by the following condition

$$\begin{aligned}
\bar{s}_t &= \arg\min_{s_t} s_t \\
s_{t+1} &= s_t + \bar{a}_{t+1} - R_t\left(\underline{u}^{\mathrm{MIO}}_t, s_t, \bar{a}_{t+1}\right) \leqslant s_t \\
\bar{a}_{t+1} &= \sigma_{t_{\mathrm{MOD}\,T}} \cdot \bar{\varepsilon}_{t+1} + \mu_{t_{\mathrm{MOD}\,T}}
\end{aligned}$$

Notation In the following, in order to simplify the notation, we will write s^i_t to denote the ith value of the state of the discretized system, i.e. we will omit the 'hat'. The context will make it clear if we are speaking of the state of the original system or of the state of the discretized one. For the same reason we will omit the 'hat' also from the control and disturbance symbols.

Chapter 8

Estimating Effects*

EB, FP, RSS and EW

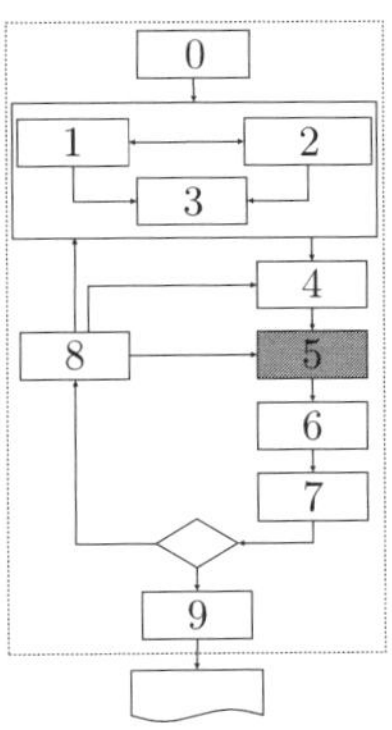

Figure 8.1: The phases of the PIP procedure. Phase 5, which is analysed in this chapter, is highlighted.

In the last three chapters we explained how to design the alternatives: in particular, in Chapter 7 we concentrated on policy design and showed how efficient (in the Pareto sense) policies can be designed by solving an (Optimal) Control Problem. We can thus associate a set of efficient policies p to each of the triples (s, r, m) of structural and normative actions that we want to consider. Each of the quadruples (s, r, m, p) that is identified in this way (structural and normative actions plus a policy) constitutes an alternative. To individuate the alternatives that obtain wide agreement from the Stakeholders, their effects on all the sectors identified in Chapter 4 must be estimated. This is Phase 5 of the PIP procedure (see Figure 8.1) which produces a matrix, called *Matrix of the Effects*, whose element (i, j) is the estimate of the value of the jth indicator produced by the ith alternative.

In Chapter 4 we said that the effects are quantified through (evaluation) indicators, whose values are estimated by simulating the system subjected to the alternative under examination. This is why we must establish a simulation procedure, which, in turn, requires the definition of the evaluation scenario and of the evaluation horizon. Once the estimates of the indicators have been obtained for all the alternatives, before moving into the *Evaluation* phase, we must make sure that each indicator is representative of the effects that we want to measure with it. These are the issues that we will study in this chapter.

8.1 Uncertainty and stochasticity

As we said in previous chapters, the policy p_i associated to alternative A_i is set-valued, in the sense that, at every time t, given the storage s_t, it proposes the set $M_t^{i,*}(s_t)$ of all the

*This chapter is based on Chapter 8 (pp. 219–238) of *Modellistica Integrata e decisione partecipata in Pratica: Il Progetto Verbano*, by R. Soncini-Sessa, McGraw-Hill, 2004.

decisions ($u_t^{\text{MIO}}, u_t^{\text{TIC}}$) that provide the 'best' system performance over the long-term (see Section 5.2.1). The effects that alternative A_i produces on the system are quantified by indicators whose value can be computed from the trajectories $\mathcal{F}$ of all the hydrological variables of the system (the lake level, the release at Miorina, the flow in the canals and in the Ticino, and the Ticino level at Pavia) over the evaluation horizon (see Section 7.1). In order to determine these trajectories, the evolution of the system must be *simulated*, by assuming that it is subject to the evaluation scenario (which will be defined in Section 8.3) and that alternative A_i has been implemented. The simulation starts from an initial time (conventionally called $t = 0$) and from a given state $\breve{s}_0$, and it should be carried out over a horizon which, to be coherent with the design phase, should be infinite. An infinite simulation horizon may cause several difficulties and so we must consider it carefully.

The reader certainly remembers that the system is characterized by the presence of uncertain and stochastic variables, from which it follows that the value of the indicators is also uncertain. Thus, it is not possible to use them directly to compare the alternatives and identify the efficient ones. In Section 7.6.1 we saw that, to overcome this difficulty, the policies can be compared on the basis of a statistic of the indicators. In the evaluation phase the same difficulty arises, but we will see that, from an algorithmic point of view, the solution adopted in Chapter 7 is not always appropriate. The reason is that, while in Chapter 7 the indicators considered were the *design indicators*, here they are the *evaluation indicators*. The difference is that the first, by construction, are always defined by separable functions, while the second may not be.

8.1.1 Uncertainty

Let us assume for the moment that only uncertain disturbances affect the system, i.e. let us assume that the inflow is deterministically known. As we have already noticed, the uncertainty about the control that the Regulator chooses at every time t from $M_t^{i,*}(s_t)$ makes the indicators uncertain, and therefore the trajectories $\mathcal{F}$ followed by the system variables cannot be univocally determined. Since this aspect is very important, let us analyse what happens during the simulation in more detail.

Given the storage $\breve{s}_0$ at time $t = 0$, the storage s_1 can be computed through equation (6.29), which becomes

$$s_1 = \breve{s}_0 + \breve{a}_1 - R_0\big(u_0^{\text{MIO}}, \breve{s}_0, \breve{a}_1\big) \tag{8.1}$$

This shows that, even if $\breve{a}_1$ were deterministically known, s_1 would not be, because we do not know the value u_0^{MIO} of the release decision, which is one of the arguments of the function $R_0(\cdot)$. We do know, however, that u_0^{MIO} is a component of the control vector ($u_0^{\text{MIO}}, u_0^{\text{TIC}}$), which belongs to the set $M_0^{i,*}(\breve{s}_0)$. Thus we can determine the set χ_1 of all the storages s_1 that would be produced as the Regulator's choice varies. We will call it set of *reachable states*. The same observation can be repeated at time $t = 1$ and at all the following instants, because the reasoning is the same if we substitute $\breve{s}_0$, which is deterministically known, with $s_1 \in \chi_1$. Therefore, the simulation of an uncertain system does not provide the trajectory of its state, which is also uncertain, but provides the trajectory $\chi_0, \chi_1, \ldots$ of the reachable states set.

Since our system is periodic with period T, and discrete (i.e. its storage can assume only n values, as shown in Section 7.8), it is possible to prove, under very broad hypotheses which are in practice always satisfied, that the trajectory of the reachable states set converges towards a cycle $\bar{\chi}_0, \bar{\chi}_1, \ldots, \bar{\chi}_{T-1}$ in a finite number of steps. From this we can conclude that it is not necessary to simulate the system over an infinite horizon, but that the simulation can be stopped when the cycle has been reached.

For every time t, we describe the set $\bar{\chi}_t$ with a vector of n Boolean variables, whose hth component equals 0 if the state s_t^h cannot occur, and 1 if it can. To keep the notation as simple as possible, we will use the symbol $\bar{\chi}_t$ to denote both the vector and the set.

Now, once the cycle $\bar{\chi}_0, \bar{\chi}_1, \ldots, \bar{\chi}_{T-1}$ of the reachable sets has been computed, i.e. once all the trajectories that the system can follow have been determined, the uncertain indicators are completely characterized and we can compute a statistic of them.

Just one statistic is of interest: the *worst case* (*Wald criterion*). All the others, such as the *expected value* (*Laplace criterion*), cannot be computed, since they require that the probability distributions of the state be known, i.e. that the state be a stochastic variable.

When the indicator i_k is a separable function and it represents a cost,[1] its worst (maximum) value over the cycle $\bar{\chi}_0, \bar{\chi}_1, \ldots, \bar{\chi}_{T-1}$ can be computed with the following expression

$$\max_{\{\mathbf{u}_t \in M_t^i(s_t)\}_{t=0,\ldots,T-1}} \left[i_k(\mathcal{F}) \right] = \max_{t=0,\ldots,T-1} \left[\bar{\chi}_t' \odot \left(\max_{\mathbf{u}_t \in M_t^i(s_t)} \mathbf{g}_t^k \right) \right] \tag{8.2a}$$

where $\mathbf{g}_t^k$ is a vector constituted by the n values that the step cost $g_t^k(\cdot)$ associated to the indicator assumes in correspondence to the n storage values of the discretized system, i.e.

$$\mathbf{g}_t^k = \left| g_t^k\left(s_t^1, \mathbf{u}_t, a_{t+1}\right), \ldots, g_t^k\left(s_t^n, \mathbf{u}_t, a_{t+1}\right) \right|' \tag{8.2b}$$

and the operator $\odot$ effects a sort of 'multiplication' between the two factors to which it is applied, whose result, denoted with b_t, is defined as follows

$$b_t = \max_h \left[\chi_t^h \max_{\mathbf{u}_t \in M_t^i(s_t^h)} g_t^k\left(s_t^h, \mathbf{u}_t, a_{t+1}\right) \right] \tag{8.3}$$

In other words, b_t is the worst step cost that can occur at time t, considering all the reachable states in $\bar{\chi}_t$ and all the controls that the Regulator can choose in correspondence to each reachable state. The values of the pair $(s_t, \mathbf{u}_t)$ that appear in equation (8.2b) are recursively determined with the simulation model, which coincides with equation (6.62), with the exception of (6.62a) and (6.62b), which are substituted by the historical time series of the inflow (a_{t+1}).

When, instead, the indicator is not a separable function, there is no algebraic formula to compute its worst case value quickly: the only way to determine it is to simulate all the possible choices that the Regulator could make, i.e. explore the entire decision tree. In our case, the number of possible combinations is enormous: for example, if at every time instant the set $M_t^{i,*}(s_t)$ contained only 10 elements, i.e. if the Regulator's choice were from only 10 controls, and if the evaluation horizon were only one year, the number of possible combinations would be 10^{365}. To appreciate the size of this number just think that the number of seconds that have passed since the Big Bang is in the order of 10^{21}. The computing time required is therefore huge (millions of billions of times the age of the universe). Therefore, when the objective is not separable, the worst case can almost never be determined.

[1] When the indicator i_k represents a benefit, the worst value is the minimum, which can be determined with a formula similar to equation (8.2) in which all the max operators are substituted by min operators.

Unfortunately, as we already observed in Section 7.3, some of the evaluation indicators in our Project are not defined by separable functions. In order to evaluate the values of all indicators, it is thus necessary to turn to a *deterministic simulation*, and to eliminate the uncertainty produced by the Regulator's choice. To do this, we must 'simulate' his choice as well, by defining a model that reproduces his behaviour. Given the function that it performs, we will call this model an *extractor*. We have thus identified an important difference with respect to the policy design phase, where we were able to deal with uncertainty explicitly, since the control indicators are, by construction, always separable.

To define the extractor we remind the reader of what we observed in Section 3.5: when it is physically possible and the lake level is not too high, the Regulator releases a volume equal to the overall water demand of the canals plus the MEF, i.e. equal to $W_t^{\text{REC}} + W_t^{\text{VC}} + W_t^{\text{IC}} + q_t^{\text{MEF}}$. The water is then allocated to the canals and the Ticino according to the distribution policy (see Section 3.4). We can thus assume that, at every time t, in the set $M_t^{i,*}(\breve{s}_t)$ the Regulator prefers, if they exist, controls that do not produce any deficit for the irrigators and hydropower producers, i.e. such that

$$u_t^{\text{MIO}} - u_t^{\text{TIC}} \geqslant W_t^{\text{REC}} + W_t^{\text{VC}} + W_t^{\text{IC}} + q_t^{\text{MEF}} \tag{8.4}$$

and that, among such controls, he chooses the control $\mathbf{u}_t = (\breve{u}_t^{\text{MIO}}, \breve{u}_t^{\text{TIC}})$ that is closest to the pair

$$\mathbf{u}_t^{\text{irr+hyd}} = \begin{bmatrix} W_t^{\text{REC}} + W_t^{\text{VC}} + W_t^{\text{IC}} + q_t^{\text{MEF}} \\ 0 \end{bmatrix} \tag{8.5}$$

The vector $\mathbf{u}_t^{\text{irr+hyd}}$ defines the *reference* which the Regulator wants to get close to, and therefore we will call this behaviour model *extractor with irrigation/hydropower reference*. With this behaviour the Regulator aims at maximizing irrigation and hydropower performances: thus he provides the water users with nothing more than their demand and leaves the least possible to the Ticino. If no controls exist in $M_t^{i,*}(\breve{s}_t)$ that satisfy condition (8.4), the one closest to the reference (8.5) will be chosen.

Note that the model proposed is extremely simple, because it assumes that the Regulator makes his decision each day without considering the decisions that were made on the previous days. The choice of this model, though realistic, is arbitrary, just like the choice of the reference $\mathbf{u}_t^{\text{irr+hyd}}$. For this reason the consequences of this choice on the indicators' estimates should be analysed in greater detail. We will deal with this further on, in Section 8.4, after the procedure for estimating the indicators has been completed. Before that, we must consider how to deal with stochastic variables.

8.1.2 Stochasticity

We have seen how to deal with uncertain variables; now we will see how to deal with stochastic ones. To keep the presentation simple, let us suppose that there is no uncertainty, i.e. let us assume that the policy is point-valued. Similarly to what we saw in the previous section, it is easy to guess that stochastic inflows make the lake storage a stochastic variable as well. Therefore at each time instant the storage is characterized by a vector $\boldsymbol{\pi}_t$ whose components represent the probability that the state s_t assumes the first, second, ..., nth of its possible values. By simulating alternative A_i, subject to a given inflow process (like, for example, the one defined by equation (6.17)), a trajectory of $\boldsymbol{\pi}_t$ is obtained.

Since the system is periodic, it is possible to prove that the trajectory $\boldsymbol{\pi}_0, \boldsymbol{\pi}_1, \ldots$ converges towards a cycle $\bar{\boldsymbol{\pi}}_0, \bar{\boldsymbol{\pi}}_1, \ldots, \bar{\boldsymbol{\pi}}_{T-1}$ as t increases. However, unlike the previous case, it is not certain that the cycle will be reached in finite time. Thus, the simulation should be stopped when the cycle is 'sufficiently well' identified (in a sense to be defined); this always requires a finite time.

Once the cycle is known, if the indicator is separable, we can easily compute a statistic of it, which can be either the worst case (the Wald criterion), or the expected value (the Laplace criterion). In the first case a form similar to equation (8.2a) is used,[2] in the second the following expression

$$\underset{\{a_t\}_{t=1,2,\ldots}}{E}\left[Z_k(\mathcal{F})\right] = \frac{1}{T}\sum_{t=0}^{T-1} \bar{\boldsymbol{\pi}}_t' \cdot \underset{a_{t+1}}{E}\left[\mathbf{g}_t^k\right] \tag{8.6}$$

where $\mathbf{g}_t^k$ is the vector defined by equation (8.2b).

However, it is not possible to compute a statistic for non-separable indicators in this way. To understand why, let us suppose for example that we must compute an indicator whose value depends on the lake storage at two consecutive time instants $t-1$ and t. To do this, knowing the two marginal probabilities $\bar{\boldsymbol{\pi}}_{t-1}$ and $\bar{\boldsymbol{\pi}}_t$ of states s_{t-1} and s_t is not sufficient: also their joint probability density $\bar{\boldsymbol{\pi}}_{t-1,t}$ must be known.

Therefore, in general, it is not possible to compute statistics of non-separable evaluation indicators with analytical formulae when the inflows are stochastic variables: just as in the previous case, we must compute them over a sample obtained through deterministic simulation. To run the simulation, the inflows must be supposed to be deterministically known in every time interval $[t, t+1)$.

Readers who are interested in studying the issues presented in this section in greater detail may refer to Chapter 19 of THEORY.

8.2 Choosing the inflow series

In order to simulate an alternative A_i we must thus have an inflow series, which, clearly, must be of a finite length. However, the finiteness of the series does not constitute a difficulty in the estimation phase. When a policy is designed, it is essential to take into account its effects over an infinite horizon: if the objectives were defined over a finite horizon, no value would be attributed to the resource after the end of the design horizon (see Section 7.1) and consequently, close to the end of the horizon, the policy might propose controls that over-exploit the resource. This risk does not exist while estimating the effects, because the policies have already been defined.

Let us choose the inflow series that is to be used. The easiest solution is to adopt a portion of the *historical series*, N years long, during which the inflow process can be reasonably regarded as cyclostationary. With this hypothesis, the N years in the series may be treated as N equiprobable one-year-long realizations, and in the estimation of the indicators, the expected value (or the maximum value, depending on the choice of the statistic) with respect to the inflow process can be substituted by the time average (or the maximum) over the N years. The estimates obtained in this way can also be regarded as the estimates of the

[2] In this case the set $\mathcal{X}_t$ must be seen as the set of state values that have a non-zero probability of being realized, namely the ones associated with a positive component of $\boldsymbol{\pi}_t$.

effects that would be produced by alternative A_i, if in the future the same inflow sequence should occur. However, they cannot be considered as estimates of the effects that would have been obtained in the past, if alternative A_i had been implemented before the series occurred. As we will see in Section 8.3, this last interpretation is precluded by the fact that in the evaluation scenario we do not consider the water demand that characterized the water users in the past, but the expected future demand. We will call this first solution *historical simulation*.

An alternative solution is more complicated: it requires the identification of a (stochastic) model of the inflow process, like, for example, the one defined by equation (6.17), which we used to design the policies. A more complicated model might be used too, given that now the complexity of the model no longer causes the computing time to increase prohibitively, as it does when designing policies. Once the model for the inflow process has been identified, L *synthetic inflow series* must be generated. Then, alternative A_i is simulated for each of them and finally the indicator values are computed, by evaluating the average (or maximum) value on the sample of the L simulations. This second solution is referred to in literature as *Monte Carlo simulation*.

In order to choose from the two solutions we will examine their advantages and disadvantages:

- First of all we observe that alternative $A0$, i.e. the one that assumes that nothing is done and everything remains as it is (Section 3.7.1), is always included among the alternatives to be simulated. However, a formal expression of the regulation policy that was adopted in the past is not available and it cannot be identified through interviews with the Regulator or the downstream water users, as we did for the distribution policy (Section 3.4). In fact, in Section 3.5 we showed that, when the lake level is close to the upper extreme of the regulation range, the Regulator no longer decides on the basis of the level alone, he considers a wider set of information, which he chooses according to the situation. It is therefore impossible to identify a policy (i.e. a function of the time and the lake level) that reproduces the historical behaviour of the Regulator. Without a policy for $A0$, it is not possible to run a Monte Carlo simulation, but it is possible to run a simulation over the historical series. If, in fact, neither the policy nor the stage–discharge relation nor the regulation range are changed, as $A0$ assumes, the historical inflows would produce the historical lake levels and from these (since the policy is unchanged) the historical releases would result. The historical simulation of $A0$ would thus provide the historical levels and releases, which therefore do not have to be computed, since they are already known. Only the series of the allotments among the canals and the Ticino would have to be computed: in fact, it might be different from the historical one, given that the users' demand values that we want to consider are those expected in the future and not the historical ones. But we have both the historical series of releases and the historical distribution policy, and so the allotments can be easily simulated.

- By adopting the historical simulation, it is easy for Experts and Stakeholders to compare the effects that the alternatives would produce in the future with what has been observed in the past. This gives significance to the indicator values and simplifies their evaluation (see Section 8.3.1 of the DVD).

- Model (6.17) does not reproduce the temporal correlations that characterize the inflow process. In Section 6.5 we showed that even the most complex models do not describe this aspect with sufficient accuracy. It follows that the Monte Carlo simulation might underestimate the probability of prolonged droughts and floods and consequently their effects might be underestimated. This risk is excluded by running a historical simulation.

- When the number N of years in the historical series is very small, the indicator values provided by the historical simulation might not be representative of what could happen in the future, because they might be strongly influenced by particular events occurring in the historical series. This does not happen with Monte Carlo simulations, because it is possible to generate an arbitrarily great number L of synthetic series, so that the indicator estimates effectively reflect the characteristics of the inflow process described by the model. This difficulty does not arise also when the historical series shows a good record of the variety of events that can occur.

From the previous points it seems that the arguments in favour of using historical simulation are more numerous and more meaningful that those against. Before definitively adopting it, however, a couple of things should be checked.

First of all, we must ascertain that the available time series contains a succession of N years, with N sufficiently large, during which the process that generates the inflow can be reasonably considered cyclostationary. In fact, only if this is true:

(a) can each year be considered as an experiment that is independent from the previous years and that can thus be used to obtain a sample estimate of the expected value (or maximum value) of the indicators;

(b) does it make sense to assume that the series being considered could represent one of the possible future realizations of the inflow process.

As we observed in Section 2.1.1, the inflow process is substantially influenced by the presence of hydropower reservoirs in the catchment. The first reservoirs were constructed in the 1920s, the majority in the 1940s and 1950s, and the last one came into service in 1973. Thus it is only since 1974 that the inflow process can be considered cyclostationary and so we cannot consider inflow data prior to that year. At the start of the Verbano Project, the last available year was 1998. Thus N is equal to 25 years, a number large enough to guarantee fairly reliable indicator estimates. Note that, by assuming that 1974–1998 is a cyclostationary period from a hydrological standpoint, we implicitly exclude the possibility that climate change (driven by the greenhouse effect) is affecting the system.

Next, we must check if in the period that has been chosen there is enough variation in the hydrological characteristics to cover the whole spectrum of events that could occur in the future. This means that it must contain particularly rainy years and dry years, as well as years that are characterized by 'average' conditions. In order to ascertain whether this variety exists, it is useful to study the inflow trajectory rather than comparing statistics, such as the average or the maximum/minimum daily inflow (both on individual days and over periods of various lengths). For example, in Figure 8.2 the inflow trajectories in three particularly interesting years are shown and the corresponding statistics are reported in Figures 8.3–8.5.

By observing the trajectories it can be noted that:

- 1976 is an extremely dry year. The inflows are exceptionally scarce until September and this could create (as, in fact, it did) a very critical situation both for the irrigators and for the downstream environment.

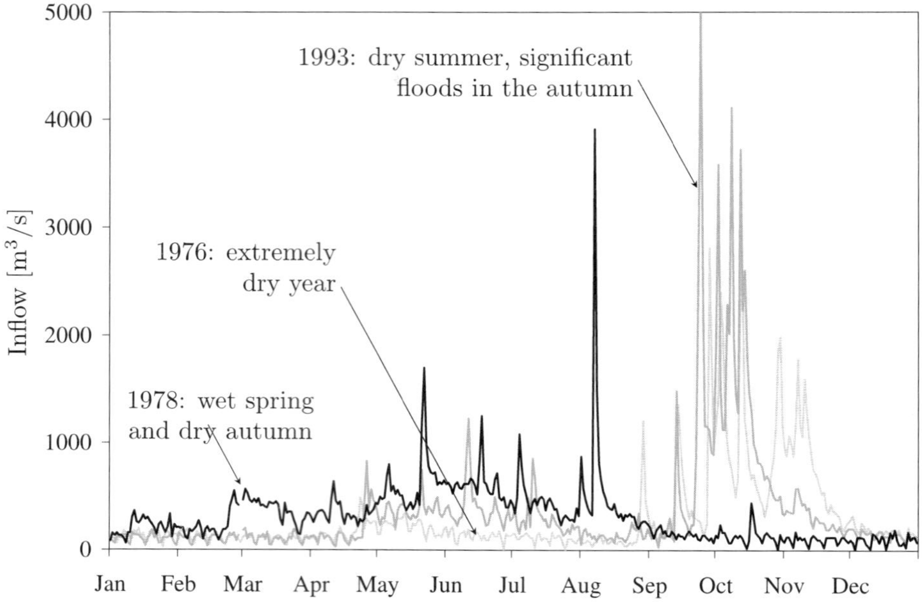

Figure 8.2: Inflow trajectories in the years 1976, 1978 and 1993.

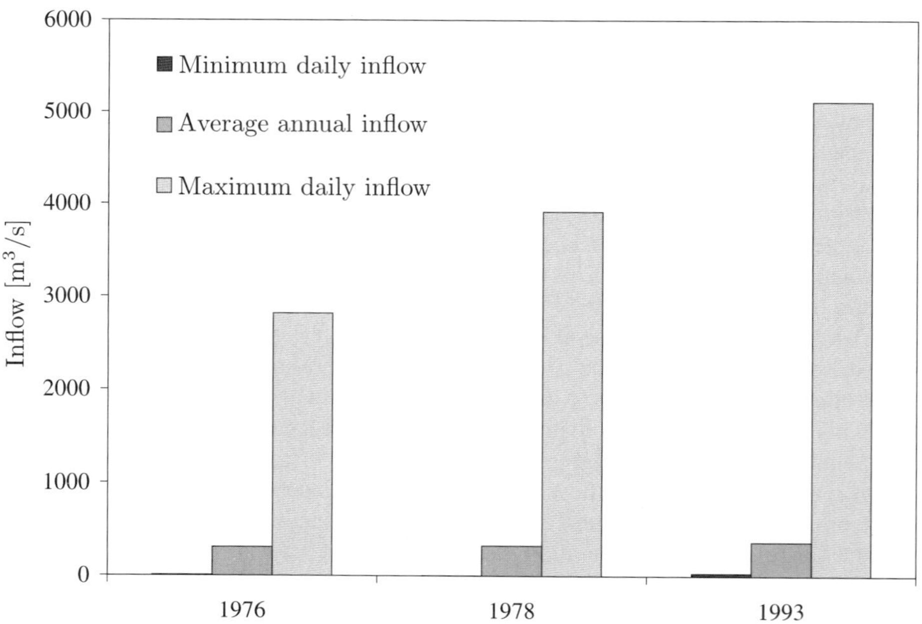

Figure 8.3: Minimum daily inflow, average annual inflow and maximum daily inflow, in the years 1976, 1978 and 1993.

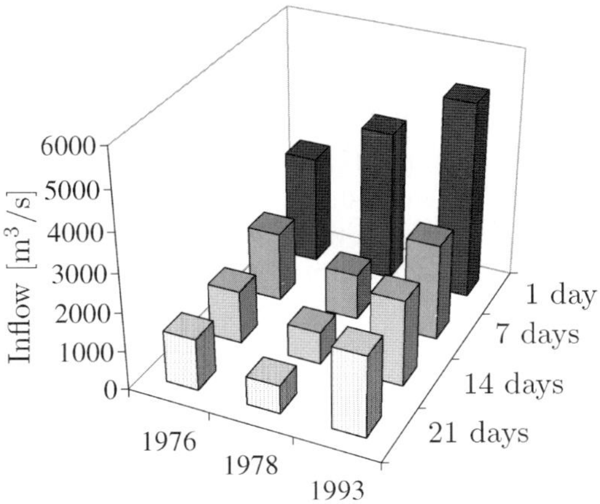

Figure 8.4: Maximum daily inflow and maximum average inflow over 7, 14 and 21 days, in the years 1976, 1978 and 1993.

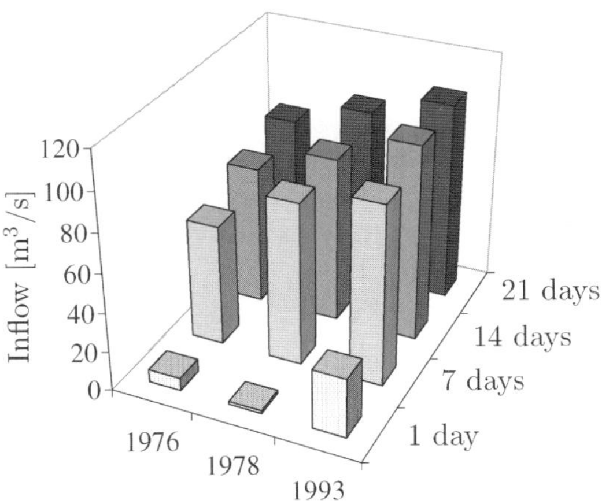

Figure 8.5: Minimum daily inflow and minimum average inflow over 7, 14 and 21 days, in the years 1976, 1978 and 1993.

- In 1978 the inflows are unusually abundant from January to September, even though there are no large floods (except for an unusual flood in mid August). Therefore in that period there are no difficulties for either the upstream or downstream Stakeholders. In the autumn months, however, the inflows are very scarce. This can make it difficult to create an adequate water reserve (lake storage) for the following winter and spring. If this storage is not formed, the hydropower producers can be damaged, because they need a significant amount of water to be stored before winter.

- In 1993 the inflow trajectory between January and August is intermediate with respect to the 1976 and 1978 trajectories. The inflows are particularly reduced in August, and this can create difficulties for the irrigators. From mid September to mid October there is a striking sequence of floods, whose return period is about 70 years. Therefore large flooding is expected both upstream and downstream.

From the statistics in Figures 8.3–8.5 it is not possible to pick out the characteristics that have now been revealed. Observe for example Figure 8.5. The minimum inflow values registered in 1976 and in 1978 might suggest that 1978 had more severe dry spells than 1976. On the other hand, from the average minimum inflows over several days it can be deduced that the most severe dry spells were in 1976. Nonetheless, the differences are so small that it is not possible to infer that 1976 actually was a much more critical year than 1978 for both the irrigators and the environment. Similar considerations are relevant to 1993 and to the other statistics that were presented. We conclude that, to judge whether there are years with sufficiently heterogenous hydrological characteristics in the period being considered, it is necessary to carefully analyse the inflow trajectories together with the Experts. This is just what we did and the Experts came to the conclusion that the years 1974–1998 can be used to generate a good estimate of the effects of the alternatives.

8.3 The evaluation scenario

At this point we have identified the inflow series to be used in the historical simulation. However, the inflow series is not the only element that defines the scenario with respect to which the effects of the alternatives are estimated. Other elements come together for its definition, part of which must coincide with the elements that define the design scenario. In fact, as mentioned in Section 7.5, the design scenario is part of the evaluation scenario: to be precise, it is the part that is required for estimating the design indicators, which are a subset of the evaluation indicators.

The evaluation scenario is completely defined by:

- The **inflow series**: we assume, for the reasons explained in the previous section, the historical series in the years 1974–1998.
- The trajectories of the **water demand**: they are assumed to be periodic, with period one year, and we adopted the patterns expected for the near future (see Figure 7.10).
- The **minimum volume** q^{Turb} that must be guaranteed each day for the Turbigo-EUROGEN thermoelectric power plant; the volume corresponding to a constant flow rate of 25 m^3/s was adopted.
- The trajectory of the **median of the Ticino flow**: the pattern estimated over the horizon 1974–1998 was adopted.
- The **flow series of the Po**, required for computing the indicators for the *Downstream Flooding* sector. In fact, as we have mentioned several times, the level of the Ticino at Pavia is not only influenced by the releases at Miorina, but also by the backflow from the Po. Releases from the lake and flows in the Po do not have a bijective relationship, because precipitation events over the Po catchment (closed at the confluence with the Ticino) are not synchronized with releases from the lake. Nevertheless, some correlation does exist between them, particularly during big floods, and so assuming the historical series from 1974 to 1998 for the flow in the Po seems reasonable.

The reader might have noticed that the indicator for *Downstream Environment* sector

$$i_{\text{Env\_D\_1}} = \sqrt{\frac{1}{N}\sum_t \left(q_t^{\text{TIC}} - r_t^n\right)^2} \tag{8.7}$$

which was introduced in Section 4.3.3, requires that also the series of natural outflows r_t^n from the lake be known (while the volume q_t^{TIC} is estimated through the simulation itself). Thus it would seem that the scenario must include the natural outflow trajectory as well, but this is not so. Such a trajectory, in fact, is univocally determined by the inflow trajectory, once the continuity equation of the natural lake is known. Let $s^n(\xi)$ be the storage in the natural lake at time instant ξ in day t; the continuity equation has the following form[3]

$$\frac{ds^n}{d\xi} = \frac{a_t}{\Delta} - N^{\text{nat}}\left(s^n(\xi)\right) \tag{8.8}$$

where $N^{\text{nat}}(\cdot)$ is the natural storage–discharge relation (derived from the natural stage–discharge relation on page 54) and Δ is the duration of the day in seconds (86 400 s). Given the initial condition $s^n(t)$, the solution $s^n_{[t,t+\Delta]}(\cdot)$ can be computed between the beginning (t) and the end $(t+\Delta)$ of the day, i.e. the storage can be estimated at every instant within the day. Thereby the storage $s^n(t+\Delta)$ at the end of the day is obtained and the volume r_t^n that would have naturally flowed from the lake that day is derived as

$$r_t^n = R_t^n\left(s_t^n, a_t\right) = \int_t^{t+\Delta} N^{\text{nat}}\left(s^n_{[t,t+\Delta)}(\xi)\right) d\xi \tag{8.9}$$

The recursive use of equations (8.8) and (8.9) over the whole evaluation horizon provides the trajectory of natural outflows.

A similar discussion is valid for indicator $i_{\text{Env\_U\_1}}$, which measures the standard deviation between the lake levels in the natural regime and in the regulated regime.

Constructing the water demand trajectories We said that it is possible to assume that water users' demands are periodic functions, with a period of one year. At first glance, one might think that the water demand trajectories over the whole evaluation horizon, i.e. 25-year-long trajectories, could be obtained by simply repeating the same sequences of 365 values 25 times and by adding a value for 29th February in leap years. Actually, this is not correct. Think, for example, of hydropower demand: it is clear that it depends on the day of the week. The demand for energy, in fact, is usually lower on Saturday and Sunday, when most of the industrial activity is suspended. To take this into account, the 365 values that define the water demand for each of the users were fixed with reference to a *standard year*, which is a non-leap-year and begins on Monday. When constructing the water trajectories over the evaluation scenario, we must consider that not all years are standard years: then, we need a rule to associate the correct demand value to each day of the evaluation horizon. The most obvious choice is to associate each day to the closest homonymous day of the standard year,

[3]Remember that the historical inflow series is composed of daily inflows a_t, which are estimated every day by closing the mass balance of the lake over the last 24 hours (see Section 6.5). To estimate the natural outflow in day t, the continuity equation must be integrated over 24 hours and this requires that the instantaneous inflow be known at every instant of the day. Since this is not the case, we are forced to assume that the instantaneous inflow is constant and equal to the average inflow rate, i.e. a_t/Δ.

i.e. the day of the standard year that has the same name (Monday, Tuesday, ...) and is the closest to the day being considered. In this way, two temporal indices are associated to each day: the first is the ordinal number (0–364/365) that labels it with respect to the first day of the year (day 0), which we call *natural date*; the second is the number that labels the closest homonymous week day in the standard year,[4] which we will call *anthropic date*.

Now we have all the necessary elements for estimating the indicators. Nevertheless, the moment has not yet arrived to do the computations: we must first assess the impact of our arbitrary assumption of the extractor (see Section 8.1.1) and estimate the overall computing time for the computation of all the indicators for all the alternatives, which may prove to be unacceptable.

8.3.1 The Alternative Zero and history

8.4 The extractor: the model of the Regulator

In Section 8.3 we called *extractor* the model that describes how the Regulator chooses a control $\mathbf{u}_t$ from the set $M_t^{i,*}(s_t)$ of controls proposed by the (set-valued) policy at time t. In Section 5.2.1.1 we qualified this set by affirming that "its controls provide performances that are *equivalent* with respect to the objectives of the Control Problem with which the policy was designed". It is now the time to make the scope of this affirmation clearer:

- the controls of $M_t^{i,*}(s_t)$ produce *equivalent* performances in the sense that they all produce the same value of the weighted sum of the objectives of the Primary Control Problem (where the weight values are the ones adopted in the policy design), and the same value of the objective of the Secondary Problem. In other words, their performances are equivalent if they are measured as they were in the Lexicographical Problem with which the policy was designed;

- it follows that the performances must be evaluated over an infinite horizon (as in the policy design); and that

- the system must be subjected to an inflow trajectory generated by the same model that was used in the policy design (in our case, the one defined by equation (6.17)).

In practice, however, the performances of the different controls in $M_t^{i,*}(s_t)$ are not all the same, and the differences increase as the distance grows between the real operational conditions and the theoretical conditions defined above. In our case, the historical inflow series that we adopted for estimating the effects is certainly not a typical example of the series that can be generated with the model (6.17) that we adopted in the policy design. Its realizations are, in fact, sequences of inflows that are not correlated in time, while the inflows in the historical series are correlated. It follows that the average length of dry spells and flood events in the historical series is longer than in the series generated by model (6.17). Furthermore, the evaluation horizon for the historical simulation is finite (25 years). Lastly, we do not estimate the performances with respect to the design indicators, but with the extended set of indicators that was defined in Chapter 4. For this reason, we must expect the estimates of such indicators to be different for different values of the control. But how much will they vary? In other words, how much does the estimate of the effects depend on the extractor adopted?

[4] The standard year is labelled using the natural date system.

8.4.1 Analysing the sensitivity of the indicators

To find an answer to this question, it is necessary to conduct a sensitivity analysis of the indicators with respect to the extractor. The analysis can be carried out by repeating the estimation with different extractors, which reflect some of the Regulator's possible selection criteria.

We can single out two categories of extractors:

1. *Reference extractors*: they operate with respect to a given reference $(W_t^{\mathrm{MIO}}, W_t^{\mathrm{TIC}})$. First of all, from the set $M_t^{i,*}(s_t)$ a subset

$$\tilde{M}_t^{i,*} = \{\mathbf{u}_t\colon \mathbf{u}_t \in M_t^{i,*}(s_t),\ u_t^{\mathrm{MIO}} - u_t^{\mathrm{TIC}} \geqslant W_t^{\mathrm{MIO}},\ u_t^{\mathrm{TIC}} \geqslant W_t^{\mathrm{TIC}}\} \tag{8.10}$$

of controls that satisfy the reference is extracted. Then, within $\tilde{M}_t^{i,*}$ the control $\mathbf{u}_t$ that is closest to the reference is the one adopted, i.e.

$$\mathbf{u}_t = \arg \min_{(u_t^{\mathrm{MIO}}, u_t^{\mathrm{TIC}}) \in \tilde{M}_t^{i,*}(s_t)} \sqrt{(u_t^{\mathrm{MIO}} - W_t^{\mathrm{MIO}})^2 + (u_t^{\mathrm{TIC}} - W_t^{\mathrm{TIC}})^2} \tag{8.11}$$

If the set $\tilde{M}_t^{i,*}$ is empty, the control that is closest to the reference is identified within the set $M_t^{i,*}(s_t)$.

The choice of the reference depends on the aim of the regulation, for example:

(a) The *irrigation/hydropower reference*

$$W_t^{\mathrm{MIO}} = W_t^{\mathrm{REC}} + W_t^{\mathrm{VC}} + W_t^{\mathrm{IC}} + q_t^{\mathrm{MEF}} \qquad W_t^{\mathrm{TIC}} = 0 \tag{8.12}$$

tends to the maximum (instantaneous) satisfaction of the irrigators and hydropower producers.

(b) The *irrigation reference*

$$W_t^{\mathrm{MIO}} = W_t^{\mathrm{REC,irr}} + W_t^{\mathrm{VC}} + W_t^{\mathrm{IC,irr}} + q_t^{\mathrm{MEF}} \qquad W_t^{\mathrm{TIC}} = 0 \tag{8.13}$$

tends to the maximum instantaneous satisfaction of the irrigators. Remember that the distribution policy allocates to the Industrial Canal a flow higher than the demand $W_t^{\mathrm{IC,irr}}$ of the Old Users only when the overall irrigation demand W_t^{MIO} is completely satisfied (Section 3.4), i.e. when the difference between the release r_{t+1} at Miorina and the flow $q_{t+1}^{\mathrm{TIC},u}$ that, following the environmental allotment decision, is left to the Ticino is higher than the reference W_t^{MIO} adopted by this extractor. It follows that, with this extractor, water volumes for hydropower production will be released from the lake only when the fear of floods makes it absolutely necessary, i.e. only when, within the set $M_t^{i,*}(s_t)$ of controls proposed by the policy, no control $(u_t^{\mathrm{MIO}}, u_t^{\mathrm{TIC}})$ exists such that $u_t^{\mathrm{MIO}} - u_t^{\mathrm{TIC}} = W_t^{\mathrm{MIO}}$. The aim of this extractor is thus the maximum satisfaction of the irrigators both in the short and the medium term.

(c) The *environmental reference*

$$W_t^{\mathrm{MIO}} = r_t^n + q^{\mathrm{Turb}} \qquad W_t^{\mathrm{TIC}} = r_t^n - q_t^{\mathrm{MEF}} \tag{8.14}$$

This extractor has the aim of maximizing the satisfaction for *Downstream Environment*. We have recently noted (see (8.7)) that the design indicator for this sector is defined as the distance between the flow left to the Ticino and the volume r_t^n that would have flowed into the river in natural conditions. In order to obtain the flow r_t^n in the Ticino, it is not enough to release an equal volume at Miorina, but it is necessary to increase it by the volume q^{Turb} needed for cooling the thermoelectric power plant Turbigo-EUROGEN. Remember, in fact, that the distribution policy makes it possible to assign volumes higher than the MEF to the Ticino only when the cooling of the Turbigo-EUROGEN plant has been assured. For the same reason, the reference W_t^{TIC} of the environmental allotment decision is the difference between r_t^n and q_t^{MEF}.

2. *Superior extractors*: they choose the control $(u_t^{\text{MIO}}, u_t^{\text{TIC}})$ that corresponds to the maximum value of u_t^{MIO} from $M_t^{i,*}(s_t)$. If more than one control satisfies this condition, the choice among such controls is made by means of a reference extractor. The superior extractors are thus differentiated by the choice of the second extractor. Some possible choices are:
   (a) *superior extractor – irrigation/hydropower reference*;
   (b) *superior extractor – irrigation reference*;
   (c) *superior extractor – environmental reference*.

   The main aim of superior extractors is to favour the *Upstream Flooding* sector as much as possible; the secondary aim is defined by the second extractor.

The sensitivity analysis was carried out by first simulating different alternatives with each of the above extractors, and then by computing the indicator values and comparing them. In order that the sensitivity analysis provide an indication about the maximum variations that can be expected as the extractor varies, it is advisable to choose alternatives characterized by sets $M_t^{i,*}(s_t)$ that contain many elements. We expect that these are the alternatives characterized by the triple ($\text{SD}_{+600}/\text{RANGE}_{\text{ENL}}/13_{\text{CON}}$), since we have already noticed that this is the one that offers the broadest space to the regulation decisions (see Figure 5.22). Since the cardinality of the set $M_t^{i,*}(s_t)$ also depends on the vector $\boldsymbol{\lambda}$ of weights used to define the Control Problem, it is advisable to consider various vectors of weights, which correspond both to extreme policies and to compromise policies. By doing so, we ascertained that the variations in the indicator values produced by changing the extractor are greater for the extreme policy, and less for the compromise policy. In both cases, however, the variations are small enough for us to conclude that the choice of the extractor to use for estimating the effects is not critical in the Verbano Project.

8.4.2 Reducing the computing time

In Section 7.6.3 we saw that the lexicographic ranking of the objectives, established by the Galli Law, obliges us to solve two Control Problems, one after the other, in order to design each policy. The first Problem, termed *Primary Problem*, considers the indicators for the *Upstream Flooding*, *Irrigation* and *Downstream Environment* sectors. The second, termed *Secondary Problem*, considers the indicator for the ENEL *Power* sector and it is constrained to choose from the set of controls suggested by the set-valued policy obtained by solving the Primary Problem. Finally, in order to estimate the effects of the policy we have obtained, we must simulate it over a horizon of 25 years. The computing time T_{tot}^c required for designing an alternative and estimating its effects is thus given by

$$T^c_{\text{tot}} = 2 \cdot T^c_{\text{con}} + T^c_{\text{sim}} \tag{8.15}$$

where T^c_{con} is the computing time required for solving a Control Problem (both Primary and Secondary Problems require the same computing time, because the number of objectives that are considered has a very negligible influence) and T^c_{sim} is the time needed to carry out the simulation. Using a Pentium III 600Mhz processor, the value of T^c_{con} varies from 35 minutes to 20 hours, according to the type of policy under design, say half a day on average, while T^c_{sim} is about 1 hour. The number of alternatives that must be designed is supposed to be of the order of two hundred. Therefore it would take about two hundred[5] days of uninterrupted computing time to generate and simulate the alternatives. It would be very useful if we could reduce this time.

There are two critical elements: the value of T^c_{con} and the fact that we have to solve two Control Problems for each alternative. We can act on the first only by improving the efficiency of the computing code or reducing the discretization, but the estimate of the time reported above was made by considering an optimized code and the discretization cannot be reduced any further without losing significance. In order to reduce the computing time, it is thus necessary to avoid solving the Secondary Problem. Is this possible?

If we want to design the policy, the answer is clearly negative: there is no way of avoiding it. Note, however, that our aim is not to design a policy for each alternative, but to have an estimate of the alternative's performance. In fact, these are the performances that the Experts and the Stakeholders must compare to identify the alternatives that obtain the widest agreement. It is essential that we know the policy only for those alternatives. This observation suggests a possibility that is worth exploring. The scope of the Secondary Problem is to individuate, for every time t and every state s_t, in the set $M^*_t(s_t)$ of equivalent controls that were obtained by solving the Primary Problem, the subset $M^{i,*}_t(s_t)$ of controls that minimize the lost revenue for ENEL's hydropower plants (Section 7.6.3). In order to identify these controls, we must take into account not only the effect that a given control has at time t, but also the effects that it produces in the future. For example, at time t there might be some controls in $M^*_t(s_t)$ that make it possible to satisfy completely the water demand of the power plants and eliminate the lost revenue on that day. However, it is not possible to conclude that those controls are optimal before assessing whether the water volume used in that day would produce a reduction for the power plants in the following days, and if in those days the price of energy would not be higher. If both these conditions are satisfied, it would be advisable to defer hydropower production. Nevertheless, cases such as these are fairly rare and so it is plausible that, if the controls in $M^*_t(s_t)$ that minimize the daily lost revenue were chosen, the error in the evaluation of the performance would not be very significant. Now, the controls that satisfy the above criterion are just the ones identified by the extractor with irrigation/hydropower reference. Therefore, if the approximation is acceptable, it is possible to avoid solving the Secondary Problem, and halve the computing time. A shift from two hundred days of computing to around a hundred is certainly not insignificant.

Before opting for this simplified procedure, one has to estimate the error that it may induce in the indicator estimates of ENEL *Power* sector. To do so, we selected a certain number of alternatives, very different from each other, and compared the estimates obtained by the application of the complete procedure with those obtained from the simplified one.

[5] Half a day on average for solving each Control Problem, and two Control Problems for each alternative, for a total of two hundred days of computing time.

The differences found were all less than 5%, so we considered the approximation acceptable and estimated the values of the indicators with the simplified procedure. Thereby we obtained the *Matrix of the Effects*, whose element (i, j) is the estimate of the jth indicator value induced by the ith alternative.

8.5 Validating the indicators

Once the indicator values have been estimated, we must verify that they are adequate for evaluating and comparing the alternatives. For this reason, two aspects must be examined for each indicator: the significance of the values obtained and its representativeness.

The necessary condition for the values to be significant, and therefore to be adequate for comparing the alternatives, is that the difference between the best and worst values of the indicator among all the alternatives be more than double the uncertainty with which the values of the indicator have been estimated. Such uncertainty can be derived from the results obtained by validating the models used to compute the indicators.

The representativeness of an indicator is verified by ascertaining, in collaboration with the Expert who defined it, whether it achieves the aim for which it was created: to quantify the criterion c_j that it is associated to and allow the ranking of the alternatives with respect to that criterion. The verification is carried out by choosing a set of alternatives and presenting the Expert with the trajectories of the hydrological variables produced by each one of them. On the basis of these trajectories, the Expert must rank the alternatives by decreasing values of satisfaction with respect to criterion c_j. The technique used to obtain the ranking is based on the pairwise comparison of the alternatives, which was already used when identifying the design indicators (Section 7.2). The Expert can only rank the alternatives if there are not too many; on the other hand, they should be numerous enough to represent most of the situations that could come about. Once the Expert's ranking has been obtained, the Analyst ranks the same alternatives, from best to worst, by increasing values[6] of the indicator that quantifies the criterion c_j. The indicator is considered to be representative when the two rankings correspond sufficiently well.

Whenever the validation of an indicator fails, the indicator should not be used to evaluate the alternatives and should be substituted by another indicator; but not always may the latter be identified. Thus, at the end of the *Estimating Effects* phase the set of criteria can be divided into two subsets: a set of criteria that are quantified by indicators and a set of criteria that have not been or cannot be quantified. With respect to the first the evaluation of the alternatives is carried out by analysing the estimated indicator values, so that the complexity of the evaluation procedure depends on the number of the criteria being considered. With respect to the second set, the evaluation requires the direct analysis of the alternatives, through the comparison of the trajectories of the hydrological variables that are produced by them; in this case the complexity of the evaluation procedure depends on the number of alternatives to evaluate.

[6]We are assuming that the indicator represents a cost; it would be of a decreasing order if the indicator expressed a benefit. Nevertheless, there are more complicated cases, in which the function that expresses the satisfaction is not monotonic in the value of the indicator. For further details see Chapter 17 of THEORY.

Chapter 9

Evaluation*

EB, FC, FP and RSS

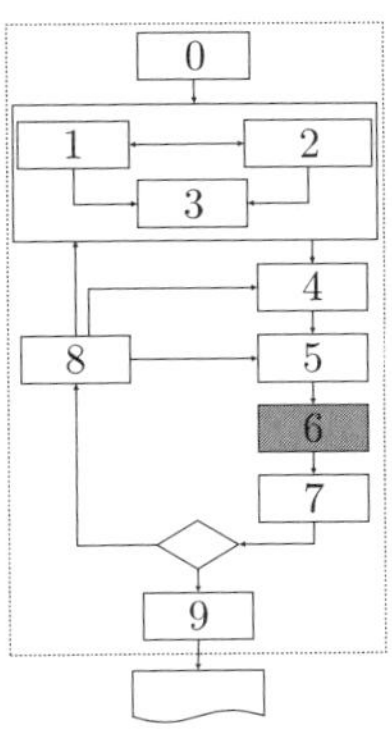

Figure 9.1: The phases of the PIP procedure. Phase 6, which is analysed in this chapter, is highlighted.

Once the estimation of effects has been completed, the alternatives must be evaluated (Phase 6 of the PIP procedure, see Figure 9.1). This means that for each sector we must identify the preference structures with which the relevant Stakeholders judge the alternatives. Then, we will be able to rank the alternatives, from best to worst, from the point of view of each sector. Generally, different rankings are obtained when considering different sectors, but this does not concern us now, because the comparison among sectors will be the subject of the next phase and therefore of the next chapter.

In Section 4.2 we identified an Expert that would work with each sector, to help the Stakeholders define their hierarchy of criteria and the indicators; the same Expert will help us now to identify their preference structures. In theory, given that the Experts are interviewed one at a time, it would be possible to use a different evaluation method for each one, to adapt as much as possible to the specific needs of each sector. However, we preferred to use the same method with all of them, in order to create a knowledge base and reference framework that is shared by all the Experts and Stakeholders. This will be of use in the next phase. Of the possible evaluation methods (some of which are presented and compared in Chapter 17 of THEORY) we chose the *Multi-Attribute Value Theory* (MAVT). It is, in fact, the most suited to our needs, for two main reasons:

- in the Verbano Project the number of alternatives to be evaluated is very high and the MAVT, unlike other methods, deals with this type of problem without difficulty (it can even be used when an infinite number of alternatives must be evaluated);
- the alternatives are designed in several steps and, in each step, an evaluation and a comparison must be carried out, in order to acquire information about the new set of

*This chapter is based on Chapter 9 (pp. 239–262) of *Modellistica Integrata e decisione partecipata in Pratica: Il Progetto Verbano*, by R. Soncini-Sessa, McGraw-Hill, 2004.

alternatives to be designed (see Chapter 5). In order that this procedure be successful, it is necessary that any alternative that is judged to be better than another at a given step, should continue to be so in the following steps. The MAVT guarantees that this condition is satisfied.

In the next two paragraphs we will illustrate the principal characteristics of this method. It will be a simple explanation strictly aimed at the Verbano Project; for a more formal description of MAVT the reader can refer to Chapter 20 of THEORY.

9.1 Multi-Attribute Value Theory

Multi-Attribute Value Theory (MAVT) (Keeney and Raiffa, 1976), is based on the identification of a *value function* $V(\cdot)$. This function has the vector of indicators that quantify the effects of an alternative as an argument, and returns a single value that expresses the satisfaction for that alternative. Since in the Verbano Project the alternatives are evaluated separately by the Experts from each sector, several value functions have been defined, one for each sector, which depend only on the indicators defined for the sector in question. Therefore the value V that the function assumes is none other than the sector index that we introduced in Chapter 4. Increasing values of V correspond to increasing levels of satisfaction. Once the value function has been identified, it is possible to rank any set of alternatives from the standpoint of the sector under examination.

Since it is extremely difficult to identify the value function directly, the problem is traced back to the definition of a *partial value function* for each one of the indicators. The global value function is then obtained from the composition of the partial value functions, usually through a weighted sum (see Figure 9.2). The value of the coefficients (*weights*) in the sum depend on the relative importance that the Expert associates to each of the indicators. Note that, by computing the value function as a weighted sum of the partial value functions, we implicitly assume that a poor performance of one of the indicators can be balanced by good performances of others. When this happens we say that *Compensation* is allowed among the criteria. The MAVT thus requires that Compensation is allowed among all the indicators (with the exception of one at most (Keeney and Raiffa, 1976), as we will see in Section 9.5).

It is important that the values to be assigned to the weights be defined on the basis of the values that each indicator can assume, using the procedure that will be defined in Section 9.4, and not on the basis of abstract considerations, far from the specific context considered. Also the definition of the partial value functions involves a number of complex operations, which are summarized in Section 9.3, and which require that the Expert express his evaluation of the effects of an alternative in a formalized way. Any attempt at simplifying these operations by adopting quick hypotheses, for example hypothesizing a priori that the partial value functions be linear, should be absolutely avoided. By doing so, in fact, the results are seriously invalidated.

The validity of the results also depends on the satisfaction of the hypotheses upon which the method is based and on the actual applicability of the approach with the Expert that it presupposes. First, we must verify that Compensation among the indicators is allowed. Note that in the *Evaluation* phase the sectors are considered one at a time and so we are not interested in Compensation among *all* the Project's indicators: Compensation must be allowed only among the indicators of the sector under examination. The interviews carried out with the Experts showed that for the sectors considered in the Verbano Project this

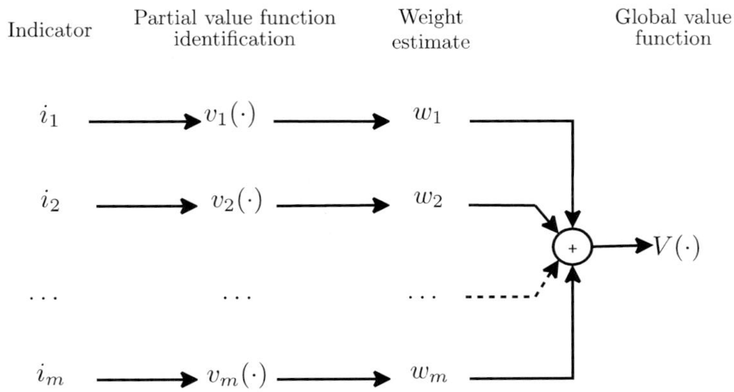

Figure 9.2: Steps in identifying the value function, given the values $i_1, \ldots, i_m$ of the indicators, according to the MAVT.

condition is satisfied, except for the ENEL *Power* sector. In Section 9.5 we will explain how to manage such a situation.

Moreover, the MAVT is a *normative method*, which means it requires that the Expert have a preference structure that satisfies three axioms: *Completeness*, *Consistency* and *Independence from Irrelevant Alternatives*. We do not want to go into the definition of those axioms here, but the interested reader can find them in Chapter 17 of THEORY. We simply say that the Experts were interviewed to see whether the axioms held and the result was positive.

Lastly, we must consider whether the Experts are able to respond to the questions we have to pose for identifying the partial value functions and the weights. We believe that response is likely to be affirmative, because the interaction that we have had with them up to this point has created a common knowledge base, which makes the MAVT approach acceptable. The complexity of this procedure, which at first glance might appear to be a limit, produces a practical advantage which comes to light in the following *Comparison* phase: getting the Experts to consider the values that the indicators can actually assume, and to contextualize each of their responses, obliges them to shift the evaluation from the level of questions of principle to the level of real problems.

9.2 The value functions

As we have seen, the central idea of the MAVT is to identify a value function for each sector, which makes it possible to compute the satisfaction (with respect to the sector criterion) for the effects that each alternative produces. The value of this function is none other than the sector index, and once this is known it is easy to obtain the ranking of the alternatives with respect to the sector criterion: just arrange them by decreasing values.

We must now define a formal procedure for the identification of the value function. Since this procedure must be independent from the sector being considered, in the following we can omit the specification of the sector to which we are referring.

For each sector, the global value V of an alternative A_i is a function of the values assumed by the r indicators $i_j(A_i)$, with $j = 1, \ldots, r$ of that sector, i.e.

$$V = V\big(i_1(A_i), i_2(A_i), \ldots, i_r(A_i)\big) = V\big(\mathbf{i}(A_i)\big) \tag{9.1}$$

where $\mathbf{i}(A_i)$ denotes the vector of the values assumed by the indicators $i_1(A_i), i_2(A_i), \ldots, i_r(A_i)$ for alternative A_i. For short, in the following we call these values *effects*. The form of the function could be extremely varied (see Keeney and Raiffa, 1976), for example

$$V = i_1^{\alpha_1} \cdot i_2^{\alpha_2} \cdot \cdots \cdot i_r^{\alpha_r} \tag{9.2}$$

or

$$V = \alpha_1 \cdot i_1 + \alpha_2 \cdot i_2 + \cdots + \alpha_r \cdot i_r \tag{9.3}$$

where $|\alpha_1, \alpha_2, \ldots, \alpha_r|$ is a vector of parameters.

Independently of its form, a value function must satisfy the following conditions:

(1) if the Expert believes the vectors of the effects produced by two alternatives are equivalent, then the values V associated to them must be the same and, on the contrary, if two vectors of effects are associated with the same value V, they must be equivalent, i.e.

$$\mathbf{i}(A1) \sim \mathbf{i}(A2) \quad \Leftrightarrow \quad V\big(\mathbf{i}(A1)\big) = V\big(\mathbf{i}(A2)\big) \tag{9.4}$$

(2) if the Expert believes that one vector of effects is better than another, then the value V associated to the first must be higher than the one associated to the second and, on the contrary, if the value V associated to one vector of effects is higher than another, then the first must be preferred over the second, i.e.

$$\mathbf{i}(A1) \succ \mathbf{i}(A2) \quad \Leftrightarrow \quad V\big(\mathbf{i}(A1)\big) > V\big(\mathbf{i}(A2)\big) \tag{9.5}$$

In order to identify the parameters that define the function $V(\cdot)$ in a given class, for example the parameters $\alpha_1, \ldots, \alpha_r$ in the class (9.3), a system of equations of the form (9.4) can be formulated and solved. To do this, the Expert is asked to identify a number of pairs of vectors $\mathbf{i}(A_i)$ that he judges to be equivalent. This procedure, however, can rarely be followed in practice, since the number r of indicators is often too high (think, for example, of the *Upstream Flooding* sector, which uses 12 indicators) and the Expert is not able to identify a sufficient number of pairs of equivalent vectors. It is therefore necessary to make the Expert's task simpler. It would certainly be easier to identify the parameters if, for example, the function could be expressed as the weighted sum of r partial value functions $v_1(\cdot), \ldots, v_r(\cdot)$, each of which having just one indicator i_j as its argument. This is possible if the following properties are satisfied:

1. *Separability*: the value V can be expressed as a function of the partial values $v_1, \ldots, v_r$ relative to the single indicators that make up the vector $\mathbf{i}(A_i)$, i.e.

$$V = V\big(v_1(i_1(A_i)), v_2(i_2(A_i)), \ldots, v_r(i_r(A_i))\big) \tag{9.6}$$

2. *Additivity*: the relation between the global value V and the partial values $v_1, \ldots, v_r$ is additive, i.e.

$$V = v_1\big(i_1(A_i)\big) + v_2\big(i_2(A_i)\big) + \cdots + v_r\big(i_r(A_i)\big) \tag{9.7}$$

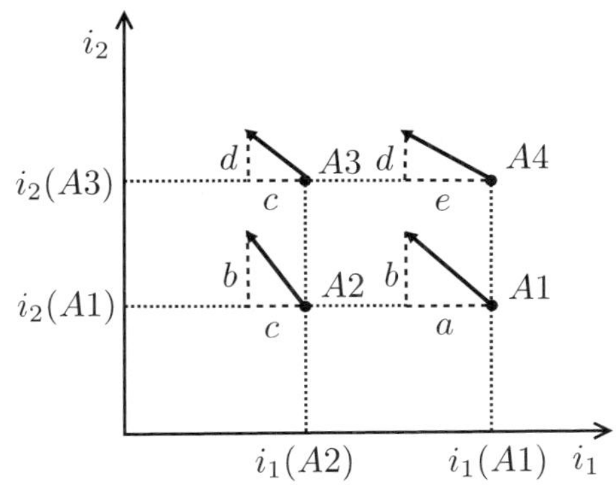

Figure 9.3: The procedure for verifying whether the *corresponding trade-off condition* holds.

Observe that if Additivity holds, then Compensation among the indicators holds as well.

These properties cannot be assumed a priori: their validity must be ascertained with an appropriate test, which is based on checking if a special property, termed *mutual preferential independence*, exists among all the pairs of indicators. Mutual preferential independence exists between two indicators when the preference order between any two values of one indicator is independent from the values assumed by the other indicator. Then the following theorem is valid (Keeney and Raiffa, 1976):

Theorem 1: *When more than two indicators are considered, the necessary and sufficient condition for Separability and Additivity is mutual preferential independence, i.e. that every subset of indicators is preferentially independent of its complementary subset.*

Thanks to this theorem, once the existence of mutual preferential independence has been confirmed, we can affirm that the properties of Separability and Additivity are valid. The situation is a little more complicated when there are only two indicators, i_1 and i_2, because in that case mutual preferential independence is a necessary but not sufficient condition for Additivity. However, we can turn to the following theorem (Keeney and Raiffa, 1976):

Theorem 2: *When only two indicators are considered, the necessary and sufficient condition for Separability and Additivity is that the so-called corresponding trade-off condition hold among the two indicators.*

The *corresponding trade-off condition* can be verified with the following procedure[1] (set out in Figure 9.3):

1. Consider a generic alternative $A1$ and ask the Expert to specify the increase[2] b of i_2 that he would be willing to accept in order to obtain a decrease a of i_1, i.e.

$$V\big(i_1(A1), i_2(A1)\big) = V\big(i_1(A1) - a, i_2(A1) + b\big)$$

[1] As usual we assume that both indicators are to be minimized.

[2] Remember that the Expert's satisfaction is assumed to decrease as the value of the indicators increases.

2. Choose an alternative $A2$, such that $i_1(A2) < i_1(A1)$ and $i_2(A2) = i_2(A1)$. Ask the Expert to specify the decrease c of i_1 that in his opinion would compensate an increase b of i_2, i.e.

$$V\big(i_1(A2), i_2(A1)\big) = V\big(i_1(A2) - c, i_2(A1) + b\big)$$

3. Then choose an alternative $A3$, such that $i_1(A3) = i_1(A2)$ and $i_2(A3) > i_2(A2)$. Ask the Expert to specify the increase d of i_2 that he would be willing to accept to obtain a decrease c of i_1, i.e.

$$V\big(i_1(A2), i_2(A3)\big) = V\big(i_1(A2) - c, i_2(A3) + d\big)$$

4. Finally, choose an alternative $A4$, such that $i_1(A4) = i_1(A1)$ and $i_2(A4) = i_2(A3)$. Ask the Expert to specify the decrease e of i_1 that in his opinion would compensate an increase d of i_2, i.e.

$$V\big(i_1(A1), i_2(A3)\big) = V\big(i_1(A1) - e, i_2(A3) + d\big)$$

5. The corresponding trade-off condition holds if and only if $e = a$.

When the properties of Separability and Additivity hold, the global value function can be expressed in the form (9.7). To identify it, we can thus start from the identification of the partial value functions $v_j(\cdot)$, with $j = 1, \dots, r$, as shown in the next section. Conversely, when the properties of Additivity and Separability do not hold, we must proceed directly with the identification of the global value function, using the more general form (9.1). In Section 9.5 we will show an example of how one can deal with the latter case, for which, however, a general identification procedure does not exist.

9.3 Identifying the partial value functions

Identifying a partial value function $v_j(\cdot)$ means passing from the value assumed by indicator i_j, which is expressed in physical units, to the satisfaction that the Expert associates to it, which is dimensionless. As an example, Figure 9.4 shows the partial value function that was identified for the indicator associated to the *Reduced access to beaches* criterion of the *Upstream Tourism* sector. Note that the function is not linear: this occurs very often because the Expert's preference structure usually is non-linear. The value $v_j(\cdot)$ is not defined on an absolute scale, but on an interval scale. This means that the intensity of preference for an alternative is not related to the absolute value of $v_j(\cdot)$, but to the differences in value with respect to other alternatives. This is not a limit for our aim, which is to establish a ranking of the alternatives.

Generally the minimum value of the function is set to 0 and the maximum value is set to 1; but this is only a convention and different choices are just as valid.

Once the minimum and maximum values for $v_j(\cdot)$ have been fixed, the following procedure (called *significant points method*) can be used to identify the function (for other possibilities see Section 20.4.2 of THEORY):

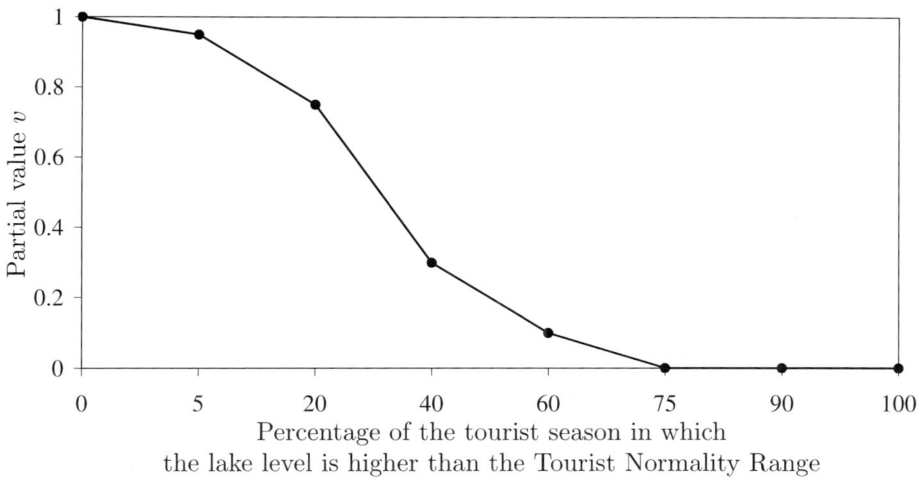

Figure 9.4: Partial value function for the indicator associated to the *Reduced access to beaches* criterion of the *Upstream Tourism* sector.

1. Determine the set $\mathcal{I}$ of values that indicator i_j can assume, i.e. establish the domain of the partial value function.

2. Ask the Expert to identify in $\mathcal{I}$ the subset $\mathcal{I}_{\min}$ of values of i_j to which the minimum value (0) is associated and the set $\mathcal{I}_{\max}$ to which the maximum value (1) is associated. Clearly, the first, or the second, or both of the subsets $\mathcal{I}_{\min}$ and $\mathcal{I}_{\max}$ could contain just a single element.

3. Ask the Expert to choose a significant point $\bar{i}_j \in \mathcal{I} - (\mathcal{I}_{\min} \cup \mathcal{I}_{\max})$ and to associate it with a value $\bar{v}_j$, taking into account the values associated to the points that have already been defined. Then ask him to choose a new point and proceed in this way until a set $\mathcal{V}$ of sufficiently numerous pairs (i_j, v_j) has been defined. Clearly, there are many ways with which the Expert could choose the new point to propose; for this reason the Analyst should guide him by asking that he first identify the points for which he believes that the function $v_j(\cdot)$, or its first derivative, would present discontinuity.

4. Interpolate the points of $\mathcal{V}$, to obtain the function $v_j^{\text{test}}(\cdot)$.

5. Verify that the function $v_j^{\text{test}}(\cdot)$ is the partial value function sought. To do this, ask the Expert to express his satisfaction for a value of i_j that he had not evaluated at previous steps, and compare his response with the value provided by function $v_j^{\text{test}}(\cdot)$. If the difference, in the Expert's opinion, is significant, $v_j^{\text{test}}(\cdot)$ is modified in order to fit with the new information. The test is repeated for other values of i_j until the Expert (not the Analyst!) is convinced that the function that has been identified mirrors his preferences.

6. Lastly, check that the Expert's preference structure is time-invariant. For this, in one or more meetings after the one in which function $v_j^{\text{test}}(\cdot)$ is identified, Step 5

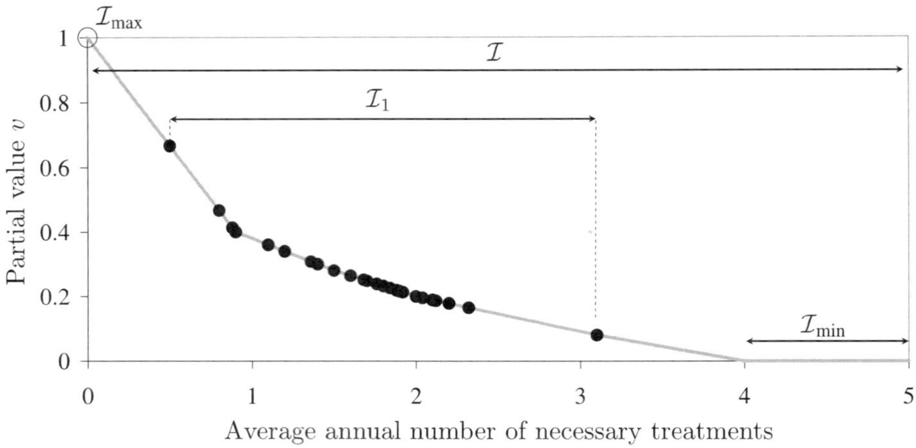

Figure 9.5: Partial value function for the indicator *Average annual number of necessary treatments* of *Upstream Tourism* sector. Sets $\mathcal{I}$, $\mathcal{I}_{min}$, $\mathcal{I}_{max}$ and the interval $\mathcal{I}_1$ are defined in the text.

is repeated. If after a number of meetings this validation test still fails, this may indicate that indicator i_j is not significant, i.e. it does not provide the information necessary for the Expert to evaluate the effects of the alternatives with respect to the leaf criterion it is associated to. If this hypothesis proves to be true, we would have to return to Step 2 of the PIP procedure and define a new indicator. Notice, however, that another reason for the Expert's preference structure being time-variant is that the Expert is not taking the work seriously, because he does not believe in it or is not motivated. Conversely, if from one meeting to the next there are no differences in the Expert's judgements, we can assume the function $v_j^{test}(\cdot)$ that was identified as the partial value function $v_j(\cdot)$.

The definition of the domain $\mathcal{I}$ of the partial value function deserves comment. Sometimes the set of all the possible values for indicator i_j is known a priori, i.e. before the alternatives are designed; at other times it can be determined only after the alternatives have been designed and their effects estimated. The first case is simpler and occurs, for example, when i_j is defined in terms of a percentage. Since the set of possible values is deterministically known, the domain of the partial value function is univocally known, too. The situation is more complicated when the set $\mathcal{I}$ is not known. In this case, after a first set of alternatives has been designed and the values of i_j have been computed, an interval $\mathcal{I}_1$ of indicator values can be defined. The width of this interval is a function of the type and number of alternatives considered: by designing new alternatives, in fact, one might get values for i_j that are not contained in it. Thus, when defining $\mathcal{I}$ it is advisable to consider an interval sufficiently larger than $\mathcal{I}_1$. Figure 9.5 shows the domain $\mathcal{I}$ that has been defined for the partial value function of indicator *Average annual number of necessary treatments*, and compares it with the indicator values obtained in correspondence to a number of alternatives. As the reader can notice, the interval $\mathcal{I}_1$ defined by these values is used only as a support for the definition of $\mathcal{I}$. Thereby, one avoids making the error of assigning zero value and unitary value to the two values of i_j in correspondence to the worst and best alternatives respectively.

9.4 Identifying the global value function

As we explained in the previous section, the partial value functions are defined on an interval scale: thereby, the function $v_j(\cdot)$ can be defined in an infinite number of equivalent ways, each related to the other by a linear transformation with positive coefficients. To better understand this statement, suppose a partial value function $v_j(\cdot)$ has been defined and derive from it the function $v'_j(\cdot)$, with the following linear transformation

$$v'_j = \alpha_j + w_j \cdot v_j \quad \text{with } \alpha_j \geqslant 0,\ w_j > 0 \tag{9.8}$$

Now consider two alternatives $A1$ and $A2$, and compute the values $v_j(i_j(A1))$, $v_j(i_j(A2))$ and $v'_j(i_j(A1))$, $v'_j(i_j(A2))$. The following variations can be defined

$$\Delta v_j(A1, A2) = v_j\big(i_j(A1)\big) - v_j\big(i_j(A2)\big) \tag{9.9}$$

$$\Delta v'_j(A1, A2) = v'_j\big(i_j(A1)\big) - v'_j\big(i_j(A2)\big) \tag{9.10}$$

By substituting definition (9.8) of v'_j into equation (9.10), the following expression is obtained

$$\begin{aligned}\Delta v'_j(A1, A2) &= \alpha_j + w_j \cdot v_j\big(i_j(A1)\big) - \alpha_j - w_j \cdot v_j\big(i_j(A2)\big) = \\ &= w_j \cdot \big[v_j\big(i_j(A1)\big) - v_j\big(i_j(A2)\big)\big] = \\ &= w_j \cdot \Delta v_j(A1, A2)\end{aligned} \tag{9.11}$$

Now repeat the same procedure for a second pair $(A3, A4)$ of alternatives. Observe that the ratio between the variation of v'_j in the passage from $A2$ to $A1$ and the variation from $A4$ to $A3$ must be equal to the ratio between the corresponding variations of v_j, i.e.

$$\frac{\Delta v'_j(A1, A2)}{\Delta v'_j(A3, A4)} = \frac{w_j \cdot \Delta v_j(A1, A2)}{w_j \cdot \Delta v_j(A3, A4)} = \frac{\Delta v_j(A1, A2)}{\Delta v_j(A3, A4)} \tag{9.12}$$

This shows that a linear transformation does not change either the ranking or the preference structure among the alternatives. Therefore, affirming that "the partial value functions contain information that is only relevant to the differences among the performances" is equivalent to affirming that "these functions are univocally known up to the coefficients α_j and w_j". Equation (9.7) should therefore be more correctly written in the following general form

$$V = \sum_{j=1}^{r} \big[\alpha_j + w_j \cdot v_j\big(i_j(A_i)\big)\big] \tag{9.13}$$

which shows that, in order to identify the global value function $V(\cdot)$, the parameters α_j and w_j must be estimated. Let us see how this is done.

In the first place, note that the scale with respect to which the global value function is defined is an arbitrary one: i.e. it is possible to fix the values 0 and 1 in an arbitrary way. A useful convention is to assume that V is 0 when all the partial value functions are 0, and that it equals 1 when they all equal 1, i.e.

$$v_j = 0 \quad \text{for } j = 1, \dots, r \qquad \Rightarrow \qquad V = 0 \tag{9.14}$$

$$v_j = 1 \quad \text{for } j = 1, \dots, r \qquad \Rightarrow \qquad V = 1 \tag{9.15}$$

By substituting equation (9.14) into equation (9.13), the following expression is obtained

$$\sum_{j=1}^{r} \alpha_j = 0 \tag{9.16}$$

Now, since all these coefficients must be positive, as required by equation (9.8), it follows that all α_j are zero. Equation (9.13) thus becomes

$$V = \sum_{j=1}^{r} \left[w_j \cdot v_j \left(i_j(A_i) \right) \right] \tag{9.17}$$

Finally, from equation (9.15) the following condition is derived

$$\sum_{j=1}^{r} w_j = 1 \tag{9.18}$$

The most general form for the global value function thus is

$$\begin{cases} V = \sum_{j=1}^{r} w_j \cdot v_j(i_j(A_i)) \\ \sum_{j=1}^{r} w_j = 1 \end{cases} \tag{9.19}$$

and the problem of identifying $V(\cdot)$ is reduced to the problem of estimating the r coefficients w_j. We call such coefficient *weights*, since they define the relative importance of the different partial value functions. In fact, the ratio w_j/w_k between two weights is equal to the ratio $\Delta v_k/\Delta v_j$ (which we will call *compensation level*), where Δv_k is the increase in satisfaction for criterion k that is required to compensate a decrease Δv_j in satisfaction for criterion j. To estimate the weights it is clearly necessary to turn once again to the Expert.

9.4.1 Estimating the weights

It is not possible to ask the Expert to just provide a vector of weights, because the values obtained in this way would not have any connection to the partial value functions. Instead, he should be asked to indicate $r-1$ pairs of vectors $\mathbf{i}$ of effects that he judges to be equivalent. In this way, a system of $r-1$ equations is obtained, each equation being of the form

$$\sum_{j=1}^{r} w_j \cdot v_j\left(i'_j\right) = \sum_{j=1}^{r} w_j \cdot v_j\left(i''_j\right) \tag{9.20}$$

where $|i'_1, i'_2, \ldots, i'_r|$ and $|i''_1, i''_2, \ldots, i''_r|$ are two equivalent vectors of effects. The solution of this system provides the weights we are looking for.

To simplify both the formulation and the resolution of the system, the Expert should not randomly search for pairs of equivalent vectors, but should choose them in such a way that the two vectors in each pair are completely identical except for the values of two components, namely the values of two indicators i_m and i_n. More precisely, the procedure is as follows. A vector $\mathbf{i}'$ of effects is chosen and a pair of indicators (i_m, i_n) is selected. The Expert is asked to derive an equivalent vector $\mathbf{i}''$ by arbitrarily assigning a new value i''_m to i_m, and looking for a value i''_n of i_n that keeps his global satisfaction the same. All the other components of vectors $\mathbf{i}'$ and $\mathbf{i}''$ are the same, i.e.

$$i'_j = i''_j \quad \text{with } j \neq m, \; j \neq n$$

Then, the term on the right-hand side of equation (9.20) can be written as

$$w_m \cdot v_m\left(i''_m\right) + w_n \cdot v_n\left(i''_n\right) + \sum_{j \neq m,n} w_j \cdot v_j\left(i'_j\right)$$

Also the term on the left-hand side can be written in a similar way, and equation (9.20) gets the following form

$$w_m \cdot v_m\left(i'_m\right) + w_n \cdot v_n\left(i'_n\right) = w_m \cdot v_m\left(i''_m\right) + w_n \cdot v_n\left(i''_n\right)$$

in which only two partial values appear. The procedure is quite laborious, but it is essential to use it if we want the value function that is identified really to reflect the Expert's preference structure.

We can conclude with the following observations:

1. If *Consistency* holds, the vector of weights identified with the above procedure does not depend on the pairs of indicators used to define the $r - 1$ equations. In fact, once we know the weight of one indicator with respect to a second indicator, and the weight of the second with respect to a third, because of Consistency we can derive the weight of the first with respect to the last, i.e.

$$w_i = a \cdot w_j \quad \text{and} \quad w_j = b \cdot w_k \qquad \Rightarrow \qquad w_i = (a \cdot b) \cdot w_k \tag{9.21}$$

   Even if the choice of the pairs does not influence the result, it is still advisable to facilitate the Expert's task by choosing pairs of indicators that are very meaningful to him.

2. In equation (9.20) we denoted the vectors of the effects with the notation $\mathbf{i}'$ and $\mathbf{i}''$, instead of the usual notation $\mathbf{i}(A')$ and $\mathbf{i}(A'')$. This is because we wanted to highlight that those vectors do not necessarily have to be produced by simulating some of the alternatives that have been designed: if we considered those alone, it might be difficult to find pairs of equivalent vectors, both because the number of designed alternatives might not be large enough, and because it might be that the values of some indicators are correlated, making it impossible to identify pairs whose components move in the desired directions. For example, as shown in Section 7.3.1, in the *Upstream Flooding* sector the average surface area and duration of floods are correlated: it is thus impossible to find alternatives in which the flooded surface increases and the duration decreases with respect to a reference alternative, if only designed alternatives are considered. Therefore, to estimate the ratio between the weights of these two indicators it is necessary to refer to hypothetical situations.

9.5 A special case: the value function for the ENEL Power sector

As explained in Section 9.2, in order that a global value function can be expressed in the form (9.7), the properties of *Separability* and *Additivity* must hold, for which the necessary condition is *mutual preferential independence* between the indicators. This implies that Compensation must be allowed among all the indicators, which is not always true. For example, Compensation is not allowed among the indicators $i_{\mathrm{EnP\_D\_1}}$ and $i_{\mathrm{EnP\_D\_2}}$ from the ENEL *Power* sector, which are respectively associated to the criteria *Lost revenue for hydropower plants* and *Continuity of the Turbigo-*EUROGEN *power plant's operation.* In fact, as we said in Section 4.2.2, when the water volume diverted into the Industrial Canal is insufficient to guarantee the operation of the Turbigo-EUROGEN thermoelectric power plant, the economic damage is so great that the value assumed by indicator $i_{EnP\_D\_1}$, relative to hydropower production, is no longer significant. It follows that the value function $V_{EnP\_D}(\cdot)$ cannot be identified by assuming the additive form (9.19).

Then, $V_{EnP\_D}(\cdot)$ should contain an 'if-then' condition. In particular, we could assume that, when the *Average annual number of days in which the thermoelectric power plant at Turbigo is out of service* exceeds a threshold $\bar{i}_{EnP\_D\_2}$, the global value V_{EnP_D} is 0; and that when that threshold is not exceeded, V_{EnP_D} depends on the value of $i_{EnP\_D\_1}$ alone, i.e.

$$V_{EnP\_D} = \begin{cases} 0 & \text{if } i_{EnP\_D\_2} > \bar{i}_{EnP\_D\_2} \\ v_{EnP\_D\_1}(i_{EnP\_D\_1}) & \text{otherwise} \end{cases} \tag{9.22}$$

The validity of this approach is confirmed by the following theorem (Keeney and Raiffa, 1976):

Theorem 3: *Given a vector composed of r indicators, if they are all mutually preferentially independent in pairs, except for one (say i_r), then a global value function exists of the form*

$$V = V(i_r, \bar{V})$$

where

$$\bar{V} = v_1(i_1) + v_2(i_2) + \cdots + v_{r-1}(i_{r-1})$$

and $V(\cdot)$ is increasing in $\bar{V}(\cdot)$.

In our case, since r is equal to 2, $\bar{V}(\cdot)$ is just the partial value function of indicator $i_{EnP\_D\_1}$.

9.6 Validating the value function

Once the value function $V(\cdot)$ has been identified, it can be used to rank any set of alternatives with respect to the sector criterion under study, provided that all the indicator values that are produced by the alternatives fall into the domain of the corresponding partial value functions (Section 9.3). Therefore, at least in theory, the Expert does not have to examine the alternatives directly in order to rank them. Nevertheless, it is good practice to validate the value function before using it. To do this, the Expert is shown a set $\mathcal{A}_{test}$ of alternatives and he is asked to order them by decreasing preference, by looking at the indicator values that they produce. Then the Analyst ascertains whether the Expert's ranking coincides with the one produced by the value function $V(\cdot)$.

If the rankings are the same, the function is validated and can be used for any set of alternatives. If they are not, it means either that the partial value functions were not correctly identified or that the weights were not correctly estimated. To ascertain whether the cause is the first, the Analyst can do nothing but run through, with the Expert, all the steps of the procedure that was used to identify the partial value functions once again. As for the second cause, a sensitivity analysis can be carried out to check the weights' values. The analysis considers one weight at a time. Let w_j be the one under examination, and compute the minimum increase Δw_j that produces a change in the ranking of the alternatives in $\mathcal{A}_{\text{test}}$. Then consider the ratios between the value of the weight w_j increased by Δw_j and the values of the other weights, i.e. the ratios $(w_j + \Delta w_j)/w_k$ for $k = 1, \ldots, r$, $k \neq j$, and ask the Expert to judge these *compensation levels* (see page 238). If he believes that they are too high, the difference between the Expert's ranking and the one produced by the value function $V(\cdot)$ cannot be corrected by increasing the weight w_j. Then a decrease in its value must be analysed, in a similar way. Otherwise, if the compensation levels are judged to be acceptable, it must be concluded that the weight w_j is badly calibrated and must be estimated again, with the procedure described in Section 9.4.

Once the value functions have been validated for all the sectors, they can be applied to the indicator values contained in the Matrix of the Effects, thus obtaining the *Matrix of the Values*, whose element (i, h) expresses the satisfaction associated to the ith alternative according to the hth sector. Let us underline once again that these values are none other than the sector index values, which will provide the basis for the Comparison of the alternatives, as illustrated in the next chapter.

Chapter 10

Comparison

EB, FC, FP and RSS

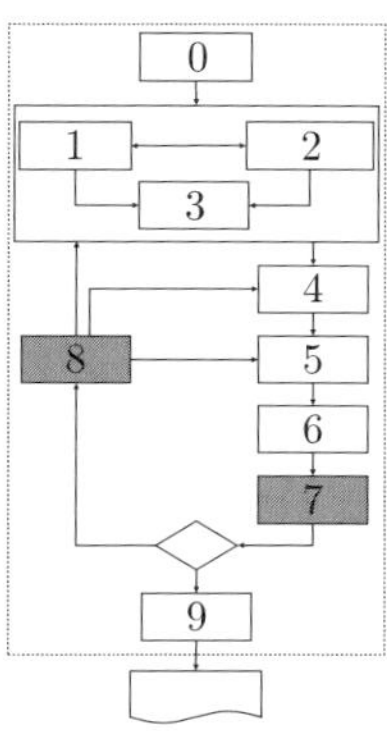

Figure 10.1: The phases of the PIP procedure. Phases 7 and 8, which are analysed in this chapter, are highlighted.

In this chapter we will deal not only with *Comparison* but also with *Mitigation*, i.e. with Phases 7 and 8 of the PIP procedure shown in Figure 10.1. Since the discussion about mitigation involves more practical than theoretic aspects, in this chapter we will just illustrate its aim and refer the reader to Chapter 14 for several examples of how it is carried out in practice. Also the presentation of the Comparison phase will be very soft with respect to theoretical aspects; for more details the reader may refer to Chapter 21 of THEORY.

The negotiation process is assisted by a *Facilitator*, a neutral third party (often the Analyst if he has the ability, as in the Verbano Project), who manages the process in such a way that the Stakeholders proceed constructively towards building of a consensus around one (or a few) alternatives. It is up to him to define the phases of the process in such a way that each of them is a step in that direction. Sometimes the Facilitator acts as a *Mediator*, which means that he assumes a more active role: he does not only facilitate, but he also structures the process, governing the agenda, using tools such as reformulating, active listening and open questions as well as his analytical abilities, and finally suggesting solutions, if desirable. The interested reader can find a detailed description of these activities in Appendix A10 of THEORY, along with a presentation of the psychological and cultural aspects of negotiations.

10.1 The Comparison Method

In the *Evaluation* phase (Chapter 9) a global value function was defined for each sector and its value V (i.e. the value of the *sector index* that we introduced in Chapter 4) was computed for all the alternatives. On the basis of these values (which, from now on, we will call sector indices) we now want to identify the reasonable alternatives, through a negotiation process involving all the Stakeholders. Clearly, the negotiation process actually involves the

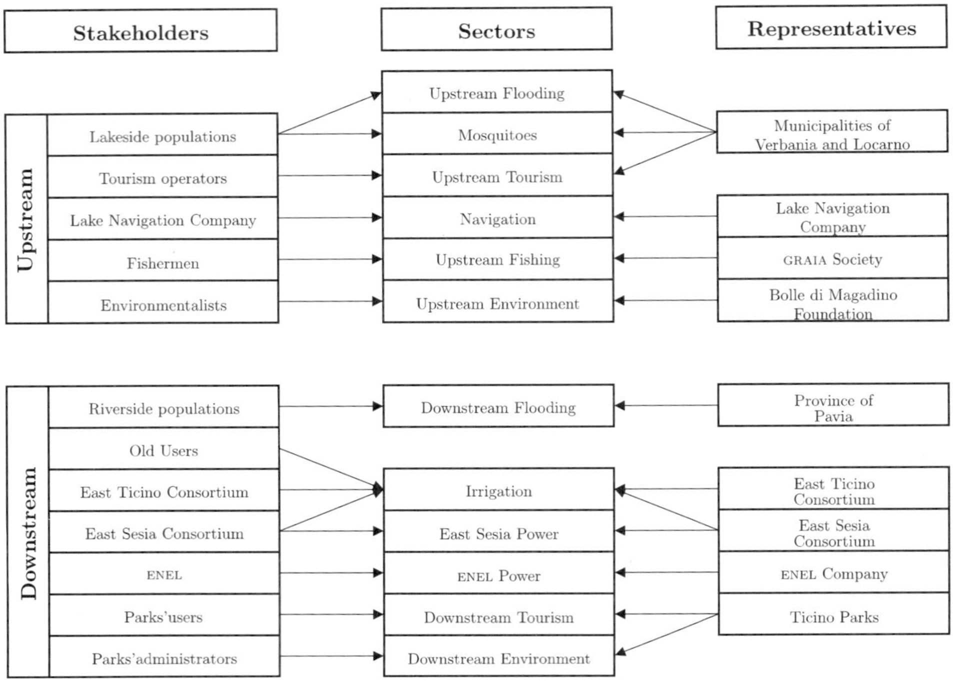

Figure 10.2: Sectors, Stakeholders and their representatives in the Verbano Project.

representatives of Stakeholders, not their totality, which would be impossible because of their number. However, for simplicity's sake, we will use the same term to refer to both the Stakeholders themselves and their representatives. The representatives that were involved in the negotiations of the Verbano Project are shown[1] in Figure 10.2.

In the *Evaluation* phase the effects of the same alternative were compared on the basis of the indicator values relevant to the sector under examination; on the other hand, in the *Comparison* phase, the alternatives are compared on the basis of the sector indices. In the *Evaluation* phase the value functions were identified by assuming that in all sectors, with the exception of the ENEL *Power* sector, Compensation was allowed. This means that, in each sector, a bad indicator performance can be compensated by a good performance from one of the other indicators. This assumption is often acceptable among the criteria relevant to the same sector, but it is not among sector criteria by the very definition of sector. In fact, a sector is a subset of evaluation criteria whose aggregation is shared by all the interested Stakeholders, which often implies that Stakeholders are willing to accept compromise among the evaluation criteria of that subset; on the contrary, the level of the sector criteria is just the one at which compensation among the criteria is no longer accepted by the interested Stakeholders (see Chapter 4). Therefore, Compensation not being allowed, in the *Comparison* phase the alternatives cannot be compared by simply summing the sector

[1] The GRAIA Society cited in the figure is a society of independent researchers in environmental engineering and ichthyology (the acronym stands for G*estione e* R*icerca* A*mbientale* I*ttica* A*cque* – Management and Research for Environment, Ichthyology and Waters). It was invited as a representative for the interests of the *Upstream Fishing* sector.

indices. It follows that in the *Comparison* phase the MAVT cannot be used, at least in its traditional formulation. This method, in fact, would require that the Stakeholders agree on a vector $\mathbf{w}$ of weights, and that this vector be used to aggregate the sector indices and define a *Project Value Function*, which could be used to rank the alternatives of set $\bar{\mathcal{A}}$. However, by doing so, we would not know if the alternatives that take the first positions in the ranking produce effects that are considered to be acceptable by all the Stakeholders, or if some of them would oppose the implementation of one or more of those alternatives.

To overcome this difficulty, we can ask each Stakeholder to specify *thresholds of acceptability* for the sector indices (s)he is interested in. If more than one Stakeholder has interests in the same sector, as for example in the case of *Irrigation* sector, the threshold with the greatest value is assumed. Once the acceptability thresholds have been defined, we can exclude all the alternatives in set $\bar{\mathcal{A}}$ for which the sector indices are below these thresholds, and use the Project Value Function to rank the remaining alternatives.

If this procedure were adopted, the discussion among the Stakeholders would be limited to the definition of the Project Value Function, namely to the definition of the vector $\mathbf{w}$ of the weights to be attributed to the sector indices. This vector should be derived from the m vectors $\mathbf{w}^i$ provided by each of the m Stakeholders ($i = 1, \ldots, m$). To do this, two procedures can be followed. In the first, the Stakeholders are shown the existing distances between the vectors $\mathbf{w}^i$ that they defined, and they are asked to modify their own position, until a vector that is shared by all of them is achieved. If this procedure should run into a conflict that cannot be overcome, a second procedure can be used. Each Stakeholder is asked to provide a second vector $\mathbf{v}^i$ of weights, whose jth element expresses the importance that Stakeholder i attributes to Stakeholder j. The m vectors $\mathbf{v}^i$ are arranged in a square matrix, from which a vector $\mathbf{v}$ is extracted that defines the relative importance of each Stakeholder. Vector $\mathbf{v}$ is then used to weight the m vectors $\mathbf{w}^i$ and obtain vector $\mathbf{w}$.

The described procedure seems weak, however, from the point of view of the interaction that is established among the Stakeholders. Therefore, it is not certain that all of them will be willing to accept the ranking that it produces. Further, it is possible that no alternative satisfies the acceptability thresholds indicated by the Stakeholders. For example, this would happen in the Verbano Project if the values of the sector indices produced by alternative $A0$ were used as thresholds, that is, if only 'win–win alternatives' were considered acceptable (remember that a 'win–win alternative' is an alternative that improves the satisfaction of *all* the Parties with respect to the current condition). Such alternatives cannot exist, since in the Verbano Project $A0$ turns out to be an efficient alternative. This consideration is very important: $A0$ is the alternative that would occur in the future if no action were implemented. Since it proves to be an efficient alternative, no alternative exists that produces greater satisfaction for *all* the sectors. In other words, whichever alternative is chosen, at least one of the sectors will be dissatisfied. This stresses the importance of establishing a thorough process of comparison that involves all the Stakeholders.

For this reason *we abandoned the search for a global ranking by means of a Project Value Function, and asked the Stakeholders to negotiate the acceptability thresholds*. To do so, we proposed a negotiation procedure in which the acceptability thresholds are not fixed a priori, but are defined in a recursive way by considering the effects of various threshold values.

To better understand the procedure that we are about to propose, it helps to study two methods, other than the weighting method presented in Section 5.2.6, that can be used to solve Multi-Objective Control Problems. The first is the *Constraint Method*, which transforms a Problem with n objectives into a (family of) Single-Objective Problems (parametric in a vector **f**), by considering only one of the objectives of the original Multi-Objective Problem, and adding the constraint that the value of the other $n - 1$ objectives must not exceed the thresholds expressed by vector **f** (see Section 18.3.4 of THEORY).

The second method is known by the name of *Pareto Race* and it was proposed in the 1980s (Korhonen and Laakso, 1986). It is a method for generating points along the Pareto Frontier (Section 5.2.1) in real time, by solving a sequence of Design Problems formulated on the basis of the preferences that the Decision Maker (termed as DM in the following) expresses as she obtains the results of the previous Problems. It is thus an interactive and iterative method defined by the following procedure:

1. the DM expresses her own vision by specifying the value that she desires for each objective (for example, 100 Mm^3 of average annual deficit and 5 km^2 of average annual flooded area). By so doing, she defines a *reference point* R in the objective space;

2. given R, a Single-Objective Problem can be formulated (see Section 18.3.3 of THEORY). Its solution is an efficient alternative, i.e. it belongs to the Pareto Frontier, which is 'close' to R, in a sense to be specified[2];

3. the values of the objectives that correspond to that alternative are shown to the DM, for example by means of a bar graph;

4. the DM is asked if she is 'satisfied' with those values, in which case the procedure is terminated, or if she wants to obtain better values for one of the objectives. In the second case, a new reference point R is defined and the procedure is repeated from Step 2.

Other methods foresee that the whole Pareto Frontier is identified in advance and that the DM is shown it and chooses what she judges to be the best compromise alternative. The Pareto Race, on the contrary, has the advantage of identifying only the alternatives that the DM thinks are interesting. As the procedure develops, the DM has the feeling of moving along the Pareto Frontier (from which the name *Pareto Race*). She thus acquires knowledge about the conflicts between the different sectors, because she sees that some of the indices increase while others decrease, and can thus appreciate the possibilities for compromise. In this way the best compromise slowly takes shape in her mind. This is why the procedure ends when the DM is 'satisfied'.

In the case of Verbano we cannot adopt the Pareto Race in its classical form for two reasons. Firstly, it is not possible to design the policies in real time: the computing time required for the resolution of one Control Problem varies from a minimum of 35 minutes to about 20 hours (on a Pentium III processor, 600 Mhz, with SCSI architecture) and if the generation of the responses is too slow, it prevents the DM from 'perceiving' the compromise. In fact, it is only when the questions and answers follow each other in quick succession that the DM can 'acquire knowledge' about the compromise, since this is an acquisition process that is based on short-term human memory. The second and more important reason is that in the Verbano Project there are two DMs (the Italian and Swiss governments). Moreover, its Goal is to identify the alternatives that get the widest agreement among the Stakeholders

[2]The adjective 'close' does not perfectly translate the mathematical criterion, for greater clarification see Section 18.3.3 of THEORY.

(the reasonable alternatives), which will be then submitted to the two DMs. Consequently, all the Stakeholders must be involved in the comparison.

The way to overcome the first difficulty is simple: it is just because of the considerable computing time required for the Design Problem that in the PIP procedure the phases of *Designing Alternatives*, *Estimating Effects* and *Evaluation* are conducted off line, i.e. before the *Comparison* (this took about 70 days of uninterrupted computing time). We can therefore overcome the difficulty if we accept that the Design Problem formulated in Step 2 is not solved by searching the whole alternative space, but by screening the alternatives that have already been designed and looking for the one that most satisfies the objectives of the Problem.

The second difficulty remains. However, we can solve it by adopting the procedure described in the following paragraph.[3]

10.1.1 The Elementary Negotiation Procedure

1. Given an initial alternative A_{chosen}, the Facilitator uses a bar chart to show the values of the indices that this alternative produces in the different sectors, and asks each of the Stakeholders to take a position: *supporting* it, *accepting* it or *opposing* it.

2. The Facilitator asks the Stakeholders to identify the sector they judge to be the most disfavoured; let us denote this sector with $Sect^d$ and the value of the corresponding sector index with V_d (the symbol V is adopted because the index value is computed by means of a value function, see Chapter 9). Note that this is not necessarily the sector whose index has the lowest value, because it makes no sense to compare the values of the indices of different sectors, both for psychological reasons and for technical reasons (the value functions are defined on arbitrary scales, see Section 9.4). Let $\mathcal{A}^c$ be the set of alternatives subject to comparison and $\mathcal{A}^d$ the subset of the alternatives that increase the index of $Sect^d$ with respect to A_{chosen}.

3. The Facilitator asks the Stakeholders, with the exception of those interested in sector $Sect^d$, if they are willing to lower the value of the sector indices with respect to the value obtained in correspondence with A_{chosen}. If the answer is positive, the Facilitator asks them to indicate the lowest values (*acceptability thresholds*) for each sector. Otherwise, a stalemate has been reached, because, since all the alternatives in $\mathcal{A}^c$ are efficient, it is not possible to improve the sector index of $Sect^d$ without worsening the value of at least one other sector index. In the case of stalemate, A_{chosen} is a *reasonable alternative*. If, instead, even one of the sector indices can be reduced, the Facilitator identifies the subset $\mathcal{A}^{\text{acc}}$ of the alternatives that satisfy the acceptability thresholds. The intersection of the sets $\mathcal{A}^d$ and $\mathcal{A}^{\text{acc}}$ provides the set $\mathcal{A}^{\text{exp}}$ of the alternatives to explore in the search for an alternative that gets a wider agreement than A_{chosen}.

4. If $\mathcal{A}^{\text{exp}}$ contains at least one alternative, proceed with the next step. If, instead, it is empty, the Facilitator proposes that the Stakeholders review the acceptability thresholds: if at least one of them can be changed, go back to Step 3; if they all refuse, A_{chosen} is a reasonable alternative and the procedure terminates.

[3]It is an ad hoc adjustment of the more general procedure presented in Section 21.5 of THEORY.

5. To search for a reasonable alternative within $\mathcal{A}^{\text{exp}}$, the Facilitator asks the Stakeholders to define a suitable rule to rank the alternatives of this set. For example, the rule could be one of the following:
   (a) *maximum value of the index*: the alternatives $A_i \in \mathcal{A}^{\text{exp}}$ are ranked by decreasing values of the index $V_d(A_i)$, i.e. the index of the most disfavoured sector;
   (b) *minimum overall dissatisfaction*: the alternatives $A_i \in \mathcal{A}^{\text{exp}}$ are ranked by decreasing values of the following function
   $$\sum_j \alpha_j \big(V_j(A_i) - V_j(A_{\text{chosen}})\big)^-$$
   where $(\cdot)^-$ is an operator that returns the value of the argument when this is negative and zero in the opposite case. The α_j coefficients define the relative weights of the sectors; in practice, they are almost always used as Boolean variables to define the sectors that are considered;
   (c) *compensation of the differences*: the alternatives are ordered by decreasing values of the following function
   $$\sum_j \alpha_j \big(V_j(A_i) - V_j(A_{\text{chosen}})\big) \tag{10.1}$$
   where the weights α_j have the same meaning as in criterion (b). Note that, unlike rule (b), with this rule the decrease of an index can be compensated by the improvement of others.

   Let A_{current} be the first alternative of the ranking obtained.
6. Using a bar graph, the Facilitator shows the values of the indices of A_{current} compared to those of A_{chosen} and asks each of the Stakeholders to declare whether (s)he *supports*, *accepts*, or *opposes* A_{current}. A_{current} is said to get a *wider agreement* than A_{chosen} if all the Stakeholders who support A_{chosen} support A_{current} and at least one of the following cases occurs:
   - one of the Stakeholders who accept A_{chosen} supports A_{current};
   - one of the Stakeholders who opposes A_{chosen} accepts or supports A_{current}.

   If A_{current} gets a *wider agreement* than A_{chosen}, replace A_{chosen} with A_{current} and go back to Step 2.
7. Otherwise, if there exists an alternative that follows A_{current} in the ranking, assume it as A_{current} and go back to Step 6. If such an alternative does not exist because the whole ranking has already been examined, A_{chosen} is a reasonable alternative, since within $\mathcal{A}^{\text{exp}}$ an alternative that gets a wider agreement has not been found. Then the procedure terminates.

For each reasonable alternative it is important to record which Stakeholders *support* it, which of them *accept* it and which *oppose* it, because this information is very useful in the phase of *Final Decision*.

Often in Step 2 it is not possible to reach an agreement among the Stakeholders to designate the most disfavoured sector $Sect^d$: if a conflict exists, in fact, it emerges right now.

When this occurs, the procedure must be split into two or more branches (we term this moment *branching* point). Denote the alternative A_{chosen}, i.e. the alternative in correspondence to which the branching occurs, with A_{bra}. Among the sectors proposed as $Sect^d$, the Facilitator chooses the one whose index has to be improved and proceeds until a reasonable alternative is found. Then, he goes back to the branching point and starts the procedure again by considering another of the sectors proposed as $Sect^d$ and setting $A_{\text{chosen}} = A_{\text{bra}}$. The procedure is repeated again and again until all the sectors proposed as disfavoured at the branching point have been considered.

Note that, by following the different pathways that come off the branching point, different reasonable alternatives can be found. An interesting case is when, by following a branch, the alternative A_{bra} itself emerges as a reasonable alternative. Note that A_{bra} can be considered reasonable only if it proves to be a reasonable alternative also by following all the other branches. In fact, if from at least one of them a different alternative emerges, by construction, this alternative achieves a wider agreement than A_{bra} and, as a consequence, A_{bra} cannot be reasonable. Therefore, in correspondence with a branching point, the above Step 4 must be substituted by the following:

4. If $\mathcal{A}^{\text{exp}}$ contains at least one alternative, proceed with the next step. If, instead, it is empty, the Facilitator proposes that the Stakeholders review the acceptability thresholds: if at least one of them accepts, go back to Step 3; if they all refuse, A_{chosen} could be a reasonable alternative. It will actually be one if, by following all the other pathways from the branching point, no other alternative emerges that gets a wider agreement. In that case the procedure terminates.

One after the other, all the branching points that are encountered are examined in this way.

10.1.2 The steps for designing the alternatives

As we explained in Section 5.4, the alternatives of the Verbano Project were designed in two steps. Let $\bar{\mathcal{A}}_1$ be the set of alternatives designed in the first step. All the alternatives in the set must be subjected to the phases of *Evaluation* and *Comparison*: therefore the ENP must be carried out on the set $\mathcal{A}^c = \bar{\mathcal{A}}_1$. By doing so, the first set $\{A^{1,i}_{\text{rea}}\}$ of reasonable alternatives is identified. On the basis of the latter, a second set $\bar{\mathcal{A}}_2$ of alternatives is identified by means of the procedure described in Section 5.4.2. The alternatives in $\bar{\mathcal{A}}_2$, together with the reasonable alternatives identified in the first step, are subjected to a second step of *Comparison*, which provides the second set $\{A^{2,i}_{\text{rea}}\}$ of reasonable alternatives. Therefore in the second step of *Comparison* the ENP is carried out on the set $\bar{\mathcal{A}}_2 \cup \{A^{1,i}_{\text{rea}}\}$.

10.1.3 Initialization of ENP

To complete the description of the ENP we must define a criterion for the choice of the alternative A_{chosen} to be considered in Step 1.

Remember that no discrimination among the sectors is allowed in the *Comparison* phase and that it is essential that both the two sets of reasonable alternatives $\{A^{1,i}_{\text{rea}}\}$ and $\{A^{2,i}_{\text{rea}}\}$ be characterized by the fact that each Stakeholder supports at least one of the alternatives in the set. To guarantee that this occurs, it is sufficient that the following conditions be satisfied

1. After the first step of the alternatives design, the ENP is repeated as many times as there are sectors, and each time the Stakeholders interested in the sector under

examination select the alternative A_{chosen} to start the procedure. The condition of *wider agreement* (see Step 6) ensures that the Stakeholders that choose A_{chosen} at Step 1 will support all the reasonable alternatives that are obtained from it.

2. After the second step of the alternatives design, the ENP is repeated, starting each time with one of the reasonable alternatives in the set $A_{\text{rea}}^{1,i}$, which was obtained at the end of the first step.

10.1.4 The least-bad alternatives

Once the second set of reasonable alternatives has been identified, it is useful to divide it into groups of alternatives that are characterized by the same pairs (SD/RANGE) of actions. For each group, if there exists at least one Stakeholder that is opposed to all the alternatives in the group, the following procedure is carried out

1. Ask each of the Stakeholders that support or accept at least one of the alternatives in the group to what extent (s)he is willing to diminish the values of the sector indices (s)he is interested in, with respect to the worst alternative (for him/her), before changing to the opposition. In this way a set of acceptability thresholds is obtained.
2. Determine the set $\mathcal{A}^s$ of the alternatives in $\mathcal{A}^c$ which satisfy the acceptability thresholds and which are characterized by the same pair (SD/RANGE) as the group of alternatives being examined. If the set $\mathcal{A}^s$ is empty, it must be enlarged by not considering in turn one or more of the acceptability thresholds defined in Step 1.
3. Each of the Stakeholders who oppose all the alternatives in the group is asked to indicate the alternatives that (s)he prefers in $\mathcal{A}^s$. These will be catalogued as *least-bad* for him/her. If the Stakeholder refuses all the alternatives in $\mathcal{A}^s$, this set must be enlarged by not considering in turn one or more of the acceptability thresholds.
4. For each least-bad alternative obtained in this way, each of the other Stakeholders is asked to state how (s)he positions him/herself with respect to it, namely whether (s)he supports, accepts, tolerates or opposes it.

By construction, the least-bad alternatives are characterized by less opposition than the reasonable alternatives identified with the ENP, since they were identified taking into consideration the preferences of the opposers as much as possible. The risk is that these alternatives might be without supporters.

10.1.5 The reasonable alternatives

To further reduce the conflict, all the reasonable alternatives that have been obtained until this point are subjected to the *Mitigation* phase, at the end of which a third and last step of *Comparison* is carried out (Section 10.4). The set of reasonable alternatives resulting from this last comparison is submitted to the DMs for the *Final Decision* (Chapter 16).

The success of the *Comparison* phase greatly depends on the ability of the Facilitator to help the Stakeholders to negotiate and to show them how to get out of a stalemate, without ever forcing them into decisions. To prepare for negotiations, three preliminary activities can be carried out between the *Evaluation* phase and the application of the ENP, which we will describe in the following section.

10.2 Preliminaries to the Comparison phase

The *Comparison* phase must be characterized by three conditions:

- *Transparency*: each Stakeholder must account for his/her own preferences;
- *Completeness of Information*: each Stakeholder must understand as much as possible how the effects of the alternatives were estimated and evaluated for each sector;
- *Availability of Information*: all information must be archived in a structured way, so that it can be accessed easily at any time.

To satisfy these conditions the negotiation process should be preceded by the following three preliminary actions:

1. *Sharing the results of the evaluation*: it helps if each Stakeholder knows how each sector index was defined (not just his/her own!) and if (s)he is able to associate some significant values of the indices with the corresponding effects on the system. The moment for sharing this knowledge is a collective meeting of all the Stakeholders, in which each in turn presents this information for the sectors (s)he is interested in.

2. *Preliminary Analysis*: the results of the sensitivity analysis used to assess the interrelations between sector indices and actions are presented to all the Stakeholders, considering one sector at a time. For example, the way that the index of the *Upstream Flooding* sector varies with the stage–discharge relation, or with the regulation range, is shown. In this way, the Stakeholders can understand which combinations of structural and normative actions are advantageous to each sector and how much the regulation policy influences the sector index. The analysis is conducted by the Facilitator and presented to the Stakeholders either collectively or individually. The Stakeholders must also have the opportunity to conduct this analysis first hand.

3. *Individual Exploration*: it is important that each Stakeholder explore the performance of the alternatives autonomously, with the aim of gaining a feel for the interconnecting links between the different sector indices. (S)he should identify the alternative that (s)he will propose at Step 1 of the ENP when it is his/her turn. In order to avoid preconceived ideas influencing this choice, it is important that the exploration of the alternatives be carried out without knowing the combination of actions that has produced each alternative.

   Examples of these three preliminary activities are provided in Chapter 12. After these activities, it is time for the

4. *Negotiations*: they are carried out through the ENP in several steps, each of which is fulfilled through a collective meeting of all the Stakeholders.

To facilitate the satisfaction of the three conditions of Transparency, Completeness and Availability, it is advisable to conduct all the interactions with the same information support system, so that they take place in the same conceptual environment, which the Stakeholders can master progressively over time. This is one of the purposes of the MODSS TWOLE, described in Chapter 24 of THEORY.

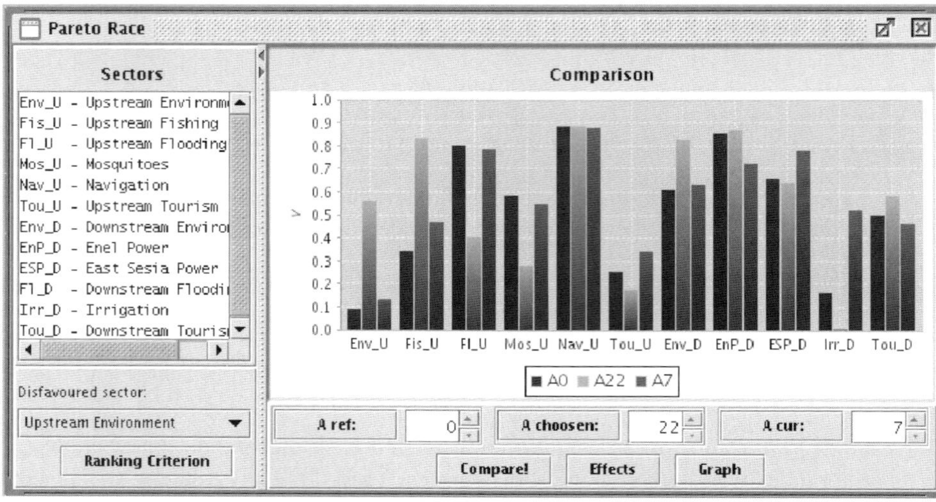

Figure 10.3: A screenshot from TWOLE, which implements the ENP.

10.3 Identification of alliances

Cognitive psychologists (Miller, 1956) have shown that, in general, human short-term memory cannot elaborate more than seven units of information at a time (in some rare individuals as many as nine units but as few as five in other rare cases). Above that threshold, a person may experience a sense of confusion and an inability to make judgements, which are manifested in asking several times for the same information, because it seems to escape him/her. For this reason, it is advisable that the number of sector indices compared in the ENP not be more than seven. The reader can experiment with this limitation to human cognitive ability by observing Figure 10.3, that shows the TWOLE's screenshot that supports the ENP. Eleven sectors are compared[4] and the effort required to express even very simple judgements is evident: for example, ascertaining if one of the three alternatives is efficient is not easy. If, as in the Verbano Project, there are more than 5–7 sectors, it is advisable to try and reduce them. Clearly, it is not possible to simply exclude several sectors (the related Stakeholders would be opposed). Instead, it is necessary to eliminate the redundant information and this is achieved by identifying the *alliances*.

Two sectors are natural allies (excluding the possibility of irrational envy) if their indices are positively correlated, i.e. if the alternatives that increase the first sector index also increase the second and vice versa. If this relationship exists between two sectors and the Stakeholders are informed about it, one of the two sector indices will not need to be visualized during the ENP without any Stakeholder interested in it feeling disfavoured: given two alternatives, it is easy to say which is preferable for the omitted sector by observing the index of the allied sector. However, if the Stakeholders need to know the value of the index of the omitted sector, for example to compare it to a threshold, it can be immediately provided.

[4]The reader might notice that the *Downstream Flooding* sector does not appear in Figure 10.3. In fact, it was excluded from the *Evaluation* and *Comparison* phases for reasons that will be explained in Chapter 12.

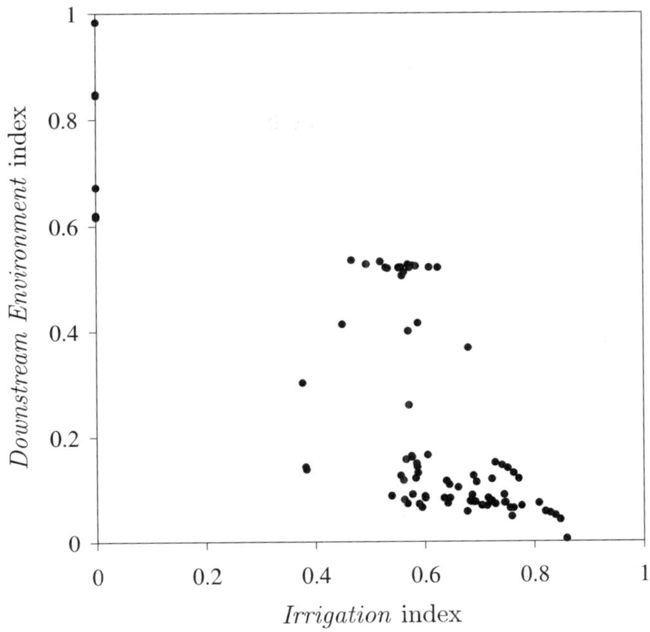

Figure 10.4: Plot of the *Irrigation* index values against the *Downstream Environment* index values. Each point represents a different alternative.

The alliances can be identified by considering one pair of sectors at a time and producing a scatter plot of their indices values for different alternatives. For example, from Figure 10.4 it can be inferred that the *Downstream Environment* and *Irrigation* sector are in strong conflict: high values for the first index are produced by alternatives that produce low values for the second and vice versa.

Therefore, these two sectors are not allies.

The following three cases can occur:

1. *Positive correlation*: the pairs of points that represent the two sector indices are close to a straight line with a positive slope. The two sectors are natural allies and it is possible not to show one of the two indices during the ENP.

2. *Negative correlation*: the regression line has negative slope. Also in this case it is not necessary to visualize both the sector indices, because if the first sector prefers one alternative over another, the preference of the second is certainly the opposite. Nevertheless, when possible in the ENP it is preferable to show both the index values for psychological reasons.

3. *Independence or uncorrelation*: if the pairs of indices are aligned along parallel horizontal or vertical lines, the two sectors are independent. They are uncorrelated if the scatter plot gives a 'cloud' of points. In both cases, knowing one sector index does not provide any information about the other. In the ENP it is therefore essential to visualize both the sectors.

10.4 Mitigation

Mitigation measures are interventions that aim at reducing the negative effects of an alternative on a particular sector, in order to reduce the opposition from the interested Stakeholders or even obtain their support. With the exception of particular or extreme cases, mitigation measures are not financial indemnities, which would simply conceal the problem without resolving it, but specific interventions designed to fit the particular conditions that are encountered. In Chapter 14 we will give two examples.

Once the mitigation measures have been identified for a given alternative, the latter should be considered to be a new alternative and thus it should be evaluated and compared to the reasonable alternatives that have already been identified. All the Stakeholders must participate in the new comparison, not only those that have interests in the concerned sector, since the new alternative could produce negative effects in other sectors. At the end of the comparison, the new alternative usually replaces the initial one, but it is also possible that both are taken forward to the phase of *Final Decision*.

10.5 Identification of the reasonable alternatives

In conclusion, the identification of the reasonable alternatives for the Verbano Project was divided into the following steps (see Figure 10.5).

- The set $\bar{\mathcal{A}}_1$ of efficient alternatives was designed and evaluated; this will be illustrated in Chapter 11.

- After the preliminary activities had been carried out, the alternatives of set $\bar{\mathcal{A}}_1$ were compared and the first set $\{A_{\text{rea}}^{1,i}\}$ of reasonable alternatives was identified, as will be reported in Chapter 12.

- Based on set $\{A_{\text{rea}}^{1,i}\}$, a new set $\bar{\mathcal{A}}_2$ of efficient alternatives was designed and then submitted to evaluation. The new alternatives, together with the alternatives of set $\{A_{\text{rea}}^{1,i}\}$, were compared in a second negotiation step, and a new set $\{A_{\text{rea}}^{2,i}\}$ of reasonable alternatives was identified. The union of the latter and of the set $\{A_{\text{lb}}\}$ of least-bad alternatives formed the set $\{A_{\text{rea}}^{3,i}\}$. All this will be described in Chapter 13.

- Mitigation measures for the alternatives in set $\{A_{\text{rea}}^{3,i}\}$ were designed, and the alternatives so obtained, together with those of set $\{A_{\text{rea}}^{3,i}\}$, were compared in the third negotiation step, as will be illustrated in Chapter 14.

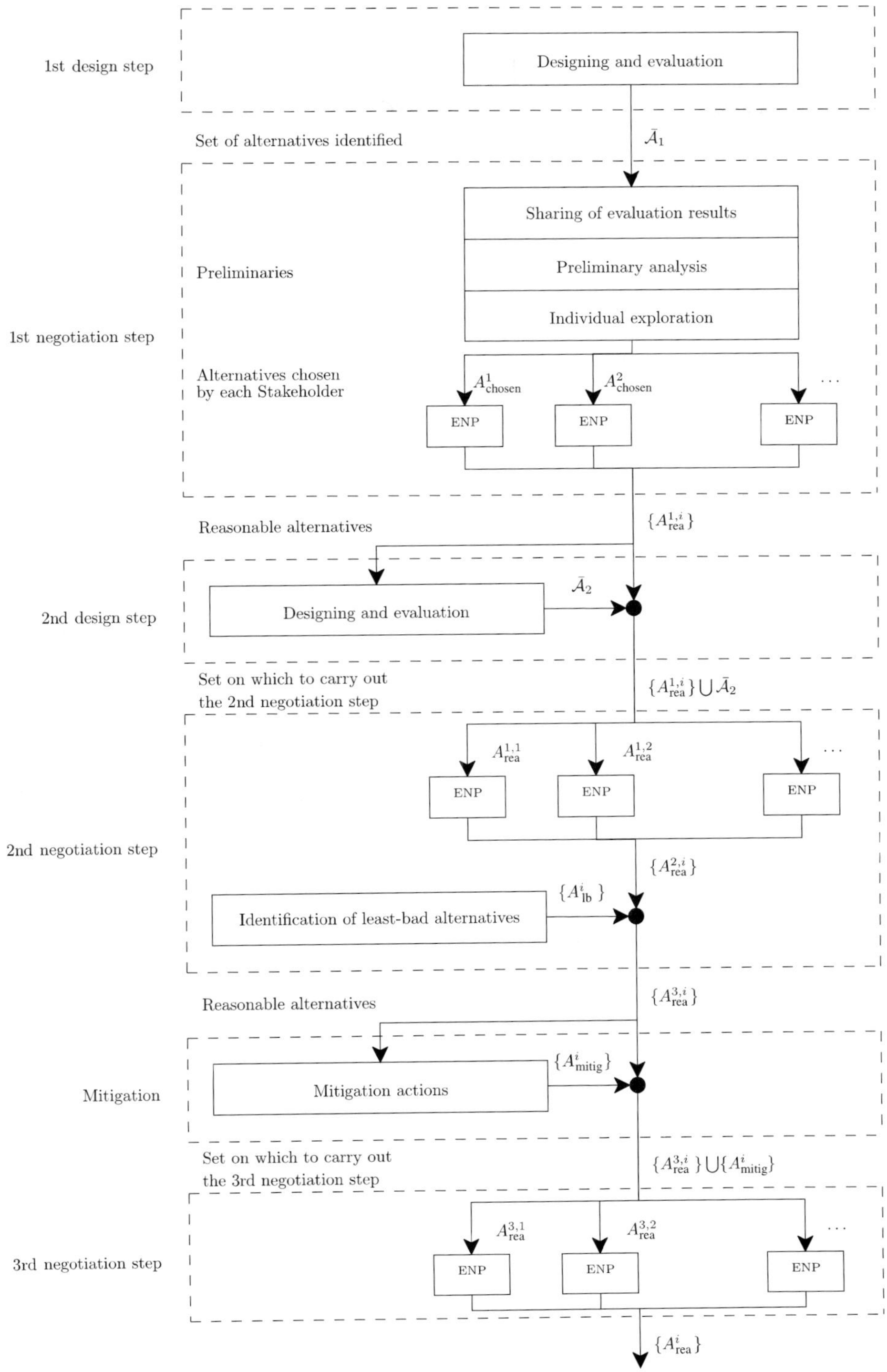

Figure 10.5: The steps for identifying the reasonable alternatives.

PART B

IDENTIFYING THE REASONABLE ALTERNATIVES

Chapter 11

Designing Alternatives and Evaluation

EB, FP, RSS and EW

This chapter presents the results of the first step of designing and evaluating the alternatives (see Figure 11.1).

11.1 Designing Alternatives

As we have said several times, an alternative A_i is defined when the actions that it is composed of (structural action: stage–discharge relation SD_i; normative actions: regulation range RANGE_i and Minimum Environmental Flow MEF_i; regulation action: policy p_i) are all specified, which is to say an alternative is defined by the following 4-tuple

$$A_i = (\text{SD}_i/\text{RANGE}_i/\text{MEF}_i, p_i) \tag{11.1}$$

The triples of structural and normative actions that we have decided to consider in the first step of designing the alternatives (see Section 5.4.1) are the following:

$$(\text{SD}_{+600}/\text{RANGE}_{\text{ENL}}, 13_{\text{CON}}) \tag{11.2}$$

$$(\text{SD}_{\text{CUR}}/\text{RANGE}_{\text{CUR}}, 13_{\text{CON}}) \tag{11.3}$$

$$(\text{SD}_{+600}/\text{RANGE}_{\text{CUR}}, 13_{\text{CON}}) \tag{11.4}$$

For each of these triples, we must design a set of efficient regulation policies, which are identified by solving an appropriate (Multi-Objective Optimal) Control Problem, whose objectives were chosen in Section 5.2.3 and quantified in Section 7.3. Among the objectives there exists a partial lexicographic order, in the sense that the use of water for the production of hydropower is subordinate to the satisfaction of human, agricultural and environmental needs. It follows that the Control Problem is a lexicographic problem and in Section 7.6.3 we showed that it is equivalent to a sequence of two Control Problems. In the first (*Primary Problem*) the objectives linked to the sectors *Upstream Flooding* ($J_{\text{fl}}(p)$), *Irrigation* ($J_{\text{irr}}(p)$) and *Downstream Environment* ($J_{\text{env}}(p)$) are considered; in the second (*Secondary Problem*), the objective linked to the ENEL *Power* sector ($J_{\text{Enel}}(p)$) is considered alone. Since the Primary Problem has three objectives, it must be solved by defining a vector $\boldsymbol{\lambda}$ of weights and combining the three objectives $J_{\text{fl}}(p)$, $J_{\text{irr}}(p)$ and $J_{\text{env}}(p)$ into a single objective $\bar{J}$. By doing so, the Multi-Objective Control Problem is traced back to a family of

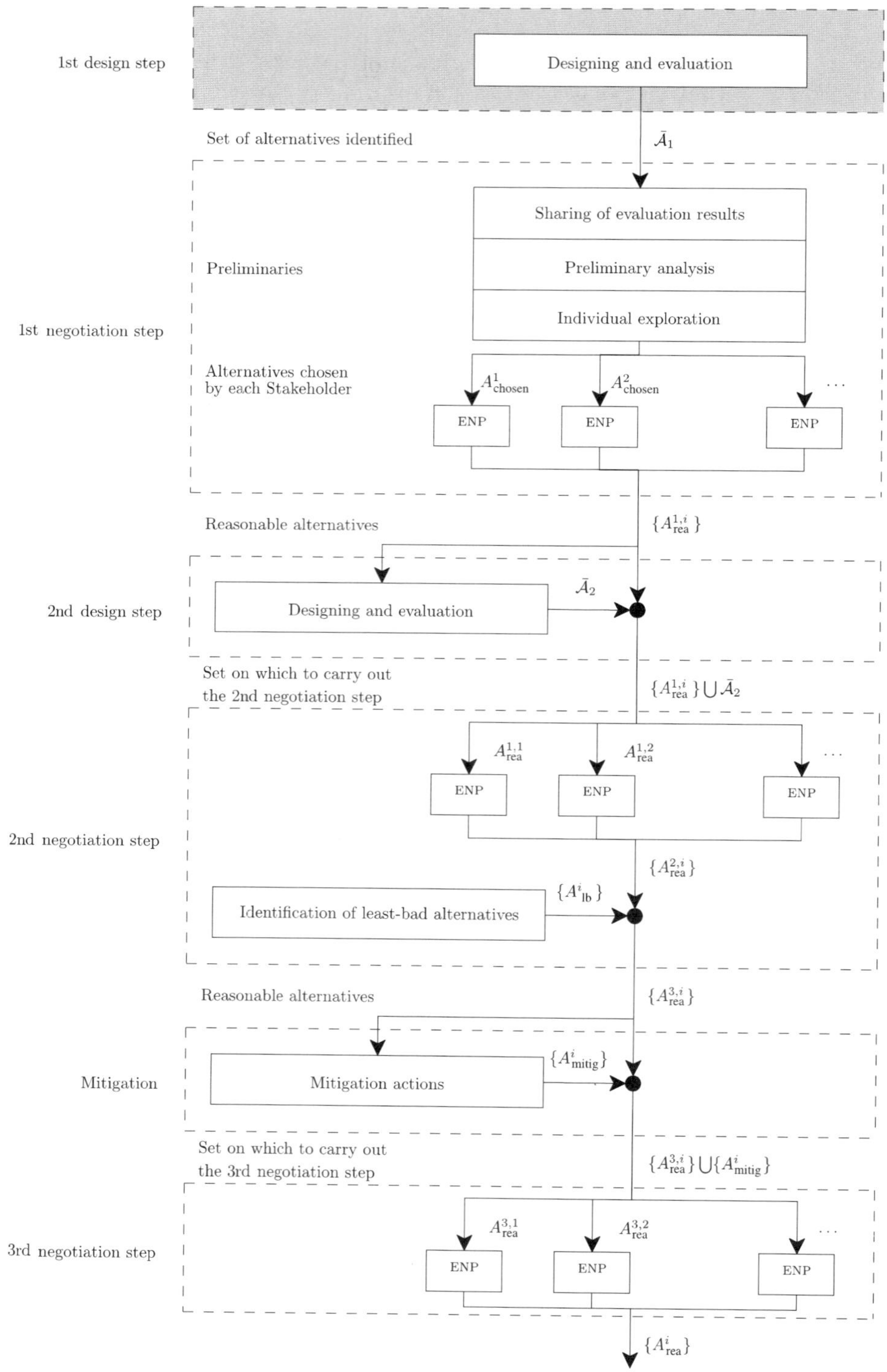

Figure 11.1: The steps for identifying the reasonable alternatives. The highlighted step is the one whose results are described in this chapter.

Single-Objective Control Problems, which are parametric in vector $\boldsymbol{\lambda}$ (see Section 5.2.6). For each value $\boldsymbol{\lambda}_i$ of $\boldsymbol{\lambda}$ two Single-Objective Control Problems (Primary and Secondary) can be defined and solved in sequence using the MODSS TWOLE (see Chapter 24 of THEORY), which provides an efficient policy p_i for the triple of structural and normative actions considered.

The above procedure allows for the exact determination of the policy p_i but in Section 8.4.2 we observed that in this phase the exact definition of the policies is less interesting than the estimation of the effects that each of them produces. In fact, the choice among the different alternatives is based on the comparison of the effects that they induce and not on the actions that define them. In that section we assessed the errors that are committed in the estimation of the effects if, instead of solving the two Control Problems in sequence, only the Primary Problem is solved and the policy obtained is simulated by using an *extractor with an irrigation/hydropower reference*. We ascertained that the errors are rather small and so we decided to adopt this shortcut, halving the computing time required for the design and evaluation of all the alternatives (from 200 to 100 days).

So far we have outlined the general framework of the operations for designing alternatives and estimating their effects. Now let us focus on the way in which the results of the simulation are presented and examined. They are arranged in a table which, for each day of the evaluation horizon, provides the value of all the hydrological variables that appear in the model of the water system. As an example, we will examine several alternatives that share the triple (11.2) of structural and normative actions and differ in the policy; in particular, we will consider the extreme policies, i.e. those that are obtained by attributing a unitary weight to one of the three objectives J_{fl}, J_{irr} or J_{env} and zero to the other two.

First we will analyse the extreme policy which favours the *Upstream Flooding* sector. If the objective is to minimize flooding, the most obvious way to achieve it would be to keep the Miorina dam completely open all the time. This is not the only way, however: in the winter months, for example, when the expected inflows are low, it might not be necessary to open the dam completely, and so it would be possible to create a storage for springtime irrigation without worsening the floods. At every time instant a number of controls thus exists that are equivalent from the *Upstream Flooding* standpoint. Furthermore, since upstream floods depend only upon the lake level and not on the water distribution downstream from the Miorina, the value of J_{fl} is influenced only by the control u_t^{MIO}, which defines the release from the lake, and thus only by the component $M_t^{\text{MIO}}(s_t)$ of the control law. Therefore minimizing J_{fl} is not sufficient for determining the second component $M_t^{\text{TIC}}(s_t)$ and thus no unique extreme policy can be defined for the flooding sector, but an infinite number of them. To completely define the policy, i.e. the sequence of pairs of control laws $(M_t^{\text{MIO}}(s_t), M_t^{\text{TIC}}(s_t))$, it is thus necessary to formulate and solve, in cascade after the first, a second Control Problem that takes account of the other two objectives (J_{irr} and J_{env}) and contains the constraint

$$u_t^{\text{MIO}} \in M_t^{\text{MIO}*}(s_t) \quad \text{for each } t$$

where $M_t^{\text{MIO}*}(s_t)$ for $t = 0, 1, \ldots$ are the control laws that solve the first Problem. Since we are searching for extreme policies, when solving the second Problem it is opportune to consider the extreme cases as well, i.e. the Problem where only the irrigation objective is considered or the Problem where only the environmental objective is considered. By doing so, the following alternatives are obtained

$$A27 = (\text{SD}_{+600}/\text{RANGE}_{\text{ENL}}, 13_{\text{CON}}, p_{27})$$
$$A240 = (\text{SD}_{+600}/\text{RANGE}_{\text{ENL}}, 13_{\text{CON}}, p_{240}) \quad (11.5)$$

The first is characterized by an extreme policy with respect to the two objectives *Upstream Flooding* and *Irrigation* (in lexicographic order), the second by an extreme policy with respect to *Upstream Flooding* and *Downstream Environment* (again in lexicographic order). The numbers attributed to the alternatives do not have any particular meaning: they serve only to distinguish them from one another.

Table 11.1 shows an extract from the results provided by the simulation of $A27$; below we describe the contents of its columns one by one.

- **Natural and anthropic date:** The second column (*Natural Date*) contains the progressive number (from 0 to 364) of the current day, i.e. the day to which the row refers (in leap year 29th February has the same number as 28th February). The third column (*Anthropic Date*) contains the number of the control law applied that day. The regulation policy is, in fact, a periodic sequence (of period 365) of set-valued control laws, and, like water demand, it depends on anthropic date and not on natural time (see page 223).

  In the Control Problem with which the policy is designed, the *standard year* is considered, namely a non-leap-year that begins on a Monday. This specification is necessary because the values of some parameters appearing in the objectives (Section 7.3), in the equations of the model (Chapter 6) and in the design scenario, depend on the day of the week: for example, the price of electricity is lower on Saturday and Sunday. When the policy is simulated over the evaluation horizon, clearly not all the years are standard years: thus we need a rule to associate control laws and days in non-standard years. The most obvious choice is to adopt the policy associated to the anthropic date of the current day: namely, to associate the current day with the control law of the closest homonymous day of the standard year, i.e. the day of the standard year that has the same name (Monday, Tuesday, ...) and is the closest to the day being considered. For example, 1st January 1974 (indexed with 0 in the natural date) falls on a Tuesday; the control law is thus the one associated to day 1 in the standard year, because 1 is the natural date closest to 0 which is associated to a Tuesday in the standard year.

- **Storage s_t:** the column contains the volume that is stored in the lake at 8 o'clock in the morning on the current day. For all the alternatives, the value reported in the first row is the same, namely the value (344.7 Mm^3) of the storage measured on 1st January 1974, the first day of the evaluation horizon H. The values in the following rows are computed with the mass balance equation (6.29). Since they depend on the release decisions assumed over the previous days, they are different for different alternatives.

- **Release and allotment decisions $u_t^{\text{MIO}}, u_t^{\text{TIC}}$:** once the control law index is defined and the lake storage is known, the set $M_0^{i*}(s_0)$ is univocally defined from which a pair $(u_0^{\text{MIO}}, u_0^{\text{TIC}})$ of controls is chosen. The choice is defined by the extractor being considered. In the example, as we said, the extractor with an irrigation/hydropower reference was adopted: therefore the pair is identified by equations (8.11) and (8.12), which choose the control that is closest to the reference $(W_t^{\text{MIO}}, W_t^{\text{TIC}})$. On day 0 the reference is the pair (210.5, 0) and, since it belongs to the set $M_0^{i*}(s_0)$, it is chosen.

Table 11.1. The results of the first steps of the simulation of alternative $A27$

Date	Natural	Anthropic	s_t [m^3]	u_t^{MIO} [m^3/s]	u_t^{TIC} [m^3/s]	a_{t+1} [m^3/s]	r_{t+1} [m^3/s]	$q_{t+1}^{TIC,u}$ [m^3/s]	h_t^{SC} [m]	$q_{t+1}^{TIC,MEF}$ [m^3/s]	$q_{t+1}^{TIC,r}$ [m^3/s]	q_{t+1}^{TIC} [m^3/s]	q_{t+1}^{REC} [m^3/s]	q_{t+1}^{IC} [m^3/s]	q_{t+1}^{VC} [m^3/s]	q_{t+1}^{PT} [m^3/s]	h_t^{PV} [m]
1/1/74	0	1	344.7	210.5	0.0	135.1	210.5	0.0	0.88	13.0	0.0	13.0	45.0	138.5	14.0	165.5	0.00
2/1/74	1	2	338.2	210.5	0.0	174.3	210.5	0.0	0.85	13.0	0.0	13.0	45.0	138.5	14.0	165.5	0.00
3/1/74	2	3	335.1	210.5	0.0	198.9	210.5	0.0	0.83	13.0	0.0	13.0	45.0	138.5	14.0	165.5	0.00
4/1/74	3	4	334.1	210.5	0.0	174.6	210.5	0.0	0.83	13.0	0.0	13.0	45.0	138.5	14.0	165.5	0.00
5/1/74	4	5	331.0	210.5	0.0	101.7	210.5	0.0	0.82	13.0	0.0	13.0	45.0	138.5	14.0	165.5	0.00
...	...	...	...	...	...	...	...	...	...	...	...	...	...	...	...	...	...
16/2/74	46	47	253.5	210.5	0.0	295.0	210.5	0.0	0.45	13.0	0.0	13.0	45.0	138.5	14.0	165.5	0.00
17/2/74	47	48	260.8	210.5	0.0	346.1	210.5	0.0	0.48	13.0	0.0	13.0	45.0	138.5	14.0	165.5	0.00
18/2/74	48	49	272.5	210.5	0.0	404.7	210.5	0.0	0.54	13.0	0.0	13.0	45.0	138.5	14.0	165.5	0.00
19/2/74	49	50	289.3	625.0	0.0	379.5	625.0	0.0	0.62	13.0	414.5	427.5	45.0	138.5	14.0	165.5	0.00
20/2/74	50	51	268.1	391.0	0.0	359.7	391.0	0.0	0.52	13.0	180.5	193.5	45.0	138.5	14.0	165.5	0.00
21/2/74	51	52	265.4	343.4	0.0	305.1	343.4	0.0	0.50	13.0	132.9	145.9	45.0	138.5	14.0	165.5	0.00

Table 11.2. The results of the first steps of the simulation of alternative $A240$

Date	Natural	Anthropic	s_t [m^3]	u_t^{MIO} [m^3/s]	u_t^{TIC} [m^3/s]	a_{t+1} [m^3/s]	r_{t+1} [m^3/s]	$q_{t+1}^{TIC,u}$ [m^3/s]	h_t^{SC} [m]	$q_{t+1}^{TIC,MEF}$ [m^3/s]	$q_{t+1}^{TIC,r}$ [m^3/s]	q_{t+1}^{TIC} [m^3/s]	q_{t+1}^{REC} [m^3/s]	q_{t+1}^{IC} [m^3/s]	q_{t+1}^{VC} [m^3/s]	q_{t+1}^{PT} [m^3/s]	h_t^{PV} [m]
1/1/74	0	1	344.7	1200.0	1187.0	135.1	760.9	747.9	0.88	13.0	0.0	760.9	0.0	0.0	0.0	760.9	0.00
2/1/74	1	2	290.7	1200.0	1187.0	174.3	648.3	635.3	0.62	13.0	0.0	648.3	0.0	0.0	0.0	648.3	0.00
3/1/74	2	3	249.7	1200.0	1187.0	198.9	564.7	551.7	0.43	13.0	0.0	564.7	0.0	0.0	0.0	564.7	0.00
4/1/74	3	4	218.1	1200.0	1187.0	174.6	496.7	483.7	0.28	13.0	0.0	496.7	0.0	0.0	0.0	496.7	0.00
5/1/74	4	5	190.3	1200.0	1187.0	101.7	432.2	419.2	0.15	13.0	0.0	432.2	0.0	0.0	0.0	432.2	0.00
6/1/74	5	6	161.7	1200.0	1187.0	101.7	371.5	358.5	0.01	13.0	0.0	371.5	0.0	0.0	0.0	371.5	0.00

- **Inflow a_{t+1}:** to simulate the alternatives, we decided to use the historical series of inflows (Section 8.3). The column shows the average daily flow rate of the inflows in the interval $[t, t+1)$, with t being the current day. For example, in the first row the average flow rate between 8 o'clock in the morning on 1st January 1974 and the same hour on 2nd January 1974 is shown: it is equal to 135.1 m^3/s.

- **Hydrological variables:** In the remaining columns the values of the hydrological variables that describe the system are reported. They are derived from equation (6.62) that describes the model of the water system, in which the data in the columns that we have described so far are replaced. For example, the level h_t^{SC} is obtained by transforming s_t with equation (6.20), while the release r_{t+1} is obtained from s_t, u_t^{MIO} and a_{t+1} by using equation (6.27).

Now let us examine the values that were obtained. An immediate idea of the greater or lesser risk of flooding on a given day t of the year is provided by Figure 11.2, which shows the pattern of the expected value μ_t^a of the average daily inflow plus or minus its standard deviation σ_t^a. The risk of flooding is clearly increasing with $\mu_t^a + \sigma_t^a$ and it is thus much higher in the spring and autumn months than in the winter months. This is why in the first days of the year policy p_{27} suggests a control that is equal to the water demand increased by the MEF (210.5 m^3/s), even if the lake level is higher than 0.80 m at Sesto Calende hydrometric station, while after 19th February much higher releases are suggested even for much lower lake levels: the reason is that, on 19th February, the inflows expected in the future are much higher than those expected on the first days of January.

Now we will consider alternative $A240$, which was obtained by solving a Secondary Problem that has J_{env}, i.e. the *Downstream Environment* quality, as its objective. The results of its simulation are reported in Table 11.2. One can observe that, unlike $A27$, in alternative $A240$ the Miorina dam is kept completely open all of the time: this is revealed by the value of the control u_t^{MIO}, which is always equal to 1200 m^3/s, a value which, given the low level of the lake, cannot actually be achieved (note the corresponding value of r_{t+1}) but which implies the complete opening of the dam. As we have already seen, opening the dam completely is one of the optimal decisions for the objective J_{fl}. Let us check whether it is optimal from the point of view of the environmental objective J_{env} as well. In Section 4.3.3 we assumed the natural regime as a reference point to define the *Downstream Environment* quality. Now, by keeping the Miorina dam completely open, we do not create natural regime conditions, but free regime conditions (with the stage–discharge relation +600), and the two regimes are not the same. Nevertheless, in the Verbano Project they are quite similar, because of the characteristics of the two stage–discharge relations: the natural one and the +600. In fact, Figure 11.3 shows that the instantaneous stage–discharge relation $N^{+600}(\cdot)$, produced by the action SD_{+600}, significantly differs from the natural stage–discharge relation ($N^{NAT}(\cdot)$). However, for sufficiently high lake levels, the second can be approximated by the first once it has been shifted by $\bar{h}$. If the approximation were exact, i.e. if

$$N^{NAT}\left(h^{SC}\right) = N^{+600}\left(h^{SC} - \bar{h}\right) \tag{11.6}$$

the free regime and the natural regime would be characterized by the same release trajectory. This is confirmed by Figure 11.4, which shows that, in free regime conditions, the release trajectory created by $A240$ tends over time to the natural outflow trajectory.[1] On the

[1] The initial difference is due to the fact that the storage on 1/1/74, which is the initial state of the simulation for $A240$, is much different from the natural storage on the same day. The difference becomes negligible after about 6 months.

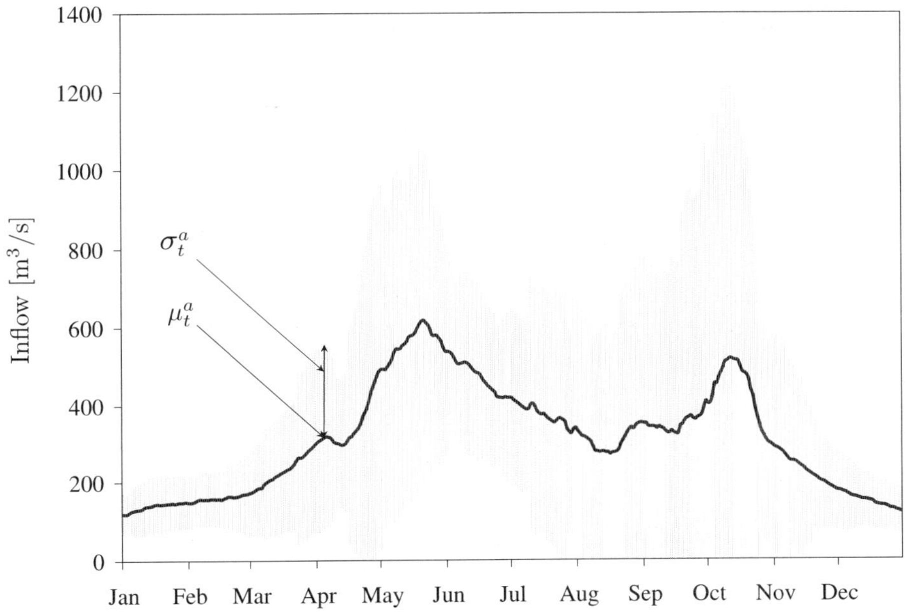

Figure 11.2: Pattern of the expected value μ_t^a of the average daily inflow $\pm$ its standard deviation σ_t^a.

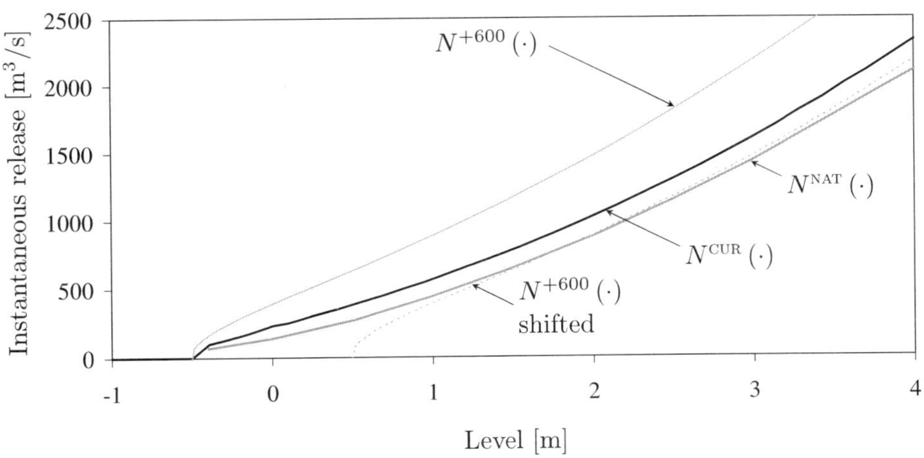

Figure 11.3: Different stage–discharge relations: natural, current, +600 and shifted +600.

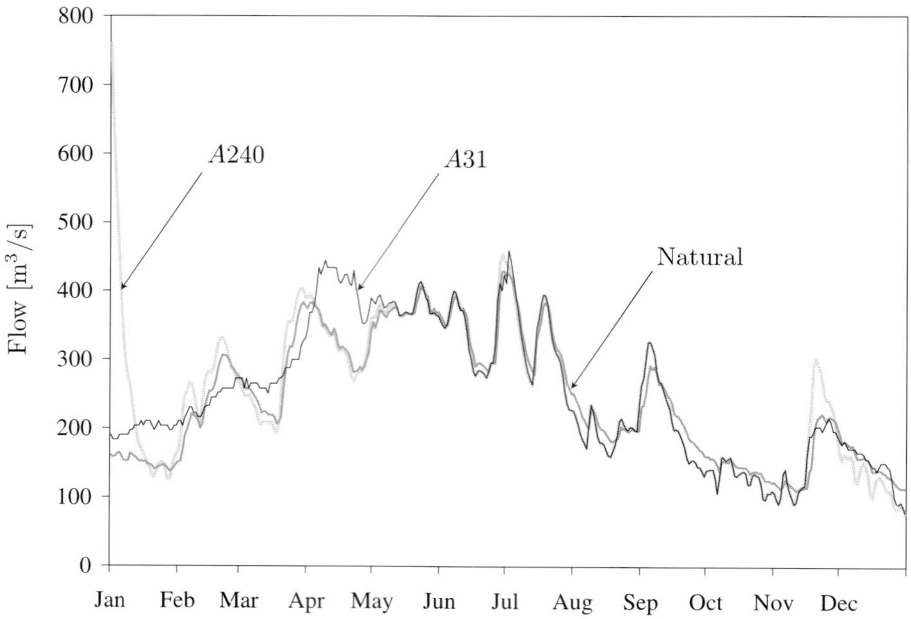

Figure 11.4: Release trajectories produced by alternatives $A240$ and $A31$, compared to natural outflow trajectory (over the year 1974).

contrary, the level trajectories would be fairly different but they would be, however, linked by a vertical shift (see Section 3.1.3 of the DVD).

In Figure 11.4 the release trajectory produced by $A31$ is shown as well. This is the extreme alternative that favours *Downstream Environment*, i.e.

$$A31 = (\text{SD}_{+600}/\text{RANGE}_{\text{ENL}}, 13_{\text{CON}}, p_{31}) \quad \text{Extreme policy for } \textit{Downstream Environment} \tag{11.7}$$

This trajectory is very similar to the other two. The small differences with respect to the natural outflow trajectory are due to the fact that, in order to reduce the computational complexity of the Control Problem, the reference value used in the definition of objective J_{env} is not the natural outflow, but its median (see Section 7.3.4).

From the above, it emerges that there is no conflict between the reduction of flooding and the improvement of the *Downstream Environment* quality. The main difference between the two extreme policies is that while the objective J_{fl} is influenced only by the release at the Miorina dam (and in some periods of the year it is almost insensitive even to this release), the objective J_{env} is influenced also by the water distribution downstream from the Miorina and is sensitive to decisions all through the year. It follows that while the *Upstream Flooding* sector can easily form alliances with other sectors (the two alternatives $A27$ and $A240$ are an example of this) the *Downstream Environment* sector has difficulties in finding allies. When compromise policies are designed, the *Upstream Flooding* sector might at first seem to be in conflict with *Downstream Environment* because the objective of the latter has a poor performance while the objective of the former does not. However, a careful analysis of the results shows that the poor performance of *Downstream Environment* is actually due to good performances of the irrigation and hydropower objectives.

Lastly, it is interesting to examine the extreme alternative that favours *Irrigation*, i.e.

$$A22 = (\text{SD}_{+600}/\text{RANGE}_{\text{ENL}}, 13_{\text{CON}}, p_{22}) \quad \text{Extreme policy for } \textit{Irrigation} \qquad (11.8)$$

The results of its simulation are reported in Table 11.3.

In order to understand the meaning of the decisions suggested by this alternative, bear in mind that the objective J_{irr} does not only consider irrigation demand, but also the risk aversion that the Expert for this sector demonstrated. The irrigation season begins in March, with the submersion of the rice paddies, but since risk aversion is high (remember exponent 12 in equation (7.22)), in January the policy is already working to guarantee that in March the lake storage will be sufficient to avoid heavy deficits. Observe that in Table 11.3 the release is strictly equal to the sum of the irrigation demands plus the MEF (104.4 m^3/s) until the lake level exceeds 0.9 m at Sesto Calende: below this level, i.e. before the water reserve is considered to be sufficient (as judged by the policy), no water is released for hydropower production. It emerges that the risk aversion of the irrigators would reduce the release for hydropower not only in the critical periods for irrigation, but also, and above all, beforehand. Since in winter the value of energy is higher than in the other months, this causes a conflict between irrigation and hydropower interests.

11.2 Estimating Effects

By simulating an alternative we obtain the trajectories of the hydrological variables: with this information, the indicator values must be computed. The operation is very simple: it just requires applying the functionals that define the indicators (see Section 4.5) to the trajectories that were obtained. The values produced in this way are arranged in a matrix (termed *Matrix of the Effects*), in which each row corresponds to an indicator and each column to an alternative. As an example, the Matrix of the Effects for the four alternatives that we discussed in the previous section is shown in Table 11.4.

11.3 Validating the indicators

As we explained in Section 8.5, once the indicator values have been estimated it is necessary to verify that they are suitable for comparing the alternatives. To achieve this, two aspects of each indicator must be examined: the significance of the values obtained and its representativeness. All the indicators proved to be significant and representative, with the exception of one relevant to *Downstream Environment*, one of *Navigation* and all the indicators of *Downstream Flooding* sector.

11.3.1 Downstream Environment

The value of the indicator $i_{\text{Env\_D\_5}}$ (*Fraction* [%] *of the reproduction period for aquatic birdlife and fish species in which the flow rate of the Ticino at Oleggio varies too rapidly*) proved to depend almost exclusively on rapid variations in the release from the Miorina dam, which are generated when the lake level reaches the upper extreme of the regulation range, with the aim of not exceeding it. The variations are very sharp because no limit was

Table 11.3. The results of the first steps of the simulation of alternative $A22$

Date	Natural	Anthropic	s_t [m^3]	u_t^{MIO} [m^3/s]	u_t^{TIC} [m^3/s]	a_{t+1} [m^3/s]	r_{t+1} [m^3/s]	$q_{t+1}^{TIC,u}$ [m^3/s]	h_t^{SC} [m]	$q_{t+1}^{TIC,MEF}$ [m^3/s]	$q_{t+1}^{TIC,r}$ [m^3/s]	q_{t+1}^{TIC} [m^3/s]	q_{t+1}^{REC} [m^3/s]	q_{t+1}^{IC} [m^3/s]	q_{t+1}^{VC} [m^3/s]	q_{t+1}^{PT} [m^3/s]	h_t^{PV} [m]
1/1/74	0	1	344.7	104.4	0.0	135.1	104.4	0.0	0.88	13.0	0.0	13.0	2.0	75.4	14.0	102.4	0.00
2/1/74	1	2	347.4	104.4	0.0	174.3	104.4	0.0	0.89	13.0	0.0	13.0	2.0	75.4	14.0	102.4	0.00
3/1/74	2	3	353.4	112.0	0.0	198.9	112.0	0.0	0.92	13.0	0.0	13.0	2.0	83.0	14.0	110.0	0.00
4/1/74	3	4	360.9	117.0	0.0	174.6	117.0	0.0	0.96	13.0	0.0	13.0	2.0	88.0	14.0	115.0	0.00
5/1/74	4	5	365.9	122.0	0.0	101.7	122.0	0.0	0.98	13.0	0.0	13.0	2.0	93.0	14.0	120.0	0.00
6/1/74	5	6	364.2	122.0	0.0	101.7	122.0	0.0	0.97	13.0	0.0	13.0	2.0	93.0	14.0	120.0	0.00

Table 11.4. The Matrix of the Effects relative to the four alternatives presented in Section 11.1

	Sectors	Indicators	A22	A27	A31	A240
Upstream	*Upstream Environment*	$i_{Env\_U\_1}$	0.74	0.67	0.53	0.67
		$i_{Env\_U\_2}$	7.52	29.54	24.02	29.54
		$i_{Env\_U\_3}$	7.76	85.15	68.37	85.15
		$i_{Env\_U\_4}$	90.35	31.85	32.38	31.85
		$i_{Env\_U\_5}$	81.66	7.79	20.76	7.79
	Upstream Fishing	$i_{Fis\_U\_1}$	1.97	8.30	5.75	8.30
		$i_{Fis\_U\_2}$	26.73	52.68	35.02	52.68
		$i_{Fis\_U\_3}$	13.44	91.29	82.85	91.29
		$i_{Fis\_U\_4}$	7.76	85.15	68.37	85.15
	Upstream Flooding	$i_{Fl\_U\_1\_Loc}$	1.10	1.05	1.05	1.05
		$i_{Fl\_U\_1\_Vb}$	0.29	0.27	0.27	0.27
		$i_{Fl\_U\_2\_Loc}$	0.49	0.23	0.29	0.23
		$i_{Fl\_U\_2\_Vb}$	0.06	0.02	0.03	0.02
		$i_{Fl\_U\_3\_Loc}$	28.	26.	26.	26.
		$i_{Fl\_U\_3\_Vb}$	12.	10.	10.	10.
		$i_{Fl\_U\_4\_Loc}$	7.36	4.08	4.92	4.08
		$i_{Fl\_U\_4\_Vb}$	1.36	0.60	0.76	0.60
		$i_{Fl\_U\_6\_Loc}$	0.00	0.00	0.00	0.00
		$i_{Fl\_U\_6\_Vb}$	0.00	0.00	0.00	0.00
	Mosquitoes	$i_{Mos\_U\_1}$	0.60	1.24	0.92	1.24
		$i_{Mos\_U\_2}$	189.93	8.81	31.65	8.81
	Navigation	$i_{Nav\_U\_1}$	21.25	9.67	12.81	9.67
		$i_{Nav\_U\_2}$	0.72	0.33	0.38	0.33
	Upstream Tourism	$i_{Tou\_U\_1}$	7.10	85.31	68.91	85.31
		$i_{Tou\_U\_2}$	48.82	2.20	7.93	2.20
		$i_{Tou\_U\_3}$	0.60	1.24	0.92	1.24
		$i_{Tou\_U\_4}$	189.93	8.81	31.65	8.81
Downstream	*Downstream Environment*	$i_{Env\_D\_1}$	247.65	209.28	94.49	97.72
		$i_{Env\_D\_5}$	25.99	10.12	3.06	5.15
	ENEL *Power*	$i_{EnP\_D\_1}$	3345.96	3439.94	20647.65	20663.38
		$i_{EnP\_D\_2}$	0.	2.	365.	365.
	East Sesia Power	$i_{ESP\_D\_1}$	1079.24	927.54	2575.35	2575.35
	Downstream Flooding	$i_{Fl\_D\_1}$	3.97	3.94	3.93	3.94
		$i_{Fl\_D\_2}$	1.71	1.51	1.57	1.51
		$i_{Fl\_D\_3\_Bt}$	2.	1.	1.	1.
		$i_{Fl\_D\_3\_Pv}$	32.	26.	26.	26.
		$i_{Fl\_D\_4\_Bt}$	0.20	0.16	0.16	0.16
		$i_{Fl\_D\_4\_Pv}$	3.52	2.84	2.88	2.84
		$i_{Fl\_D\_6}$	0.72	0.60	0.64	0.60
	Irrigation	$i_{Irr\_D\_1}$	147.66	387.72	4707.64	4707.64
		$i_{Irr\_D\_2}$	14.90	53.13	100.00	100.00
		$i_{Irr\_D\_3\_IC}$	4.61	12.87	100.00	100.00
		$i_{Irr\_D\_3\_REC}$	21.84	27.45	100.00	100.00
		$i_{Irr\_D\_3\_VC}$	8.19	35.04	100.00	100.00
	Downstream Tourism	$i_{Tou\_D\_1}$	257.51	242.20	2.72	2.72
		$i_{Tou\_D\_2}$	8.11	4.76	7.40	7.71

put on the speed at which the dam is manoeuvred, but the analysis of the individual events showed that slower and more cautious manoeuvres would have eliminated the problem, without compromising the values of the other indicators. In other words, the negative effects measured by the indicator can be eliminated by acting at the operational control level (see Section 1.4). The indicator is thus not representative, because it signals damage which, in fact, does not exist, and so it will not be considered in the *Evaluation* phase.

11.3.2 Navigation

The values of the indicator $i_{\mathrm{Nav\_U\_3}}$ (*Average annual lost revenue* [k€/year] *due to the inability to load heavy cargo when the lake level is low*) proved to be so small that they had no real economic significance. Therefore, the indicator will not be used in the *Evaluation* phase.

11.3.3 Downstream Flooding

An indicator is significant when it makes it possible to compare the alternatives on the basis of its value. This means that the difference between the best and worst values of the indicator among all the alternatives that had been designed must be more than double the uncertainty with which the values of the indicator have been estimated. To evaluate the uncertainty in the estimation of the *Downstream Flooding* indicators, we first considered the errors committed in estimating the flood peaks of the River Ticino at Pavia with the neural network described in Section 6.9.1. It was found out that the errors have zero mean and a standard deviation of 31 cm (Table 6.6). This value provides an estimate of the uncertainty σ_i in the estimation of indicator $i_{\mathrm{Fl\_D\_1}}$ (*Maximum peak flood level in Pavia*). The uncertainty in the estimation of indicator $i_{\mathrm{Fl\_D\_2}}$ (*Average annual flooded area in Pavia*) can be derived from $\sigma_{i_{\mathrm{Fl\_D\_1}}}$ also: it is given by the minimum increment that the flooded area can take following an increase in the flood peak equal to $\sigma_{i_{\mathrm{Fl\_D\_1}}}$. That value is 0.76 Km^2.

As shown in Table 11.5, the difference between the minimum and maximum values of $i_{\mathrm{Fl\_D\_1}}$ among all the alternatives is only 8 cm: a value that is clearly less than the uncertainty in the estimation. Therefore the indicator is not significant. One might suspect, however, that the judgement is polarized: the value of $i_{\mathrm{Fl\_D\_1}}$ depends on a single flood event, which is the same for both alternatives. In other words, the lack of discrimination might be due to the particular characteristics of that event, rather than to the inaccuracy of the model that is used to compute the indicator. This suspicion weakens when one notes that the condition of significance is not satisfied by indicator $i_{\mathrm{Fl\_D\_2}}$ either (Table 11.6), which

Table 11.5. *Maximum peak flood level in Pavia*: comparison between the performances of the best and worst alternatives and the accuracy of the estimate

	Alternative	SD	RANGE	MEF	Policy	Indicator value
Best	*A*20	ACT	ACT	13	Extreme for floods	3.89
Worst	*A*22	600	ENL	13	Extreme for irrigation	3.97
Significance			Difference			0.08
			Uncertainty			0.31
			Difference/Uncertainty			0.27

Table 11.6. *Average annual flooded area in Pavia*: comparison between the performances of the best and worst alternatives and the accuracy of the estimate

	Alternative	SD	RANGE	MEF	Policy	Indicator value
Best	*A*20	ACT	ACT	13	Extreme for floods	1.40
Worst	*A*22	600	ENL	13	Extreme for irrigation	1.71
			Difference			0.31
Significance			Uncertainty			0.76
			Difference/Uncertainty			0.40

considers the *Average annual flooded area in Pavia* and depends on all the events rather than just one.

To ascertain whether the suspicion is founded and above all to decide how to solve this difficulty, we must consider the cause of flooding at Pavia. Floods depend on the Ticino level, which in turn depends both on the release from the Miorina dam and on the flow in the Po (Section 6.9.1). Therefore we must examine these two factors.

11.3.4 The release from Miorina dam

The release from the Miorina dam can influence the flooding in Pavia only when it is significantly high. This occurs when the lake is in free regime conditions, and in such circumstances the release is univocally defined by the stage–discharge relation. However, as we have already observed, the time instant $\bar{t}$ and the level $h_{\bar{t}}^{SC}$ at which the lake passes from regulated to free regime conditions depend on the policy adopted and thus on the alternative being considered.

The stage–discharge relation and the regulation range being equal, the extreme alternative that favours the *Downstream Flooding* sector is alternative *A*27, which adopts an extreme policy that favours *Upstream Flooding*; the one against *Downstream Flooding* is *A*22, which uses an extreme policy that favours *Irrigation*. Alternative *A*27 tends to switch to free regime conditions for intermediate or even low lake levels, so that at each time instant the lake level is the lowest possible; the release into the downstream Ticino is thus reduced to the minimum and so is the flow that reaches Pavia. On the contrary, alternative *A*22 does not consider the periods of high inflows as dangerous, but as an opportunity for increasing the lake storage. Therefore it passes into free regime only when the regulation range imposes it and produces flood waves characterized by a peak flow rate much higher than the one produced at the same time by *A*27, because the lake level with *A*22 is higher than it is with *A*27. The validity of this deduction is confirmed by Figure 11.5, which compares the release and level trajectories produced by the two alternatives in the period between 20th and 23rd of September during the flood event in September–October 1993.

To understand the influence of the stage–discharge relation, which determines the value of the outflow in free regime conditions, let us compare the lake level and release trajectories produced by the current stage–discharge relation $N^{\text{CUR}}(\cdot)$ and by the one $(N^{+600}(\cdot))$ that will be obtained once the action SD_{+600} is implemented. We will denote the quantities obtained with the first relation with the superscript CUR and those obtained with the second with +600. For simplicity's sake we suppose that, when a flood occurs, the time $\bar{t}$ at which the lake passes into free regime conditions is the same for both relations, but that the

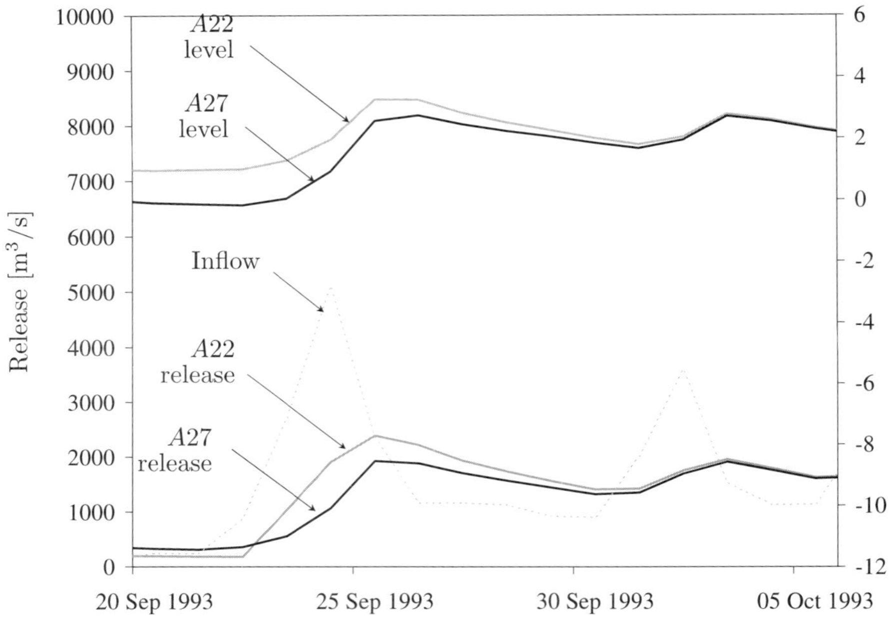

Figure 11.5: Lake level and release trajectories produced by alternatives $A22$ and $A27$ in some days during the flood event of September–October 1993.

corresponding lake level $h_{\bar{t}}^{SC}$ could be different. As we observed on page 264 for the +600 and the natural stage–discharge relations, we can consider, with a good approximation, that these two relations are linked by the following relation

$$N^{\text{CUR}}(h^{SC}) \cong N^{+600}(h^{SC} - \bar{h}) \quad \text{with } \bar{h} = 0.75 \text{ m} \tag{11.9}$$

It follows that:

- At time $\bar{t}$ the release produced by $N^{+600}(\cdot)$ is higher than the one produced by $N^{\text{CUR}}(\cdot)$ if

$$h_{\bar{t}}^{+600} > h_{\bar{t}}^{\text{CUR}} - \bar{h} \tag{11.10}$$

  This inequality is likely to hold in most cases, since there will be only few policies that reduce the lake level trajectory by 75 cm with respect to the current trajectory.

- If the lake remains in the free regime conditions for a long enough period of time (see Section 6.6.7) the two lake levels tend to satisfy the following relation

$$h_{\bar{t}}^{+600} = h_{\bar{t}}^{\text{CUR}} - \bar{h} \tag{11.11}$$

  When this relation is satisfied, from equation (11.9) it follows that the two flow rates are equal.

We can thus conclude that:

- passing from SD_{CUR} to SD_{+600} does not necessarily cause a reduction in the release in free regime conditions, especially if at the same time the regulation range is widened;

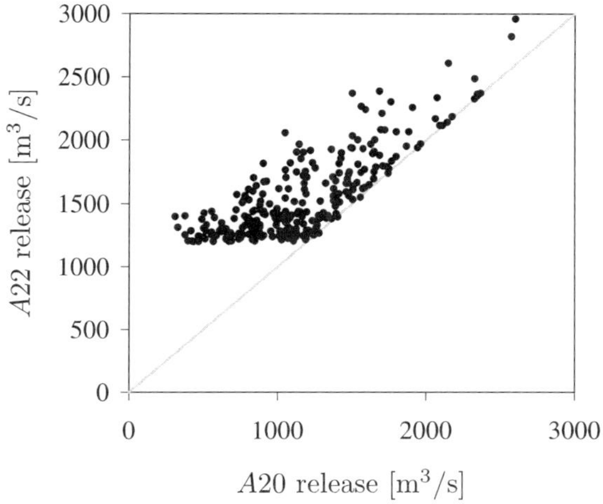

Figure 11.6: Release (in free regime conditions) produced by $A20$ against release produced by $A22$, in the flood events of the evaluation horizon.

- the major differences in the release are expected to occur between alternative $A20$ (SD_{CUR}, $\text{RANGE}_{\text{CUR}}$, extreme policy that favours flooding) and $A22$ (SD_{+600}, $\text{RANGE}_{\text{ENL}}$, extreme policy that favours irrigation). This intuition is confirmed by Figure 11.6, which compares the releases produced by $A22$ and by $A20$ in free regime conditions in the flood events of the evaluation horizon. Note that all the pairs of values lie above the bisector, which means that in free regime conditions the release produced by $A22$ is never inferior to the release produced by $A20$. Note also that as the release increases the pairs get closer to the bisector, which indicates that the difference between the two releases tends to diminish as their value increases: this is what one expects from equation (11.11), because very high release values are reached only when the lake has been in free regime for some time.

11.3.5 Flow in the Po

Now we will consider the effect of the flow in the Po. We said in Section 8.3 that the time series of the Po flow is a component of the *evaluation scenario*: therefore, all the alternatives are compared based on the same flow values. A query springs to mind: if the level at Pavia were much more sensitive to the flow in the Po than to the release at the Miorina, the small differences in the flood peaks at Pavia produced by the different alternatives could be explained by the fact that all of them are influenced by the same time series of Po flow.

To check whether this is actually the cause, we constructed the *abacus*[2] shown in Figure 11.7. The curves express the level of the Ticino at Pavia as a function of the release at the Miorina dam, the Po flow being equal. Note that these curves become increasingly horizontal as the flow in the Po increases: this confirms that the hypothesis is more and more acceptable as the flow rate in the Po increases. We can thus affirm that, if most of the flood events at Pavia occurred in correspondence to high flows in the Po, floods would not be very

[2]The abacus was computed with the neural network described in Section 6.9.1, by adopting the hypothesis that the flow of the Po were stationary.

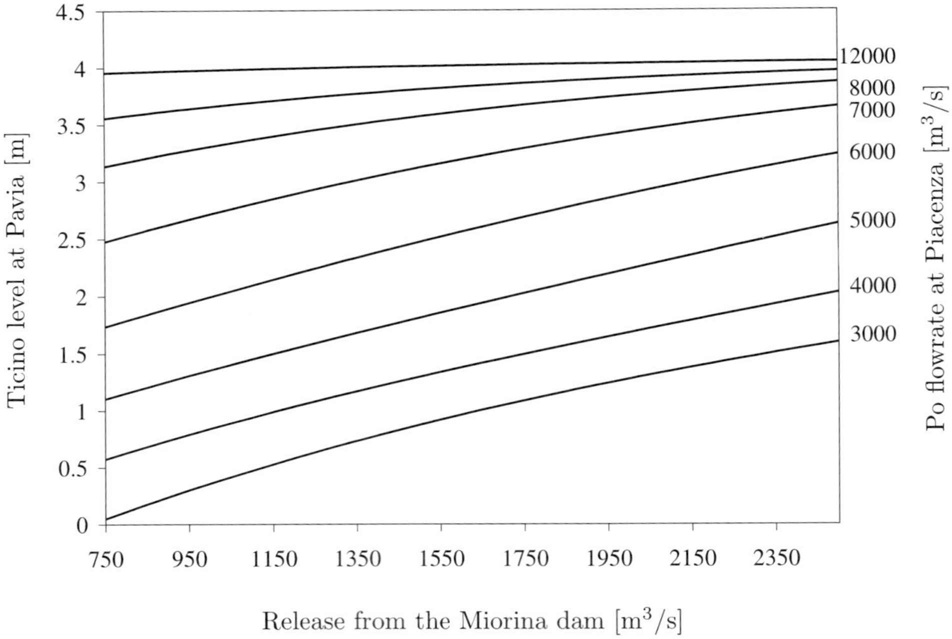

Figure 11.7: The abacus with which the level of the Ticino at the Covered Bridge is computed as a function of the release from the Miorina and the flow in the Po.

sensitive to the alternatives. In the event that produced the highest peak at Pavia (November 1994) the flow rate in the Po was about 10 000 m^3/s, and for this value the curve of the abacus is almost horizontal. In this way we explained the result shown in Table 11.5: the insensitivity of the maximum peak at Pavia with respect to the alternative.

This conclusion is confirmed by the results shown in Figure 11.8, which compares the levels at Pavia produced by alternatives *A*20 and *A*22 in the flood events of the evaluation horizon. Most of the differences in level that exceed $\sigma_{i_{\mathrm{Fl\_D\_1}}}$ (i.e. the uncertainty in level estimation) are observed when the levels produced by *A*22 are lower than the flood threshold at Pavia. These are the very events that now we know are most sensitive to the release from the Miorina, also because, as we previously showed, *A*22 is the worst alternative from the point of view of *Downstream Flooding*, in the sense that it is the one that produces the highest flows in the Ticino. At high levels the differences are less significant and drop below the threshold of uncertainty, since the flow in the Po becomes the dominant factor.

11.3.6 Conclusions

The previous observations lead us to conclude that, if the indicator values for the *Downstream Flooding* sector are not significant, it is not only because of the inaccuracy of the model that describes the propagation of the wave along the stretch of the Ticino, but also and above all because of the nature of the system, in which the floods depend on the backflow from the Po. We also confirmed that several alternatives, in particular those based on the SD_{+600} relation, like *A*22, for example, can actually increase the levels at Pavia.

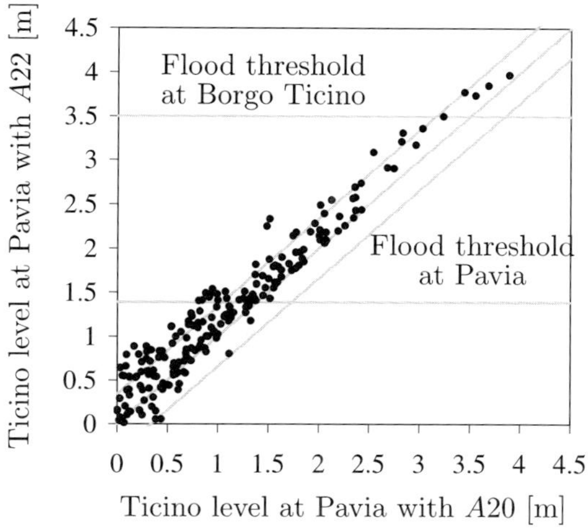

Figure 11.8: Level of the Ticino at the Covered Bridge (Pavia) produced by $A20$ versus level produced by $A22$, in the flood events of the evaluation horizon. The band around the bisector has a width equal to twice the uncertainty $\sigma_{i_{\text{Fl\_D\_1}}}$ in the level estimates.

Therefore, the fear expressed by the population of Pavia about the excavation is not unfounded. It follows that we cannot avoid considering the effects of the alternatives on the floods at Pavia but, at the same time, we are not able to quantify the effects with the indicators.

In Chapter 9 we explained that, in these cases, the comparison must be made by examining the effects of the alternatives directly, i.e. without using the indicators. However, it is not completely clear how to apply the suggestion to our case given that, to compute the effect (Ticino level at Pavia), we must necessarily use a model and the one that we have is affected by an error that does not permit us to discriminate between the alternatives. Moreover, even if this difficulty were overcome, there would still be the problem of how the *Comparison* could be carried out: if the alternatives are compared on the basis of the effects and not on the basis of the indicators, the ENP procedure cannot be used.

To overcome these obstacles, with the agreement of the Stakeholders who represent the *Downstream Flooding* sector, the following decisions were taken:

1. The first and second negotiation steps (see Figure 11.1) are completed without taking their sector into account. By doing so, a set of reasonable alternative is identified.

2. The Stakeholders interested in the *Downstream Flooding* sector are asked to examine the effects that each reasonable alternative produces, in order to:
   (a) check if it is likely that the alternative increases the flood peak levels at Pavia;
   (b) identify, if necessary, appropriate mitigation measures.

3. The third negotiation step is carried out without taking account of the *Downstream Flooding* sector. The reasonable alternatives that come out of it will be submitted to the *Final Decision* phase only if they prove not to be worse than the current situation from the *Downstream Flooding* standpoint, i.e. if the reasonable alternatives have the support of the Stakeholders interested in flooding in Pavia.

4. If, when operating as above, no reasonable alternatives are found that gain their support, the decision-making process will re-start with a direct pairwise comparison of all the alternatives that have been designed in the various steps.

The reader might object that this procedure gives the Stakeholders that represent the *Downstream Flooding* sector a sort of veto power. We can reply by observing that if the clause for re-examination at step 4 were used, it would mean carrying out a long pairwise comparison of all the alternatives, at the conclusion of which, if no better alternatives had emerged, the reasonable alternatives previously found could, in any case, be subjected to the *Final Decision* phase. The proposed procedure thus does not constitute, in practice, a real right of veto for the Stakeholders interested in the *Downstream Flooding* sector. It is actually only a short cut, which makes it possible to avoid, as long as it is not strictly necessary, the direct pairwise comparison of the alternatives that have been designed.

For the operational applicability of the procedure, we should also define how the alternatives are compared from the point of view of the *Downstream Flooding* sector. We postpone the description of this topic to Section 14.1, even though it was actually discussed before the procedure was proposed, since otherwise the procedure itself would make no sense.

11.4 Evaluating the alternatives

Once the effects of the alternatives have been estimated and the significance of the indicators has been verified, the next step in the decision-making procedure is the *Evaluation* phase, in which the value functions are identified for each sector (Chapter 9). They are identified separately for each sector and so we can show, as an example, how to proceed with one sector: *Upstream Environment*. For brevity's sake we will omit the results relative to the other sectors. We will only include a short comment for the *Downstream Environment* sector, for which the MAVT hypotheses proved not to be satisfied.

11.4.1 The value function of the Upstream Environment sector

11.4.1.1 Identifying the partial value functions

The first step of the MAVT evaluation method requires the identification of a partial value function for each of the indicators (Section 9.3). In the case of *Upstream Environment* the indicators are the following:

- $i_{\text{Env\_U\_1}}$: *Distance* [m] *from natural regime*;
- $i_{\text{Env\_U\_2}}$: *Average annual number of days* [d/year] *in which the lake level is in the range of reed bed erosion*;
- $i_{\text{Env\_U\_3}}$: *Fraction* [%] *of the egg-laying period in which the Cyprinids cannot access the reed beds*;
- $i_{\text{Env\_U\_4}}$: *Fraction* [%] *of the nesting period in which the lake level is higher than the threshold of the nests*;
- $i_{\text{Env\_U\_5}}$: *Fraction* [%] *of the normal shelf emergence period in which the lake level is higher than the emergence threshold.*

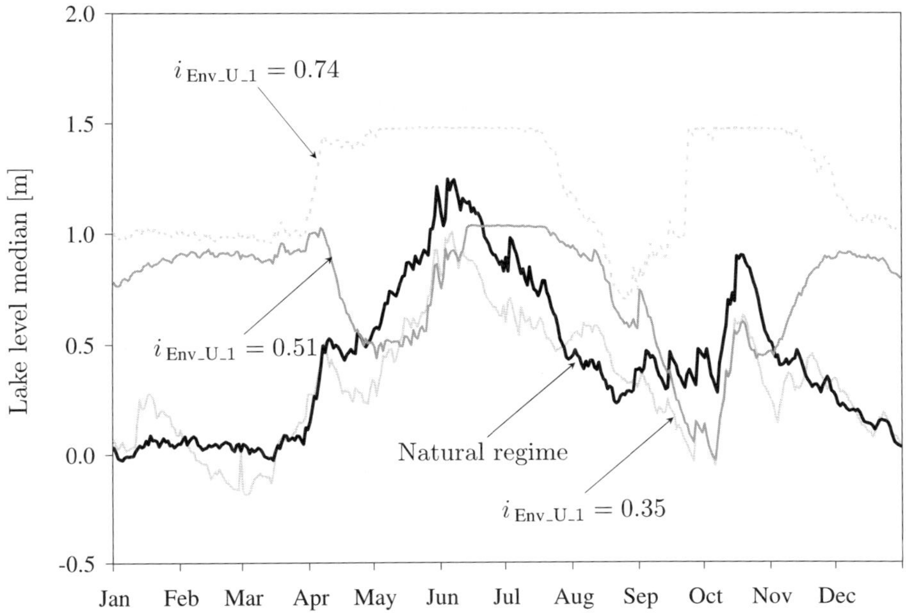

Figure 11.9: The trajectory of the lake level median in natural regime conditions and for different alternatives.

Let us consider the first, for example: $i_{\text{Env\_U\_1}}$. In order to identify a partial value function for it, the Expert must be able to associate mentally every value that the indicator can assume with the corresponding effect. Having this ability is essential to be able to evaluate the alternatives correctly. To help the Expert acquire or improve this ability, it is useful to show him the trajectories of the lake level (or of some suitable statistics of it) for different alternatives, and to accompany each trajectory with the corresponding indicator value. In the case of $i_{\text{Env\_U\_1}}$ the Expert asked to see the trajectories of the lake level median, such as those in Figure 11.9.

The choice of the alternatives that will be shown to the Expert is quite important. In order not to make the interaction with the Expert too complicated, the Analyst cannot show him too many alternatives. On the other hand, the Expert's exploration should cover evenly the entire interval $\mathcal{I}_{\text{alt}}$ of the values that the indicator assumes for the designed alternatives (for $i_{\text{Env\_U\_1}}$ the interval of variation is between 0.31 and 0.74 m). It is advisable, even if it is not always possible, that the Expert define the partial value function over an interval that is larger than $\mathcal{I}_{\text{alt}}$, because in the following we will design new alternatives and we cannot exclude the possibility that one of them might produce a value of $i_{\text{Env\_U\_1}}$ that does not belong to $\mathcal{I}_{\text{alt}}$. In the case that we are examining, the Expert was able to define the partial value function over the entire feasible interval [0, 1]. It is a piecewise linear function, obtained by interpolating several significant points, as shown in Figure 11.10.

In the same way, the Expert defined the partial value functions for the other four indicators; they are shown in Figure 11.11.

Given the partial value functions, the sub-Matrix of the Effects relative to the *Upstream Environment* sector alone can be transformed into the (sub)matrix of partial values (Figure 11.12).

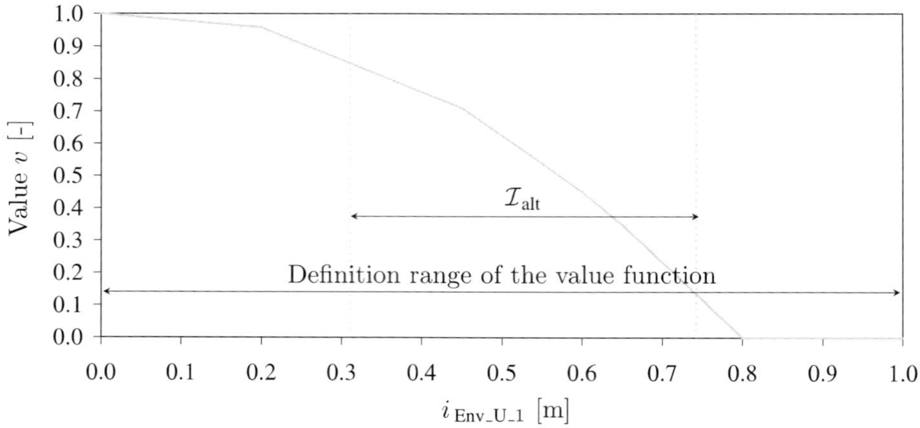

Figure 11.10: The partial value function for the indicator $i_{Env\_U\_1}$: *Distance* [m] *from natural regime.*

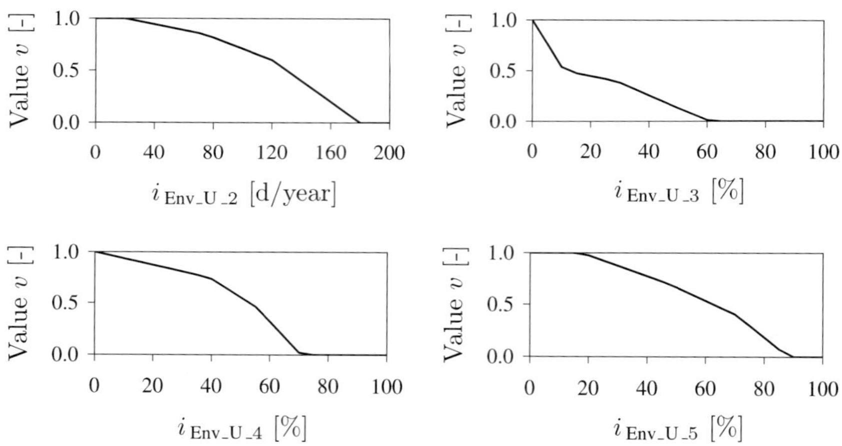

Figure 11.11: The partial value functions for the indicators $i_{Env\_U\_2}$–$i_{Env\_U\_5}$ of the *Upstream Environment* sector.

11.4.1.2 Defining the weights

Once the partial value functions have been identified, it is necessary to specify the value of the weights with which they must be aggregated into a global value function, which returns the sector index. To do this, as we explained in Section 9.4.1, for each possible pair of indicators the Expert is asked to indicate two pairs of values that he judges to be equivalent, in order to determine the increment of the first indicator that would compensate the corresponding decrease of the second. The pairs of values might also be specified by choosing two pairs of alternatives that he considers to be equivalent (from the point of view expressed by the two indicators under examination). It is not necessarily true, however, that in the set of the designed alternatives such pairs exist, for two reasons: the set is finite and the two indicators might be positively correlated.

Matrix of the Effects

Sector	Indicators		*A*22	*A*27	*A*31	*A*240
Upstream Environment	$i_{\mathrm{Env\_U\_1}}$	[m]	0.74	0.67	0.53	0.67
	$i_{\mathrm{Env\_U\_2}}$	[d/year]	7.52	29.54	24.02	29.54
	$i_{\mathrm{Env\_U\_3}}$	[%]	7.76	85.15	68.37	85.15
	$i_{\mathrm{Env\_U\_4}}$	[%]	90.35	31.85	32.38	31.85
	$i_{\mathrm{Env\_U\_5}}$	[%]	81.66	7.79	20.76	7.79

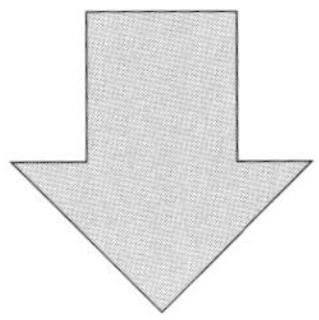

Matrix of partial values

Sector	Indicators	*A*22	*A*27	*A*31	*A*240
Upstream Environment	$v_{\mathrm{Env\_U\_1}}$	0.13	0.31	0.57	0.31
	$v_{\mathrm{Env\_U\_2}}$	1.00	0.97	0.99	0.97
	$v_{\mathrm{Env\_U\_3}}$	1.00	0.07	0.43	0.07
	$v_{\mathrm{Env\_U\_4}}$	0.00	0.80	0.79	0.80
	$v_{\mathrm{Env\_U\_5}}$	0.00	0.64	0.44	0.64

Figure 11.12: From the Matrix of the Effects to the matrix of partial values for the *Upstream Environment* sector and the four alternatives presented in Section 11.1.

In the example that we are considering, all the indicators are negatively correlated: see, for example, the pair ($i_{\mathrm{Env\_U\_2}}$, $i_{\mathrm{Env\_U\_5}}$) in Figure 11.13. The cause of this is understood from Figure 11.14, which shows the trajectories of the median lake level produced by several alternatives and the thresholds and periods that appear in the definition of the indicators (see Section 4.5.1 of the DVD). An indicator (for example, $i_{\mathrm{Env\_U\_2}}$) assumes a 'good' value for an alternative if the latter produces a trajectory that remains only temporarily in the zone identified by the relevant threshold and period (in the example, the area termed *reed bed erosion*). Since these areas cover almost all the space, it is evident that an alternative that improves one indicator will inevitably worsen another.

The vector of weights shown in Table 11.7 was estimated on the basis of the pairs of values provided by the Expert, with the procedure defined in Section 9.4.1. Once this vector is known, the global value function is univocally defined and computing the global value

Table 11.7. The vector of weights for the *Upstream Environment* sector

Sector	Indicators	Weight
Upstream Environment	$i_{\mathrm{Env\_U\_1}}$	0.244
	$i_{\mathrm{Env\_U\_2}}$	0.070
	$i_{\mathrm{Env\_U\_3}}$	0.318
	$i_{\mathrm{Env\_U\_4}}$	0.297
	$i_{\mathrm{Env\_U\_5}}$	0.071

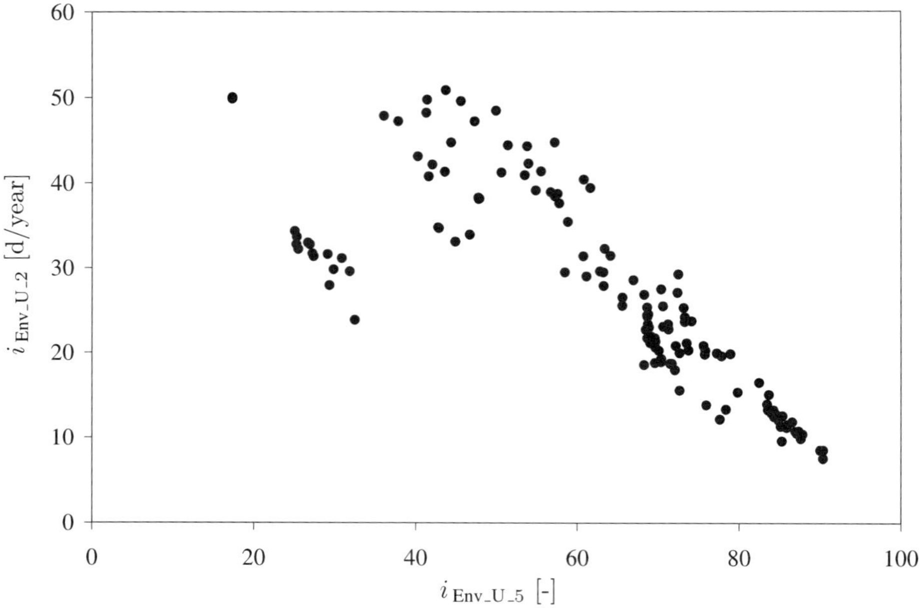

Figure 11.13: Values of indicator $i_{\text{Env\_U\_5}}$ against those of indicator $i_{\text{Env\_U\_2}}$ for the designed alternatives. The first indicator is the *Fraction of the normal shelf emergence period in which the lake level is higher than the emergence threshold*, the second is the *Average annual number of days in which the lake level is in the range of reed bed erosion.*

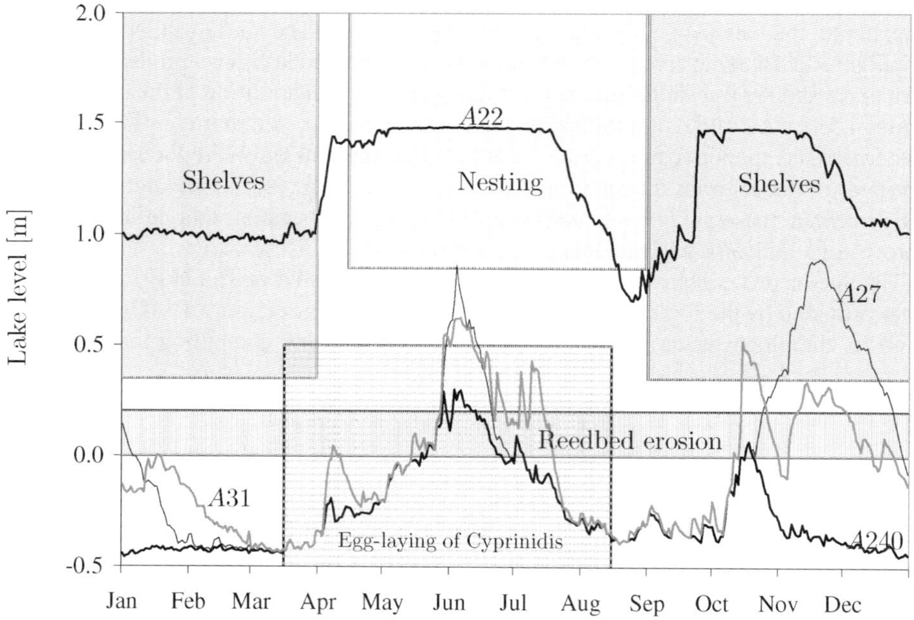

Figure 11.14: Lake level medians produced by several alternatives. The threshold values and the periods that appear in the definition of the indicators of the *Upstream Environment* sector are shown as well.

Table 11.8. The sector index values of the alternatives in Figure 11.12 for the *Upstream Environment* sector

Sector	A22	A27	A31	A240
Upstream Environment	0.17	0.59	0.62	0.59

(i.e. the value of the sector index) associated to an alternative is immediate: for example, from the (sub)matrix of partial values in Figure 11.12 one obtains the matrix (row vector) of the sector indices in Table 11.8.

11.4.2 The value function of the Downstream Environment sector

In Section 4.3.3 we said that the Expert for the *Downstream Environment* sector uses two indicators to evaluate the *Overall condition of the ecosystem*:

- $i_{\text{Env\_D\_1}}$: *Distance from the natural regime*;
- $i_{\text{Env\_D\_2}}$: *Value of the* MEF.

By showing the Expert pairs of alternatives, we found that when the differences in the value of $i_{\text{Env\_D\_1}}$ are sufficiently high, the Expert always attributes a higher global value to the alternative for which that indicator is the lowest. When, instead, the differences in the value of $i_{\text{Env\_D\_1}}$ are small, for example less than 2 m^3/s, the Expert attributes a higher global value to the alternative for which $i_{\text{Env\_D\_2}}$ (i.e. the MEF) is higher, regardless of the value of indicator $i_{\text{Env\_D\_1}}$, because he does not perceive the differences in the value of the latter. In other words, he demonstrates a limited discrimination capacity. In this case, according to MAVT, it is no longer possible to identify a value function and, to proceed, we should turn to a different method.[3] In the introduction of Chapter 9, however, we said that the MAVT is particularly well suited for the Verbano Project, since it allows us progressively to enlarge the set of alternatives that are considered, as is required by the procedure that we have adopted (see Figure 10.5). Therefore, it was necessary to force the Expert to adopt the behaviour model required by MAVT, by assuming that the global value function for the *Downstream Environment* sector could be defined on the basis of the indicator $i_{\text{Env\_D\_1}}$ alone. In the *Comparison* phase the Expert will compare the alternatives on the basis of that global value function but, if the resulting values are similar, he will base his judgement on the value of $i_{\text{Env\_D\_2}}$.

11.4.3 The matrix of partial values

By proceeding in a similar way with the Experts from all the other sectors, the partial value functions for all the indicators and the weights for every sector were obtained. Then, from the Matrix of the Effects (Table 11.4), the matrix of partial values (Table 11.9) and thus the matrix of global values (sector indices) (Table 11.10) were derived.

[3] For example, a descriptive method like those of ELECTRE family, described in Section 17.5 of THEORY.

Table 11.9. The matrix of partial values for all the sectors and the four alternatives presented in Section 11.1

	Sectors	Indicators	$A22$	$A27$	$A31$	$A240$
Upstream	*Upstream Environment*	$v_{Env\_U\_1}$	0.13	0.31	0.57	0.31
		$v_{Env\_U\_2}$	1.00	0.97	0.99	0.97
		$v_{Env\_U\_3}$	1.00	0.07	0.43	0.07
		$v_{Env\_U\_4}$	0.00	0.80	0.79	0.80
		$v_{Env\_U\_5}$	0.00	0.64	0.44	0.64
	Upstream Fishing	$v_{Fis\_U\_1}$	0.98	0.92	0.94	0.92
		$v_{Fis\_U\_2}$	0.73	0.47	0.65	0.47
		$v_{Fis\_U\_3}$	0.95	0.16	0.31	0.16
		$v_{Fis\_U\_4}$	1.00	0.07	0.43	0.07
	Upstream Flooding	$v_{Fl\_U\_1\_Loc}$	0.75	0.80	0.80	0.80
		$v_{Fl\_U\_1\_Vb}$	0.58	0.64	0.63	0.64
		$v_{Fl\_U\_2\_Loc}$	0.71	0.98	0.94	0.98
		$v_{Fl\_U\_2\_Vb}$	0.34	0.95	0.80	0.95
		$v_{Fl\_U\_3\_Loc}$	0.20	0.35	0.35	0.35
		$v_{Fl\_U\_3\_Vb}$	0.41	0.45	0.45	0.45
		$v_{Fl\_U\_4\_Loc}$	0.82	0.98	0.94	0.98
		$v_{Fl\_U\_4\_Vb}$	0.34	0.95	0.83	0.95
		$v_{Fl\_U\_6\_Loc}$	1.00	1.00	1.00	1.00
		$v_{Fl\_U\_6\_Vb}$	1.00	1.00	1.00	1.00
	Mosquitoes	$v_{Mos\_U\_1}$	0.60	0.33	0.40	0.33
	Navigation	$v_{Mos\_U\_2}$	0.00	1.00	0.93	1.00
		$v_{Nav\_U\_1}$	0.65	0.84	0.79	0.84
		$v_{Nav\_U\_2}$	0.93	0.97	0.96	0.97
	Upstream Tourism	$v_{Tou\_U\_1}$	0.97	0.08	0.37	0.08
		$v_{Tou\_U\_2}$	0.22	0.98	0.91	0.98
		$v_{Tou\_U\_3}$	0.60	0.33	0.40	0.33
		$v_{Tou\_U\_4}$	0.00	1.00	0.93	1.00
Downstream	*Downstream Environment*	$v_{Env\_D\_1}$	0.01	0.12	0.62	0.60
		$v_{Env\_D\_5}$	–	–	–	–
	ENEL *Power*	$v_{EnP\_D\_1}$	0.86	0.86	0.16	0.16
		$v_{EnP\_D\_2}$	–	–	–	–
	East Sesia Power	$v_{ESP\_D\_1}$	0.58	0.64	0.00	0.00
	Downstream Flooding	$v_{Fl\_D\_1}$	–	–	–	–
		$v_{Fl\_D\_2}$	–	–	–	–
		$v_{Fl\_D\_3\_Bt}$	–	–	–	–
		$v_{Fl\_D\_3\_Pv}$	–	–	–	–
		$v_{Fl\_D\_4\_Bt}$	–	–	–	–
		$v_{Fl\_D\_4\_Pv}$	–	–	–	–
		$v_{Fl\_D\_6}$	–	–	–	–
	Irrigation	$v_{Irr\_D\_1}$	0.74	0.05	0.00	0.00
		$v_{Irr\_D\_2}$	0.85	0.47	0.00	0.00
		$v_{Irr\_D\_3\_IC}$	0.95	0.87	0.00	0.00
		$v_{Irr\_D\_3\_REC}$	0.78	0.73	0.00	0.00
		$v_{Irr\_D\_3\_VC}$	0.92	0.65	0.00	0.00
	Downstream Tourism	$v_{Tou\_D\_1}$	0.29	0.34	0.99	0.99
		$v_{Tou\_D\_2}$	0.98	0.99	0.98	0.98

Table 11.10. The matrix of sector index values for the four alternatives presented in Section 11.1; the index of the *Downstream Flooding* sector cannot be estimated

Sectors	Indicators	*A*22	*A*27	*A*31	*A*240
Upstream Environment	$v_{\text{Env\_U\_1}}$	0.17	0.59	0.62	0.59
Upstream Fishing	$v_{\text{Fis\_U\_1}}$	0.88	0.41	0.60	0.41
Upstream Flooding	$v_{\text{Fl\_U\_1\_Loc}}$	0.56	0.73	0.70	0.73
Mosquitoes	$v_{\text{Mos\_U\_1}}$	0.28	0.69	0.68	0.69
Navigation	$v_{\text{Nav\_U\_1}}$	0.83	0.92	0.90	0.92
Upstream Tourism	$v_{\text{Tou\_U\_1}}$	0.40	0.68	0.71	0.68
Downstream Environment	$v_{\text{Env\_D\_1}}$	0.01	0.12	0.62	0.60
ENEL *Power*	$v_{\text{EnP\_D\_1}}$	0.86	0.86	0.16	0.16
East Sesia Power	$v_{\text{ESP\_D\_1}}$	0.58	0.64	0.00	0.00
Downstream Flooding	$v_{\text{Fl\_D\_1}}$	–	–	–	–
Irrigation	$v_{\text{Irr\_D\_1}}$	0.82	0.37	0.00	0.00
Downstream Tourism	$v_{\text{Tou\_D\_1}}$	0.64	0.66	0.99	0.99

Chapter 12

Comparison: the first negotiation step

EB, FC, FP and RSS

This chapter presents the preliminaries to the *Comparison* phase and the first negotiations step (Figure 12.1).

12.1 Sharing the results of the evaluation

The aim of *sharing the results of the evaluation* (Section 10.2) is to get each Stakeholder to understand how the value functions were defined for all the sectors and which effects are evaluated. To achieve this, an open meeting was organized for all the Stakeholders and the members of the Italian–Swiss Commission. In the meeting the Stakeholders (or rather, the Experts they commissioned) took turns in presenting the sector(s) that interested them. We will outline the structure of the presentation from the *Upstream Flooding* sector as an example:

- Explanation of each indicator and presentation of the partial value function associated to it.

  For example, Figure 12.2 shows the partial value function associated to the indicator *maximum flooded area* [km^2] *in Locarno* ($i_{\text{Fl\_U\_1\_Loc}}$). The non-linearity of the function derives from the fact that no substantial damage is recorded for small floods, and so the value function is equal to one. Once a threshold value (0.7 km^2) has been exceeded, the flood begins to cause damage and it is necessary to take defence measures, which have a cost; this causes a decrease in the value function, which is increasingly marked as the flooded area grows. When the flooded area exceeds 2.5 km^2, the greatest possible damage is recorded: from this threshold upwards the value function will thus be zero. On the same graph the points that represent the performances of the alternatives designed in the first step are marked. Note that they are divided into two groups. To understand the reason for this, observe that in all the alternatives the worst event is the flood in October 1993, which was produced by a sequence of three flood waves entering the lake, of which the biggest was the last. In practice, what counts in these conditions is only the stage–discharge relation that is adopted; the effect of

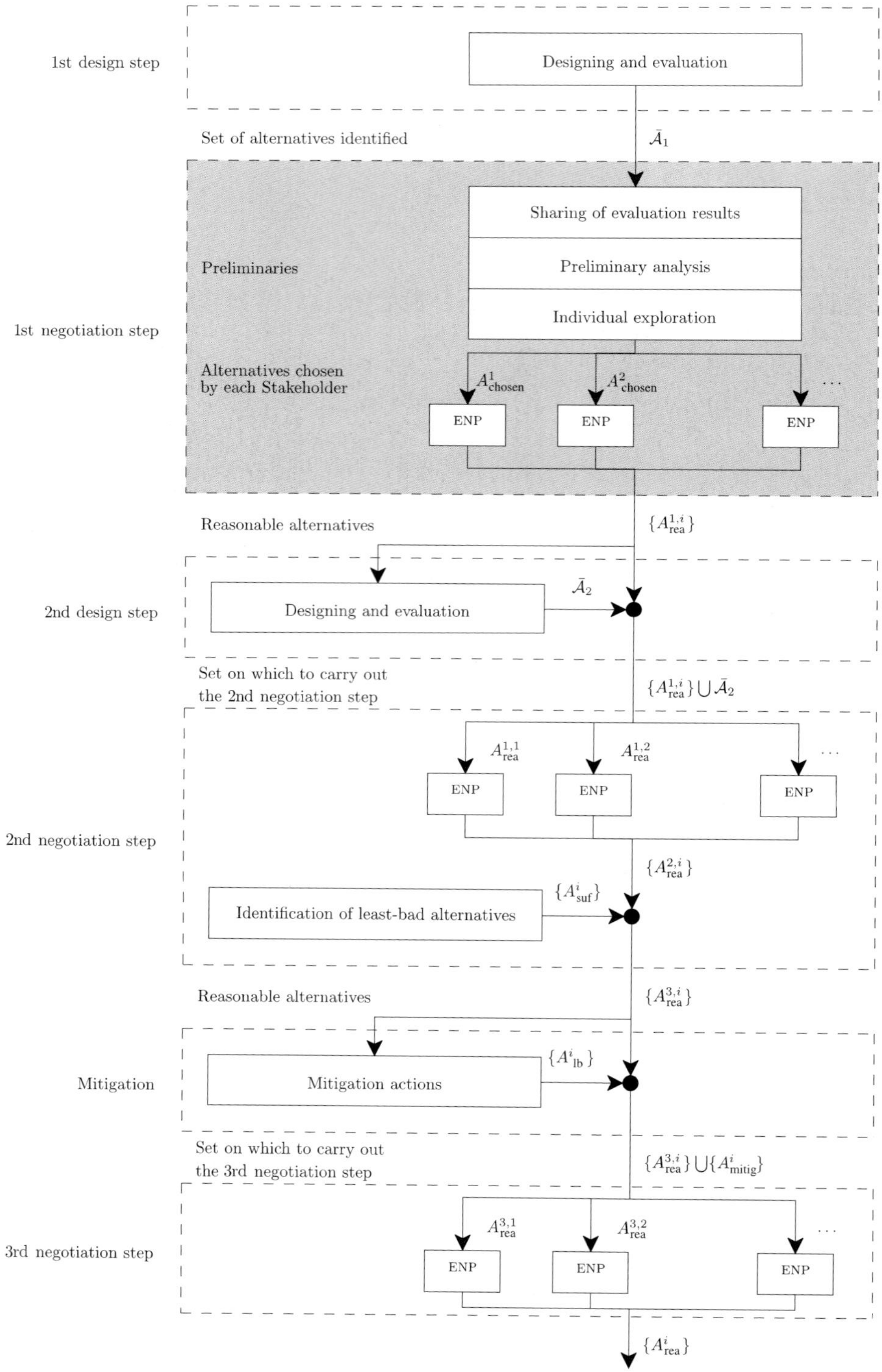

Figure 12.1: The steps for identifying the reasonable alternatives. The highlighted steps are the ones whose results are described in this chapter.

Figure 12.2: The images that were used to explain the meaning of the partial value function for the indicator *maximum flooded area* [km^2] *in Locarno* ($i_{\mathrm{Fl\_U\_1\_Loc}}$).

the regulation policy is secondary. The two groups (from left to right) thus correspond to the values that follow from the actions SD_{+600} and $\mathrm{SD}_{\mathrm{CUR}}$ respectively. To help the participants get an idea of what the different values represent, the Expert showed two images of the same street and the water level that would be reached in each case.

- Presentation of the vector of weights used to aggregate the partial value functions: this shows the relative importance of the indicators (and so of the effects that they quantify) in defining the global value function.

  The vector of the weights for the *Upstream Flooding* sector is shown in Table 12.1. Note above all that the overall weight of the indicators for the township of Locarno is higher than that of the indicators for Verbania. There are two reasons for this: on the one hand, Locarno represents a larger fraction of lakeside towns; on the other hand, in these towns there are more valuable structures and activities that can be damaged than in Verbania. Next observe that the Expert gave greater importance to the extreme event (flooded area and duration) than the average behaviour. Lastly, zero weight was given to the two indicators of interrupted traffic because no alternative produced, according to the Expert's judgement, significant effects on the traffic; in other words, the partial value functions for both the indicators are equal to one for all the alternatives. It follows that the weight attributed to them is irrelevant.

Table 12.1. The vector of weights used in the *Upstream Flooding* sector to aggregate the partial value functions

Indicator		Weight
$i_{\text{Fl\_U\_1\_Loc}}$	Maximum flooded area in Locarno	0.21
$i_{\text{Fl\_U\_1\_Vb}}$	Maximum flooded area in Verbania	0.08
$i_{\text{Fl\_U\_2\_Loc}}$	Average annual flooded area in Locarno	0.18
$i_{\text{Fl\_U\_2\_Vb}}$	Average annual flooded area in Verbania	0.03
$i_{\text{Fl\_U\_3\_Loc}}$	Maximum number of consecutive flooding days in Locarno	0.25
$i_{\text{Fl\_U\_3\_Vb}}$	Maximum number of consecutive flooding days in Verbania	0.05
$i_{\text{Fl\_U\_4\_Loc}}$	Average annual number of flooding days in Locarno	0.18
$i_{\text{Fl\_U\_4\_Vb}}$	Average annual number of flooding days in Verbania	0.02
$i_{\text{Fl\_U\_5\_Loc}}$	Maximum number of days of interrupted traffic in Locarno	0.00
$i_{\text{Fl\_U\_5\_Vb}}$	Maximum number of days of interrupted traffic in Verbano	0.00
$i_{\text{Fl\_U\_6\_Loc}}$	Average annual number of days of interrupted traffic in Locarno	0.00
$i_{\text{Fl\_U\_6\_Vb}}$	Average annual number of days of interrupted traffic in Verbania	0.00

12.2 Preliminary analysis

The aim of this analysis (Section 10.2) is to show the Stakeholders the sensitivity of the sector index values, computed by means of the global value function, to the actions. To do this, graphs like the one shown in Figure 12.3 (again referring to the *Upstream Flooding* sector) were used. Remember that the alternatives designed up to this point are all characterized by a MEF equal to 13 m^3/s (Sections 5.4.1 and 11.1), so that, apart from the policy, they are differentiated by the pair ($\text{SD}_i/\text{RANGE}_i$). As shown in the figure, three possible pairs were considered. For each pair, the bar represents the interval of variation of the index value produced by changing the policy. Observe that, for the same stage–discharge relation, the size of the interval increases as the regulation range is widened; conversely, for the same regulation range, the size decreases when passing from SD_{CUR} to SD_{+600}, because the latter, by allowing for higher releases, reduces the preventative role of regulation. The difference ΔF between the sector index values associated to the worst alternative with ($\text{SD}_{+600}/\text{RANGE}_{\text{ENL}}$) and to the worst with ($\text{SD}_{+600}/\text{RANGE}_{\text{CUR}}$) represents the maximum decrease in the index

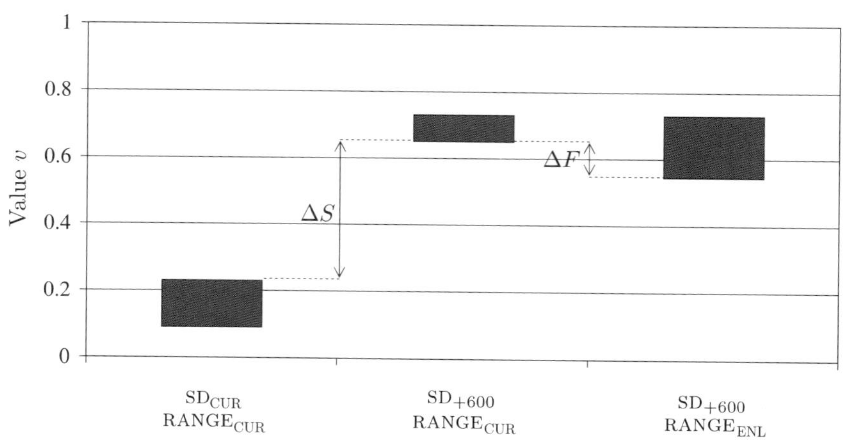

Figure 12.3: Sensitivity of the *Upstream Flooding* sector index to actions.

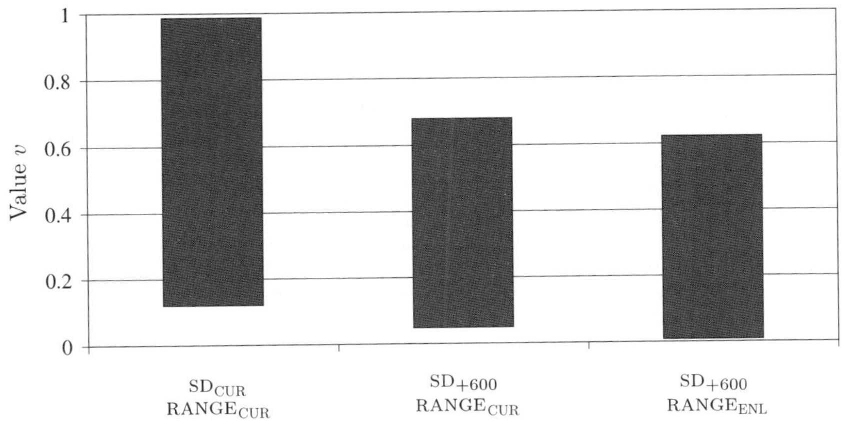

Figure 12.4: Sensitivity of the *Downstream Environment* sector index to actions.

value that the shore-dwellers can fear as a consequence of widening the regulation range. The difference ΔS provides information similar to that given by ΔF, in relation to a change in the stage–discharge relation.

The information provided by this type of graph is very important for the development of the decision-making process. On the one hand, it confirms the rationality of the Swiss request to excavate the lake outlet (ΔS has a high value); on the other hand, it shows that they should accept the widening of the regulation range, if this is necessary to obtain the excavation. In fact, even if the most unfavourable policy for *Upstream Flooding* were chosen, the corresponding index value would be much greater than the maximum index value that the sector could obtain in the current situation, i.e. with the pair (SD_{CUR}/$RANGE_{CUR}$).

The considerations presented clearly pertain to the *Upstream Flooding* sector only. If another sector is examined, e.g. *Downstream Environment* (Figure 12.4), it can be noticed that the sensitivity to the policy is much higher than the sensitivity to the pair (SD_i/$RANGE_i$), which means that this sector should concentrate its attention on the choice of the first, rather than the second.

12.3 Individual exploration

Each Stakeholder compares all the alternatives that have been designed, with the aim of identifying the one that (s)he prefers and that (s)he will propose when her/his turn to start the ENP comes. Each Stakeholder can compare the alternatives by considering the sector index values (s)he is interested in and, if necessary, the indicator values and the trajectories of several hydrological variables.

In Figure 12.5 the list of the alternatives chosen by the representatives of the Stakeholders is shown.

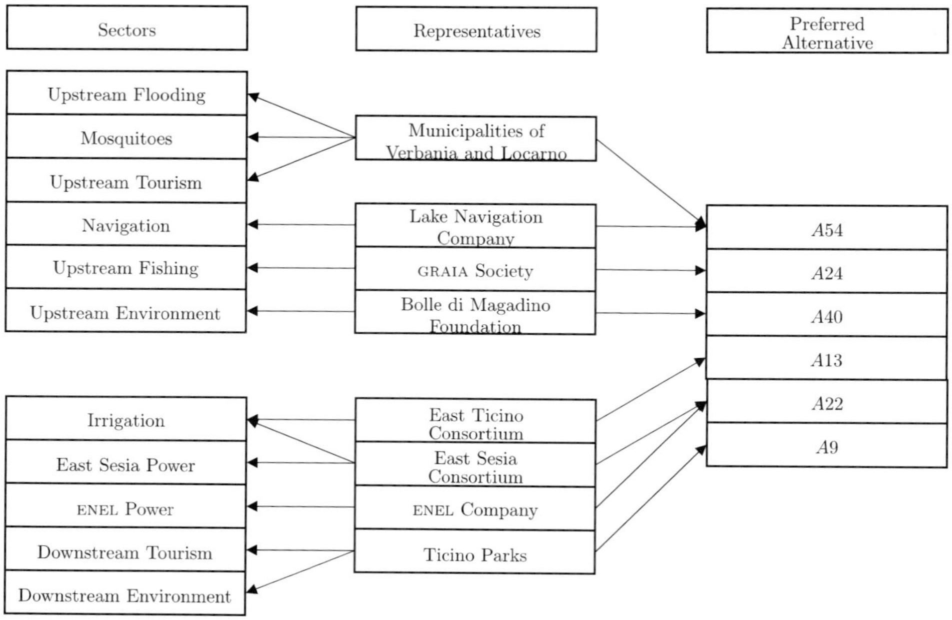

Figure 12.5: The alternatives preferred by the Stakeholders.

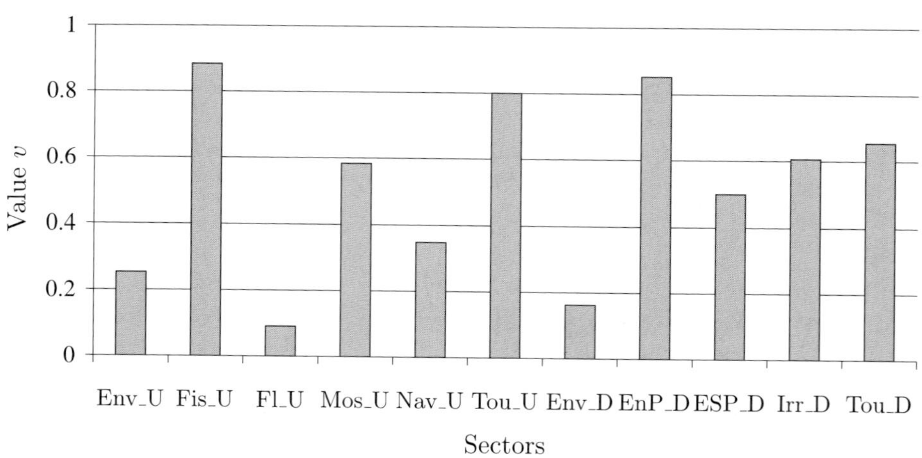

Figure 12.6: $A0$'s performance.

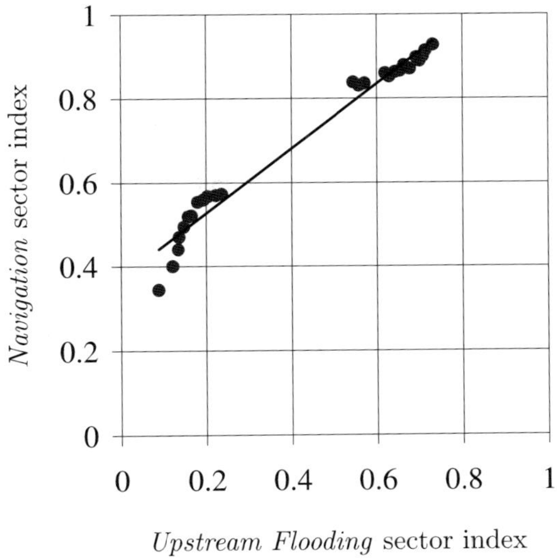

Figure 12.7: *Upstream Flooding* sector index vs *Navigation* sector index.

12.4 *A*0's performance

The alternative *A*0 assumes that nothing is changed with respect to the current situation (Section 3.7.1): it is thus a natural reference point for the Stakeholders during the individual exploration and the negotiations.

In Figure 12.6 the performance of *A*0 is shown for each of the sectors. We already know that the sector index values for *Upstream Flooding*, *Irrigation* and *Downstream Environment* are held to be unsatisfactory by the corresponding Stakeholders, and it is for this very reason that Switzerland and Italy put forward the proposals and counterproposals described in Chapter 3. We do not know, however, how the other sectors judge A_0's performance: it is worth finding this out because it will make it possible to identify any possible alliance among the sectors.

Navigation By observing Figure 12.7 we can see that there is a strong positive correlation between this sector's index and the *Upstream Flooding* index. As we noted in Section 4.2.1, the navigability of the lake is limited when the lake levels are either too high or too low. None of the alternatives that were designed creates problems due to low levels, and so the satisfaction for the *Navigation* sector depends only on the need to close some of the stops or to put up raised walkways to allow the passengers access to the ferries when the lake level is too high. Since the heights of the stops are higher than the elevation at which flooding begins at Locarno and Verbania, it is easy to understand why the two sectors are positively correlated. However, it cannot be concluded that, if the Stakeholders who are interested in the *Upstream Flooding* sector are dissatisfied, then those interested in the *Navigation* sector are as well. It was necessary to interview the second group, who confirmed that they too are dissatisfied with *A*0.

Upstream Tourism Table 12.2 shows that the index of this sector is more influenced by *reduced landscape aesthetics* and by *reduced access to beaches* than by *discomfort produced*

Table 12.2. The vector of weights that defines the global value function of the *Upstream Tourism* sector

Indicator		Weight
$i_{Tou\_U\_1}$	Percentage of the tourist season in which the lake level is lower than the Tourist Normality Range	0.21
$i_{Tou\_U\_2}$	Percentage of the tourist season in which the lake level is higher than the Tourist Normality Range	0.44
$i_{Tou\_U\_3}$	Average annual number of necessary treatments	0.16
$i_{Tou\_U\_4}$	Average duration of the periods in which treatments are probable	0.19

by mosquitoes. The first two criteria are quantified by indicators that consider the percentage of the tourist season in which the lake level is higher and lower than the *Tourist Normality Range* (Section 4.5.10 of the DVD). The latter is defined as a sequence of level ranges, the tth of which is the range of levels that 'usually' occurred in the past on day t. It follows that *Upstream Tourism* rates the alternative $A0$ very highly.

Mosquitoes The historical regulation of Verbano produced an increase in the average level of the lake with respect to the natural situation (Figure 12.8). In this way the probability that the areas where the mosquitoes lay their eggs will be flooded has increased (see Section 2.2.1.4 of the DVD), thus increasing their proliferation, which is displeasing for the population and the tourist operators. Since the lake level trajectory for alternative $A0$ coincides with the historical trajectory by definition (see Section 3.7.1), alternative $A0$ is judged to be unsatisfactory by this sector.

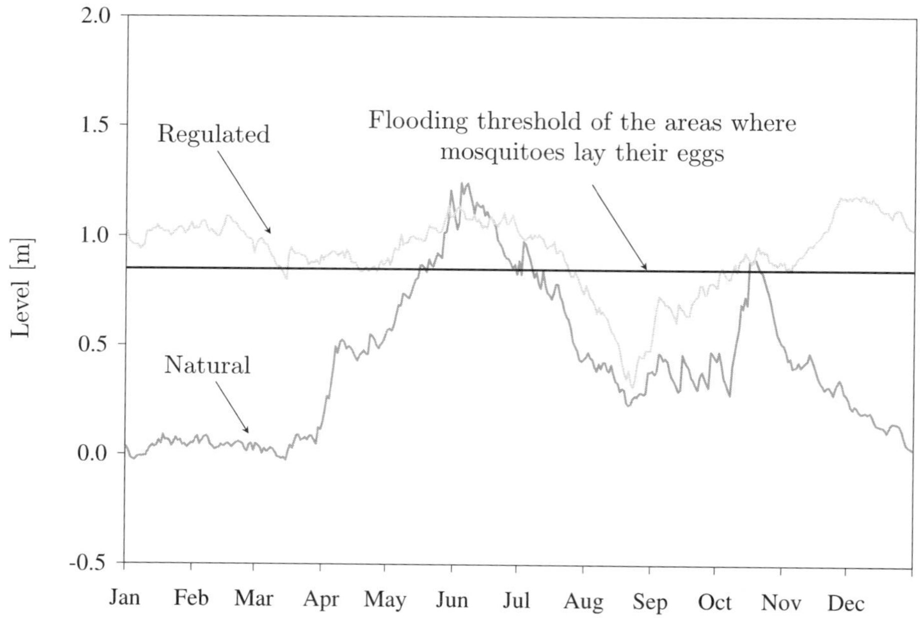

Figure 12.8: Trajectories of the lake level median in natural and regulated regime conditions.

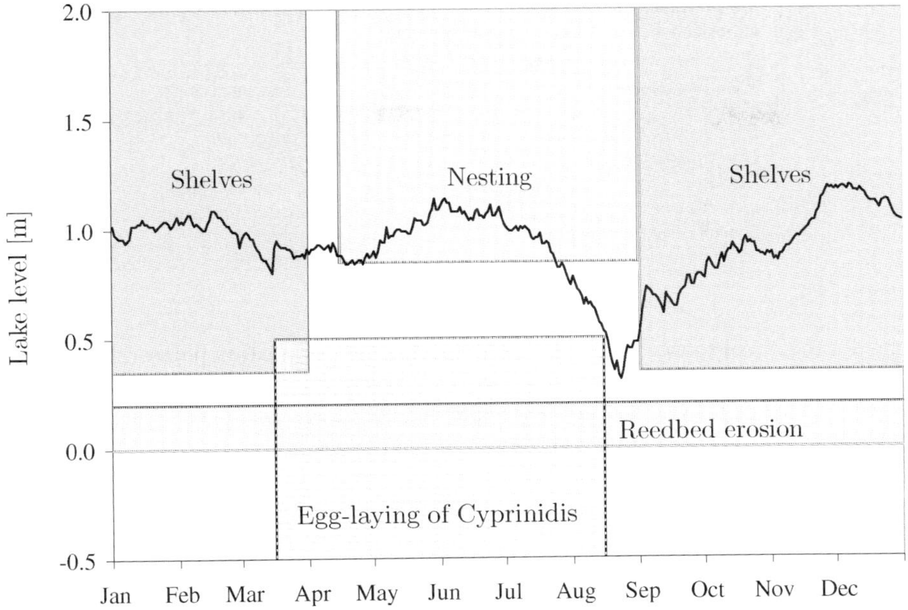

Figure 12.9: Trajectory of the median lake level produced by alternative *A*0 and the threshold values and periods that appear in the definition of the indicators of the *Upstream Environment* sector.

Upstream Fishing Alternative *A*0 is evaluated positively by *Upstream Fishing* for several reasons. For example, *A*0 does not produce sudden falls in the lake level in the month of December, which would have a negative impact on the reproductive cycle of the White Fish; this behaviour was not dictated by attention for the concerns of the fishermen, but by the need to create a storage for irrigation in spring. Surprisingly, we discovered that two sectors, *Upstream Fishing* and *Irrigation*, which one might have expected to be in conflict, are potentially allied in practice, at least for the protection of White Fish. A second example: the lake level trajectory produced by alternative *A*0 favours access to the reed beds for the Pike and the Cyprinids, which go there to lay their eggs.

Upstream Environment The *Upstream Environment* sector, on the other hand, is not at all satisfied with the performance of *A*0. The reason is evident: the lake ecosystem is damaged when the level is in either of the ranges highlighted in Figure 12.9 and this occurs frequently, as the figure shows.

ENEL Power The situation of this sector is quite particular: the introduction of the Galli Law (Section 7.6.2), which subordinates the use of water for hydropower production to domestic use and agricultural and environmental purposes, makes it very difficult, if not impossible, to maintain the past production rate into the future. The provisions of the Galli Law are totally satisfied by the formulation of the (Optimal) Control Problem (Section 7.6.3), while it was only partially possible to take it into account in the definition of alternative *A*0: the distribution policy takes account of the law,[1] but the release from the lake, which is

[1] In fact, when the release does not satisfy the overall water demand, the first reduction is to the supply for hydropower plants (Section 6.7.3).

Table 12.3. The sectors that were explicitly considered in the negotiations and those associated to each of them

Visualized sector	Associated sector
Upstream Flooding	*Navigation*
Upstream Environment	*Mosquitoes*
Irrigation	*Upstream Fishing*
ENEL *Power*	*East Sesia Power*
Downstream Environment	–

assumed to be equal to the historical release over all the evaluation horizon (1974–1998), is influenced by the law only in the last five years, because it came into force only in 1994. Alternative $A0$ thus provides a very high value for the ENEL *Power* sector index, but ENEL knows full well that this performance cannot be maintained.

East Sesia Power The above considerations for the ENEL Power sector are valid also for this sector.

Downstream Tourism There are no particular observations: alternative $A0$ is considered to be fairly satisfactory.

12.5 Negotiations

In Section 10.3 we explained that it is not advisable to visualize more than 5÷7 sector indices in the negotiations based on ENP. Thus, in order to reduce the number of indices, only one sector index can be visualized from any set of positively correlated indices. By means of scatter plots like the one shown in Figure 12.7, pairwise positive correlation among the sectors was examined and it was discovered that four sectors can be selected to represent eight (see the first four rows in Table 12.3). Moreover, the Stakeholders interested in the *Upstream Tourism* sector, who are also interested in the *Upstream Flooding* sector (see Figure 10.2), declared that mediocre performances for *Upstream Tourism* can be compensated by good performances for *Upstream Flooding*, because this is the sector that interests them the most. For this reason, the sector index of *Upstream Tourism* is not visualized and it will be shown on request only. A similar decision was made for *Downstream Tourism* too. Remember finally that, for the reason presented in Section 11.3.6, the *Downstream Flooding* sector is not considered in these negotiations.

In conclusion, during negotiations the sector indices for *Upstream Flooding*, *Upstream Environment*, *Irrigation*, ENEL *Power* and *Downstream Environment* were visualized. To keep the presentation simple, in the following we will focus our attention on the positions assumed by the representatives of these five sectors, even if representatives of the other sectors actively participated in the negotiations as well.

As we already said (page 249), the negotiations are carried out by applying the Elementary Negotiation Procedure (ENP) as many times as there are Stakeholders and each time, at Step 1, alternative A_{chosen} is proposed by a different Stakeholder. We will use the East Ticino Consortium's turn to exemplify.

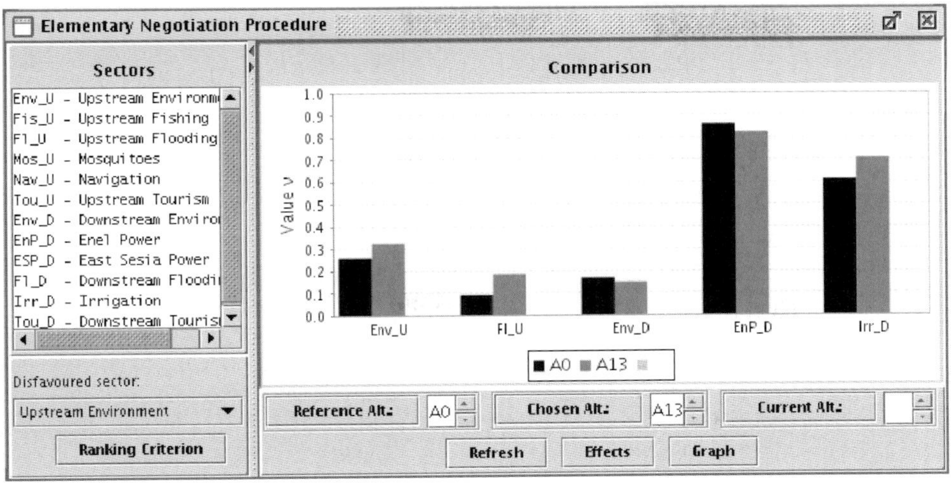

Figure 12.10: The sector indices for alternative $A13$ compared with those for alternative $A0$.

At Step 1 the East Ticino Consortium proposes their preferred alternative, namely $A13$, which thus becomes A_{chosen}. Its performance is shown in Figure 12.10. The Stakeholders take the following positions:

- *Support*: East Ticino Consortium;
- *Accept*: ENEL, East Sesia Consortium;
- *Oppose*: Municipalities of Verbania and Locarno, Bolle di Magadino Foundation, Ticino Parks.

The position of the East Sesia Consortium deserves some comment: it might be surprising that they do not support the alternative proposed by the East Ticino Consortium, since both Consortia are interested in *Irrigation*. The difference in the position of the two Consortia can be explained by observing that they are in different conditions. The East Sesia Consortium's situation is more critical, because their Abstraction Licence provides less than their current water needs (Section 6.9.2 of the DVD), and so it is difficult to guarantee each year that at the end of March there will be enough water to submerge the rice paddies. On the other hand, the East Ticino Consortium's Abstraction Licence currently provides for more than the Consortium's current water demand, which has gradually decreased over time with the industrialization of its territory. Furthermore, the East Sesia Consortium is not interested in irrigation alone, but also in the hydropower production, which, as shown in Figure 12.10, diminishes from $A0$ to $A13$.

Then the Stakeholders are asked to identify the sector $Sect^d$ that seems to be the most disfavoured. It is at this point that the conflict emerges, because each of the opposers believes that they are the most disfavoured. We are thus at a *branching point* of the procedure: let us call it BP_1. The Facilitator arbitrarily chooses *Upstream Flooding* as the disfavoured sector, promising to return later and consider both *Upstream Environment* and *Downstream Environment* (Step 2).

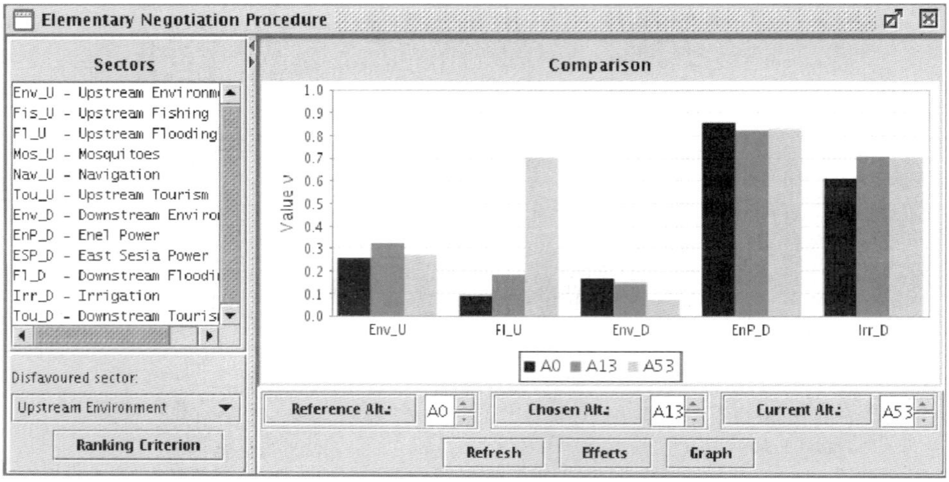

Figure 12.11: The sector indices for alternative $A53$ compared with those for $A13$ and $A0$.

Then the Stakeholders are asked to fix the acceptability thresholds for the sector indices they are interested in. The Stakeholders interested in *Upstream Environment* and *Downstream Environment*, after some discussion, decide not to fix any threshold, because they want to reserve their judgement until they are treated as disfavoured sectors. Those interested in *Irrigation* and ENEL *Power* establish their threshold at 0.7: a value that is a little higher than that produced by $A0$ for *Irrigation* and a little lower for ENEL *Power* (Step 3).

Since the set $\mathcal{A}^{\text{exp}}$ of the alternatives that satisfy the acceptability thresholds is not empty (Step 4), the Facilitator proceeds with Step 5.

Once the alternatives in $\mathcal{A}^{\text{exp}}$ have been ordered with the rule of *minimum overall dissatisfaction*, as the Stakeholders requested, the first alternative on the list is $A53$, which thus becomes alternative A_{current} (Step 5). Its performance is shown in Figure 12.11. Note the strong increase in value for the *Upstream Flooding* sector, a slight increase in ENEL *Power*, while the value for *Irrigation* is unchanged. The sectors *Upstream Environment* and *Downstream Environment* experience a decrease. Given its performance, the Stakeholders take the following positions in relation to this alternative:

- *Support*: East Ticino Consortium, Municipalities of Verbania and Locarno;
- *Accept*: ENEL, East Sesia Consortium;
- *Oppose*: Bolle di Magadino Foundation, Ticino Parks.

Alternative $A53$ thus gets a wider agreement than alternative $A13$: then the ENP procedure continues from Step 2, with $A53$ as A_{chosen} (Step 6).

Given the large increase of the *Upstream Flooding* index produced by $A53$, compared to $A13$ and $A0$, the East Sesia Consortium proposes that the lakeside towns should be satisfied with that and that an effort should be made to find an alternative that increases the satisfaction for *Irrigation*. The proposal is accepted by all, with the abstention of the Stakeholders interested in *Upstream* and *Downstream Environment*, who are already opposed to the alternative and reserve their judgement for the time when the branching point BP_1 is re-

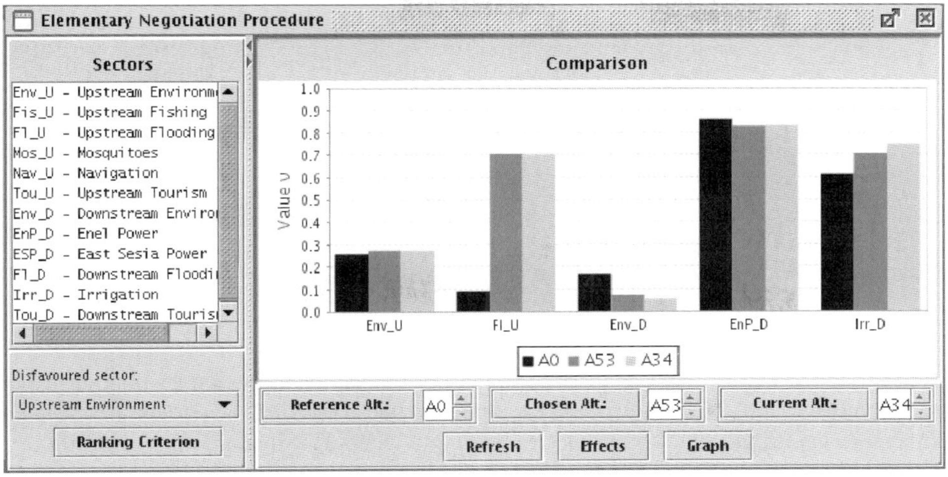

Figure 12.12: The sector indices for alternative $A34$ compared with those for $A53$ and $A0$.

visited. Thus the sector considered as disfavoured is *Irrigation*, which is not the sector with the lowest index: this is not surprising because, as we have already noted, such a selection criterion would be unjustified since the indices are defined on an arbitrary scale (Step 2). The interested Stakeholders pose a threshold of 0.6 for *Upstream Flooding* and maintain 0.7 for ENEL *Power* (Step 3); clearly, a threshold is no longer necessary for *Irrigation*, given that we are looking for alternatives that increase the index of this sector. Since $\mathcal{A}^{\text{exp}}$ is not empty, we may proceed (Step 4); once the alternatives are ranked with the rule *minimum overall dissatisfaction* (the Stakeholders do not want to change it), alternative $A34$ comes up first and so becomes alternative A_{current} (Step 5). Its performance is shown in Figure 12.12. The index of *Irrigation* grows, while the other sector indices remain unchanged (or almost unchanged). The East Sesia Consortium declares that it *supports* alternative $A34$, while the others judge it to be equivalent to $A53$: the agreement has thus been further broadened. Since A_{current} gets a wider agreement than A_{chosen}, A_{chosen} is replaced with A_{current} and the negotiators return to Step 2 (Step 6).

The lakeside towns and the Irrigation Consortia are satisfied with alternative A_{chosen}: $Sect^d$ must thus be chosen from the other three sectors. Since there are no specific requests from the Stakeholders, it is decided to look for alternatives that increase the index of *Upstream Environment*, and then move to the other two sectors. This is a new *branching point* of the procedure: let us call it BP_2 (Step 2). The Stakeholders interested in *Upstream Flooding* and ENEL *Power* keep the thresholds that they fixed previously, while those interested in *Irrigation* set it at 0.74. The set $\mathcal{A}^{\text{acc}}$ of alternatives that satisfy these thresholds is not empty but none of those alternatives improves the index of *Upstream Environment* (Step 3). Thus the set $\mathcal{A}^{\text{exp}}$ is empty: the Stakeholders are asked to lower their thresholds but they refuse. Thus the Facilitator comes back to the branching point BP_2 (end of Step 4 for the branching case, see page 249). By maintaining the same thresholds, the Stakeholders look for alternatives that increase the index of *Downstream Environment* with respect to $A34$; but without success (Steps 2, 3 and 4). The only thing left to do is to explore the possibility of increasing the index of ENEL *Power* (last pathway from BP_2). Both the Bolle di Magadino Foundation and the Ticino Parks affirm that they are not willing to lower their index any

Table 12.4. The reasonable alternatives identified in the first negotiation step

Alternative	SD	RANGE	MEF	Weights		
				λ_{fl}	λ_{irr}	λ_{env}
A0	CUR	CUR	13_{CON}	–	–	–
A9	CUR	CUR	13_{CON}	0.00	0.00	1.00
A18	CUR	CUR	13_{CON}	0.01	0.98	0.01
A54	+600	CUR	13_{CON}	0.99	0.01	0.00
A34	+600	ENL	13_{CON}	0.99	0.01	0.00
A36	+600	ENL	13_{CON}	0.98	0.01	0.01

further; the Facilitator thus sets their acceptability thresholds at the values that correspond to alternative $A34$: 0.26 for *Upstream Environment* and 0.05 for *Downstream Environment*. With these values the set $\mathcal{A}^{\text{exp}}$ proves to be empty (Step 3). Since no one is willing to reduce his/her own threshold, the Facilitator concludes that alternative $A34$ is a reasonable alternative (Step 4).

Now, as promised, the branching point BP_1 is revisited: thereby $A13$ is assumed once more as A_{chosen} and one of the two sectors that were proposed earlier, but not considered, is chosen as $Sect^d$. The Facilitator begins with *Upstream Environment* (Step 2). The Stakeholders from *Irrigation* and ENEL *Power* maintain their threshold at 0.7, while the representatives for *Upstream Flooding* decide not to pose any threshold. The resulting set $\mathcal{A}^{\text{exp}}$ turns out to be empty (Step 3). Since none of the Stakeholders is willing to reduce their acceptability thresholds, it is concluded that it is impossible to find an alternative which gets a wider agreement than $A13$, by moving in the direction of *Upstream Environment* (Step 4). The result is the same when *Downstream Environment* is considered. All the pathways from the branching point BP_1 have now been examined. Note that, even if with the last two attempts the agreement was not broadened with respect to $A13$, this alternative is not a reasonable one, because when exploring the first pathway from BP_1 we found an alternative that obtains a wider agreement: $A34$. Therefore East Ticino Consortium's turn concludes with the identification of this alternative as a reasonable alternative.

12.6 The results of the first negotiations step

By proceeding as detailed in the previous section with one Stakeholder at a time, the six reasonable alternatives listed in Table 12.4 were identified. In Table 12.5 the positions of the Stakeholders with respect to them are shown.

The Ticino Parks' choice to support only alternative $A9$, which is characterized by the extreme policy in favour of the environment, namely no water supply to the canals, must be considered only as a provocative choice, while they are waiting to analyse alternatives with a MEF higher than 13 m^3/s. For this reason, none of the other Stakeholders support this alternative, not even the Bolle di Magadino Foundation, even if the *Upstream Environment* index of alternative $A9$ is particularly high.

Table 12.5. The positions of the Stakeholders with respect to the reasonable alternatives identified in the first negotiation step

SD	RANGE	MEF	Alternative	Stakeholders		
				Support	Accept	Oppose
CUR	CUR	13_{CON}	*A0*	GRAIA Society ENEL Company		Municipalities of Verbania and Locarno Lake Navigation Company Bolle di Magadino Foundation East Sesia Consortium East Ticino Consortium Ticino Parks
CUR	CUR	13_{CON}	*A9*	Ticino Parks		Municipalities of Verbania and Locarno Lake Navigation Company Bolle di Magadino Foundation GRAIA Society East Sesia Consortium East Ticino Consortium ENEL Company
CUR	CUR	13_{CON}	*A18*	Bolle di Magadino Foundation	GRAIA Society Ticino Parks	Municipalities of Verbania and Locarno Lake Navigation Company East Sesia Consortium East Ticino Consortium ENEL Company
+600	CUR	13_{CON}	*A54*	Lake Navigation Company Municipalities of Verbania and Locarno	GRAIA Society East Sesia Consortium East Ticino Consortium ENEL Company	Ticino Parks Bolle di Magadino Foundation
+600	ENL	13_{CON}	*A34*	Municipalities of Verbania and Locarno Lake Navigation Company East Sesia Consortium East Ticino Consortium	ENEL Company GRAIA Society	Ticino Parks Bolle di Magadino Foundation
+600	ENL	13_{CON}	*A36*	Municipalities of Verbania and Locarno Lake Navigation Company Bolle di Magadino Foundation	GRAIA Society	East Sesia Consortium East Ticino Consortium ENEL Company Ticino Parks

Chapter 13

Comparison: the second negotiation step

EB, FC, FP and RSS

This chapter describes the second step in designing the alternatives and the negotiations that follow (Figure 13.1).

13.1 Designing the alternatives

The aim of the second design step (Section 5.4.2) is to generate new alternatives that could get a wider agreement than the reasonable alternatives identified so far; this will be ascertained through a second step of negotiations. The new alternatives (black dots in Figure 13.2) are sought in the neighbourhood of the reasonable alternatives A_{rea} that were identified in the first step (grey dots surrounded by a circle in Figure 13.2). Two types of alternatives are designed in the second step:

- alternatives characterized by the same combination of structural and normative actions as A_{rea};
- alternatives characterized by the same stage–discharge relation and regulation range as A_{rea}, but with a higher MEF value.

When designing the first group of alternatives, we explore the same decision space that was explored in the first design step but with a higher resolution. Thereby these alternatives are obtained by solving the same (Optimal) Control Problem as the one used to design the alternatives of A_{rea}, except for the weights assigned to the objectives, which are slightly different. They will be chosen by taking account of information that has been acquired in the first negotiation step. Let us consider, for example, alternative $A34$, which is one of the reasonable alternatives identified in the first step and was designed with the weights

$$\lambda_{\text{fl}} = 0.99 \qquad \lambda_{\text{irr}} = 0.01 \qquad \lambda_{\text{env}} = 0$$

We know that the Ticino Parks and the Bolle di Magadino Foundation are opposed to it. In order to find an alternative in the neighbourhood of $A34$ that might be accepted by these opposers, the value of λ_{env} must be increased, but this is only possible if the weight of another sector is reduced.[1] To choose the new value for λ_{env}, remember that in the first design step

[1] The sum of the weights must always be equal to one.

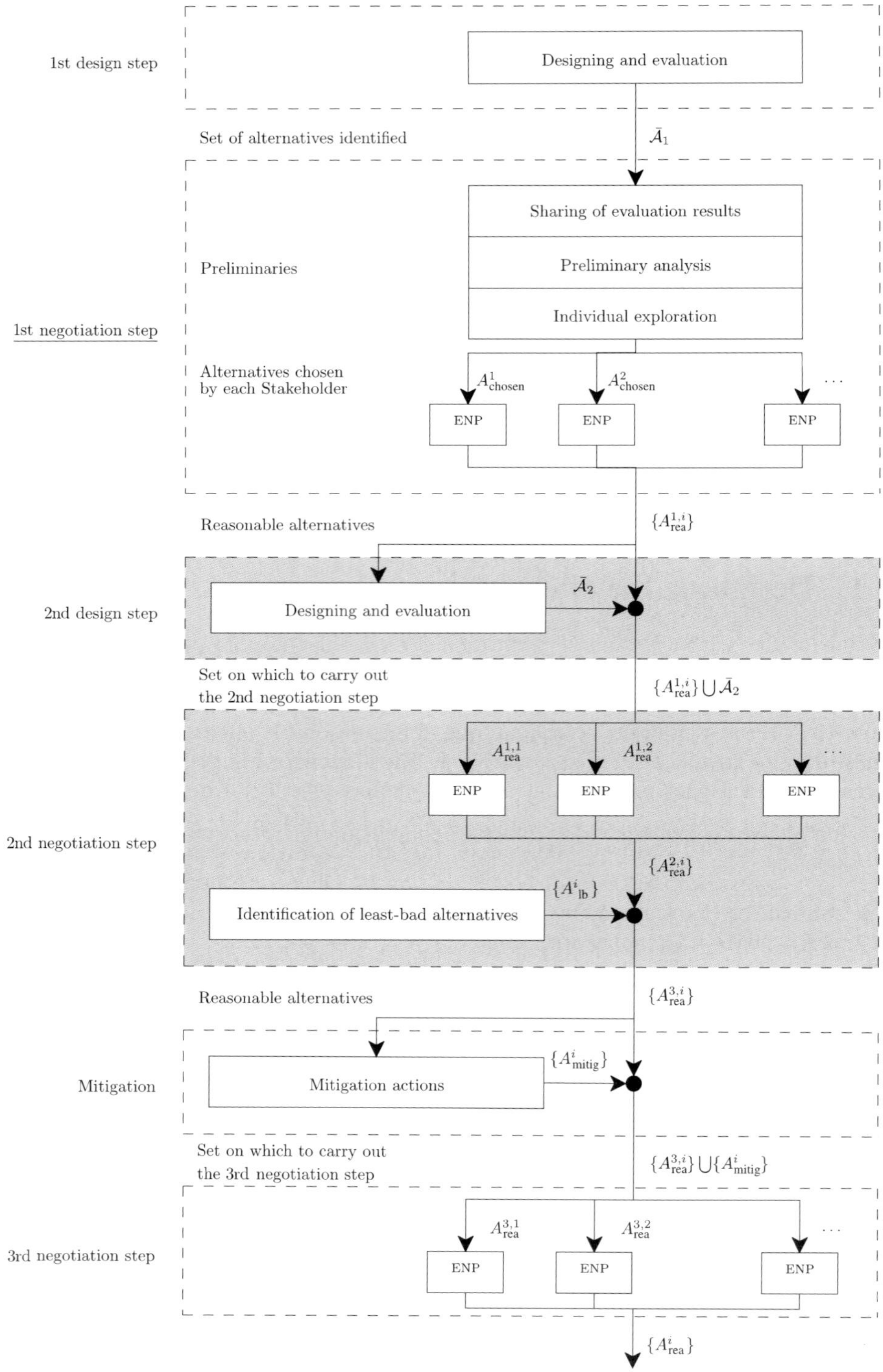

Figure 13.1: The steps for identifying the reasonable alternatives. The highlighted steps are the ones whose results are described in this chapter.

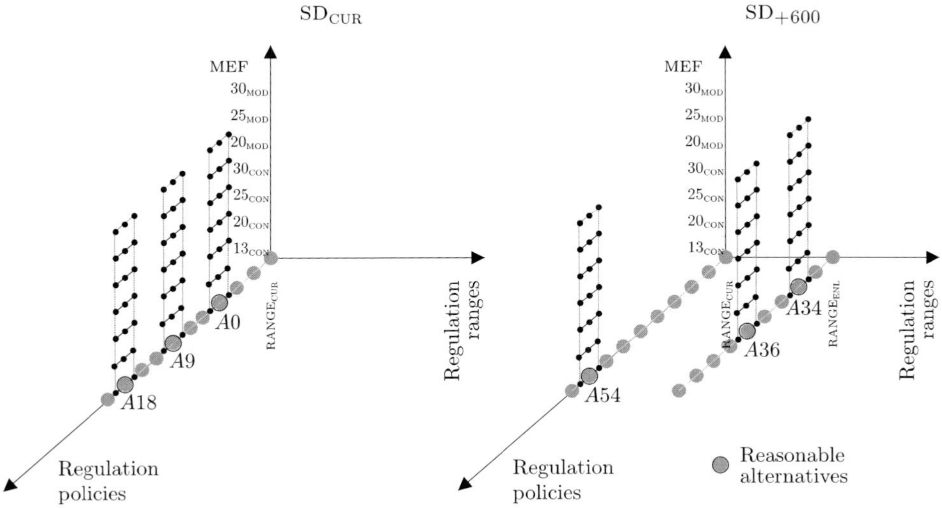

Figure 13.2: The alternatives designed in the second step are indicated by black dots.

the same combination that characterizes $A34$, namely (SD_{+600}/$\text{RANGE}_{\text{ENL}}$/$13_{\text{CON}}$) was also used to design alternative $A36$, with the following weights

$$\lambda_{\text{fl}} = 0.98 \qquad \lambda_{\text{irr}} = 0.01 \qquad \lambda_{\text{env}} = 0.01$$

Since the Irrigation Consortia (and ENEL) are opposed to it, we understand that, in order to maintain the agreement from the Consortia that was achieved with $A34$, the increase in λ_{env} cannot be compensated with a reduction in λ_{irr} but only in λ_{fl}.

As for the second group of alternatives, it might be surprising that they have to be characterized by a value of MEF higher than the one of A_{rea} even when that alternative does not arouse opposition from *Downstream Environment*. Observe, however, that the decision space that was not explored in the first design step is the very space that corresponds to the MEF values: an increase in the MEF value might make it possible to modify an alternative that pleases *Downstream Environment* in the direction required by the opposers, without loosing the agreement of the Ticino Parks. Alternatives of this type are designed by solving Control Problems that were not considered in the previous step and so, in contrast to what happens for the first group, we do not have explicit references for the choice of the weights. However, indications can still be derived from the information that we have already acquired. For example, considering once more the case of alternative $A34$, one can observe that since the Ticino Parks are opposed to it and they are in favour of an increase in the MEF, it might be reasonable to keep the same weights as the $A34$, or at the most to vary them very little. The validity of these guesses must still be checked with an analysis of the results obtained.

Once the new alternatives have been designed, it is advisable to identify all the alternatives that, from the information in our possession, we know will not be interesting to the opposers, or will be such as to lose the supporters. These alternatives should be immediately eliminated and should not be subjected to the second negotiation step. From the results obtained in the first negotiation step we can, in fact, deduce several *conditions for acceptability* of the new alternatives. For example, in the neighbourhood of *A*34, a new alternative *A* can gain a wider agreement only if:

(1) the value of the *Downstream Environment* sector index that it produces is higher than the one produced by *A*34, i.e.

$$V_{\text{Env\_D}}(A) > V_{\text{Env\_D}}(A34)$$

(2) the value of the *Irrigation* sector index that it produces is higher than the one produced by *A*0, i.e.

$$V_{\text{Irr\_D}}(A) > V_{\text{Irr\_D}}(A0)$$

If one or the other of these conditions were not respected, we are sure that the Stakeholders who are interested in those two sectors would oppose *A*.

Just as we did in the previous chapters, we will describe this procedure in practice, considering the reasonable alternative *A*34 as an example. We chose it because it allows us to make some interesting observations. For didactic reasons, we will begin the presentation with the alternatives of the second group, starting with the alternatives with time-invariant MEF, and then passing to those with time-variant MEF. In the analysis attention will be paid mostly to the conflict between *Irrigation* and *Downstream Environment*, because this conflict must be solved or at least mitigated in order to get a wider agreement. Once this first analysis is concluded we will deal with the alternatives of the first group.

13.1.1 Alternatives with time-invariant MEF

The reader might wonder why it is necessary to design new policies in order to assess the effects of an increase in the MEF: it might seem sufficient that the policies that have already been designed be simulated with different MEF values. Such a procedure would not be correct, however, because the policy obtained as a solution to a Control Problem is efficient for the system subject to the constraints that appear in the formulation of the Problem, but it cannot be if the constraints are modified. Since the MEF is a parameter that defines one of the constraints, when it is varied the policy must be re-designed.

Given the pair of actions ($\text{SD}_{+600}/\text{RANGE}_{\text{ENL}}$) with which alternative *A*34 was designed and the discretization that we adopted for the MEF (Section 5.3.3), we designed the six alternatives described in Table 13.1. Table 13.2 shows the sector index values they produce and, as a comparison, also those produced by the three reasonable alternatives *A*0, *A*34 and *A*36. The first is shown because it is the Stakeholders' point of reference, the second because it is the alternative whose agreement we hope to broaden, and the third because it is associated to the same pair of actions as *A*34.

Table 13.2 shows that, as we expected, all the new alternatives either increase the *Downstream Environment* index with respect to *A*34, or at least do not worsen it. None of the new alternatives is dominated by the others because as the MEF increases the *Irrigation* index decreases.

Alternative *A*66 produces a value for *Irrigation* that is lower than the value produced by *A*0; thus *A*66 violates the second condition of accessibility and it must be discarded. Since *A*66 is characterized by a MEF value equal to 30_{CON} and since, as we have just seen, the value for *Irrigation* decreases as the MEF increases, the unacceptability of *A*66 reveals that it would be useless to consider MEF values that are higher than 30_{CON}.

Table 13.1. Characteristics of the alternatives with time-invariant MEF higher than 13 m^3/s

Alternative	SD	RANGE	MEF	Weight for *Irrigation*
*A*56	+600	ENL	20_{CON}	$\lambda_{irr} > \lambda_{irr}^{A34}$
*A*58	+600	ENL	20_{CON}	$\lambda_{irr} = \lambda_{irr}^{A34}$
*A*60	+600	ENL	25_{CON}	$\lambda_{irr} > \lambda_{irr}^{A34}$
*A*62	+600	ENL	25_{CON}	$\lambda_{irr} = \lambda_{irr}^{A34}$
*A*64	+600	ENL	30_{CON}	$\lambda_{irr} > \lambda_{irr}^{A34}$
*A*66	+600	ENL	30_{CON}	$\lambda_{irr} = \lambda_{irr}^{A34}$

Table 13.2. Sector indices associated to the alternatives with time-invariant MEFs

Sector	MEF $[m^3/s]$								
	13_{CON}			20_{CON}	25_{CON}	30_{CON}	20_{CON}	25_{CON}	30_{CON}
	*A*0	*A*34	*A*36	*A*58	*A*62	*A*66	*A*56	*A*60	*A*64
Downstream Environment	0.16	0.06	0.15	0.07	0.07	0.08	0.06	0.06	0.06
Irrigation	0.61	0.74	0.60	0.68	0.62	0.56	0.73	0.68	0.61
ENEL *Power*	0.85	0.83	0.70	0.82	0.81	0.80	0.83	0.82	0.81
Upstream Environment	0.26	0.27	0.47	0.26	0.26	0.26	0.22	0.22	0.22
Upstream Flooding	0.09	0.70	0.69	0.70	0.70	0.70	0.68	0.68	0.66

Alternatives *A*58 and *A*62 bring about a decrease in the *Irrigation* index with respect to *A*34. In particular, the value produced by *A*62 is just a little higher than the value produced by *A*0, despite the fact that *A*62 was designed giving *Irrigation* the same weight λ_{irr} that was used when designing *A*34. We inferred that if a lower weight were used, alternatives that violate the second condition of acceptability would be designed and so no attempts were made in this direction. The three remaining alternatives (*A*56, *A*60 and *A*64) were designed giving *Irrigation* a higher weight than in *A*34. All three are dominated by alternative *A*34 and so they are not interesting, as well as being unacceptable because they violate the first condition (the *Downstream Environment* index is not higher than that produced by *A*34).

By observing the release trajectories produced by the new alternatives, we understood that an increase in the weight λ_{irr} causes an increase in the number of days that the flow in the Ticino is equal to the MEF. Since the latter is time-invariant, this means that the flow in the Ticino stays the same for long periods and the seasonal fluctuations that characterize the natural regime are absent. The distance between regulated and natural regime thus remains the same as for *A*34, even though the value of the MEF was increased.

In conclusion, by adopting time-invariant MEFs, alternatives that increase the *Downstream Environment* index are obtained, but at the expense of the *Irrigation* index. Of the alternatives designed, only *A*58 and *A*62 are acceptable, even if the *Downstream Environment* index remains lower than that produced by *A*0 and its increase with respect to *A*34 is very small. As we have just noted, this might be due to the fact that the presence of a time-invariant MEF induces long periods of constant flow in the Ticino. A more significant

Table 13.3. Characteristics of the alternatives with time-variant (modulated) MEF

Alternative	SD	RANGE	MEF	Weight for *Irrigation*
*A*72	+600	ENL	20_{MOD}	$\lambda_{\text{irr}} > \lambda_{\text{irr}}^{A34}$
*A*74	+600	ENL	20_{MOD}	$\lambda_{\text{irr}} = \lambda_{\text{irr}}^{A34}$
*A*76	+600	ENL	25_{MOD}	$\lambda_{\text{irr}} > \lambda_{\text{irr}}^{A34}$
*A*78	+600	ENL	25_{MOD}	$\lambda_{\text{irr}} = \lambda_{\text{irr}}^{A34}$

Table 13.4. Sector indices associated to the alternatives with time-variant (modulated) MEF

Sector	MEF [m^3/s]						
	13_{CON}			20_{CON}	20_{MOD}	20_{CON}	20_{MOD}
	*A*0	*A*34	*A*36	*A*58	*A*74	*A*62	*A*78
Downstream Environment	0.16	0.06	0.15	0.07	0.07	0.07	0.07
Irrigation	0.61	0.74	0.60	0.68	0.68	0.62	0.63
ENEL *Power*	0.85	0.83	0.70	0.82	0.82	0.81	0.81
Upstream Environment	0.26	0.27	0.47	0.26	0.26	0.26	0.26
Upstream Flooding	0.09	0.70	0.69	0.70	0.70	0.70	0.70

increase in the *Downstream Environment* index could possibly be obtained with time-variant MEFs, so it is time to put them to the test.

13.1.2 Alternatives with time-variant MEF

To assess the effects of a time-variant MEF (defined in Section 5.3.3 and which we will also term *modulated* MEF) we designed the four alternatives in Table 13.3.

Alternatives *A*72 and *A*76 produce values for *Downstream Environment* that are lower than the value produced by *A*34 and so they are discarded. This is why they do not appear in Table 13.4, where the indices relative to the other two alternatives (*A*74 and *A*78) are reported, along with the indices produced by the two alternatives (*A*58 and *A*62) that were introduced in the previous section and the usual three reference alternatives (*A*0, *A*34 and *A*36). From the table it seems that the *Downstream Environment* index does not increase if we pass from time-invariant to modulated MEF and that the *Irrigation* index either increases slightly (*A*78) or does not change (*A*74). This is certainly not the result that we expected. Nevertheless, a moment's thought shows why this can happen: if it is possible to increase environmental satisfaction without lowering irrigation satisfaction, then it must also be possible to increase irrigation satisfaction whilst maintaining the same level of environmental satisfaction. For some reason, which we will try to identify, we only found pairs of alternatives that produce the second effect. Alternatives that produce the first effect could certainly be found as well with further explorations. Nevertheless, since the increases in the index that we obtained up to now are very small, it seemed useless to continue the exploration.

We will simply note that alternatives *A*58 and *A*74 are indistinguishable, while alternative *A*78 dominates *A*62. Thus *A*58 and *A*62, which are characterized by a time-invariant MEF, are discarded because the alternatives with a modulated MEF are preferable. The reason for this is simple. For each possible MEF value, the water volume that must be released

Table 13.5. Sector indices associated to the alternatives with 13_{CON} and different weights for *Downstream Environment*

Sector	MEF [m^3/s]							
	13_{CON}						20_{MOD}	25_{MOD}
	*A*0	*A*34	*A*36	*A*71	*A*81	*A*82	*A*74	*A*78
Downstream Environment	0.16	0.06	0.15	0.15	0.14	0.14	0.07	0.07
Irrigation	0.61	0.74	0.60	0.63	0.70	0.67	0.68	0.63
ENEL *Power*	0.85	0.83	0.70	0.71	0.72	0.71	0.82	0.81
Upstream Environment	0.26	0.27	0.47	0.40	0.26	0.29	0.26	0.26
Upstream Flooding	0.09	0.70	0.69	0.69	0.69	0.70	0.70	0.70

into the Ticino over the course of a year is, by construction, the same for both the time-invariant and the modulated MEF values (see Section 5.3.3). It follows that the modulated MEF guarantees a performance for *Downstream Environment* that is no worse than the one that would be obtained with the corresponding time-invariant MEF. A time-variant MEF imposes higher releases in the very periods when, in natural regime conditions, more water would be available, which is advantageous with view to a compromise between irrigation and environmental needs. However, since the alternatives that we are considering were all designed giving zero weight to *Downstream Environment* and a high weight to *Irrigation*, the entire advantage is acquired by the latter sector. Now it is clear why we did not obtain the result that we expected: the surprising result is produced by the weights, which evidently play a very important role. Therefore, it is mandatory to explore the possibility of increasing the value for *Downstream Environment* by acting on the weights, i.e. we must explore the alternatives belonging to the first group that was defined in the introduction.

13.1.3 Alternatives oriented to Downstream Environment

The interesting alternatives from the first group are characterized by values of the *Downstream Environment* weight between the one used for *A*34 and the one used for *A*36. In Table 13.5 we show the indices produced by three alternatives (*A*71, *A*81 and *A*82) with these characteristics. In the same table the indices produced by alternatives *A*74 and *A*78 are also reported (these are the only acceptable alternatives that have been designed so far), as well as the usual three reference alternatives. The table reveals that if we considered only the *Irrigation* and *Downstream Environment* sectors, alternatives *A*74 and *A*78 would be dominated by *A*81 and *A*82, but such dominance no longer exists if we consider all five sectors.

This result definitively confirms that in order to diminish the conflict between *Downstream Environment* and *Irrigation* it is more fruitful to act upon the regulation policy (by carefully selecting the weights), rather than introducing normative actions (i.e. imposing a particular MEF). In fact, on the one hand the adoption of a time-invariant MEF imposes an increase in the release at the most critical times for irrigation, and on the other the adoption of a time-variant MEF reduces the conflict but does not completely eliminate it.[2] Instead, acting on the regulation policy makes it possible to increase the flow in the Ticino especially in the autumn and winter months, as demonstrated, for example, by Figure 13.3, which compares

[2]In the summer period a time-variant MEF imposes lower releases than the corresponding time-invariant MEF, but in the period when the rice paddies are being flooded the time-variant MEF imposes higher releases.

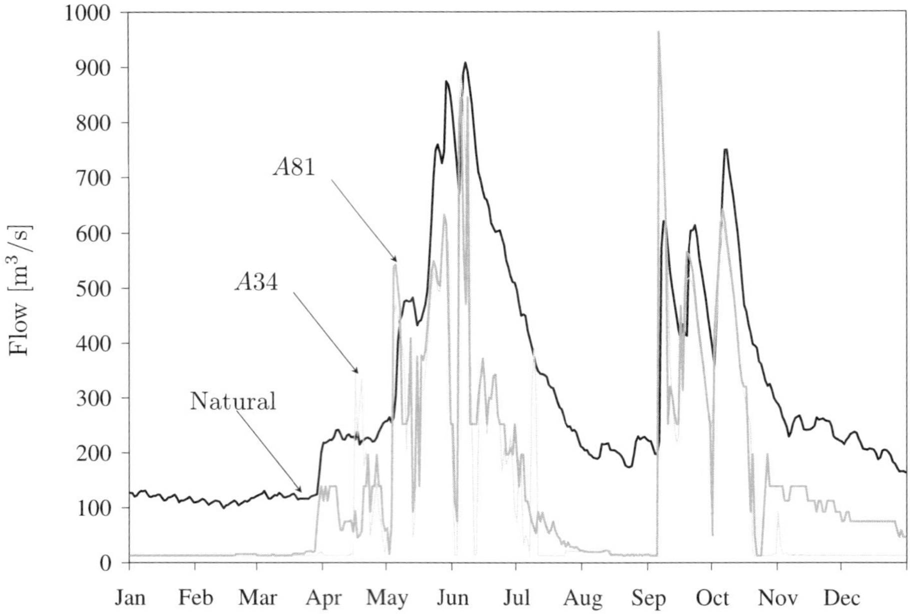

Figure 13.3: Release trajectories produced by $A34$ and $A81$ and natural outflow trajectory over one year (1984).

the release trajectories produced by $A34$ and $A81$ with the natural outflow trajectory over one year (1984).

Thus, acting on the regulation policy design through the weights is the best way to mitigate the conflict. Note, however, that the weights should not be modified too much, lest a new conflict with ENEL be generated: the winter period is, in fact, the most critical one for ENEL, because in these months the value of hydropower is higher. Table 13.5 shows that both $A81$ and $A82$ produce a value for ENEL *Power* that is clearly inferior to the ones produced by $A74$ and $A78$.

In conclusion: starting from the reasonable alternative $A34$, we obtained five new alternatives ($A71$, $A74$, $A78$, $A81$ and $A82$), which will be presented to the second negotiation step. The set of new alternatives is completed by repeating the procedure for all the other reasonable alternatives.

13.1.4 A comment

In this section we have seen that the conflict between *Downstream Environment* and *Irrigation* is structural and cannot be solved by increasing the MEF or by modifying the weights in the regulation policy design. The five alternatives that were designed starting from $A34$ increase the *Downstream Environment* index a little, but without bringing it up to the value produced by $A0$. Moreover, these small increments bring about a decrease in the index values for *Irrigation* and ENEL *Power*. The space for further compromise is reduced to the slight margin among these five alternatives. The advent of irrigation and hydropower production in an era when very little attention was paid to environmental problems relegated *Downstream Environment* to a condition of poor satisfaction, from which it is unlikely to

Table 13.6. The reasonable alternatives identified in the second negotiation step

Alternative	SD	RANGE	MEF	Weights		
				λ_{fl}	λ_{irr}	λ_{env}
*A*0	ACT	ACT	13_{CON}	–	–	–
*A*119	ACT	ACT	25_{MOD}	0.01	0.98	0.01
*A*54	+600	ACT	13_{CON}	0.99	0.01	0.00
*A*34	+600	ENL	13_{CON}	0.99	0.01	0.00
*A*36	+600	ENL	13_{CON}	0.98	0.01	0.01

escape without lowering the levels of satisfaction to which the Stakeholders interested in *Irrigation* and ENEL *Power* have by now become accustomed.

The only way to increase the satisfaction of *Downstream Environment* without disturbing *Irrigation* and ENEL *Power* would be to include a modification of irrigation techniques among the actions considered in the Project. For example, it would be interesting to explore the effects of changing from submersion to sprinklers (or to drip irrigation where possible). Alternatives with these characteristics might make it possible to satisfy the Stakeholders interested in *Irrigation* and, at the same time, to allow for more water to be available to increase the *Downstream Environment* index, but it is unlikely that they would satisfy ENEL. Clearly, the cost of the transformation would be included among the indicators for *Irrigation*, and we should also take into account the impact on the landscape that an alternative of this type would provoke, since it would reduce the flow in the canals and in the irrigation ditches. However, the study of these actions exceeds the limits imposed on the Verbano Project and so we will not consider it.

13.2 The second negotiation step

Just as in the first step, negotiations are carried out according to the ENP (page 247). The only difference is that the procedure is no longer re-started as many times as there are Stakeholders, each time beginning from the alternative that one of them prefers, but it is carried out as many times as there are reasonable alternatives identified in the first negotiation step, each time beginning with a different one. In this way the set of alternatives in Table 13.6 is obtained, with respect to which the Stakeholders assume the positions presented in Table 13.7.

By comparing these two tables with Tables 12.4 and 12.5, which describe the reasonable alternatives identified in the first negotiation step, one notes that they differ only in the fact that alternative *A*9 is replaced by alternative *A*119, which is still supported by the Ticino Parks but, unlike the first, is supported also by the Bolle di Magadino Foundation and accepted by the GRAIA Society (*Upstream Fishing* sector). This replacement is thus a step forward towards a compromise.

13.3 The least-bad alternatives

As Figure 13.1 shows, the second negotiation step foresees identifying not only the reasonable alternatives but also, if necessary, the least-bad alternatives.

Table 13.7. The positions of the Stakeholders with respect to the reasonable alternatives identified in the second negotiation step

SD	RANGE	MEF	Alternative	Stakeholders		
				Support	Accept	Oppose
CUR	CUR	13_{CON}	$A0$	GRAIA Society ENEL Company		Municipalities of Verbania and Locarno Lake Navigation Company Bolle di Magadino Foundation East Sesia Consortium East Ticino Consortium Ticino Parks
CUR	CUR	25_{MOD}	$A119$	Bolle di Magadino Foundation Ticino Parks	GRAIA Society	Municipalities of Verbania and Locarno Lake Navigation Company East Sesia Consortium East Ticino Consortium ENEL Company
+600	CUR	13_{CON}	$A54$	Municipalities of Verbania and Locarno Lake Navigation Company	GRAIA Society East Sesia Consortium East Ticino Consortium ENEL Company	Bolle di Magadino Foundation Ticino Parks
+600	ENL	13_{CON}	$A34$	Municipalities of Verbania and Locarno Lake Navigation Company East Sesia Consortium East Ticino Consortium	GRAIA Society ENEL Company	Bolle di Magadino Foundation Ticino Parks
+600	ENL	13_{CON}	$A36$	Municipalities of Verbania and Locarno Lake Navigation Company Bolle di Magadino Foundation	GRAIA Society	East Sesia Consortium East Ticino Consortium ENEL Company Ticino Parks

Table 13.8. The acceptability thresholds expressed by the Stakeholders when identifying the least-bad alternatives for the Ticino Parks. The search starts from the reasonable alternatives characterized by the pair (SD_{+600}/$\text{RANGE}_{\text{ENL}}$)

Sectors	Acceptability threshold
Upstream Environment	0.40
Upstream Fishing	0.80
Upstream Flooding	0.60
Mosquitoes	0.55
Navigation	0.80
Upstream Tourism	0.60
Downstream Environment	0.00
ENEL *Power*	0.83
East Sesia Power	0.40
Irrigation	0.70
Downstream Tourism	0.00

More precisely, for each class of alternatives, defined by the same pair (stage–discharge relation/regulation range), we must ascertain whether there are Stakeholders who are opposed to all the alternatives in that class. If there are, the procedure described on page 250 is used to search for an alternative that might be 'tolerated', even reluctantly, by the opponent(s). This least-bad alternative will be included among those submitted to the *Final Decision* phase.

As a practical example of how to proceed, we consider the pair (SD_{+600}/ $\text{RANGE}_{\text{ENL}}$), for which two reasonable alternatives were identified: *A*34 and *A*36. Each of them is opposed by several Stakeholders, but only the Ticino Parks are opposed to both, and so it is for these Stakeholders that least-bad alternatives must be identified. Following the procedure, first the Stakeholders that support these two alternatives are asked to define their acceptability thresholds (Step 1). In response, the values reported in Table 13.8 are obtained. On the basis of these thresholds, the set $\mathcal{A}^s$ of the alternatives that satisfy them is identified: it proves to be empty. To identify the least-bad alternatives it is thus necessary to enlarge the set $\mathcal{A}^s$; this is done by not considering in turn one or more of the most critical acceptability thresholds, which are those for the *Upstream Environment*, *Irrigation* and ENEL *Power* sector. The aim is to identify at least one set $\mathcal{A}^s$ that is not empty (Step 2). When only one of the three thresholds is not considered, the set $\mathcal{A}^s$ remains empty. Thus it is necessary not to consider at least two of them and the following cases are obtained:

- When the *Upstream Environment* and ENEL *Power* thresholds are not considered, the set $\mathcal{A}^s$ contains the alternative *A*81. After examining its performances, the Ticino Parks declare that they would tolerate this alternative (Step 3). This alternative is supported by the Municipalities of Verbania and Locarno and the Lake Navigation Company; accepted by the Irrigation Consortiums and the GRAIA Society; and opposed by the Bolle di Magadino Foundation and ENEL (Step 4).

- When the *Irrigation* and ENEL *Power* thresholds are not considered, the set $\mathcal{A}^s$ contains several alternatives, all of which are characterized by a regulation policy obtained with the same weights as *A*36 and by MEF values higher than 13 m^3/s. Of these alternatives, the Parks prefer *A*112 (Step 3). This alternative is supported by the Municipalities of Verbania and Locarno, the Lake Navigation Company and the

Table 13.9. The characteristics of the least-bad alternatives

Alternative	SD	RANGE	MEF	Weights		
				λ_{fl}	λ_{irr}	λ_{env}
$A7$	ACT	ACT	13_{CON}	0.00	0.99	0.01
$A18$	ACT	ACT	13_{CON}	0.01	0.98	0.01
$A168$	+600	ACT	25_{MOD}	0.99	0.01	0.00
$A81$	+600	ENL	13_{CON}	0.60	0.38	0.02
$A122$	+600	ENL	20_{MOD}	0.60	0.39	0.01
$A110$	+600	ENL	25_{MOD}	0.55	0.35	0.10
$A112$	+600	ENL	25_{MOD}	0.98	0.01	0.01

Bolle di Magadino Foundation; accepted by the GRAIA Society, and is opposed by the Irrigation Consortiums and ENEL (Step 4).

- When the *Irrigation* and *Upstream Environment* thresholds are not considered, the set $\mathcal{A}^s$ is empty.
- When all three of the thresholds are not considered, the set $\mathcal{A}^s$ contains various alternatives, among which the Parks prefer $A110$ and $A122$. Both these alternatives are supported by the Municipalities of Verbania and Locarno and the Lake Navigation Company; accepted by the GRAIA Society; and opposed by the Bolle di Magadino Foundation and ENEL. The Irrigation Consortia are opposed to the first and accept the second (Step 4).

We proceeded in a similar way for the other two classes of reasonable alternatives, obtaining the alternatives $A7$, $A18$ and $A168$. The least-bad alternatives are those listed in Table 13.9 and the Stakeholders position themselves with respect to them as described in Table 13.10. These alternatives will be subjected to the phase of *Mitigation*, described in the next chapter, along with the reasonable alternatives listed in Table 13.6.

Table 13.10. The positions of the Stakeholders with respect to the least-bad alternatives

SD	RANGE	MEF	Alternative	Stakeholders			
				Support	Accept	Oppose	Tolerate
CUR	CUR	13_{CON}	$A7$	Bolle di Magadino Foundation	GRAIA Society Ticino Parks	Municipalities of Verbania and Locarno Lake Navigation Company ENEL Company	East Sesia Consortium East Ticino Consortium
		13_{CON}	$A18$	Bolle di Magadino Foundation	GRAIA Society Ticino Parks	East Sesia Consortium East Ticino Consortium ENEL Company	Municipalities of Verbania and Locarno Lake Navigation Company
+600	CUR	25_{MOD}	$A168$	Municipalities of Verbania and Locarno Lake Navigation Company	GRAIA Society ENEL Company	Bolle di Magadino Foundation East Sesia Consortium East Ticino Consortium	Ticino Parks
	ENL	13_{CON}	$A81$	Municipalities of Verbania and Locarno Lake Navigation Company	GRAIA Society East Sesia Consortium East Ticino Consortium	Bolle di Magadino Foundation ENEL Company	Ticino Parks
		20_{MOD}	$A122$	Municipalities of Verbania and Locarno Lake Navigation Company	GRAIA Society	Bolle di Magadino Foundation East Sesia Consortium East Ticino Consortium ENEL Company	Ticino Parks
		25_{MOD}	$A110$	Municipalities of Verbania and Locarno Lake Navigation Company	GRAIA Society	Bolle di Magadino Foundation East Sesia Consortium East Ticino Consortium ENEL Company	Ticino Parks
		25_{MOD}	$A112$	Municipalities of Verbania and Locarno Lake Navigation Company Bolle di Magadino Foundation	GRAIA Society	East Sesia Consortium East Ticino Consortium ENEL Company	Ticino Parks

Chapter 14

Mitigation measures

EB, FC, RSS and EW

Once they have been identified, the reasonable alternatives must be analysed with the aim of identifying mitigation measures that can bring about a wider agreement (see Figure 14.1). This must be done before they are subjected to the *Final Decision* phase. A *mitigation measure* is an intervention (namely an action or a mix of actions) specifically aimed at mitigating the effects of an alternative on a particular sector, in order to reduce the opposition of some of the Stakeholders or, better still, to transform it into acceptance or support. Unlike the actions that have been considered until now, a mitigation action should be designed in such a way that its effects are limited, as much as possible, to the sector it was designed for.

Mitigation measures are sought for each of the reasonable alternatives. As usual, we will provide only one example, with reference to an alternative ($A34$) that is supported by most of the Stakeholders and has few opponents, because there seems to be real hope of reaching consensus on it.

Before proceeding, however, we must free ourselves from the commitment that we made in Section 11.3.6: as the reader will remember, we agreed with the Stakeholders to suspend judgement about the effects of the alternatives on the *Downstream Flooding* sector until the end of the comparison. Now, as promised, we must examine the reasonable alternatives from that point of view. We will discover that mitigation measures must be sought for that sector as well.

14.1 The effects on Downstream Flooding

As foreseen in the procedure defined in Section 11.3.6, we must examine the effects produced by each of the reasonable alternatives on the *Downstream Flooding* sector, to identify which ones have the greatest probability of increasing the flood peak levels at Pavia and thus the risk (or the consequences) of flooding. For these alternatives we must then look for appropriate mitigation measures.

To estimate the effect of the alternatives on the water level at Pavia, we must necessarily use the neural network described in Section 6.9.1. However, we discovered (Section 11.3.3) that the neural network is affected by an estimation error that is so great that it prevented us from comparing the alternatives on the basis of the indicators computed with the network output. It would thus seem impossible to use the network to ascertain whether an alternative

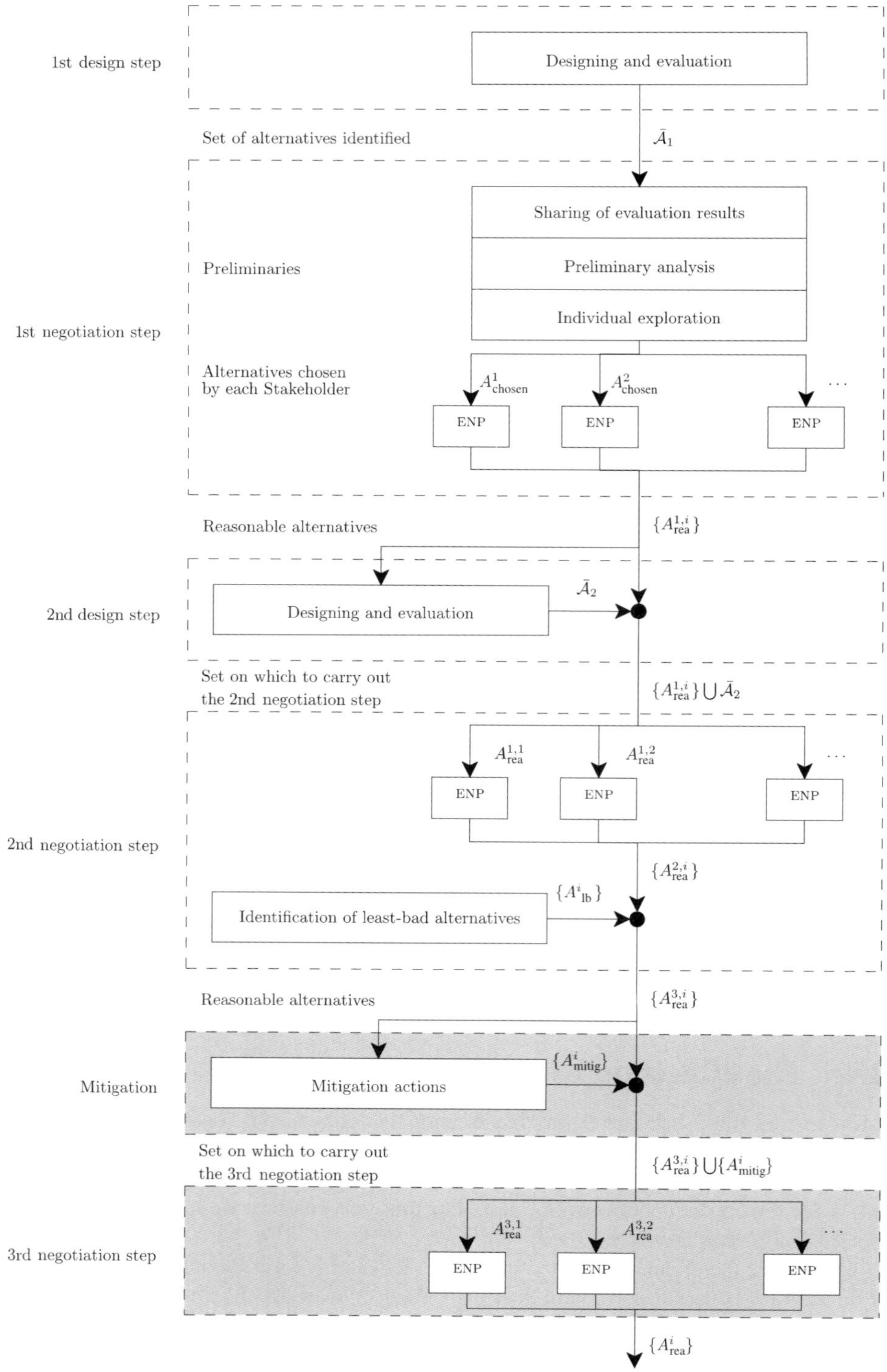

Figure 14.1: The steps for identifying the reasonable alternatives. The highlighted steps are the ones whose results are described in this chapter.

Table 14.1. Flood peak levels (h^{Pv}) at Pavia produced by alternative $A119$ and level differences (Δh^{A0}) with respect to $A0$

Starting date of the event	Alternative $A119$			
	h_t^{Pv} [m]	r_t [m^3/s]	q_{t+1}^{Po} [m^3/s]	Δh^{A0} [m]
4 Oct 1976	1.78	1475	4680	0.02
14 Oct 1976	2.20	1472	5080	0.08
28 Oct 1976	2.96	1149	7220	0.07
8 Nov 1976	2.08	1260	5780	0.00
2 May 1977	2.81	1760	6840	−0.08
21 May 1977	1.63	877	5360	0.04
30 Aug 1977	2.34	1490	5620	0.10
8 Oct 1977	**3.70**	**1660**	**8370**	**−0.04**
15 Oct 1979	2.92	1883	6300	0.06
2 Apr 1981	1.46	790	5160	−0.05
28 Sep 1981	2.08	2169	4470	−0.07
17 May 1983	1.64	1728	4160	0.12
26 Apr 1986	2.08	1522	5140	−0.06
7 Oct 1992	1.52	860	5890	0.01
25 Sep 1993	**3.11**	**1816**	**6930**	**−0.07**
6 Nov 1994	**3.91**	**1056**	**9774**	**0.00**
Peaks lowered with respect to $A0$				38%
Peaks increased with respect to $A0$				50%

increases the peak of the flood or not. The difficulty can be solved by comparing the levels produced by an alternative with those produced by $A0$, instead that with the historically recorded levels. By doing so, both the terms of the comparison are estimated with the same model and are thus affected by similar errors; these errors will therefore have little influence on the results of the comparison.

In other words, for each reasonable alternative we will use the neural network to estimate the flood peak level that it produces in each flood event of the simulation horizon, and we will compare these levels with those produced by alternative $A0$ in the same events. With this information, the Stakeholders interested in *Downstream Flooding* sector will be able to define their position with respect to each of the reasonable alternatives (i.e. whether they support it, accept it or oppose it).

For example, Table 14.1 shows this information for alternative $A119$: h_t^{Pv} is the flood peak level estimated for the event whose starting date is shown in the first column; r_t and q_{t+1}^{Po} are the flows in the Ticino and in the Po respectively, on the same day; the value Δh^{A0} reported in the last column is the difference between h_t^{Pv} and the estimated flood peak level for the same event with $A0$. To understand the meaning of the values in Table 14.1, bear in mind that when h_t^{Pv} exceeds 1.39 m the flood plains are flooded, while at 3.50 m flooding starts at Borgo Ticino (see Section 4.5.6 of the DVD). The data show that 38% of the peaks are lowered, while 50% are increased. However, note that the increases are only of a small amount and that they all occur for levels that are very far from the flood level at Borgo Ticino. The three events in which the level at Pavia exceeded 3 m are highlighted in bold: in none of these is the difference Δh^{A0} positive. In the light of these estimates, the Stakeholders (the Province of Pavia) declared that it would accept $A119$.

Table 14.2. Flood peak levels (h^{Pv}) at Pavia produced by alternative $A34$ and level differences (Δh^{A0}) with respect to $A0$

Starting date of the event	Alternative $A34$			
	h_t^{Pv} [m]	r_t [m^3/s]	q_{t+1}^{Po} [m^3/s]	Δh^{A0} [m]
4 Oct 1976	1.83	1543	4680	0.08
14 Oct1976	2.22	1494	5080	0.09
28 Oct 1976	**3.08**	**1312**	**7220**	**0.19**
8 Nov 1976	2.10	1279	5780	0.02
2 May 1977	2.85	1804	6840	−0.05
21 May 1977	1.63	880	5360	0.04
30 Aug 1977	2.45	1616	5620	0.21
8 Oct 1977	**3.75**	**1822**	**8370**	**0.01**
15 Oct 1979	**3.01**	**2009**	**6300**	**0.16**
2 Apr 1981	1.65	980	5160	0.14
28 Sep 1981	2.13	2233	4470	−0.02
17 May 1983	1.70	1815	4160	0.19
26 Apr 1986	2.12	1566	5140	−0.02
7 Oct 1992	1.57	900	5890	0.05
25 Sep 1993	**3.21**	**1963**	**6930**	**0.03**
6 Nov 1994	**3.96**	**1446**	**9774**	**0.05**
Peaks lowered with respect to $A0$				19%
Peaks increased with respect to $A0$				81%

Alternatives $A7$ and $A18$, which are the other two alternatives characterized by SD$_{\mathrm{CUR}}$, produce similar effects, so they were also accepted.

Now let us consider the alternatives characterized by SD$_{+600}$. Table 14.2 shows the effects of alternative $A34$: 81% of the peaks were increased and in two out of five events in which the peak exceeded 3 m the increments are significant (19 and 16 cm). The effects of the other alternatives with SD$_{+600}$ are similar and so the Stakeholders were opposed to all of them.

It is thus necessary to identify any existing mitigation measure that reduces the negative effects of these alternatives. We describe them for alternative $A34$, but we will see that the measures that are identified can be applied to all the reasonable alternatives characterized by SD$_{+600}$.

14.2 Mitigation measures for Downstream Flooding

To identify measures that mitigate the effects of $A34$ on the *Downstream Flooding* sector, we analysed in detail what happened during the flood event that started on 28th October 1976, which is characterized by an increase of 19 cm in the flood peak level with respect to $A0$ (Table 14.2). Figure 14.2 shows the trajectories of lake inflow (a), lake level, release at Miorina, flow in the Po (q^{Po}) and Ticino level at Pavia, over the whole event. The trajectories of Ticino level at Pavia are drawn only for a part of the event, because the neural network was calibrated, and therefore can be used, only for releases from the Miorina higher than 750 m^3/s (see Section 6.9.1 of the DVD).

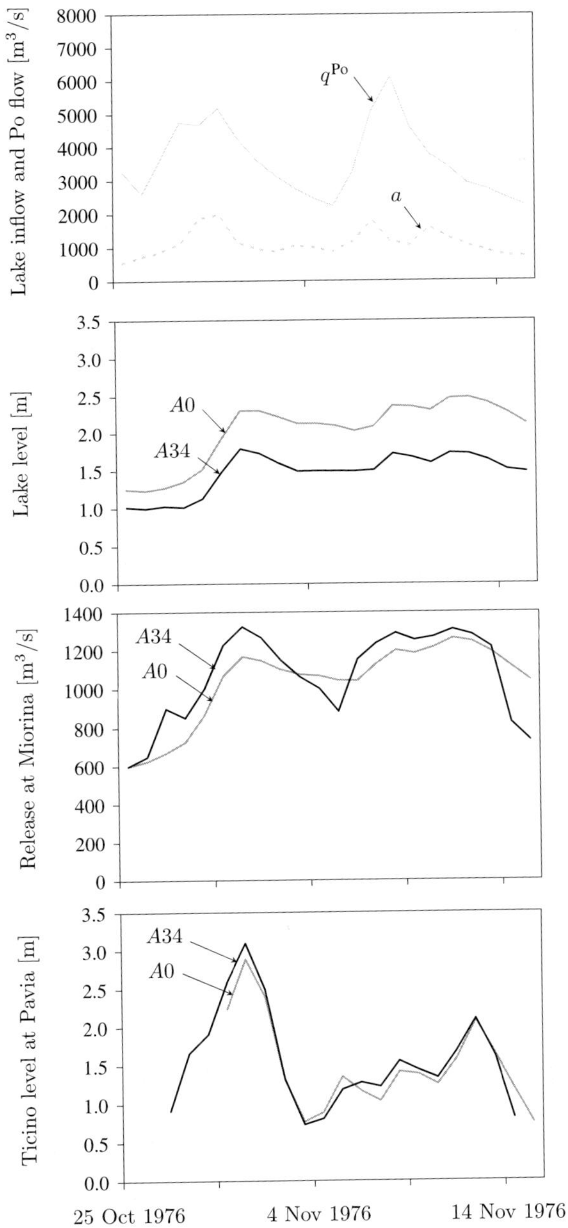

Figure 14.2: The flood event of October 1976: a comparison of alternatives $A34$ and $A0$.

Because of the different policies that the two alternatives adopt, on 25th October, the eve of the flood, the lake is at different levels in the two alternatives: 1.02 m with $A34$, and 1.25 m with $A0$. Since with $A0$ the level is higher than the upper extreme of its regulation range (1.00 m), the lake is in free regime conditions and thus the value of the release is exclusively dictated by SD_{CUR}. On the contrary, with alternative $A34$, on 25th October the regime is still regulated and the passage to free regime takes place only on 30th October. The delay is due to two reasons: first, $\text{RANGE}_{\text{ENL}}$, which is adopted by $A34$, imposes passing to free regime at 1.50 m and, second, SD_{+600} makes the lake level rise more slowly, because, for the same level, it makes it possible to release greater volumes than SD_{CUR}. Despite the fact that the level with $A34$ is always lower than the level with $A0$, from 25th October to 2nd November the release with the first alternative is higher than with the second. This is why $A34$ floods the flood plains first and produces a flood peak level at Pavia that is 19 cm higher than the one produced by $A0$. The higher peak is thus caused by the increase in the release which is allowed by SD_{+600} with respect to SD_{CUR}, and not by the adopted policy which, on the contrary, produces lower lake levels during the event. To lower the levels at Pavia, it is thus necessary to introduce a norm that limits the flow that can be released from the lake even when the upper extreme of the regulation range is exceeded.

14.2.1 The REDUCTION action

As soon as Switzerland proposed excavating the outlet, it was suggested that the intervention would probably increase the flood peak levels at Pavia. To limit this unwanted effect and increase the acceptability of the proposal, Switzerland accompanied it with a willingness to consider a norm that would make it possible to limit the release in free regime conditions. More precisely, the proposal might be that the release be driven by SD_{+600} only until it is lower than a threshold $\bar{r}$ (to be established) and that, once that level has been reached, the dam be progressively closed so as to keep the release equal to $\bar{r}$ until the lake reaches the level at which SD_{CUR} would produce $\bar{r}$. Above this level, the release is once again determined by SD_{CUR} rather than SD_{+600}.

In other words, the idea is that SD_{+600} be used to favour *Upstream Flooding* sector only when this does not bring about an increase in the risk for *Downstream Flooding*. The proposal implicitly assumes that this risk is linked to the release at Miorina alone and that it is zero until the release exceeds the threshold $\bar{r}$. This proposal could thus be the mitigation measure that we are looking for.

The technical feasibility of this measure has never been evaluated. Given the structure of the dam, an accurate study would be needed to ascertain whether the reduction in the release would be technically possible in flood conditions. It is probable that a norm of this type could only be applied if the dam were modified, and so the relevant costs would have to be considered. Nevertheless, before undertaking such a study, it would be a good idea to know whether such a norm would produce the hoped-for advantages for the *Downstream Flooding* sector, without causing significant disadvantages for the *Upstream Flooding* sector or any of the others. Thus, it is interesting to analyse the effects of this mitigation measure, which we will term as $\text{REDUCTION}_{\bar{r}}$.

$\text{REDUCTION}_{\bar{r}}$ is a normative action and so the alternatives that we want to examine are associated to the pair ($\text{SD}_{+600}/\text{RANGE}_{\text{ENL}}$ and $\text{REDUCTION}_{\bar{r}}$). However, this pair is equivalent in all effects to the pair ($\text{SD}_{\bar{r}}/\text{RANGE}_{\text{ENL}}$), where $\text{SD}_{\bar{r}}$ is the following equiva-

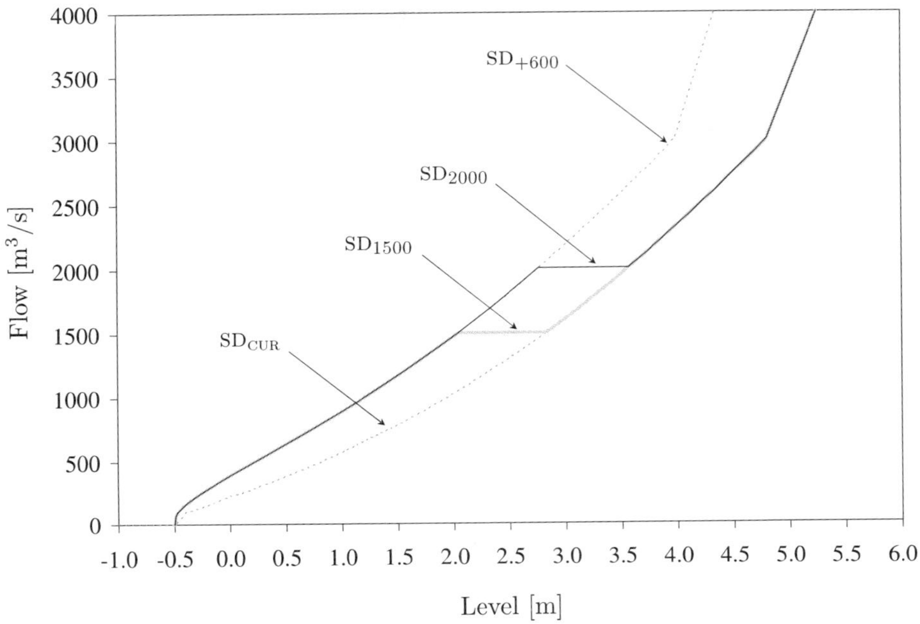

Figure 14.3: The equivalent stage–discharge relations ($\text{SD}_{\bar{r}}$) that result from the realization of the actions SD_{+600} and $\text{REDUCTION}_{\bar{r}}$, for different values of the threshold $\bar{r}$.

lent stage–discharge relation

$$r = \begin{cases} N^{+600}(h^{\text{SC}}) & \text{for } h^{\text{SC}} \leqslant \bar{h}^{+600} \\ \bar{r} & \text{for } \bar{h}^{+600} < h^{\text{SC}} < \bar{h}^{\text{CUR}} \\ N^{\text{CUR}}(h^{\text{SC}}) & \text{for } h^{\text{SC}} \geqslant \bar{h}^{\text{CUR}} \end{cases} \tag{14.1}$$

where h^{SC} and r are the instantaneous level and release, and $\bar{h}^{\text{CUR}}$ and $\bar{h}^{+600}$ denote the levels at which the release equals $\bar{r}$, with the current instantaneous stage–discharge relation $N^{\text{CUR}}(\cdot)$ and with the one ($N^{+600}(\cdot)$) that will be obtained once the action SD_{+600} is implemented, i.e.

$$N^{+600}\left(\bar{h}^{+600}\right) = \bar{r} \qquad N^{\text{CUR}}\left(\bar{h}^{\text{CUR}}\right) = \bar{r}$$

The choice of the threshold $\bar{r}$ is the critical element of the proposal, because, according to the value adopted, both the upstream and the downstream effects could vary noticeably. Thus we decided to carry out two trials, using the following values:

(1) $\bar{r} = 1500 \text{ m}^3/\text{s}$, which corresponds to $\bar{h}^{+600} = 2.04$ m and $\bar{h}^{\text{CUR}} = 2.83$ m;

(2) $\bar{r} = 2000 \text{ m}^3/\text{s}$, which corresponds to $\bar{h}^{+600} = 2.76$ m and $\bar{h}^{\text{CUR}} = 3.56$ m.

The corresponding equivalent relations are shown in Figure 14.3. Since they provide the instantaneous relation between level and outflow, while the model that we adopted describes the water system with a daily time step (see Section 6.3), the equivalent relations must be integrated over 24 hours, as already explained in Section 6.6.5. When examining the figures

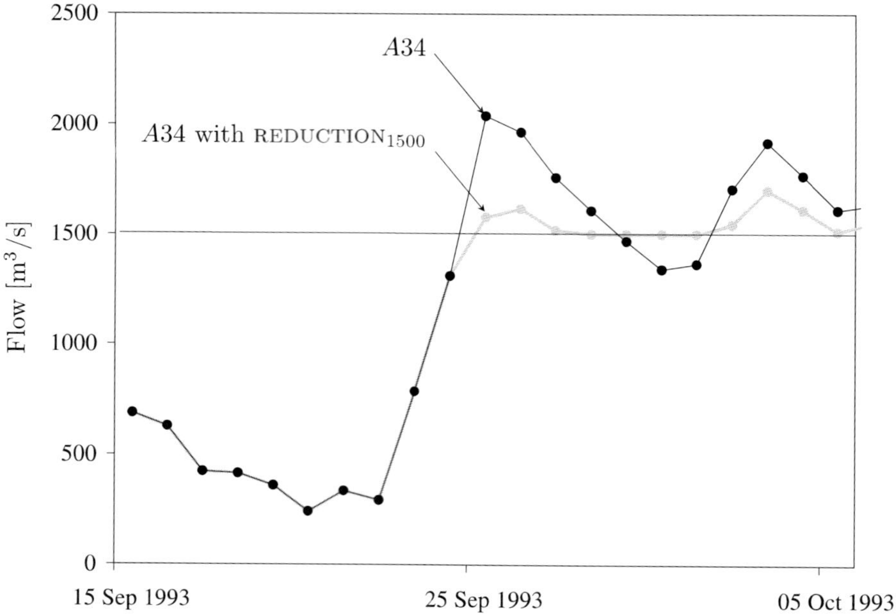

Figure 14.4: Release trajectories produced by $A34$, with and without REDUCTION_{1500}, during the flood event of September 1993.

that are presented in the following, bear in mind that the average daily flow is shown, not the instantaneous flow. It follows that when the trajectories produced by $A34$ are compared with those obtained by applying the mitigation measure REDUCTION_{1500}, no clean break can be seen in correspondence to $r = 1500\ \text{m}^3/\text{s}$, as the reader might expect. Instead there is a gradual separation of the trajectories, as shown in Figure 14.4.

14.2.1.1 Design or simulation?

Passing from $A34$ to $A34$ with $\text{REDUCTION}_{\bar{r}}$ is the same as substituting the structural action SD_{+600} with the action $\text{SD}_{\bar{r}}$. Note, however, that this does not mean that the mitigated alternative would be the following

$$(\text{SD}_{\bar{r}}/\text{RANGE}_{\text{ENL}},\ 13_{\text{CON}},\ p_{34}) \tag{14.2}$$

In fact, as we explained in Section 5.1, if the structural and normative actions are changed, then the regulation policy must also be changed, so that it is still efficient with respect to the new actions. Thus the policy that defines the mitigated alternative is an efficient solution to the Multi-Objective (Optimal) Control Problem whose constraints are defined by the triple

$$(\text{SD}_{\bar{r}}/\text{RANGE}_{\text{ENL}},\ 13_{\text{CON}})$$

This policy must be such to guarantee the same effects as p_{34} for all the sectors, with the exception of *Upstream Flooding*, *Navigation* and *Downstream Flooding*. In fact, this is the condition that makes $\text{REDUCTION}_{\bar{r}}$ a mitigation measure rather than a design action. Identifying this policy requires a careful examination of the Pareto boundary, because it is certainly not defined by the same weights that were used to identify p_{34}. In fact, when the constraints of the Problem vary, the values of the objectives vary as well, so that if the weights (i.e. the ratios between the objectives) are kept the same, all the objective values must vary, not just the one for *Upstream Flooding*. Since we want only that one to vary, the weights must be modified. Searching for the mitigated alternative is thus long and complicated.

To find a shortcut that would allow us to reduce the research time, we observe that the three sectors *Upstream Flooding*, *Navigation* and *Downstream Flooding* are influenced almost exclusively by what happens in free regime conditions, while the other sectors are influenced also, or above all, by what happens in the regulated regime. Therefore, in order that the effects for these sectors change very little, the policy must be practically the same. It follows that the policy that we are looking for will not be very different from policy p_{34} so that, even if we know that the mitigated alternative is not the one defined by (14.2), the effects that the latter induces will not be very different from the effects produced by the alternative we are looking for. Therefore, even if we know that this is not perfectly correct from the theoretical standpoint, in order to shorten the computing time we decided to estimate the effects of the mitigation measure by simulating alternative (14.2). In conclusion we will simulate the two new alternatives

$$A400 = (\text{SD}_{+600}/\text{RANGE}_{\text{ENL}} \text{ and REDUCTION}_{1500}, 13_{\text{CON}}/p_{34})$$

$$A401 = (\text{SD}_{+600}/\text{RANGE}_{\text{ENL}} \text{ and REDUCTION}_{2000}, 13_{\text{CON}}/p_{34})$$

that can be defined more briefly as

$$A400 = (\text{SD}_{1500}/\text{RANGE}_{\text{ENL}}, 13_{\text{CON}}, p_{34})$$

$$A401 = (\text{SD}_{2000}/\text{RANGE}_{\text{ENL}}, 13_{\text{CON}}, p_{34})$$

14.2.1.2 Threshold $\bar{r}$ equal to 1500 m^3/s

Table 14.3 shows the flood peak levels produced by alternative $A400$ and the differences Δh^{A0} and Δh^{A34} between these levels and those produced by $A0$ and $A34$.

It is evident that $A400$ significantly diminishes the peak levels with respect to $A34$ (69% are lowered), but too many of them (56%) are still increased with respect to $A0$. In particular, the peak of 28th October 1976 is still 18 cm higher than the one that occurs with $A0$. The Stakeholders interested in *Downstream Flooding* are thus opposed to alternative $A400$, even though they recognize that it is better than $A34$. From the point of view of the *Upstream Flooding* sector, alternative $A400$ has a decidedly worse performance than alternative $A34$ (see Table 14.4): the maximum values ($i_{\text{Fl\_U\_1\_Loc}}$ and $i_{\text{Fl\_U\_1\_Vb}}$) of flooded area are equal to those produced by $A0$.

The mitigation measure REDUCTION$_{1500}$ thus does not produce the hoped-for results downstream and at the same time worsens the situation upstream: it is thus completely uninteresting.

14.2.1.3 Threshold $\bar{r}$ equal to 2000 m^3/s

The effects of raising the threshold to 2000 m^3/s (alternative $A401$) are shown in Tables 14.5 and 14.6, for the *Downstream Flooding* and *Upstream Flooding* sectors respectively.

Raising the threshold makes the measure less effective for Pavia. The percentage of peaks that are increased with respect to $A0$ goes from 56% (for $A400$) to 75%, which is close to the percentage (81%) produced by $A34$. On the other hand, the maximum flooded

Table 14.3. Flood peak levels (h^{Pv}) at Pavia produced by alternative $A400$ and level differences (Δh^{A0} and Δh^{A34}) with respect to $A0$ and $A34$

Starting date of the event	Alternative $A400$				
	h_t^{Pv} [m]	r_t [m^3/s]	q_{t+1}^{Po} [m^3/s]	Δh^{A0} [m]	Δh^{A34} [m]
4 Oct 1976	1.80	1500	4680	0.04	−0.04
14 Oct 1976	2.21	1487	5080	0.09	−0.01
28 Oct 1976	**3.07**	**1292**	**7220**	**0.18**	**−0.01**
8 Nov 1976	2.10	1279	5780	0.02	0.00
2 May 1977	2.65	1576	6840	−0.24	−0.20
21 May 1977	1.65	897	5360	0.06	0.02
30 Aug 1977	2.35	1500	5620	0.11	−0.10
8 Oct 1977	**3.65**	**1500**	**8370**	**−0.10**	**−0.11**
15 Oct 1979	**2.70**	**1618**	**6300**	**−0.15**	**−0.31**
2 Apr 1981	1.65	983	5160	0.14	0.00
28 Sep 1981	1.89	1741	4580	−0.26	−0.24
17 May 1983	1.46	1500	4160	−0.05	−0.24
26 Apr 1986	2.09	1500	5600	−0.06	−0.03
7 Oct 1992	1.57	900	5890	0.05	0.00
25 Sep 1993	**2.96**	**1616**	**6930**	**−0.22**	**−0.25**
6 Nov 1994	**3.96**	**1446**	**9774**	**0.05**	**0.00**
Peaks lowered with respect to				44%	69%
Peaks increased with respect to				56%	6%

Table 14.4. The performances of alternative $A400$ for the *Upstream Flooding* sector and the differences (Δi^{A0} and Δi^{A34}) with respect to alternatives $A0$ and $A34$

Indicator		Alternative $A400$		
		Value	Δi^{A0}	Δi^{A34}
$i_{\mathrm{Fl\_U\_1\_Loc}}$	[km^2]	2.04	−0.00	0.99
$i_{\mathrm{Fl\_U\_1\_Vb}}$	[km^2]	0.51	−0.01	0.24
$i_{\mathrm{Fl\_U\_2\_Loc}}$	[km^2/year]	0.41	−0.30	0.09
$i_{\mathrm{Fl\_U\_2\_Vb}}$	[km^2/year]	0.05	−0.07	0.03
$i_{\mathrm{Fl\_U\_3\_Loc}}$	[d]	29	−7	3
$i_{\mathrm{Fl\_U\_3\_Vb}}$	[d]	22	−4	11
$i_{\mathrm{Fl\_U\_4\_Loc}}$	[d/year]	5.88	−7.85	0.60
$i_{\mathrm{Fl\_U\_4\_Vb}}$	[d/year]	1.88	−1.66	1.12
$i_{\mathrm{Fl\_U\_5\_Loc}}$	[d]	0	−3	0
$i_{\mathrm{Fl\_U\_5\_Vb}}$	[d]	0	−3	0
$i_{\mathrm{Fl\_U\_6\_Loc}}$	[d/year]	0.00	−0.28	0.00
$i_{\mathrm{Fl\_U\_6\_Vb}}$	[d/year]	0.00	−0.28	0.00
Sector index		Alternative $A400$		
		Value	ΔV^{A0}	ΔV^{A34}
$V_{\mathrm{Fl\_U}}$	[−]	0.37	0.24	−0.33

Table 14.5. Flood peak levels (h^{Pv}) at Pavia produced by alternative $A401$ and level differences (Δh^{A0} and Δh^{A34}) with respect to $A0$ and $A34$

Starting date of the event	Alternative $A401$				
	h_t^{Pv} [m]	r_t [m^3/s]	q_{t+1}^{Po} [m^3/s]	Δh^{A0} [m]	Δh^{A34} [m]
4 Oct 1976	1.82	1527	4680	0.06	−0.01
14 Oct 1976	2.22	1491	5080	0.09	0.00
28 Oct 1976	**3.07**	**1292**	**7220**	**0.18**	**−0.01**
8 Nov 1976	2.10	1279	5780	0.02	0.00
2 May 1977	2.84	1801	6840	−0.05	0.00
21 May 1977	1.63	878	5360	0.04	0.00
30 Aug 1977	2.45	1615	5620	0.21	0.00
8 Oct 1977	**3.75**	**1804**	**8370**	**0.00**	**−0.01**
15 Oct 1979	**3.00**	**1998**	**6300**	**0.15**	**−0.01**
2 Apr 1981	1.65	983	5160	0.14	0.00
28 Sep 1981	2.09	1999	4580	−0.06	−0.04
17 May 1983	1.70	1815	4160	0.19	0.00
26 Apr 1986	2.12	1566	5140	−0.03	0.00
7 Oct 1992	1.57	900	5890	0.05	0.00
25 Sep 1993	**3.21**	**1958**	**6930**	**0.03**	**0.00**
6 Nov 1994	**3.96**	**1446**	**9774**	**0.05**	**0.00**
Peaks lowered with respect to				19%	31%
Peaks increased with respect to				75%	0%

Table 14.6. The performances of alternative $A401$ for the *Upstream Flooding* sector and the differences (Δi^{A0} and Δi^{A34}) with respect to alternatives $A0$ and $A34$

Indicator		Alternative $A401$		
		Value	Δi^{A0}	Δi^{A34}
$i_{Fl\_U\_1\_Loc}$	[km^2]	1.66	−0.39	0.61
$i_{Fl\_U\_1\_Vb}$	[km^2]	0.45	−0.05	0.19
$i_{Fl\_U\_2\_Loc}$	[km^2/year]	0.35	−0.35	0.02
$i_{Fl\_U\_2\_Vb}$	[km^2/year]	0.03	−0.09	0.01
$i_{Fl\_U\_3\_Loc}$	[d]	27	−9	1
$i_{Fl\_U\_3\_Vb}$	[d]	14	−12	3
$i_{Fl\_U\_4\_Loc}$	[d/year]	5.32	−7.60	0.08
$i_{Fl\_U\_4\_Vb}$	[d/year]	0.84	−2.68	0.16
$i_{Fl\_U\_5\_Loc}$	[d]	0	−3	0
$i_{Fl\_U\_5\_Loc}$	[d]	0	−3	0
$i_{Fl\_U\_6\_Loc}$	[d/year]	0.00	−0.28	0.00
$i_{Fl\_U\_6\_Vb}$	[d/year]	0.00	−0.28	0.00
Sector index		**Alternative $A401$**		
		Value	ΔV^{A0}	ΔV^{A34}
V_{Fl_U}	[–]	0.52	0.39	−0.19

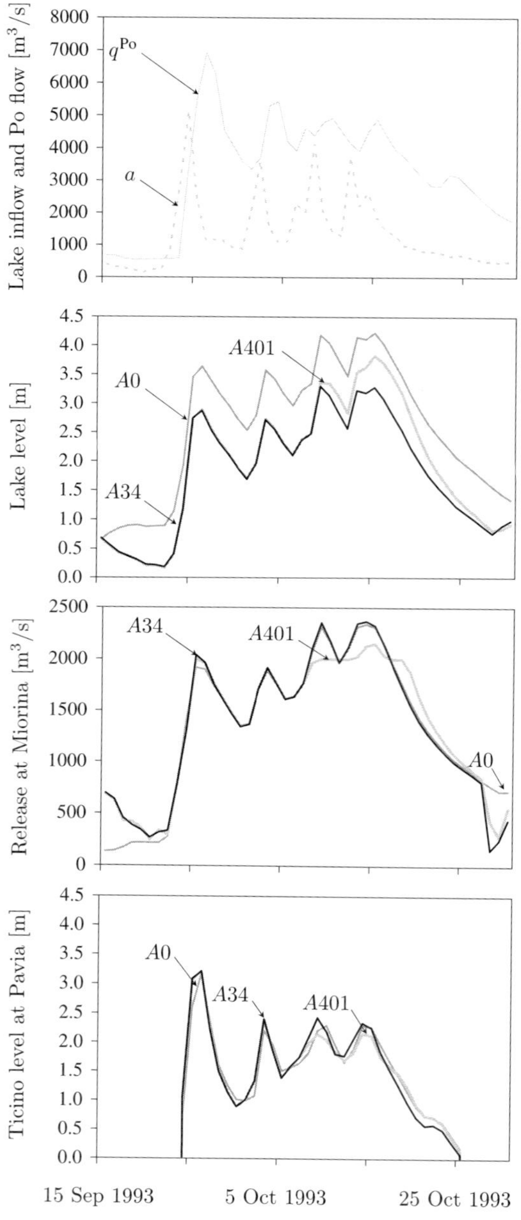

Figure 14.5: The flood event of October 1993: a comparison of alternatives $A34$, $A0$ and $A401$.

areas diminish and the satisfaction for the *Upstream Flooding* sector rises from 0.37 to 0.52. The conflict appears to be evident: an increase in the satisfaction upstream is paid for with an increase in the flood peaks downstream and vice versa. It seems that there is no way out. But is this really the case?

In order to answer this intriguing question, consider Figure 14.5, in which the effects of alternatives $A0$, $A34$ and $A401$ are compared for the flood event of October 1993. The downstream peak level was recorded on 26th September, in correspondence to the flood peak in the River Po. Until that date the release had never exceeded 2000 m^3/s (except for 25th September, when it reached the value of 2039 m^3/s), so that the REDUCTION2000 measure does not produce any effect for *Downstream Flooding*. Its effects become evident from 8th October, when the release at Miorina definitely exceeds 2000 m^3/s in correspondence to the last two inflow peaks. We can see that, with respect to alternative $A34$, $A401$ produces a slight decrease in the last two peak levels at Pavia, accompanied, however, by a significant increase in the lake levels. This is particularly serious because the highest lake level is reached in the last peak, and so this last increase in level is particularly damaging. Conversely, the maximum level at Pavia occurs in correspondence to the first peak and it is the same as with $A34$. It follows that alternative $A401$ produces more damage upstream than $A34$, without any significant advantages downstream. The reason for this *non-sense* is that the trend of the release in the Ticino is not synchronized with the trend of the flow in the Po. The first and highest peak level at Pavia is not associated to the maximum release from the Miorina dam, but to the peak flow in the Po (almost 7000 m^3/s); while the maximum release from the lake, which is clearly synchronous with the maximum lake level, occurs when the flow in the Po is less than 5000 m^3/s and thus the level at Pavia is not critical. The fact that the level at Pavia depends both on the release from the lake and on the backflow from the Po is well known, and we took account of this in the identification of the neural network that describes the propagation of flood waves in the Ticino (Section 6.9.1), but until now we have not considered this fact in the definition of the mitigation measure.

14.2.2 The REDUCTION WITH PO action

The simplest way to take account of the Po is to modify the value of the threshold $\bar{r}$ that appears in equation (14.1), according to the forecasted average daily flow $\hat{q}_{t+1}^{\text{Po}}$ in the Po, i.e. the forecast for the interval $[t, t+1)$. The threshold $\bar{r}$ is thus time-variant and takes the form

$$\bar{r}_t = \bar{r}\left(\hat{q}_{t+1}^{\text{Po}}\right) \tag{14.3}$$

The choice of the function $\bar{r}(\cdot)$ is the element that distinguishes the new mitigation measure.

The correct development of this idea requires the definition of the function $\bar{r}(\cdot)$ and the identification of a model which provides the forecast $\hat{q}_{t+1}^{\text{Po}}$ of the flow in the Po as a function of rainfall and previous flows. ARX models and neural networks are suitable for this purpose. Nevertheless, as we have already seen, before we set about defining function $\bar{r}(\cdot)$ and identifying the model, it is advisable to make sure that a measure of this type might have a reasonable probability of achieving the mitigation objective.

Therefore, as a first attempt, the function $\bar{r}(\cdot)$ should be fixed only on the basis of 'good engineering sense'. If we obtain promising results, it will be worthwhile to devise a procedure to identify the function $\bar{r}(\cdot)$ that provides the best performances; if we do not, we can heuristically state that taking the Po into account in the regulation of Verbano is probably useless.

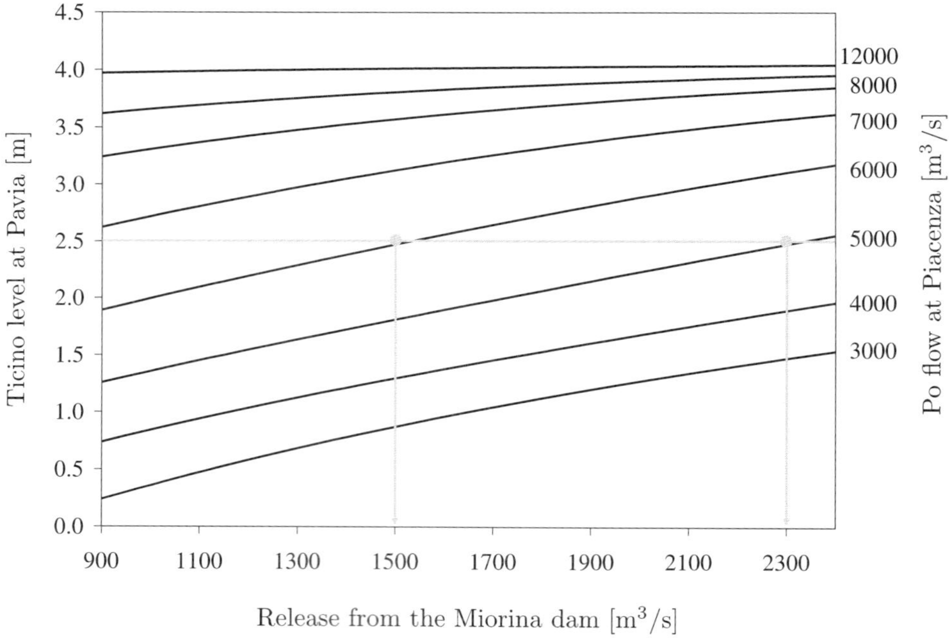

Figure 14.6: Abacus that relates the release at Miorina and the Po flow at Piacenza to the Ticino level at Pavia. The abacus was derived from the neural network described in Section 6.9.1.

For the same reason, before developing a predictor of the flow in the Po, it is opportune to try with a simple *persistent predictor* (which assumes $\hat{q}^{\mathrm{Po}}_{t+1} = q^{\mathrm{Po}}_t$) and with an *ideal predictor* ($\hat{q}^{\mathrm{Po}}_{t+1} = q^{\mathrm{Po}}_{t+1}$). The latter cannot actually be used but, given that we are working with historical data, it can be easily simulated. If the persistent predictor provides interesting results, we may hope that even better results would be provided by any other good predictor and we know for sure that they will not be worst; on the other hand, we know that no predictor could do better than the ideal predictor. The two test cases thus allow us to evaluate the extent to which the result depends on the choice of the predictor and how much attention should be devoted to this.

Now let us go back to the definition of the function $\bar{r}(\cdot)$. Given the forecast $\hat{q}^{\mathrm{Po}}_{t+1}$, the most reasonable choice is to assume that $\bar{r}_t$ is equal to a flow rate value which, according to the neural network, produces a level at Pavia of 2.50 m: this value guarantees a margin of 1.00 m with respect to the flooding level at Borgo Ticino, which is certainly sufficient to absorb the estimation errors produced by the network. With this hypothesis, the pairs $(\bar{r}_t, \hat{q}^{\mathrm{Po}}_{t+1})$ are derived directly from the neural network, which is visualized in the form of an abacus[1] in Figure 14.6. By discretizing the Po flow into classes of 1000 m^3/s and approximating the function $\bar{r}(\cdot)$ as a piecewise constant function, the mitigation measure, which we will call REDUCTION WITH PO, is defined by the following equivalent stage–discharge relation:

(1) if $\hat{q}^{\mathrm{Po}}_{t+1} < 5000\ \mathrm{m}^3/\mathrm{s}$ the stage–discharge relation is SD_{+600};

[1]This is the same abacus that we encountered on page 274.

(2) if $\hat{q}_{t+1}^{\text{Po}} \in [5000, 6000)$ the stage–discharge relation is given by equation (14.1) with $\bar{r} = 2300 \text{ m}^3/\text{s}$;

(3) if $\hat{q}_{t+1}^{\text{Po}} \in [6000, 7000)$ the stage–discharge relation is given by equation (14.1) with $\bar{r} = 1500 \text{ m}^3/\text{s}$;

(4) if $\hat{q}_{t+1}^{\text{Po}} > 7000 \text{ m}^3/\text{s}$ the stage–discharge relation is SD_{CUR}.

We can associate the REDUCTION WITH PO measure with both the ideal predictor (REDUCTION WITH PO-IDE) and the persistent predictor (REDUCTION WITH PO-PER), thus obtaining the following two alternatives

$$A402 = (\text{SD}_{+600}/\text{RANGE}_{\text{ENL}} \text{ and REDUCTION WITH PO-IDE}, 13_{\text{CON}}, p_{34})$$

$$A403 = (\text{SD}_{+600}/\text{RANGE}_{\text{ENL}} \text{ and REDUCTION WITH PO-PER}, 13_{\text{CON}}, p_{34})$$

It is opportune to remember that the first cannot actually be realized, because an ideal predictor does not exist. The second, on the contrary, can be realized because the use of a persistent predictor is equivalent to assuming that the reduction is decided on the basis of the Po flow on the previous day.

14.2.2.1 The ideal predictor

Let us examine the behaviour of $A402$ during the flood event of October 1993 (Figure 14.7). In correspondence to the first flood peak level of the lake, the flow in the Po exceeds $6000 \text{ m}^3/\text{s}$ and REDUCTION WITH PO-IDE intervenes by limiting the release at Miorina to $1500 \text{ m}^3/\text{s}$. It follows that the first flood peak level at Pavia is lowered, not only with respect to $A34$, but also to $A0$. After October 5th, REDUCTION WITH PO-IDE is fairly inactive in practice, so the level and flow trajectories tend to be similar to those produced by alternative $A34$. Since the effects on *Upstream Flooding* are essentially due to the last two peaks, alternative $A402$ produces the same performances as $A34$ in this sector, thus obtaining the brilliant result of reducing the peak downstream without increasing the peak upstream. We can conclude that for the 1993 event the REDUCTION WITH PO-IDE action produces the results that we hoped for. This result is not fortuitous, it occurs for other events as well, as shown in Tables 14.7 and 14.8. From the first we deduce that the performance of alternative $A402$ is worse than $A0$ only in events in which the peak level at Pavia is low (less than 2.50 m). Among the most relevant events, the only peak not lowered is the one beginning on 28th October 1976. From the second table we see that the effects on the *Upstream Flooding* sector are very similar to those produced by $A34$.

14.2.2.2 Persistent predictor

Unfortunately alternative $A402$ uses the ideal predictor and so in practice it cannot be realized. Now we must check if these results, which are so satisfying, are attainable with a real predictor. To do so, let us analyse alternative $A403$, which uses the persistent predictor. In Table 14.9 the effects that it has on the peak levels at Pavia are shown: they appear to be identical to those produced by alternative $A34$. Thus it must be concluded that with the persistent predictor all the advantages gained by the ideal predictor are lost. But the persistent predictor is the simplest predictor that can be imagined; more complex predictors will probably provide performances included between those produced by $A401$ and those produced by $A402$.

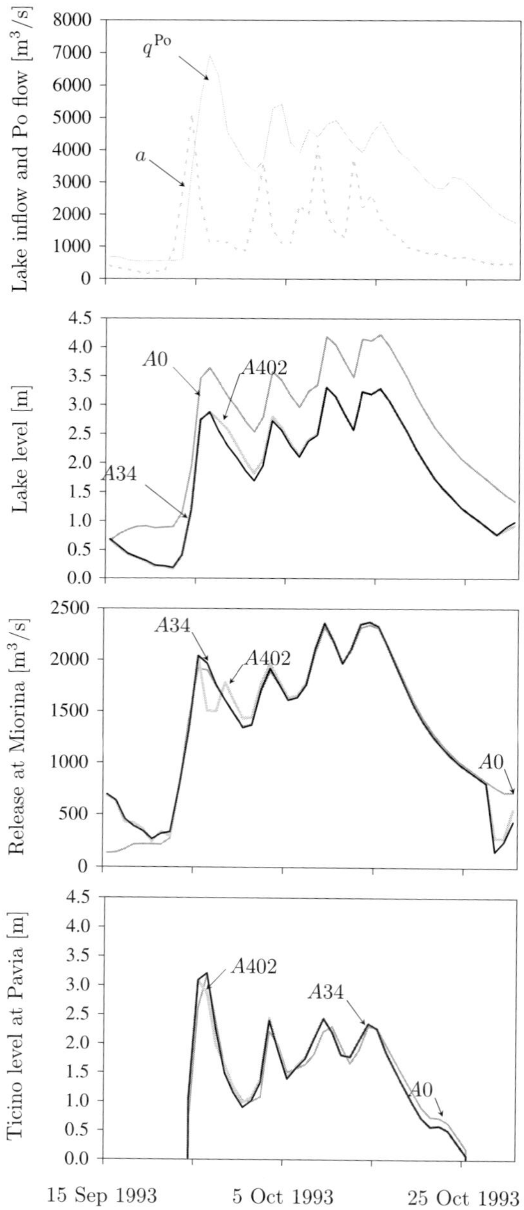

Figure 14.7: The flood event of October 1993: a comparison of alternatives $A34$, $A0$ and $A402$.

Table 14.7. Flood peak levels (h^{Pv}) at Pavia produced by alternative $A402$ and the differences (Δh^{A0}, Δh^{A34} and Δh^{A400}) with respect to alternatives $A0$, $A34$ and $A400$

Starting date of the event	Alternative $A402$					
	h_t^{Pv} [m]	r_t [m³/s]	q_{t+1}^{Po} [m³/s]	Δh^{A0} [m]	Δh^{A34} [m]	Δh^{A400} [m]
4 Oct 1976	1.82	1527	4680	0.06	−0.01	0.02
14 Oct 1976	2.22	1491	5080	0.09	0.00	0.00
28 Oct 1976	**3.07**	**1292**	**7220**	**0.18**	**−0.01**	**0.00**
8 Nov 1976	2.10	1279	5780	0.02	0.00	0.00
2 May 1977	2.82	1955	5240	−0.07	−0.02	0.17
21 May 1977	1.63	881	5360	0.04	0.00	−0.02
30 Aug 1977	2.45	1615	5620	0.21	0.00	0.10
8 Oct 1977	**3.61**	**1406**	**8370**	**−0.13**	**−0.14**	**−0.04**
15 Oct 1979	2.80	1872	5700	−0.05	−0.21	0.10
2 Apr 1981	1.65	983	5160	0.14	0.00	0.00
28 Sep 1981	2.13	2233	4470	−0.02	0.00	0.24
17 May 1983	1.70	1815	4160	0.19	0.00	0.24
26 Apr 1986	2.12	1566	5140	−0.03	0.00	0.03
7 Oct 1992	1.57	900	5890	0.05	0.00	0.00
25 Sep 1993	**3.09**	**2038**	**5620**	**−0.09**	**−0.12**	**0.13**
6 Nov 1994	**3.91**	**1045**	**9774**	**0.00**	**−0.05**	**−0.05**
Peaks lowered with respect to				38%	44%	19%
Peaks increased with respect to				56%	0%	44%

Table 14.8. The performances of alternative $A402$ in the *Upstream Flooding* sector and the differences (Δi^{A0}, Δi^{A34} and Δi^{A400}) with respect to alternatives $A0$, $A34$ and $A400$

Indicator		Alternative $A402$			
		Value	Δi^{A0}	Δi^{A34}	Δi^{A400}
$i_{Fl\_U\_1\_Loc}$	[km²]	1.06	−1.00	0.01	−1.00
$i_{Fl\_U\_1\_Vb}$	[km²]	0.27	−0.23	0.00	−0.23
$i_{Fl\_U\_2\_Loc}$	[km²/year]	0.33	−0.37	0.00	−0.09
$i_{Fl\_U\_2\_Vb}$	[km²/year]	0.03	−0.09	0.01	−0.03
$i_{Fl\_U\_3\_Loc}$	[d]	27	−9	1	−2
$i_{Fl\_U\_3\_Vb}$	[d]	12	−14	1	−10
$i_{Fl\_U\_4\_Loc}$	[d/year]	5.40	−7.52	0.16	−0.44
$i_{Fl\_U\_4\_Vb}$	[d/year]	0.84	−2.68	0.16	−0.96
$i_{Fl\_U\_5\_Loc}$	[d]	0	−3	0	0
$i_{Fl\_U\_5\_Loc}$	[d]	0	−3	0	0
$i_{Fl\_U\_6\_Loc}$	[d/year]	0.00	−0.28	0.00	0.00
$i_{Fl\_U\_6\_Vb}$	[d/year]	0.00	−0.28	0.00	0.00
Sector index		Alternative $A402$			
		Value	ΔV^{A0}	ΔV^{A34}	ΔV^{A400}
V_{Fl_U}	[–]	0.67	0.55	−0.03	0.30

Table 14.9. Flood peak levels (h^{Pv}) at Pavia produced by alternative $A403$ and the differences (Δh^{A0}, Δh^{A34} and Δh^{A400}) with respect to alternatives $A0$, $A34$ and $A400$

Starting date of the event	Alternative $A403$					
	h_t^{Pv} [m]	r_t [m^3/s]	q_{t+1}^{Po} [m^3/s]	Δh^{A0} [m]	Δh^{A34} [m]	Δh^{A400} [m]
4 Oct 1976	1.82	1527	4680	0.06	−0.01	0.00
14 Oct 1976	2.22	1491	5080	0.09	0.00	0.00
28 Oct 1976	**3.07**	**1292**	**7220**	**0.18**	**−0.01**	**0.00**
8 Nov 1976	2.10	1279	5780	0.02	0.00	0.00
2 May 1977	2.84	1802	6840	−0.05	0.00	0.02
21 May 1977	1.63	881	5360	0.04	0.00	0.00
30 Aug 1977	2.46	1623	5620	0.22	0.01	0.01
8 Oct 1977	**3.75**	**1804**	**8370**	**0.01**	**0.00**	**0.14**
15 Oct 1979	**3.01**	**2009**	**6300**	**0.16**	**0.00**	**0.21**
2 Apr 1981	1.58	912	5160	0.07	−0.07	−0.07
28 Sep 1981	2.13	2233	4470	−0.02	0.00	0.00
17 May 1983	1.71	1820	4160	0.20	0.01	0.01
26 Apr 1986	2.12	1566	5140	−0.03	0.00	0.00
7 Oct 1992	1.57	900	5890	0.05	0.00	0.00
25 Sep 1993	**3.21**	**1964**	**6930**	**0.03**	**0.00**	**0.12**
6 Nov 1994	**3.91**	**1045**	**9774**	**0.00**	**−0.05**	**0.00**
Peaks lowered with respect to				19%	25%	6%
Peaks increased with respect to				75%	3%	38%

14.2.2.3 The flood of October 2000

To check the real validity of the REDUCTION WITH PO mitigation measure it would be necessary to design a real predictor of the flow in the Po. Unfortunately this was not possible owing to the time and the budget constraints posed on Project Verbano.

To test the validity of the REDUCTION WITH PO mitigation measure in another way, we decided to evaluate its effects also on the flood of October 2000. In that event, in fact, the Ticino exceeded the level +3.50 m at the Covered Bridge hydrometric station and the city of Pavia was flooded, which had happened only two times over the evaluation horizon (1974–1998).

Furthermore, by simulating the 2000 event the policy is validated, since that time series was not used in the design. The validation is significant because the event is exceptional: it has a return time of about 70 years and produced greater damage than any of the other events considered until now (in Switzerland alone the damage amounted to about 110 million euro). The event took place between 10th and 20th October 2000 and in that period the Po reached a peak flow of 8500 m^3/s (Cattaneo et al., 2000). The results of the simulation are shown in Tables 14.10, 14.11 and 14.12. As the reader may notice, in Table 14.12 the historical levels at the Covered Bridge are not included, only those simulated with alternative $A0$ (which assumes the historical releases at Miorina) are shown; the reason was explained on page 317.

Let us analyse these results. The flow rate in the Po stayed below 5000 m^3/s until 15th October (Table 14.10) so that, up to that date, the decisions taken with alternative $A403$ do not differ from those taken with alternative $A34$: the trajectories of both lake level and release are identical. Note that for both alternatives the lake level at the beginning of the

Table 14.10. Releases from the Miorina dam during the flood of October 2000 with alternatives $A0$, $A403$ and $A34$, and the contemporary flows in the Po

Date	Release [m^3/s]			Flow in the Po
	$A0$	$A403$	$A34$	[m^3/s]
13 Oct 2000	773	775	775	720
14 Oct 2000	1280	1642	1642	1232
15 Oct 2000	1821	2283	2283	2860
16 Oct 2000	2472	2300	2711	5021
17 Oct 2000	2632	2204	2649	8503
18 Oct 2000	2447	2068	2373	7237
19 Oct 2000	2206	2405	2090	4456

Table 14.11. Lake levels during the flood of October 2000 with alternatives $A0$, $A403$ and $A34$

Date	Lake level [m]		
	$A0$	$A403$	$A34$
13 Oct 2000	1.45	0.68	0.68
14 Oct 2000	2.51	1.83	1.83
15 Oct 2000	3.40	2.67	2.67
16 Oct 2000	4.36	3.57	3.57
17 Oct 2000	4.59	3.90	3.73
18 Oct 2000	4.33	3.78	3.43
19 Oct 2000	3.98	3.54	3.06

Table 14.12. Ticino levels at the Covered Bridge (Pavia) during the flood of October 2000 with alternatives $A0$, $A403$ and $A34$

Date	Level at Pavia [m]		
	$A0$	$A403$	$A34$
15 Oct 2000	0.01	0.36	0.36
16 Oct 2000	1.16	1.44	1.44
17 Oct 2000	3.07	3.00	3.16
18 Oct 2000	3.86	3.76	3.86
19 Oct 2000	3.08	2.78	3.03
20 Oct 2000	1.73	1.89	1.64

flood event is lower than it is with $A0$ (Table 14.11), since they both adopt a policy that suggests leaving a margin for flood detention. On the morning of 16th October the flow in the Po exceeds 5000 m^3/s and consequently the normative action shows its effects: for $A34$ the average release in the following 24 hours is 2711 m^3/s, while for $A403$ it is only 2300 m^3/s (Table 14.10). Consequently, the next day the lake level with $A403$ is higher than with $A34$. On 17th and 18th October the flow rate of the Po is higher than 7000 m^3/s and so the release with $A403$ is determined by SD_{CUR}; it follows that the divergence between the levels produced by $A403$ and those produced by $A34$ increases. Nonetheless, when the lake level reaches its peak on 17th October, alternative $A403$ shows a level which is still 69 cm lower than the historical one (see $A0$). On the following days, the flow rate in the Po drops below 5000 m^3/s and decisions are taken with reference to SD_{+600} once again, both for alternatives $A403$ and $A34$. Thereby the deviations between the level and release trajectories produced by the two alternatives progressively diminish.

From 17th to 19th October, as a consequence of the reduction in the release, the levels at the Covered Bridge at Pavia produced by $A403$ are lower than those produced by $A34$ and by $A0$ (Table 14.12) and the peak level is lowered by 10 cm. The reduction in the release is significant, both with respect to $A34$ (17%) and with respect to $A0$ (13%), but it has only a slight effect on the peak level at Pavia because the latter is strongly influenced by the backflow from the Po. This is deduced by observing that the peak level occurs on 18th October with all three alternatives, despite the fact that they produce the peak release on different days (Table 14.10).

From the above discussion, it may be concluded that even in the October 2000 event the REDUCTION WITH PO action would have proved to be effective, and so it is interesting to ask whether it could be designed in a rational way and not just in an intuitive way.

14.2.3 Designing mitigation measures

The measures that we have examined so far for mitigating the effects on *Downstream Flooding* have been defined by trial and error, by analysing actions that are intuitively reasonable. This is the limit of the approach that has been followed so far; it would be more rational to design the mitigation measure by modifying the (Optimal) Control Problem in the following way:

1. A fifth objective that expresses the damage for the *Downstream Flooding* sector is added.

2. The normative action

$$r = N^{+600}\left(h^{\text{SC}}\right) \quad \text{for } h^{\text{SC}} > 1.50 \text{ m} \tag{14.4}$$

which orders that the dam be completely opened and the release be determined by SD_{+600} when the lake level is higher than the regulation range, is replaced with a normative action that allows for lower releases. For example, the following norm could be used

$$N^{\text{CUR}}\left(h^{\text{SC}}\right) \leqslant r \leqslant N^{+600}\left(h^{\text{SC}}\right) \quad \text{for } h^{\text{SC}} > 1.50 \text{ m} \tag{14.5}$$

which allows the regulation of the lake in the space between SD_{+600} and SD_{CUR} when the upper extreme of the regulation range is exceeded.[2] More precisely, while

[2] Note that (14.4) and (14.5) are defined with respect to the instantaneous level and flow.

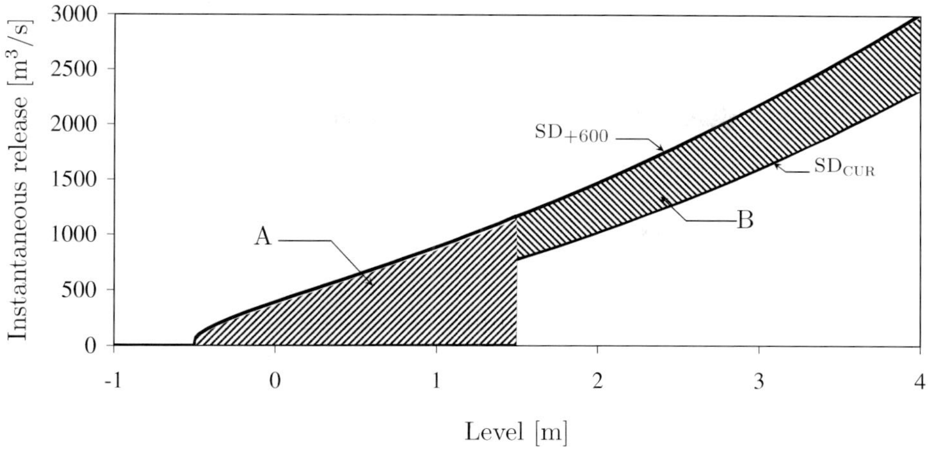

Figure 14.8: With the normative action RANGE$_{\text{ENL}}$ the control laws that define a policy must be contained in the set A. With the normative action proposed in this section they must be contained in the set A + B.

the controls suggested by the policies designed until this point must be contained in set A in Figure 14.8, the new normative action makes it possible for them to be contained in the set A + B. Furthermore, it is necessary that the control laws in set B be *point-valued*, for the reasons that will be explained later.

3. The average flow q_t^{Po} in the Po on the previous day is assumed to be known at time t, and possibly the forecast $\hat{q}_{t+1}^{\text{Po}}$ as well. By doing so, the policy will no longer be a function of the lake storage s_t alone but of the pair (s_t, q_t^{Po}) or the triple $(s_t, q_t^{\text{Po}}, \hat{q}_{t+1}^{\text{Po}})$.

The third point makes it difficult to solve the Control Problem with the Stochastic Dynamic Programming (SDP) approach that has been used until now. In fact, in order to use SDP, all the exogenous variables must be *white noises*, namely sequences of random variables that are all uncorrelated with each other. If this condition is not satisfied, SDP cannot manage the exogenous variable in the recursive Bellman equation. Nevertheless, assuming that the flow q_t^{Po} is generated by a white process means losing all the information that it provides. Alternatively, in order to use SDP, the dynamics of q_t^{Po} could be described with a model, but by doing so the computing time required for the design of each policy would be significantly increased: as we have seen, the computing time increases exponentially with the dimension of the state. Furthermore, the model adopted to describe q_t^{Po} should not contain any exogenous variables; otherwise we would face again the same difficulty we are trying to solve by adopting a model for the Po flow. However, assuming that the flow in the Po does not depend either on the precipitation in its catchment or on the flows recorded further upstream is an unacceptable hypothesis.

The difficulty can be overcome by abandoning SDP and using very recent algorithms based on Reinforcement Learning (Barto and Sutton, 1998; Kaelbling et al., 1996), which are able to manage *model-free* Problems, namely Problems formulated without modelling the system. In particular, the *Q-learning planning* algorithm (Castelletti et al., 2001; Section 13.4 of THEORY) might be quite suitable, as it makes it possible to deal with the process of the flow formation in the Po with a *model-free* approach, while the rest of the system is still described by the same model that we have used until now.

Aside from this technical difficulty, this design framework may also encounter some difficulties with social acceptability. To the lakeside population it might seem

(a) that giving permission to regulate the release beyond the regulation range might introduce too much discretion into the Regulator's behaviour;

(b) that not releasing the maximum amount possible, as both the normative actions RANGE$_{CUR}$ and RANGE$_{ENL}$ order, will be to their detriment.

To reply to point (a), we note that it is certainly true that the regulation range was introduced with the aim of avoiding the possibility that, during a flood event, the Regulator would receive contradictory orders from different authorities (for example, the Prefect of Verbania might order him to release the maximum amount possible, while the one of Pavia might order him to reduce the release); it is not true, however, that norm (14.5) would open the Regulator's decision to this kind of conflict. It authorizes a search for a policy which in flood conditions could adopt controls included in the set B in Figure 14.8, but it requires also that, in that set, the policy be point-valued. It follows that, once the policy has been defined in the set B, it is none other than a lookup table which, for each pair (s_t, q_t^{Po}) or for every triple $(s_t, q_t^{\text{Po}}, \hat{q}_{t+1}^{\text{Po}})$, provides the decision u_t^{MIO} that the Regulator *must* adhere to. If this table is made an integral part of the Regulation Licence, the certainty of the Regulator's behaviour is guaranteed.

Regarding point (b), one must observe that reducing the release is the condition that makes the excavation of the lake outlet acceptable to the Stakeholders interested in *Downstream Flooding* and that excavating can produce a significant reduction in lake floods, even if at times the release might not be the maximum possible.

The solution to the new (Optimal) Control Problem is completely reopened (as we have already underlined, the decision-making process is full of recursions!). Given the time and the costs that this interaction requires, we decided that it was not advisable to take it on before the final Decision Makers (DMs) had shown interest in this possibility. In fact, they are able to evaluate it, since the alternatives $A402$ and $A403$ provide indications about what the achievable results might be.

On the basis of these results, the Stakeholders interested in *Downstream Flooding* had already declared that they would accept alternatives that include normative actions such as REDUCTION WITH PO, while they would be opposed to all the reasonable alternatives based on SD$_{+600}$ alone. Finally, they would accept alternatives $A0$, $A7$, $A18$ and $A119$.

14.3 Mitigation for Upstream Environment

The reasonable alternative that is most favourable for the *Upstream Environment* sector is $A18$, whose index scores 0.44; the worst is $A0$ (0.26), which is only just bettered by $A34$ (0.27). The difference between the first and these last two is so significant that the representative of the interested Stakeholders, the Bolle di Magadino Foundation, is opposed to $A0$ and $A34$. This means that in order to broaden the agreement it is necessary to identify mitigation measures for this sector. We will describe the search for these measures with reference to alternative $A34$, which is the one that receives the widest agreement at the moment. However, the analysis can be easily repeated for all the other alternatives to which the Foundation is opposed.

Table 14.13. The effects of alternatives *A*34, *A*36 and *A*0 on the *Upstream Environment* sector

Indicator			*A*34	*A*36	*A*0
$i_{\text{Env\_U\_1}}$	Distance from natural regime	[m]	0.62	0.50	0.59
$i_{\text{Env\_U\_2}}$	Average annual number of days in which the lake level is in the range of reed bed erosion	[d/year]	19.	41.	13.
$i_{\text{Env\_U\_3}}$	Fraction of the egg-laying period in which the Cyprinids cannot access the reed beds	[%]	22.	21.	10.
$i_{\text{Env\_U\_4}}$	Fraction of the nesting period in which the lake level is higher than the threshold of the nests	[%]	72.	51.	84.
$i_{\text{Env\_U\_5}}$	Fraction of the normal shelf emergence period in which the lake level is higher than the emergence threshold	[%]	54.	54.	62.

The indicators for *Upstream Environment* were defined with explicit reference to the Bolle di Magadino Reserve (Section 4.5.1 of the DVD). There are two reasons for this: the first is that the Reserve is by far the most prestigious naturalistic area along the shores of the lake, for its size, the variety of biotypes involved and the importance of the migratory bird life. The second reason is that the area is representative of other areas, such as the Fondo Toce area near Verbania and the small areas of reed bed that are still present here and there along the shores of Verbano. Since the lake is actually a single environmental system and it is almost in equilibrium, favourable conditions for the Bolle di Magadino Reserve are also favourable for the other areas, and vice versa. The satisfaction for *Upstream Flooding* was thus measured in relation to the Reserve, but it is not confined to that area alone.

The mitigation actions cannot be designed with reference to all these areas at the same time: we must work individually with each one of them. In fact, even if the effects produced by an alternative are the same in all the areas, the possibilities for mitigation depend on the characteristics of each one. Given that the Reserve is the area with the highest environmental value, we will describe the mitigation in relation to it.

Table 14.13 shows the values of the indicators for the *Upstream Environment* sector produced by alternatives *A*34, *A*36 (which is one of the alternatives that is supported by the Bolle di Magadino Foundation) and *A*0. With respect to *A*36, alternative *A*34 increases the value of indicator $i_{\text{Env\_U\_4}}$ by 42% (from 51 to 72), so that the corresponding partial value function (Figure 11.11) shifts from 0.54 to 0. Given the high weight (0.29) that this partial value has in the definition of the global value function, which is used to compute the sector index, a sharp decrease in the index value follows.

On the other hand, the average annual number of days in which the lake level is in the range of reed bed erosion ($i_{\text{Env\_U\_2}}$) decreases by 54%. In this case, however, both because of the form of the partial value function and because of the low weight associated to it, the increase in the sector index is very small. The main difference between the two alternatives is the increase in the distance from the natural regime ($i_{\text{Env\_U\_1}}$), as shown by the increase in the lake level medians (Figure 14.9).

These effects, which are worrying in the short-term, would not be so in the long-term if the shores of the lake were not inhabited by human populations: if they were not, with the passage of time not only the edge of the reed bed on the lakeside would withdraw, but also the edge on the land side. Thus there would be a transfer of all the reed beds and the ecosystem that it supports towards the interior. In current conditions this readjustment is prevented by the presence of cultivated land in the interior (Figure 14.10). An interesting

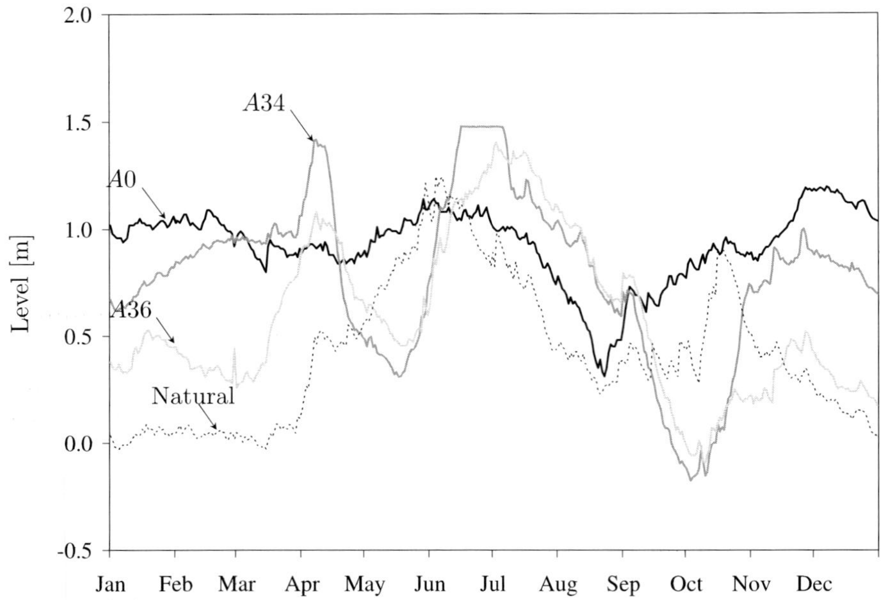

Figure 14.9: Trajectories of the lake level median produced by alternatives $A34$, $A36$, $A0$ and in the natural regime.

Figure 14.10: The Bolle di Magadino Reserve: the area that is currently occupied by the reed beds is highlighted in light grey. Zone A is completely protected while zones B and C are Respected Areas which today are partially occupied by agricultural land (source: Bolle di Magadino Foundation).

mitigation measure might be to incorporate this land into the Reserve and let it return to the natural state. An experiment of this type had already been conducted by the Bolle di Magadino Foundation, which purchased some agricultural land about ten years ago and then left it uncultivated. As the years passed the reed bed, which was free to expand into the area, began to show the characteristic compactness of those areas that traditionally belong to the wetland and the colonization process took place spontaneously and rapidly.

More precisely, the mitigation measure, which we will call BOLLE RESERVE measure, could be subdivided in the following interventions:

- the creation of shelves at higher levels than the current ones;
- acquisition of the 40 to 60 hectares of land that is currently used for agriculture behind the marsh area, in zones B and C in Figure 14.10;
- reactivation of the delta dynamics, by removing of a part of the Ticino banks at its mouth, so that the river would again influence the lands acquired.

A broad estimate of the overall cost of these interventions indicates approximately 10 to 15 million euros. Clearly, to define thoroughly the action the financing body would have to be identified: it could either be an external body, for example the Swiss government, or a body that would be benefitted by the implementation of alternative $A34$: the insurance companies, for example, who would have much fewer claims for flooding damages if that alternative were carried out.

Before dealing with these aspects, however, it is advisable to assess whether the measure effectively brings about an increase in the *Upstream Environment* sector index. If the conclusion is positive one can look for similar measures for the other areas of the lake that have environmental importance.

14.3.1 The effects of BOLLE RESERVE measure

The task is now to define how to evaluate the effects that would be obtained from the realization of the BOLLE RESERVE measure. We will only examine the steady-state effects. If the measure proves to be interesting, its effects in the transient period would have to be studied as well, because if this were too fast it might provoke serious damage, given that the ecosystem would not have time to adapt to the new condition. Such a detailed analysis exceeds the limits of the Verbano Project, however.

Figure 14.9 shows that, following the implementation of alternative $A34$, the lake level at steady-state would often be higher than it is today. Consequently, the lakeside edge of the reed beds would withdraw to a higher altitude. The average rise in the reed bed would change the levels at which the bird life can nest. Therefore, one of the effects would be a variation in the threshold values that appear in the indicators of *Upstream Environment* sector (see Section 4.5.1 of the DVD). After a careful study of the lake level trajectory over the entire evaluation horizon, Mr. Patocchi, scientific director of the Bolle di Magadino Foundation, proposed that the indicators be modified in the following way:

- $i_{\text{Env\_U\_1}}$: since this indicator measures the distance from the natural regime, it should not be modified.
- $i_{\text{Env\_U\_2}}$: the erosion of the roots of the reed bed, which currently provokes its withdrawal, would no longer be a significant problem, because the reed bed is allowed to

Table 14.14. The effects of alternatives $A34$ and $A505$ on the *Upstream Environment* sector

Indicator			$A34$ Upstream Environment		$A505$ Bolle di Magadino	
			i	v	i	v
$i_{Env\_U\_1}$	Distance from natural regime	[m]	0.62	0.42	0.70	0.23
$i_{Env\_U\_2}$	Average annual number of days in which the lake level is in the range of reed bed erosion	[d/year]	18.67	1.00	0.00	1.00
$i_{Env\_U\_3}$	Fraction of the egg-laying period in which the Cyprinids cannot access the reed beds	[%]	54.22	0.09	2.11	0.90
$i_{Env\_U\_4}$	Fraction of the nesting period in which the lake level is higher than the threshold of the nests	[%]	71.64	0.00	0.00	1.00
$i_{Env\_U\_5}$	Fraction of the normal shelf emergence period in which the lake level is higher than the emergence threshold	[%]	22.09	0.95	20.68	0.97
Sector index			–	0.27	–	0.85

withdraw. The partial value of this indicator in the mitigated alternative is thus equal to 1.

- $i_{Env\_U\_3}$: the possibilities for Cyprinids to access the reed bed would not be influenced by the proposed mitigation measures. Thus, the indicator does not have to be modified.

- $i_{Env\_U\_4}$: due to the withdrawal of the reed bed, the critical threshold for nesting would shift from +0.85 m to +1.55 m above zero at Sesto Calende hydrometric station.

- $i_{Env\_U\_5}$: the banks of wet sand and gravel (*shelves*), on which annual wetland vegetation (*Nanocyperion*) grows, would remain submerged for most of the year, but new areas would be made available with the same characteristics, thanks to the creation of artificial shelves. The partial value of this indicator in the mitigated alternative can thus be assumed to be equal to 1.

The partial value functions corresponding to the indicators that have not been modified can also remain the same, and the weights used to define the global value function do not have to be modified either.

The alternative that is obtained by applying the BOLLE RESERVE measure to $A34$ will be termed $A505$. The indicator values that it produces are shown in Table 14.14 along with those produced by $A34$. It is evident that the BOLLE RESERVE measure has a strong positive effect on the *Upstream Environment*: the problem of nesting ($i_{Env\_U\_4}$) and the loss of the shelves environments ($i_{Env\_U\_5}$) would, in fact, largely be solved.

Note that even the hight cost of the measure – which, as noted above, can be estimated at about 10 to 15 million euro – is not a large sum when one considers the reduction in the flooding damage and in the irrigation deficit that $A34$ would allow. For example, the damage produced by the October 2000 flood event was estimated at 110 million euro in the Switzerland alone.

Once these results had been examined, the Bolle di Magadino Foundation judged the effects of *A*505 to be very good and declared that, if the effects in the transient period were not dangerous or if they could be satisfactorily managed by a time-variant policy, they would support the alternative. In order to consider this discussion definitively concluded, the possibility of carrying out similar measures in other areas of environmental importance would have to be considered, but this task goes beyond the limits of the Verbano Project and so it must be postponed to a future Project.

Note that the BOLLE RESERVE measure can be applied not only to alternative *A*34, but to all the alternatives that include SD_{+600}. By its nature, it is evident that in all of them it would give optimal results. It follows that *Upstream Environment* is potentially supportive of any reasonable alternative, as long as it is accompanied by the BOLLE RESERVE measure and the study of the transient period has satisfactory results. Since such a study would be long and costly, it would be advisable to carry it out for just one alternative, the one that obtained the widest agreement. This alternative will emerge in the *Final Decision* phase, when the Decision Makers compare the reasonable alternatives that are identified at the end of the third negotiation step. Therefore, the decision-making process should proceed in the following way:

(1) the phase of *Final Decision* is conducted under the hypothesis that all the reasonable alternatives with SD_{+600} are accepted or supported by *Upstream Environment*, provided that they are mitigated;

(2) if the *best compromise alternative* A_{com} chosen by the final DMs is an alternative with action SD_{CUR}, the decision-making process is concluded; otherwise the effects on *Upstream Environment* in the transient period must be studied, looking for a time-variant policy to be applied in the transition from the current policy to the policy included in A_{com};

(3) the results of the latter study will be presented to the Bolle di Magadino Foundation; if the Foundation accepts the result, the decision-making process will be concluded; otherwise we must return to point (1), taking account of the opposition from *Upstream Environment* to alternative A_{com}.

14.4 Mitigation for the other sectors

We must still look for measures that mitigate the effects of the reasonable alternatives on the other sectors. In the preceding chapter (Section 13.1.4) we already revealed that the spaces for an increase in the sector indices for *Downstream Environment* and for *Irrigation* are, in fact, nonexistent, except if we consider modifying the irrigation technique, which is not a mitigation measure, but a design action. Mitigation measures for the other sectors either were not found (this is the case for ENEL *Power* sector, for example) or the Stakeholders did not ask for them, which is the case for *Upstream Tourism*, whose Stakeholders (or rather, their representatives, i.e. the Municipalities of Locarno and Verbania) are satisfied with the compensation that exists with the *Upstream Flooding* sector (see page 294).

Lastly, there is still the possibility to evaluate localized mitigation measures that satisfy only small groups of Stakeholders: this is the case, for example, for the lakeside protective barrier for Locarno (see Section 3.6.1 of the DVD), which could defend the city against floods. Given the local nature of these interventions, we did not include them among those

Table 14.15. The final set of reasonable alternatives

Alternative	SD	RANGE	MEF	Mitigation measure	Weights		
					λ_{fl}	λ_{irr}	λ_{env}
$A0$	ACT	ACT	13_{CON}	–	–	–	–
$A7$	ACT	ACT	13_{CON}	–	0.00	0.99	0.01
$A18$	ACT	ACT	13_{CON}	–	0.01	0.98	0.01
$A119$	ACT	ACT	25_{MOD}	–	0.01	0.98	0.01
$M54$	+600	ACT	13_{CON}	REDUCTION WITH PO BOLLE RESERVE	0.99	0.01	0.00
$M168$	+600	ACT	25_{MOD}	REDUCTION WITH PO BOLLE RESERVE	0.99	0.01	0.00
$M34$	+600	ENL	13_{CON}	REDUCTION WITH PO BOLLE RESERVE	0.99	0.01	0.00
$M81$	+600	ENL	13_{CON}	REDUCTION WITH PO BOLLE RESERVE	0.60	0.38	0.02
$M122$	+600	ENL	20_{MOD}	REDUCTION WITH PO BOLLE RESERVE	0.60	0.39	0.01
$M110$	+600	ENL	25_{MOD}	REDUCTION WITH PO BOLLE RESERVE	0.55	0.35	0.10

examined in this Project, but the construction of such barriers is a mitigation measure that Locarno could take into consideration.

14.5 The third negotiation step

The reasonable alternatives and the least-bad alternatives that emerge from the second negotiation step, together with the mitigated alternatives, were subjected to a third negotiation step, in order to identify the final set of reasonable alternatives.

The Bolle di Magadino Foundation proposed excluding alternatives $A36$ and $A112$ from the reasonable alternatives, because they consider them to be dominated by alternative $A34$ mitigated with the BOLLE RESERVE measure; all the other Stakeholders agreed to this.

As a result, the set of reasonable alternatives described in Table 14.15 was obtained, with respect to which the Stakeholders took the positions reported in Table 14.16. The mitigated alternatives were labelled keeping the same number of the alternative that they derive from, preceded by an M, in place of the A. For example, the alternative that was derived from $A34$ by adopting the REDUCTION WITH PO and BOLLE RESERVE mitigation measures is denoted as $M34$.

Table 14.16. The positions of the Stakeholders with respect to the reasonable alternatives of Table 14.15

SD	RANGE	MEF	Alternative	Stakeholders			
				Support	Accept	Oppose	Tolerate
CUR	CUR	13_{CON}	*A*0	GRAIA Society ENEL Company	Province of Pavia	Municipalities of Verbania and Locarno Lake Navigation Company Bolle di Magadino Foundation East Sesia Consortium East Ticino Consortium Ticino Parks	
		13_{CON}	*A*7	Bolle di Magadino Foundation	GRAIA Society Ticino Parks Province of Pavia	Municipalities of Verbania and Locarno Lake Navigation Company ENEL Company	East Sesia Consortium East Ticino Consortium
		13_{CON}	*A*18	Bolle di Magadino Foundation	GRAIA Society Ticino Parks Province of Pavia	East Sesia Consortium East Ticino Consortium ENEL Company	Municipalities of Verbania and Locarno Lake Navigation Company
		25_{MOD}	*A*119	Bolle di Magadino Foundation Ticino Parks	GRAIA Society Province of Pavia	Municipalities of Verbania and Locarno Lake Navigation Company East Sesia Consortium East Ticino Consortium ENEL Company	

(*continued on next page*)

Table 14.16. *(Continued)*

SD	RANGE	MEF	Alternative	Stakeholders			
				Support	Accept	Oppose	Tolerate
+600	CUR	13_{CON}	$M54$	Municipalities of Verbania and Locarno Lake Navigation Company Bolle di Magadino Foundation	GRAIA Society East Sesia Consortium East Ticino Consortium ENEL Company Province of Pavia	Ticino Parks	
		25_{MOD}	$M168$	Municipalities of Verbania and Locarno Lake Navigation Company Bolle di Magadino Foundation	GRAIA Society ENEL Company Province of Pavia	East Sesia Consortium East Ticino Consortium	Ticino Parks
	ENL	13_{CON}	$M34$	Municipalities of Verbania and Locarno Lake Navigation Company Bolle di Magadino Foundation East Sesia Consortium East Ticino Consortium	GRAIA Society ENEL Company Province of Pavia	Ticino Parks	
		13_{CON}	$M81$	Municipalities of Verbania and Locarno Lake Navigation Company Bolle di Magadino Foundation	GRAIA Society East Sesia Consortium East Ticino Consortium Province of Pavia	ENEL Company	Ticino Parks
		20_{MOD}	$M122$	Municipalities of Verbania and Locarno Lake Navigation Company Bolle di Magadino Foundation	GRAIA Society East Sesia Consortium East Ticino Consortium Province of Pavia	ENEL Company	Ticino Parks
		25_{MOD}	$M110$	Municipalities of Verbania and Locarno Lake Navigation Company Bolle di Magadino Foundation	GRAIA Society Province of Pavia	East Sesia Consortium East Ticino Consortium ENEL Company	Ticino Parks

PART C

THE FINAL DECISION

Chapter 15

Summary of the Project

FC, FP and RSS

An IWRM project should conclude with a summary document that provides the final Decision Makers (DMs) with all the information necessary for them to make the decision. It is essential that this document be concise, which means that it should not report the whole study, but should contain:

- a description of the steps that were taken to individuate the *reasonable alternatives*, with references to the technical documents that describe in detail how those steps were carried out;
- a description of the reasonable alternatives (including any *mitigation measure*) and their effects.

In this chapter we present such a summary with two main aims: to provide an example of how the document can be drawn up, and to give the reader a concise description of the reasonable alternatives that were identified. Because of the way it is written, this chapter could be easily skipped – with the exception of Section 15.3, which describes the set of reasonable and least-bad alternatives – by a reader who has already read the preceding chapters. For the reader who has not read the rest of the book, it provides sufficient information for reading the two concluding chapters, where we will see how the Project concluded.

15.1 The Problem

Several years ago Switzerland proposed that Verbano's lake outlet be excavated to increase the outflow capacity and thus reduce the intensity and frequency of lake flooding (Section 3.1). The proposal stimulated great interest from the inhabitants of the Italian towns along the shores of Verbano, since their interests are analogous to those of the Swiss shoreline dwellers. On the contrary, it was not well received by the populations along the River Ticino and the River Po, who were afraid that an increase in the lake's outflow capacity might increase the probability of flooding in their areas. To evaluate the effects of the Swiss proposal, the present Project (*Verbano Project*) was financed by the EU INTERREG II program. Meantime, the two Regional Parks in the Ticino valley, one in Lombardy and one in Piedmont, requested an increase in the *Minimum Environmental Flow* (MEF) in the Ticino downstream from the lake (Section 3.4). For this reason, when the Project began, it was decided to evaluate this request as well.

During the process of formalizing the subject of the study, it was understood right from the beginning that the Swiss request would be more acceptable for the downstream Italian population[1] if it were accompanied by some modifications to the system that would be to its advantage. This modification was identified as an increase in the active storage of the lake (Section 3.3), which would allow a greater water volume to be available for satisfying irrigation and hydropower demand. The active storage of Verbano is defined by the *regulation range*, namely a range of levels within which the Ticino Consortium (the body that is entrusted with the regulation of the lake) can choose the daily release according to its own judgement. The range is dictated by the Regulation Licence (see Section 2.3.3), which sets the norms for the regulation of the lake. In 1943 the Miorina dam, which is located a little downstream from Sesto Calende, came into service, and since then Verbano has been a regulated lake (Section 2.3). The upper extreme of the range is actually set at +1.00 m above zero at the Sesto Calende hydrometric station, in the period from 15th March to 31st October, and at +1.50 m during the rest of the year. When the lake level exceeds these levels, the Ticino Consortium must open the dam completely, allowing the most rapid outflow. When this happens, that water is lost from the point of view of the downstream users, both the irrigators and the hydropower producers. Maintaining the upper extreme of the regulation range at +1.50 m all year would permit an increase in the active volume to be regulated. On the other hand, this could also produce an increase in the extension or the frequency of the lake flooding events, because the Regulator would begin reducing the lake level at levels that are higher than the ones currently imposed. Raising the upper extreme of the regulation range is thus a modification that the lakeside population does not like a priori.

A change in the stage–discharge relation and in the regulation range cannot be the fruit of a unilateral Italian or Swiss decision. Instead, it requires the signing of an international agreement. The modification of the MEF, however, concerns Italy alone, because the users that support it and those that oppose it are all Italians. In any case, it must be borne in mind that, until a truly shared decision is reached, it is essential to promote the participation of all of the *Stakeholders*. A careful evaluation of the different aspects of the problem convinced us that the framework most suited for the Verbano Project was *Multi-Attribute Value Theory* and that a consultative role had to be given to the Stakeholders: their opinions influence and guide the choice, but the real decision is the responsibility of the Italian and Swiss governments. Given that the requests expressed by the Stakeholders guide the governments' decision, it is advisable to begin the study with an examination of their viewpoints and start the negotiations between the two nations only when these positions are well understood. This is why only the next chapter of this book is dedicated to the analysis of the final decision, while all the previous chapters consider the problem from the Stakeholders' viewpoint.

Accepting or rejecting the proposal from which the Project was born means deliberating some *actions*. Each one of them can assume different values, which we denote with conventional names in capital letters. The first is a *structural action* which can assume the following values:

- SD$_{\text{CUR}}$ (*current stage–discharge relation*): the outlet is left unchanged and thus the stage–discharge relation is unaltered;

[1] Here and in the following, the terms *upstream* and *downstream* should be understood in reference to the lake outlet.

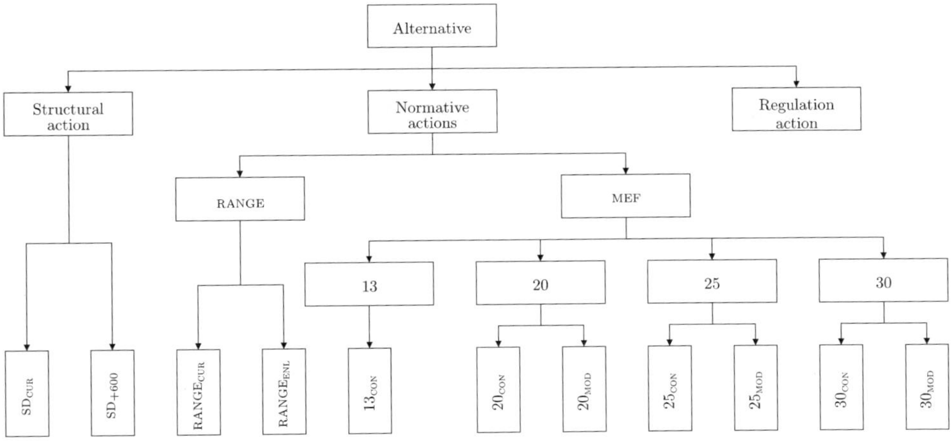

Figure 15.1: The tree of actions that compose an alternative.

- SD_{+600}: the lake outlet is excavated, so that the maximum outflow that can be released when a flood starts is greater than the current maximum outflow of about 600 m^3/s.

The second is a *normative action*, more precisely a pair of actions, which concern two different elements: the regulation range and the Minimum Environmental Flow (MEF) in the Ticino effluent. With regard to the range, two values have been considered:

- $\mathrm{RANGE}_{\mathrm{CUR}}$ (*current range*): the upper extreme of the range is unchanged, i.e. it follows the current time-variant trajectory;

- $\mathrm{RANGE}_{\mathrm{ENL}}$ (*enlarged range*): the upper extreme of the range is set over all the year at the level +1.50 m above zero at the Sesto Calende hydrometric station.

With regard to the MEF, the actions are differentiated by the values adopted (13, 20, 25 or 30 m^3/s) and by behaviour over time, which can be constant (CON) or modulated (MOD), to reproduce the median of the natural flow in the Ticino (Section 5.3.3).

Once the structural and normative actions have been fixed, the behaviour of the system, and so the effects that it produces, depend on the way in which the dam is regulated. Therefore it is necessary to consider also the regulation action which is expressed by the adopted *policy*[2] (Section 3.5).

A (project) *alternative* is defined by a combination of structural, normative and regulation actions (Figure 15.1). The alternatives to examine were carefully chosen (Chapter 5), but whichever criterion is used to select them, it is absolutely necessary that the so-called *Alternative Zero* ($A0$) be included among them, which means keeping the system unchanged ($\mathrm{SD}_{\mathrm{CUR}}/\mathrm{RANGE}_{\mathrm{CUR}}/13_{\mathrm{CON}}$, the historical regulation policy). This is the alternative that would be automatically achieved if Italy and Switzerland did not reach a new agreement and thus it constitutes a reference point for the comparison of the effects produced by the other alternatives (Section 3.7.1).

[2]With the term *policy* we refer to a function which, given the water volume stored in the lake, suggests a set of release volumes from which the Regulator of the lake may freely choose the daily release decision. This type of policy is termed *set-valued* policy.

The aim of the Verbano Project is to identify all the possible alternatives and then find the *reasonable alternatives*, which are those that are supported or accepted by a good number of, if not all, Stakeholders (Section 1.3). These alternatives will be presented to the *final Decision Makers* (DMs), who must choose the *best compromise alternative*.

15.2 The development of the Project

The Project analyses planning interventions that are suitable to solve the Verbano problem and it is carried out according to a *Participatory and Integrated Planning* (PIP) procedure, which is shown in Figure 15.2. Some considerations about the subsequent implementation of the results, at the management level, are presented in Chapter 17. Phase 1 (*Defining Actions*) of PIP procedure was briefly described in the previous section (for more details see Chapter 3), so we will begin with a description of Phase 2: *Defining Criteria and Indicators* (Chapter 4).

To identify the effects that the Stakeholders are sensitive to, we questioned their representatives. With their help, we identified a number of *evaluation criteria*, i.e. those attributes or factors with which the Stakeholders judge the performances of the alternatives. Identifying all the evaluation criteria is not an easy task: thus we proceeded by assuming a set of high-level criteria and then by breaking each of them down into lower-level criteria. This hierarchy is developed until the *leaf criteria* are identified: these are the criteria whose satisfaction can be estimated directly from the trajectories of the hydrological variables that the alternative produces, with a procedure which is called *indicator*.

The criteria which express homogeneous interests were grouped to form the so-called *sectors*, each of which is associated with a sector criterion which 'synthesizes' the point of view that the sector represents (Section 4.1). The Stakeholders, sectors and criteria from the Verbano Project are presented in Figure 15.3, while the list of indicators is reported in Tables 15.1 and 15.2.

Once the indicators had been defined, the model of the water system was identified (Phase 3: *Identifying the Model*, Chapter 6): it will be used later on to estimate the effects induced by the alternatives, i.e. to compute the indicator values for each alternative.

The alternatives were designed in Phase 4 (*Designing Alternatives*, see Chapter 5). To do this, for each possible combination of structural and normative actions, *efficient* policies were designed by solving suitable *Multi-Objective* (*Optimal*) *Control Problems*. The formulation of these Problems requires defining the *objectives* and the *constraints* of the Problem.

To define the objectives, the Experts were involved again and, by means of ad hoc procedures (Section 7.2), first the following four *design indicators* were selected from the set of indicators defined in Phase 2 (Section 7.3):

- average annual flooded area [km^2/year] in Locarno and Verbania;
- average annual potential damage from stress [–] to the irrigated crops;
- average annual lost revenue [M€] for the ENEL hydropower plants;
- distance [m^3/s] from the natural regime in the River Ticino.

Then, from these indicators, the objectives were derived by specifying how to deal with the uncertainty and the stochasticity that affect the water system (Section 7.6.1).

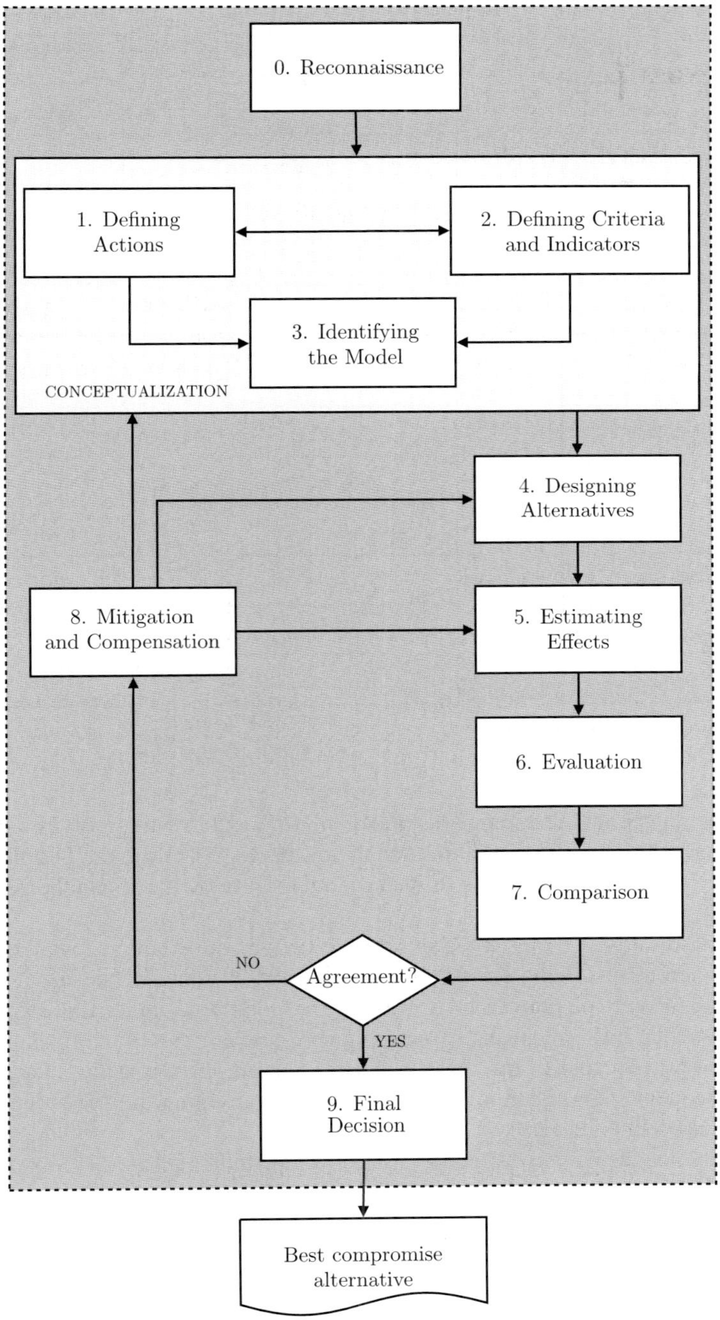

Figure 15.2: The PIP procedure, which defines the phases in which the Verbano Project is articulated.

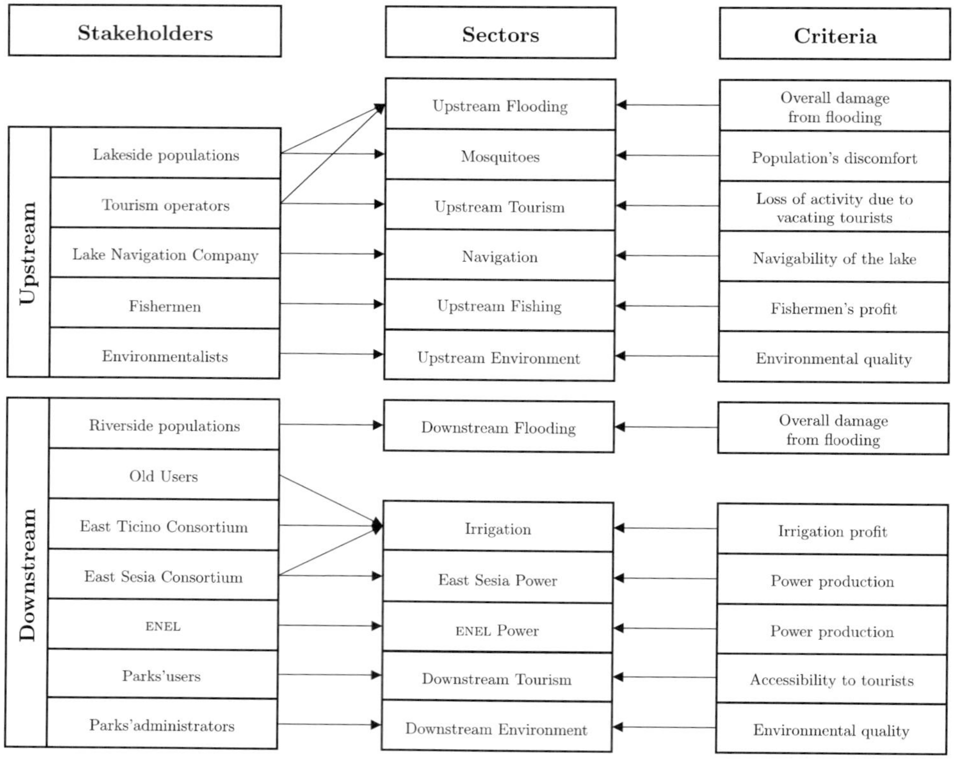

Figure 15.3: Stakeholders, sectors and sector criteria.

As for the constraints of the Control Problem, they are the union of the physical constraints, expressed by the water system model, and the legal constraints that the regulation of the lake must respect. The first are influenced by the structural action adopted, the second by the normative actions.

By solving the Multi-Objective Control Problems a set of policies was designed, and thus a set of alternatives, each of which is *efficient* (in Pareto sense), because it minimizes a certain combination of the objectives. Since there are infinite ways of combining the objectives and in practice it is only possible to consider a finite number of them, and this number cannot be too large because of the computing time needed, the alternatives design was split into two subsequent steps (Section 5.4). This led to subdividing the following phases into several steps as well.

Each of the alternatives so designed was then simulated of (Phase 5: *Estimating Effects*, see Chapter 8) in order to estimate the value of all the indicators. The simulations were made over a horizon of 25 years, and the inflow time series in the years 1974–1998 was adopted as inflow scenario (Section 8.3). By analysing the effects so computed, it became apparent that for the *Downstream Flooding* sector the differences in the indicator values between the best alternative and the worst alternative are lower than the estimation uncertainty (Section 11.3), so that these indicators cannot be used to compare the alternatives. Therefore it was decided to postpone judgement relative to this sector and to first identify a set of reasonable alternatives by comparing the alternatives from all standpoints but *Downstream*

Table 15.1. The list of indicators of the upstream sectors

Sector	Indicator		
Upstream Environment	$i_{\text{Env\_U\_1}}$	Distance from natural regime	[m]
	$i_{\text{Env\_U\_2}}$	Average annual number of days in which the lake level is in the range of reed bed erosion	[d/yr]
	$i_{\text{Env\_U\_3}}$	Fraction of the egg-laying period in which the Cyprinids cannot access the reed beds	[%]
	$i_{\text{Env\_U\_4}}$	Fraction of the nesting period in which the lake level is higher than the threshold of the nests	[%]
	$i_{\text{Env\_U\_5}}$	Fraction of the normal shelf emergence period in which the lake level is higher than the emergence threshold	[%]
Upstream Fishing	$i_{\text{Fis\_U\_1}}$	Percentage of the Bleak eggs laid that do not hatch	[%]
	$i_{\text{Fis\_U\_2}}$	Percentage of the White Fish eggs laid that do not hatch	[%]
	$i_{\text{Fis\_U\_3}}$	Fraction of the egg-laying period in which the Pike cannot access the reed beds	[%]
	$i_{\text{Fis\_U\_4}}$	Fraction of the egg-laying period in which the Cyprinids cannot access the reed beds	[%]
Upstream Flooding	$i_{\text{Fl\_U\_1\_Loc}}$	Maximum flooded area in Locarno	[km^2]
	$i_{\text{Fl\_U\_1\_Vb}}$	Maximum flooded area in Verbania	[km^2]
	$i_{\text{Fl\_U\_2\_Loc}}$	Average annual flooded area in Locarno	[km^2/yr]
	$i_{\text{Fl\_U\_2\_Vb}}$	Average annual flooded area in Verbania	[km^2/yr]
	$i_{\text{Fl\_U\_3\_Loc}}$	Maximum number of consecutive flooding days in Locarno	[d]
	$i_{\text{Fl\_U\_3\_Vb}}$	Maximum number of consecutive flooding days in Verbania	[d]
	$i_{\text{Fl\_U\_4\_Loc}}$	Average annual number of flooding days in Locarno	[d/yr]
	$i_{\text{Fl\_U\_4\_Vb}}$	Average annual number of flooding days in Verbania	[d/yr]
	$i_{\text{Fl\_U\_5\_Loc}}$	Maximum number of days of interrupted traffic in Locarno	[d]
	$i_{\text{Fl\_U\_5\_Vb}}$	Maximum number of days of interrupted traffic in Verbania	[d]
	$i_{\text{Fl\_U\_6\_Loc}}$	Average annual number of days of interrupted traffic in Locarno	[d/yr]
	$i_{\text{Fl\_U\_6\_Vb}}$	Average annual number of days of interrupted traffic in Verbania	[d/yr]
Mosquitoes	$i_{\text{Mos\_U\_1}}$	Average annual number of necessary treatments	[treatments/yr]
	$i_{\text{Mos\_U\_2}}$	Average duration of the periods in which treatments are probable	[d/yr]
Navigation	$i_{\text{Nav\_U\_1}}$	Average annual cost of the construction and dismantling of raised walkways	[k€/yr]
	$i_{\text{Nav\_U\_2}}$	Average annual lost revenue from the closure of ports	[k€/yr]
	$i_{\text{Nav\_U\_3}}$	Average annual lost revenue from the inability to load heavy cargo when the lake level is low	[k€/yr]
Upstream Tourism	$i_{\text{Tou\_U\_1}}$	Percentage of the tourist season in which the lake level is lower than the Tourist Normality Range	[%]
	$i_{\text{Tou\_U\_2}}$	Percentage of the tourist season in which the lake level is higher than the Tourist Normality Range	[%]
	$i_{\text{Tou\_U\_3}}$	Average annual number of necessary treatments	[treatments/yr]
	$i_{\text{Tou\_U\_4}}$	Average duration of the periods in which treatments are probable	[d/yr]

Table 15.2. The list of indicators of the downstream sectors

Sector	Indicator		
Downstream Flooding	$i_{\mathrm{Fl\_D\_1}}$	Maximum peak flood level in Pavia	[m]
	$i_{\mathrm{Fl\_D\_2}}$	Average annual flooded area in Pavia	[km²/yr]
	$i_{\mathrm{Fl\_D\_3\_Bt}}$	Maximum number of consecutive flooding days in Borgo Ticino (Pavia)	[d]
	$i_{\mathrm{Fl\_D\_3\_Pv}}$	Maximum number of consecutive flooding days in the flood plains (Pavia)	[d]
	$i_{\mathrm{Fl\_D\_4\_Bt}}$	Average annual number of flooding days in Borgo Ticino (Pavia)	[d/yr]
	$i_{\mathrm{Fl\_D\_4\_Pv}}$	Average annual number of flooding days in the flood plains (Pavia)	[d/yr]
	$i_{\mathrm{Fl\_D\_5}}$	Maximum number of days of interrupted traffic in Pavia	[d]
	$i_{\mathrm{Fl\_D\_6}}$	Average annual number of days of interrupted traffic in Pavia	[d/yr]
Irrigation	$i_{\mathrm{Irr\_D\_1}}$	Average annual volume of total deficit	[Mm³/yr]
	$i_{\mathrm{Irr\_D\_2}}$	Average annual potential damage from stress	[–]
	$i_{\mathrm{Irr\_D\_3\_IC}}$	Variability of supply to the Industrial Canal	[–]
	$i_{\mathrm{Irr\_D\_3\_REC}}$	Variability of supply to the Regina Elena Canal	[–]
	$i_{\mathrm{Irr\_D\_3\_VC}}$	Variability of supply to the Villoresi Canal	[–]
ENEL Power	$i_{\mathrm{EnP\_D\_1}}$	Average annual lost revenue for the ENEL hydropower plants	[k€/yr]
	$i_{\mathrm{EnP\_D\_2}}$	Average annual number of days in which the thermoelectric power plant at Turbigo is out of service	[d/yr]
Downstream Tourism	$i_{\mathrm{Tou\_D\_1}}$	Average annual number of days in which the flow rate of the Ticino at Panperduto is higher than the Tourist Normality Range	[d/yr]
	$i_{\mathrm{Tou\_D\_2}}$	Average annual number of days in which the flow rate of the Ticino at Panperduto is lower than the Tourist Normality Range	[d/yr]
Downstream Environment	$i_{\mathrm{Env\_D\_1}}$	Distance from the natural regime	[m³/s]
	$i_{\mathrm{Env\_D\_2}}$	Value of the MEF	[–]
	$i_{\mathrm{Env\_D\_3}}$	Average annual number of days in which the flow rate of the Ticino at Panperduto is lower than or equal to the MEF	[d/yr]
	$i_{\mathrm{Env\_D\_4}}$	Average annual number of consecutive days in which the flow rate of the Ticino at Panperduto is lower than or equal to the MEF	[d/yr]
	$i_{\mathrm{Env\_D\_5}}$	Fraction of the reproduction period for aquatic birdlife and fish species in which the flow rate of the Ticino at Oleggio varies too rapidly	[%]
East Sesia Power	$i_{\mathrm{ESP\_D\_1}}$	Average annual lost revenue for the East Sesia hydropower plants	[k€/yr]

Flooding, and to finally return to estimate the effects on this sector only for those reasonable alternatives.

In the phase of *Evaluating Alternatives* (Phase 6, see Chapter 9) we collaborated with the Experts once again in order to associate each indicator with a *partial value function*. This function transforms the value of the indicator, which is expressed in physical units, into the level of satisfaction that it produces (value). By combining the partial value functions of the indicators for a sector, using weights to express their relative importance, the *sector indices* were obtained (Sections 9.2–9.4).

Finally we moved to the phase of *Comparison* (Phase 7, see Chapter 10), which is articulated and interlaced with the others according to the diagram in Figure 15.4. After some preliminary activities, to which the Stakeholders as well as some members of the

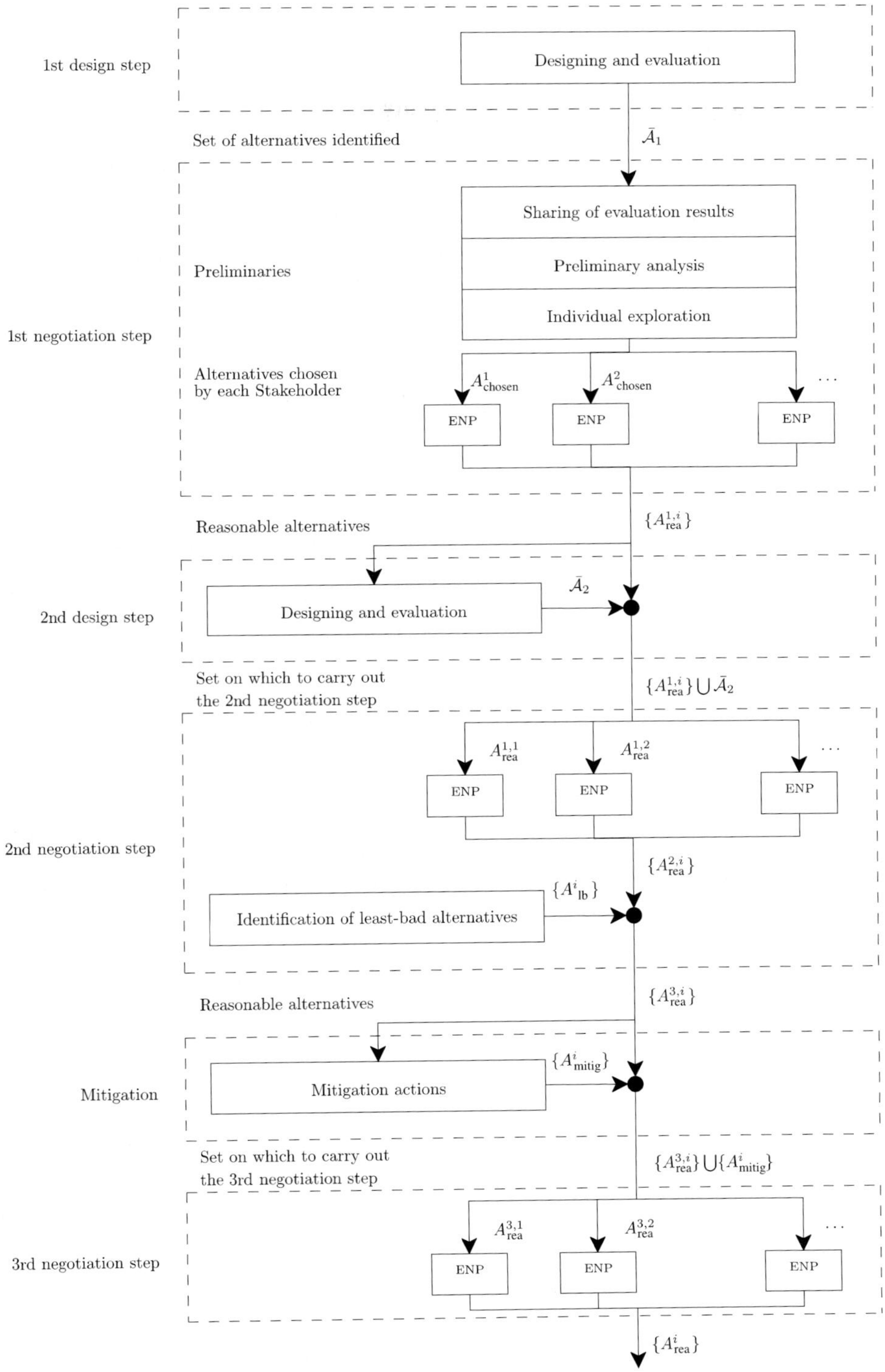

Figure 15.4: The steps in which the phase of *Comparison* is articulated.

Italian–Swiss Commission participated (Sections 12.1–12.2), the first negotiation step was carried out, and a first set of *reasonable alternatives* was identified (Chapter 12). Starting from the characteristics of these alternatives and the needs expressed by the Stakeholders, new alternatives were designed. The second negotiation step led to the identification of a second set of reasonable alternatives and of a set of *least-bad alternatives* (Chapter 13). All the alternatives of these two sets were carefully analysed to understand the effects that they produce on the *Downstream Flooding* sector. We learned that all the alternatives in which the structural action SD_{+600} is considered produce effects that the Stakeholders interested in *Downstream Flooding* are not willing to accept (Section 14.1). For this reason, suitable *mitigation measures* were studied (Section 14.2). Other mitigation measures were designed for *Upstream Environment*, particularly for the Bolle di Magadino area (Section 14.3). Then we moved to the last step of the negotiations, which led to the identification of the final set of reasonable alternatives (Section 14.5).

All computations were carried out by means of the *Multi-Objective Decision Support System* TWOLE, which also supported the decision-making procedure by providing tools for the interaction with the Stakeholders (see Section 1.5). Details about the software architecture of TWOLE MODSS can be found in Chapter 24 of THEORY.

In the next section, we will describe the characteristics of the reasonable and least-bad alternatives that were obtained at the end of the second negotiation step; in other words, we will describe the effects that the reasonable alternatives would produce if the mitigation measures were not realized. The following two sections will be dedicated to the description of those measures. Thereby the importance of the mitigation measures will be more evident.

15.3 The reasonable and least-bad alternatives

Table 15.3 reports the list of reasonable and least-bad alternatives identified in the second negotiation step (see Sections 13.2 and 13.3), from which the final set of reasonable alternatives was derived (Section 14.5). For the sake of brevity, in the following we will refer to them as 'alternatives' without further specification: the reader should remember that from now on we will consider only the alternatives listed in Table 15.3. Table 15.4 shows the Stakeholders' positions with respect to them.

Table 15.3. The reasonable and least-bad alternatives identified in the second negotiation step, from which the final set of reasonable alternatives was derived

Alternative	SD	RANGE	MEF	Weights		
				λ_{fl}	λ_{irr}	λ_{env}
*A*0	ACT	ACT	13_{CON}	–	–	–
*A*7	ACT	ACT	13_{CON}	0.00	0.99	0.01
*A*18	ACT	ACT	13_{CON}	0.01	0.98	0.01
*A*119	ACT	ACT	25_{MOD}	0.01	0.98	0.01
*A*54	+600	ACT	13_{CON}	0.99	0.01	0.00
*A*168	+600	ACT	25_{MOD}	0.99	0.01	0.00
*A*34	+600	ENL	13_{CON}	0.99	0.01	0.00
*A*81	+600	ENL	13_{CON}	0.60	0.38	0.02
*A*122	+600	ENL	20_{MOD}	0.60	0.39	0.01
*A*110	+600	ENL	25_{MOD}	0.55	0.35	0.10

Table 15.4. The positions of the Stakeholders with respect to the alternatives of Table 15.3. The precise meaning of the terms is defined in Section 10.1

SD	RANGE	MEF	Alternative	Stakeholders			
				Support	Accept	Oppose	Tolerate
CUR	CUR	13_{CON}	*A0*	GRAIA Society ENEL Company		Municipalities of Verbania and Locarno Lake Navigation Company Bolle di Magadino Foundation East Sesia Consortium East Ticino Consortium Ticino Parks	
		13_{CON}	*A7*	Bolle di Magadino Foundation	GRAIA Society Ticino Parks	Municipalities of Verbania and Locarno Lake Navigation Company ENEL Company	East Sesia Consortium East Ticino Consortium
		13_{CON}	*A18*	Bolle di Magadino Foundation	GRAIA Society Ticino Parks	East Sesia Consortium East Ticino Consortium ENEL Company	Municipalities of Verbania and Locarno Lake Navigation Company
		25_{MOD}	*A119*	Bolle di Magadino Foundation Ticino Parks	GRAIA Society	Municipalities of Verbania and Locarno Lake Navigation Company East Sesia Consortium East Ticino Consortium ENEL Company	

(*continued on next page*)

Table 15.4. *(Continued)*

SD	RANGE	MEF	Alternative	Stakeholders			
				Support	Accept	Oppose	Tolerate
+600	CUR	13_{CON}	*A*54	Municipalities of Verbania and Locarno Lake Navigation Company	GRAIA Society East Sesia Consortium East Ticino Consortium ENEL Company	Bolle di Magadino Foundation Ticino Parks	
+600	CUR	25_{MOD}	*A*168	Municipalities of Verbania and Locarno Lake Navigation Company Bolle di Magadino Foundation	GRAIA Society ENEL Company	Bolle di Magadino Foundation East Sesia Consortium East Ticino Consortium	Ticino Parks
+600	ENL	13_{CON}	*A*34	Municipalities of Verbania and Locarno Lake Navigation Company East Sesia Consortium East Ticino Consortium	GRAIA Society ENEL Company	Bolle di Magadino Foundation Ticino Parks	
+600	ENL	13_{CON}	*A*81	Municipalities of Verbania and Locarno Lake Navigation Company	GRAIA Society East Sesia Consortium East Ticino Consortium	Bolle di Magadino Foundation ENEL Company	Ticino Parks
+600	ENL	20_{MOD}	*A*122	Municipalities of Verbania and Locarno Lake Navigation Company	East Sesia Consortium East Ticino Consortium GRAIA Society	Bolle di Magadino Foundation ENEL Company	Ticino Parks
+600	ENL	25_{MOD}	*A*110	Municipalities of Verbania and Locarno Lake Navigation Company	GRAIA Society	Bolle di Magadino Foundation East Sesia Consortium East Ticino Consortium ENEL Company	Ticino Parks

Table 15.5. The performances of the alternatives of Table 15.3 in the upstream sectors

Sector	Indicator		$A0$	$A7$	$A18$	$A119$	$A54$	$A168$	$A34$	$A81$	$A122$	$A110$
Upstream Environment	$i_{Env\_U\_1}$	[m]	0.59	0.45	0.40	0.42	0.57	0.60	0.62	0.62	0.63	0.63
	$i_{Env\_U\_2}$	[d/yr]	13.20	28.50	44.73	32.22	18.67	21.11	18.67	20.19	18.75	21.07
	$i_{Env\_U\_3}$	[%]	10.14	11.03	14.90	14.41	22.80	23.53	22.09	13.93	20.89	20.65
	$i_{Env\_U\_4}$	[%]	83.55	66.91	57.20	63.34	71.45	68.98	71.64	70.03	69.56	69.13
	$i_{Env\_U\_5}$	[%]	61.69	68.57	60.84	65.60	51.11	48.49	54.22	59.18	54.11	54.76
Upstream Fishing	$i_{Fis\_U\_1}$	[%]	7.75	7.67	7.38	7.67	3.55	3.29	3.93	3.39	3.74	3.56
	$i_{Fis\_U\_2}$	[%]	23.26	20.98	23.65	20.79	40.78	31.77	36.49	11.84	11.02	14.62
	$i_{Fis\_U\_3}$	[%]	15.80	34.50	57.80	46.18	16.68	22.21	16.41	16.48	20.53	21.94
	$i_{Fis\_U\_4}$	[%]	10.14	11.03	14.90	14.51	22.80	23.53	22.09	13.93	20.89	20.65
Upstream Flooding	$i_{Fl\_U\_1\_Loc}$	[km^2]	2.28	2.06	2.06	2.06	1.05	1.05	1.05	1.05	1.05	1.05
	$i_{Fl\_U\_1\_Vb}$	[km^2]	0.52	0.50	0.50	0.50	0.27	0.27	0.27	0.27	0.27	0.27
	$i_{Fl\_U\_2\_Loc}$	[km^2/yr]	0.71	0.65	0.62	0.64	0.28	0.28	0.32	0.34	0.32	0.33
	$i_{Fl\_U\_2\_Vb}$	[km^2/yr]	0.12	0.10	0.10	0.10	0.02	0.02	0.02	0.02	0.02	0.02
	$i_{Fl\_U\_3\_Loc}$	[d]	36.	35.	35.	35.	26.	26.	26.	26.	26.	26.
	$i_{Fl\_U\_3\_Vb}$	[d]	26.	25.	24.	24.	11.	11.	11.	11.	11.	11.
	$i_{Fl\_U\_4\_Loc}$	[d/yr]	13.72	12.67	12.11	12.47	4.44	4.44	5.28	5.40	5.24	5.28
	$i_{Fl\_U\_4\_Vb}$	[d/yr]	3.54	3.20	3.00	3.12	0.72	0.72	0.76	0.84	0.76	0.76
	$i_{Fl\_U\_5\_Loc}$	[d]	6.	6.	6.	6.	0.	0.	0.	0.	0.	0.
	$i_{Fl\_U\_5\_Vb}$	[d]	6.	6.	6.	6.	0.	0.	0.	0.	0.	0.
	$i_{Fl\_U\_6\_Loc}$	[d/yr]	0.28	0.28	0.28	0.28	0.00	0.00	0.00	0.00	0.00	0.00
	$i_{Fl\_U\_6\_Vb}$	[d/yr]	0.28	0.28	0.28	0.28	0.00	0.00	0.00	0.00	0.00	0.00
Mosquitoes	$i_{Mos\_U\_1}$	[treat./yr]	2.05	1.92	2.28	2.12	2.12	2.16	1.80	1.84	1.80	1.88
	$i_{Mos\_U\_2}$	[d/yr]	35.35	47.14	33.86	38.98	31.88	29.28	41.93	44.80	41.84	40.55
Navigation	$i_{Nav\_U\_1}$	[k€/yr]	25.66	22.78	21.77	22.41	12.26	12.17	14.98	15.28	14.89	14.99
	$i_{Nav\_U\_2}$	[k€/yr]	7.66	6.03	5.37	5.66	0.35	0.35	0.36	0.38	0.37	0.37
	$i_{Nav\_U\_3}$	[k€/yr]	.	.	.	.	.	.	.	.	.	.
Upstream Tourism	$i_{Tou\_U\_1}$	[%]	10.82	8.96	12.22	10.40	19.02	19.29	18.56	15.72	18.22	17.88
	$i_{Tou\_U\_2}$	[%]	8.58	10.29	10.17	10.06	4.37	4.21	27.45	30.68	21.15	23.5
	$i_{Tou\_U\_3}$	[treat./yr]	2.05	1.92	2.28	2.12	2.12	2.16	1.80	1.84	1.80	1.88
	$i_{Tou\_U\_4}$	[d/yr]	35.35	47.14	33.86	38.98	31.88	29.28	41.93	44.80	41.84	40.55

In Tables 15.5 and 15.6 the performances (i.e. the indicator values) of these alternatives are reported, for the upstream sectors and the downstream sectors respectively. The index values reported in Figure 15.5 were computed on the basis of these performances.

We may observe that:

- The Stakeholders interested in the *Upstream* and *Downstream Environment* sectors, represented by the Bolle di Magadino Foundation and the Ticino Parks, prefer the alternatives in the class (SD_{CUR}/$\text{RANGE}_{\text{CUR}}$). In fact, the only alternative in this class to which they are opposed is $A0$. This reveals that these Stakeholders are only interested in a change in the MEF and/or in the regulation policy.
- The Stakeholders interested in *Downstream Flooding*, represented by the Province of Pavia, are not willing to accept alternatives with SD_{+600}; instead they do accept all the alternatives with SD_{CUR}.

Table 15.6. The performances of the alternatives of Table 15.3 in the downstream sectors

Sector	Indicator		$A0$	$A7$	$A18$	$A119$	$A54$	$A168$	$A34$	$A81$	$A122$	$A110$
Downstream Flooding	$i_{\text{Fl\_D\_1}}$	[m]	3.91	3.91	3.90	3.91	3.95	3.95	3.96	3.96	3.96	3.96
	$i_{\text{Fl\_D\_2}}$	[km^2/yr]	1.45	1.45	1.44	1.44	1.55	1.55	1.54	1.56	1.54	1.54
	$i_{\text{Fl\_D\_3\_Bt}}$	[d]	2.	2.	2.	2.	2.	2.	2.	2.	2.	2.
	$i_{\text{Fl\_D\_3\_Pv}}$	[d]	18.	18.	18.	18.	19.	19.	19.	19.	19.	19.
	$i_{\text{Fl\_D\_4\_Bt}}$	[d/yr]	2.72	2.72	2.68	2.72	2.92	2.92	3.00	2.96	3.00	3.00
	$i_{\text{Fl\_D\_4\_Pv}}$	[d/yr]	0.16	0.16	0.12	0.12	0.16	0.16	0.16	0.16	0.16	0.16
	$i_{\text{Fl\_D\_5}}$	[d]	3.	3.	3.	3.	3.	3.	3.	3.	3.	3.
	$i_{\text{Fl\_D\_6}}$	[d/yr]	0.52	0.52	0.52	0.52	0.60	0.60	0.60	0.68	0.64	0.64
Irrigation	$i_{\text{Irr\_D\_1}}$	[Mm^3/yr]	255.61	277.42	314.23	390.10	221.40	284.75	197.32	233.36	288.45	319.99
	$i_{\text{Irr\_D\_2}}$	[–]	26.19	17.23	17.94	25.98	20.25	25.61	16.92	15.96	19.88	22.48
	$i_{\text{Irr\_D\_3\_IC}}$	[–]	4.44	4.90	5.57	2.42	6.69	4.78	6.88	5.90	3.89	4.44
	$i_{\text{Irr\_D\_3\_REC}}$	[–]	21.50	23.50	24.86	20.69	20.58	15.73	22.07	18.44	14.71	16.86
	$i_{\text{Irr\_D\_3\_VC}}$	[–]	12.15	8.01	8.56	5.59	8.08	7.49	8.73	7.90	7.01	8.08
ENEL Power	$i_{\text{EnP\_D\_1}}$	[k€/yr]	3668.50	6814.06	6919.53	7074.59	4239.22	4667.02	4172.67	6982.67	7079.75	7180.81
	$i_{\text{EnP\_D\_2}}$	[d/yr]	0.36	0.00	0.00	0.08	0.00	0.16	0.00	0.00	0.00	0.12
Downstream Tourism	$i_{\text{Tou\_D\_1}}$	[d/yr]	248.42	156.30	152.98	151.02	247.44	250.20	255.31	158.98	160.69	160.61
	$i_{\text{Tou\_D\_2}}$	[d/yr]	4.55	4.44	4.12	4.24	5.32	5.32	6.00	6.20	6.04	6.12
Downstream Environment	$i_{\text{Env\_D\_1}}$	[m^3/s]	195.69	183.92	181.70	180.08	223.75	218.59	231.05	204.63	202.25	201.49
	$i_{\text{Env\_D\_2}}$	[–]	13	13	13	25_{MOD}	13	25_{MOD}	13	13	20_{MOD}	25_{MOD}
	$i_{\text{Env\_D\_3}}$	[d/yr]	156.14	18.67	14.71	9.67	148.74	79.99	161.17	35.46	9.35	15.23
	$i_{\text{Env\_D\_4}}$	[d/yr]	10.28	2.06	1.80	1.50	7.41	2.15	8.42	3.76	1.41	1.66
	$i_{\text{Env\_D\_5}}$	[%]	11.95	17.75	18.67	15.27	30.56	27.06	29.80	21.41	22.98	22.93
East Sesia Power	$i_{\text{ESP\_D\_1}}$	[k€/yr]	1283.47	1389.05	1398.24	1440.89	1086.72	1114.46	1098.51	1501.21	1404.95	1428.93

- The Stakeholders interested in the remaining sectors prefer alternatives that include SD_{+600}.
- The Stakeholders interested in the *Irrigation* and ENEL *Power* sectors, that is the Irrigation Consortia and ENEL, would like $\text{RANGE}_{\text{ENL}}$ to be adopted. On the contrary, the Stakeholders interested in *Upstream Flooding* and *Navigation*, represented by the Municipalities of Locarno and Verbania and the Lake Maggiore Navigation Company, are indifferent to the regulation range that is adopted.
- The Stakeholders interested in the *Fishing* sector, i.e. the professional fishermen represented by the GRAIA Society, accept all the alternatives.

Now we will look at the effects produced by each alternative, comparing the trajectories of the median lake level (estimated over the period 1974–1998) and that of the release into the Ticino in 1984 with the analogous trajectories produced by alternative $A0$. The *median* lake level was chosen instead of the *average* lake level, because the first is less polarized by extreme flood events, thus being more appropriate for representing 'normal conditions'.

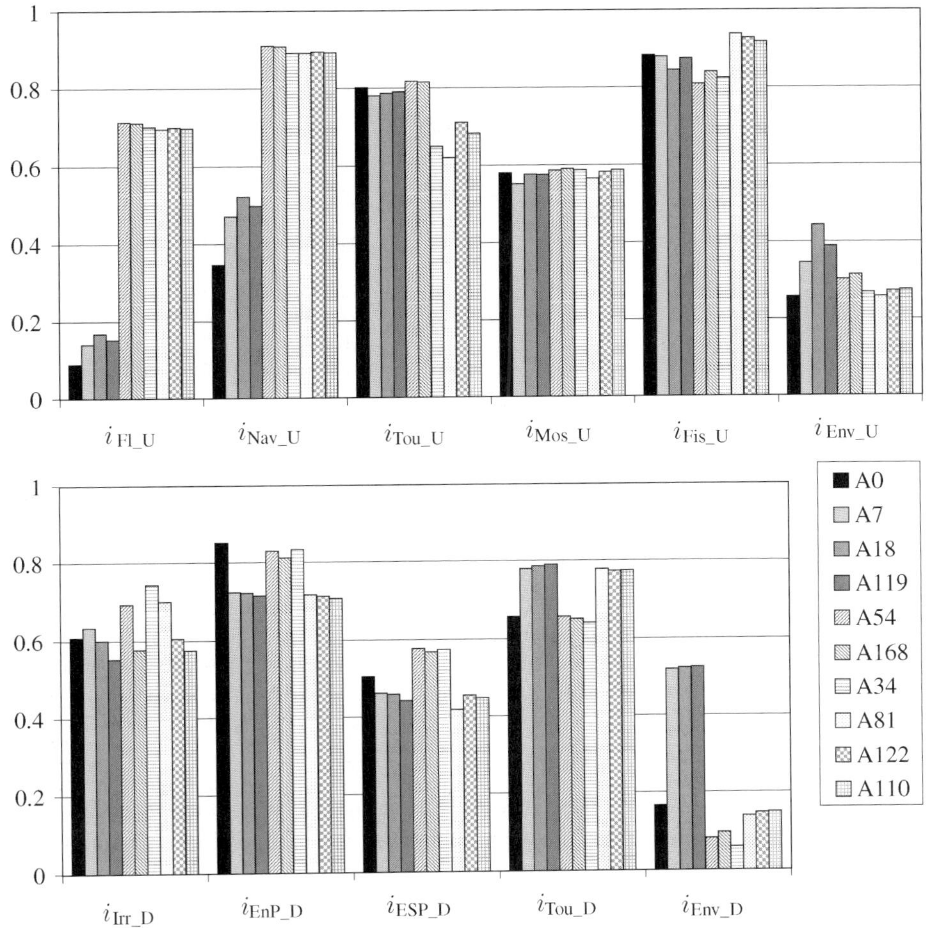

Figure 15.5: The indices produced by the alternatives of Table 15.3 in the upstream (top panel) and in the downstream (bottom) sectors.

Year 1984 was chosen to represent the release trajectory because of the peculiarity of the inflow trajectory in that year, which helps highlighting the characteristics of each alternative.

The description will be aimed at pointing out the sectors that are at an advantage and those that are at a disadvantage because of each alternative. For simplicity's sake, we will speak mostly of five sectors: *Upstream Flooding*, *Upstream Environment*, *Irrigation*, ENEL *Power* and *Downstream Environment*. This will not cause a loss of generality because the satisfaction of each of the other sectors is correlated to the satisfaction of one of these five, as shown by Table 12.3 and explained in Section 10.3. Firstly, we describe the three alternatives that obtain the widest agreement,[3] presenting them in the following order: $A34$, $A54$ and $A119$. Then we take the other alternatives into consideration and illustrate the differences with respect to the first three.

[3]Which is 'measured' by the number of sectors that support or accept the alternative.

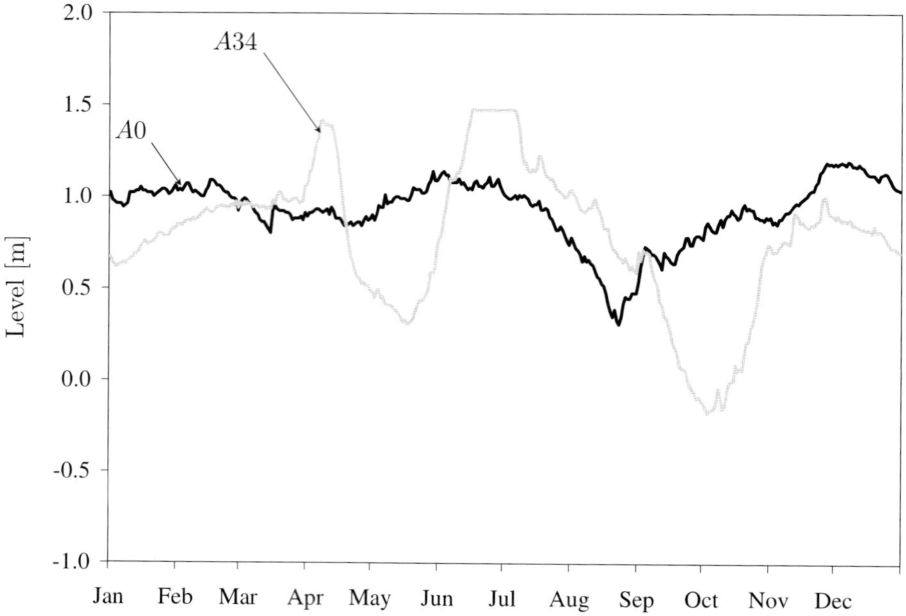

Figure 15.6: The trajectories of the median lake level at Sesto Calende with alternatives *A*0 and *A*34.

15.3.1 Alternative *A*34

The regulation policy of *A*34 was designed by attributing zero weight to the environmental objective: the release decisions are thus guided only by the need to reduce flooding around the lake and to maximize the satisfaction of the irrigation and hydroelectric users. The distribution of the flow downstream from Miorina, once the Ticino has been guaranteed the MEF, is carried out according to the current distribution policy (Section 6.7.3).

Considering Figure 15.6, which shows the trajectories of the median lake level, Figure 15.7, which compares the natural outflow trajectory with the release trajectories produced by *A*34 and *A*0, and the performances and indices reported in the previous tables and figures, the following observations can be made:

- From November until the beginning of March, *A*34 keeps the lake at levels that are lower than *A*0 levels, due to a higher release aimed at maximizing the satisfaction of the hydropower producers (the price of electricity is particularly high in those months). The loss for the ENEL *Power* sector caused by the introduction of the Galli Law is minimal: in fact, the index of this sector drops only to 0.83, from the 0.85 value associated to *A*0 (Figure 15.5, bottom).

- From March to mid-April, *A*34 usually produces a lake storage that is greater than the one resulting from *A*0. In no less than 50% of the years, the full storage is reached at the beginning of April, i.e. at the beginning of the period that the rice paddies are submerged. Therefore *A*34 is more reliable than *A*0 for satisfying the irrigation demand in this period.

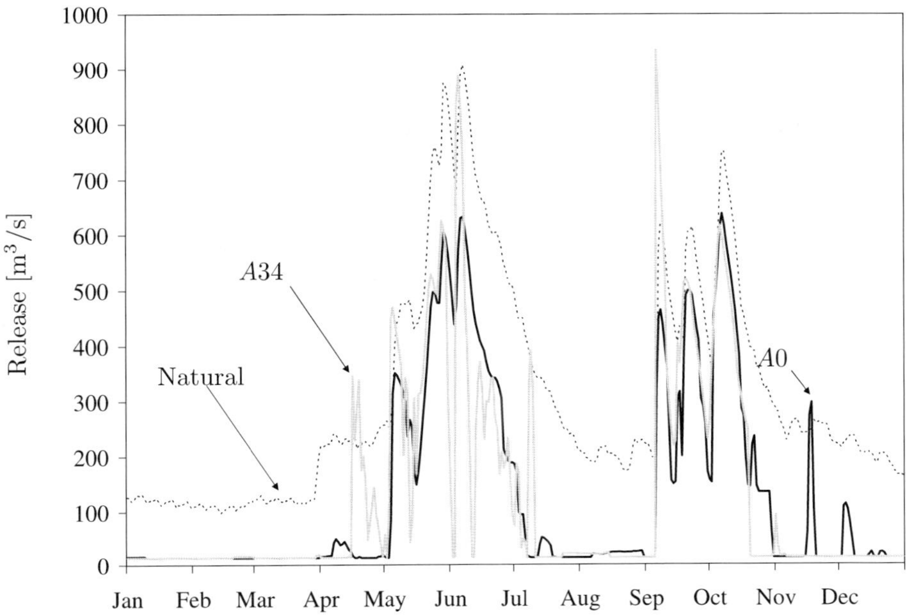

Figure 15.7: Trajectories of the release into the Ticino in 1984 produced by alternatives $A0$ and $A34$ and natural outflow trajectory in the same year.

- From mid-April to mid-May, $A34$ produces a higher release and thus reduces the risk of flooding: notice that the median lake level with $A34$ stays below the $A0$ median until mid-June. It is in this period, in fact, that springtime floods occur because of the snow melt in the lake catchment. An analogous pattern occurs in September. Thanks to the higher release and the greater outflow capacity offered by the SD_{+600}, the *Upstream Flooding* index increases to 0.70 with $A34$, compared to 0.09 achieved by $A0$.

- From mid-June until mid-July, in at least 50% of the years $A34$ achieves the full storage. This favours *Irrigation* in the summer period, when the irrigation demand reaches its maximum. In this period the median lake level produced by $A34$ remains higher than the one obtained with $A0$, because the policy takes into account the Irrigation Consortia's aversion to risk: by conserving water volumes in the lake, small deficits may occur each day but the probability of producing severe deficits in critical periods is reduced. Consequently, the indicator for the potential damage from stress ($i_{\text{Irr\_D\_2}}$), which is equal to 26.19 with $A0$, is reduced to 16.92 with $A34$ (Table 15.6) and the index of the sector rises from 0.61 ($A0$) to 0.74 ($A34$) (Figure 15.5, bottom).

- The index of the *Upstream Environment* sector is almost the same for both alternatives (0.27 with $A0$, 0.26 with $A34$). In fact, with respect to $A0$, $A34$ produces an increase in the distance from the natural regime and a worsening in the erosion of the reed beds and in the reproduction of the Cyprinids, but these negative effects are almost completely compensated by an improvement in the nesting conditions for the bird life and in the increase in the shelf emergence period (see Table 15.5).

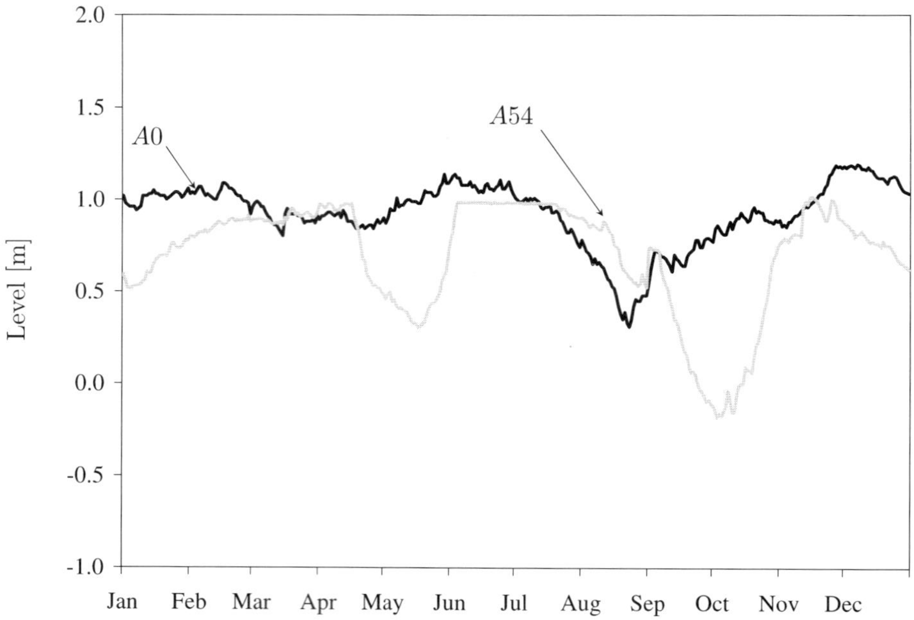

Figure 15.8: The trajectories of the median lake level at Sesto Calende with alternatives $A0$ and $A54$.

- Alternative $A34$ is seen to be disadvantageous by the *Downstream Environment* sector for two reasons. The first is that the trajectory of the release into the Ticino shows sudden peaks and violent oscillations, which are very dangerous for the river ecosystem. Both phenomena can be seen in Figure 15.7, where the release trajectories produced by $A34$ and $A0$ are shown and compared with the natural outflow trajectory. Notice that the $A0$ trajectory has more accentuated peaks than the natural outflow trajectory, but it has no oscillations. The latter are a direct consequence of excavating the lake outlet and enlarging the regulation range: maintaining the maximum storage for prolonged periods requires a timely opening of the dam when the inflow exceeds the demand, to prevent the level from rising out of the regulation range. The sudden peaks produced by the opening are higher than those produced with $A0$, both because the lake level is higher (1.50 m instead of 1.00 m) and because excavating permits a greater outflow for the same level. The second reason why *Downstream Environment* considers $A34$ disadvantageous is that, for a good six months each year, namely from November to April, $A34$ keeps the release into the Ticino almost constantly equal to the MEF (13 m^3/s) so that in that period the Ticino is similar to a canal, losing the natural fluctuations that characterize a river. Both of these reasons come together to bring the index of *Downstream Environment* down from 0.16 ($A0$) to 0.06 ($A34$).

15.3.2 Alternative $A54$

Alternative $A54$ has very similar characteristics to $A34$: as we can see in Figure 15.8, the trajectory of the median lake level has a similar form to $A34$, except for the fact that the maximum storage is limited to 1.00 m, since $A54$ adopts $\text{RANGE}_{\text{CUR}}$. From the analysis of the performances and of the indices it emerges that:

- With *A*54 the *Upstream Flooding* index is practically the same (0.71) as the one produced by *A*34 (0.70). This shows that, with an efficient regulation policy, raising the higher extreme of the range does not bring an increase in the risk of flooding.

- The effect of the different regulation range is felt in the *Upstream Tourism* sector, whose index value rises to 0.82 from 0.65 associated to *A*34. Therefore, while *A*34 is better than *A*0 for the *Upstream Flooding* sector but worse for the *Upstream Tourism* sector, *A*54 is preferable for both sectors. This is the reason why the representatives of the *Upstream Flooding* sector, i.e. the Municipalities of Locarno and Verbania, prefer *A*54 to *A*34, even if in order to reduce the damage from lake flooding they are willing to accept a reduction in the *Upstream Tourism* index (see page 294).

- The index of the ENEL *Power* sector remains the same (0.83) as for *A*34 and so only a little lower than for *A*0 (0.85). We could have expected this result, since *A*34 satisfies the hydropower demand especially in the winter and thus replacing $\text{RANGE}_{\text{CUR}}$ with $\text{RANGE}_{\text{ENL}}$ cannot provoke significant variations in the index value of this sector.

- However, the choice of the regulation range has effects on the *Irrigation* sector, whose index falls to 0.69 from 0.74 with *A*34; nevertheless, it remains greater than the value (0.61) associated to *A*0.

- The regulation range also influences the satisfaction of the *Upstream Environment* sector, but in the opposite way. The sector index rises to 0.30, compared to 0.27 for *A*34 and 0.26 for *A*0.

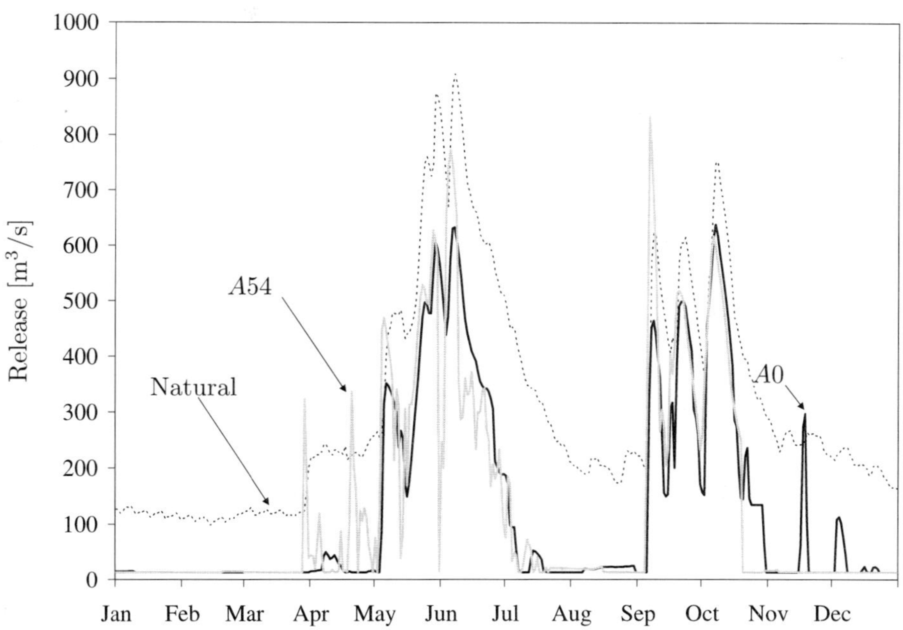

Figure 15.9: Trajectories of the release into the Ticino in 1984 produced by alternatives *A*0 and *A*54 and natural outflow trajectory in the same year.

- Alternatives $A54$ and $A34$ are very similar also in their effects on the *Downstream Environment*, as is shown in Figure 15.9: in fact, $A54$ also produces very high peaks and long periods in which the flow remains constant and equal to the MEF. The value of the sector index is 0.08 and so much less than the value (0.16) associated to $A0$.

15.3.3 Alternative $A119$

Unlike $A34$ and $A54$, the regulation policy for $A119$ was designed considering the environmental objective as well. Moreover, the MEF used is 25_{MOD} rather than 13_{CON}. These differences must be borne in mind when analysing the graphs presented in this section, especially when comparing them to Figures 15.6–15.9. Look, for example, at Figure 15.10, which compares the trajectories of the median lake level generated by $A119$ and $A0$. From November to the end of March the trajectory generated by $A119$ is lower than that produced by $A0$. However, one can no longer conclude a priori, as we did with $A34$, that this is due to the satisfaction of hydropower production. The difference could also be caused by a greater release for environmental purposes. Figure 15.11 shows that this is precisely what happens.

As for the indices:

- With $A119$ the value of the *Upstream Flooding* sector index is 0.15, and thus it is greater than the value (0.09) associated to $A0$ but much less that those associated to $A34$ (0.70) and $A54$ (0.71). The improvement with respect to $A0$ is due to the increase in the autumn release; the worsening with respect to the other two alternatives is due to the lack of a significant springtime release and the adoption of SD_{CUR}, which results in less outflow for the same lake level.

- The value of the ENEL *Power* sector index falls to 0.71 from the value of 0.85 associated to $A0$. As we have seen, the reason is that in the winter period alternative $A119$ increases the flow in the Ticino to the detriment of the Industrial Canal.

- The index of *Irrigation* diminishes, with respect to $A0$, passing from 0.61 to 0.55. The increase in the flow into the Ticino in the spring–summer period (Figure 15.11) imposes a reduction in the flow diverted into the canals, so that the average annual supply deficit rises from 255 Mm^3/year ($A0$) to 390 Mm^3/year ($A119$). This notwithstanding, the potential damage from stress ($i_{\text{Irr\_D\_2}}$) is just a little bit less, passing from 26.19 ($A0$) to 25.61 ($A119$).

- The index of *Upstream Environment* increases, passing from 0.26 ($A0$) to 0.38.

- The index of *Downstream Environment* has a remarkable improvement, from 0.16 ($A0$) to 0.52, thanks to the change in the MEF value and in the regulation policy, which explicitly considers this sector.

15.3.4 Other alternatives in the ($\text{SD}_{+600}/\text{RANGE}_{\text{ENL}}$) class: $A81$, $A110$ and $A122$

The other three alternatives in the ($\text{SD}_{+600}/\text{RANGE}_{\text{ENL}}$) class are different from $A34$ because the environmental objective was taken into account in the design of the regulation policies and because the value of the MEF passes from 13_{CON} in $A34$ to 20_{MOD} in $A122$ and to 25_{MOD} in $A110$. Figure 15.12 shows that consequently the maximum storage is not reached in either April or in June–July.

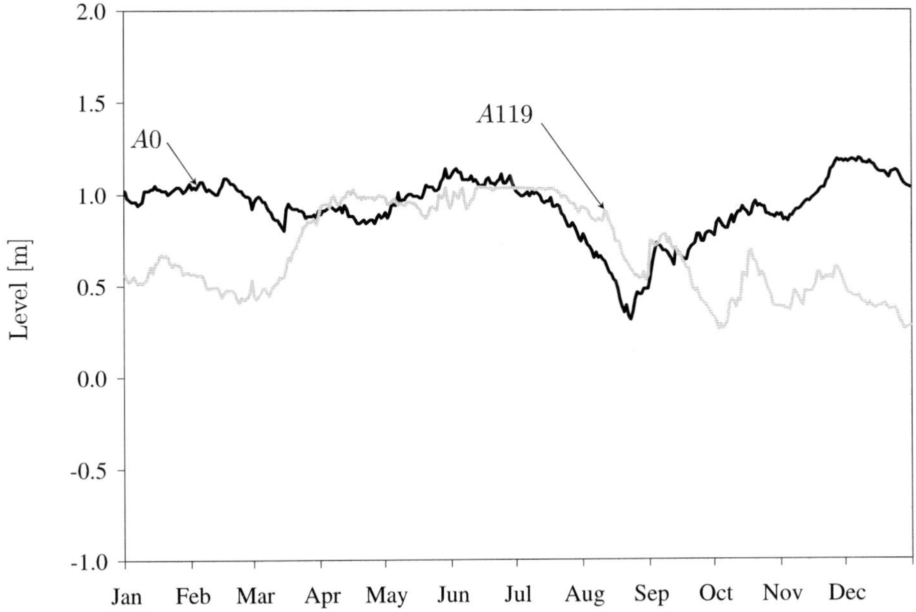

Figure 15.10: The trajectories of the median lake level at Sesto Calende with alternatives $A0$ and $A119$.

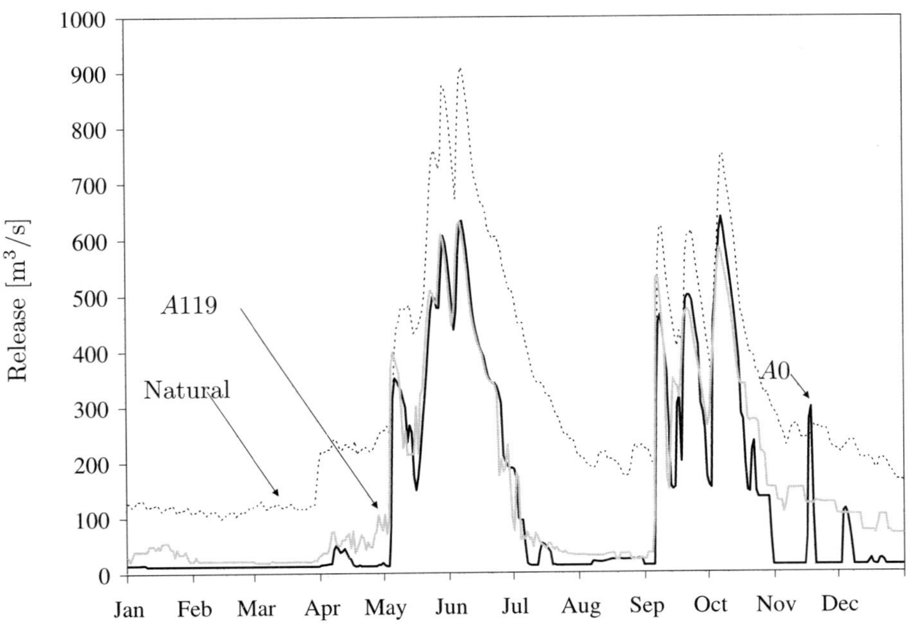

Figure 15.11: Trajectories of the release into the Ticino in 1984 produced by alternatives $A0$ and $A119$ and natural outflow trajectory in the same year.

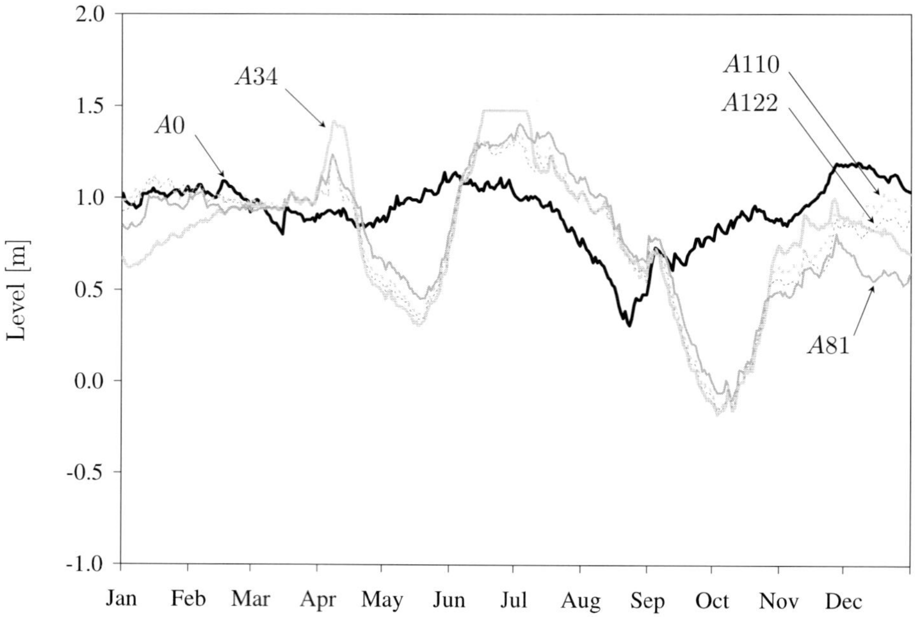

Figure 15.12: The trajectories of the median lake level at Sesto Calende with alternatives in the ($\text{SD}_{+600}/\text{RANGE}_{\text{ENL}}$) class and with $A0$.

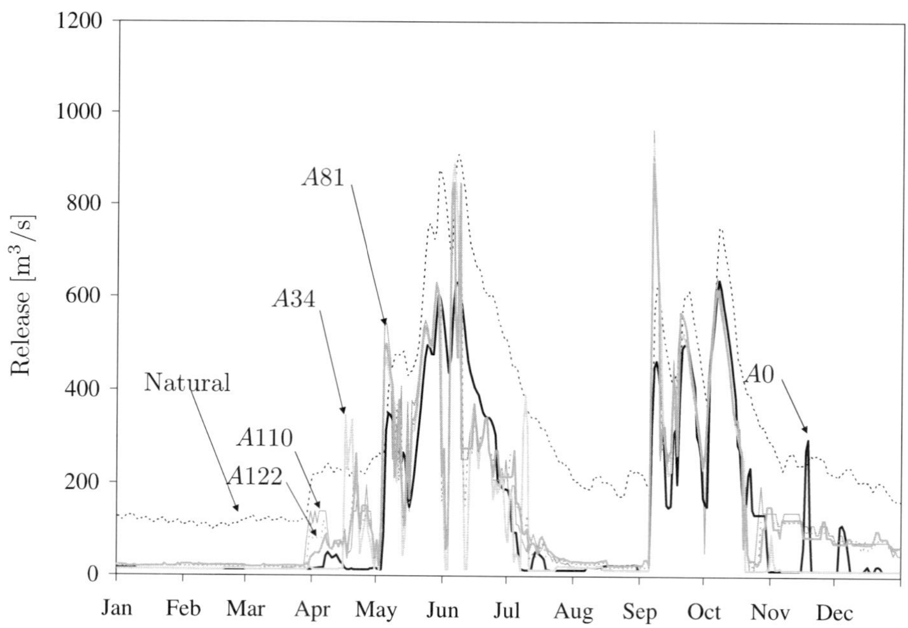

Figure 15.13: Trajectories of the release into the Ticino in 1984 produced by the alternatives in the ($\text{SD}_{+600}/\text{RANGE}_{\text{ENL}}$) class and by $A0$, and natural outflow trajectory in the same year.

As for the indices:

- The index of the *Downstream Environment* sector increases from 0.06 for *A*34 to 0.14 for *A*81 and *A*110, until it reaches 0.15 for *A*122, but it remains less than the index for *A*0 (0.16), even if only a little. The improvements produced by these alternatives, with respect to *A*34, can be appreciated by looking at Figure 15.13, which shows that the trajectories of the release into the Ticino in the periods November–December and July–August are higher and much more variable than with *A*34. In contrast, in the first months of the year the differences among the alternatives are not significant. This is due to the fact that when inflows are scarce, all the alternatives produce a release equal to the MEF. Given that in winter the difference between the MEF values of the various alternatives is reduced, the trajectories are very similar in this period. All the alternatives produce peaks and violent oscillations in the release into the Ticino, as we already noted when analysing *A*34, which are due to SD_{+600}. It is their presence that determines the low index of *Downstream Environment* and the consequent opposition from the Ticino Parks.

- The index of the *Upstream Environment* sector does not undergo significant variations with respect to *A*34: it remains unchanged at 0.27, except for *A*81, which is associated with a value of 0.26 (the same as for *A*0).

- The index of *Irrigation* falls from 0.74 for *A*34 to 0.70 for *A*81, to 0.61 for *A*122, and reaches 0.57 for *A*110: a value that is less than the value for *A*0, which is equal to 0.60.

- The index of ENEL *Power* is equal to 0.71 in all three of the alternatives and so much lower than for *A*34 (0.83) and *A*0 (0.85).

- The index of *Upstream Flooding* is very high for all three alternatives: 0.69 for *A*81, and 0.70 for *A*110 and *A*122.

15.3.5 Other alternatives in the (SD_{+600}/$\text{RANGE}_{\text{CUR}}$) class: *A*168

Alternative *A*168 produces behaviours and performances similar to *A*54. This is not surprising, since they practically differ only in the value of the MEF (13_{CON} in *A*54, 25_{MOD} in *A*168), while their policies were designed with the same weights for the objectives and, in particular, with zero weight given to the environmental objective. Figure 15.14 shows that, regarding the median lake level, the difference between *A*168 and *A*54 is concentrated in the late autumn and winter months, in which *A*168 maintains a higher storage than *A*54. Since there is no risk of flooding in this period, the only active objective is irrigation, and so both policies tend to fill the lake to guarantee that the rice paddies can be submerged in spring. Since *A*168 has a higher MEF, it must maintain a higher storage level to have the same reliability as *A*54, but this brings about a reduction in the supply to the hydropower producers. The index of the ENEL *Power* sector thus falls to 0.81 from 0.83, which is its value for *A*54.

In the summer period, however, it is not possible to fill the lake any further because *A*54 already achieves full storage in at least 50% of the years. It follows that, in order to increase the flow into the Ticino (see Figure 15.15), as imposed by the MEF, it is no longer sufficient to reduce the hydropower supply, but it is necessary to cut into the irrigation supply as well.

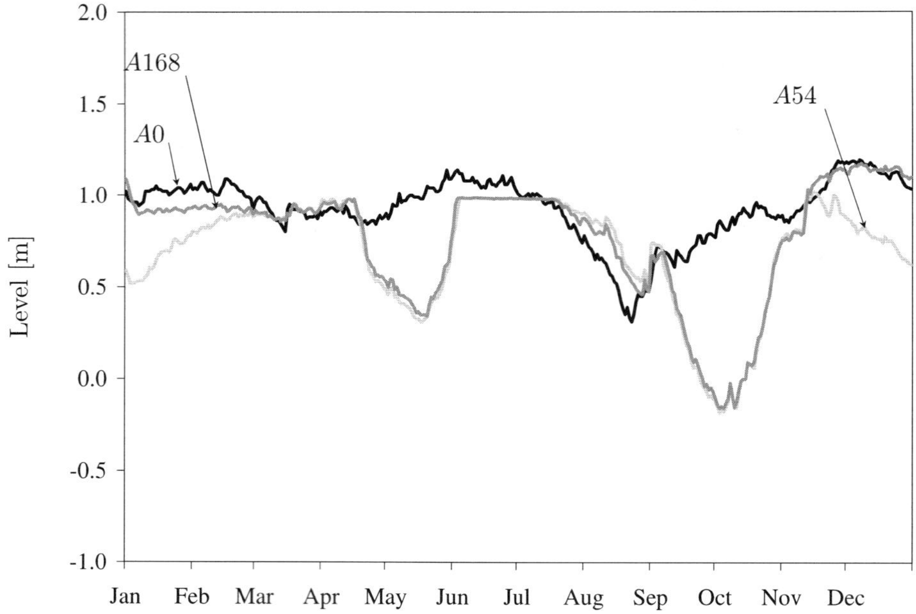

Figure 15.14: Trajectories of the median lake level at Sesto Calende with alternatives in the (SD_{+600}/$RANGE_{CUR}$) class and with $A0$.

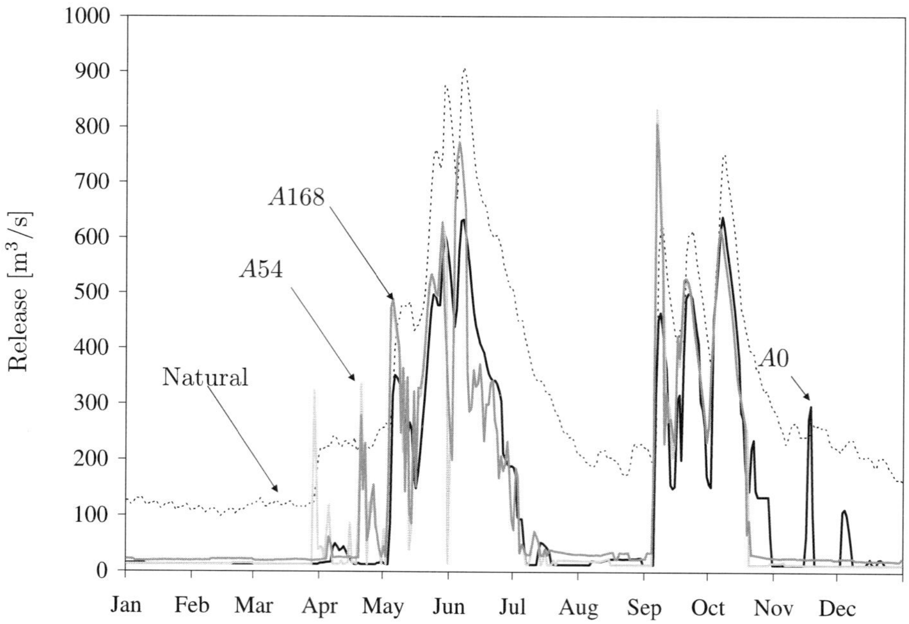

Figure 15.15: Trajectories of the release into the Ticino in 1984 produced by the alternatives in the (SD_{+600}/$RANGE_{CUR}$) class and by $A0$, and natural outflow trajectory in the same year.

The value of the *Irrigation* sector index thus falls from 0.69, which is the value for *A*54, to 0.57.

In conclusion, passing from *A*54 to *A*168 the indices for the ENEL *Power* and *Irrigation* sectors diminish, the index of the *Downstream Environment* sector increases (from 0.08 to 0.09) and the indices for the other sectors remain unchanged.

15.3.6 Other alternatives in the (SD_{CUR}/$RANGE_{CUR}$) class: *A*7 and *A*18

Alternatives *A*7 and *A*18 share with *A*119 the fact that they were designed giving some weight to the environmental objective, but they differ in their MEF value (13_{CON} in both, while *A*119 adopts 25_{MOD}). From Figure 15.16 it can be noted that all three alternatives produce very similar trajectories of the median lake level; the only difference is that, with respect to *A*7 and *A*119, *A*18 produces slightly higher releases in the spring and in the autumn (more pronounced). This is why the index of the *Upstream Flooding* sector increases slightly with respect to *A*119, passing from 0.15 to 0.16. However, the difference is so modest that the Stakeholders interested in this sector are indifferent to it.

Figure 15.17 shows that there are no significant differences in the release into the Ticino either, with the exception of the summer period, in which the effects of the different MEF values can be appreciated. All three alternatives produce that same index value (0.52) for *Downstream Environment*, but since *A*119 has a higher MEF, the Ticino Parks support it, while only accepting the other two.

15.4 Mitigation measures

By introducing mitigation measures, the alternatives of Table 15.3 are transformed into the reasonable alternatives of Table 15.7, and the positions of the Stakeholders become those in Table 15.8.

The positions are noticeably changed: only the Ticino Parks continue to prefer the alternatives of the class (SD_{CUR}/$RANGE_{CUR}$), while the Bolle di Magadino Foundation now supports all the alternatives with SD_{+600}, which are accepted by the Province of Pavia as well. The mitigation measures thus play a fundamental role in the decision-making process and so careful attention must be paid to them. We will briefly describe them (the reader can find more details in Chapter 14), beginning with the one that was studied for the *Downstream Flooding* sector, and then passing on to the one dedicated to *Upstream Environment*. For the other sectors, either no mitigation measures were identified or they were not sought, because the interested Stakeholders were satisfied with the performances that were obtained.

15.4.1 Mitigation for the Downstream Flooding sector

For each of the reasonable alternatives and for each flood event in the evaluation horizon, the peak flood level at Pavia was estimated using the neural network described in Section 6.9.1, and the estimates were compared with those produced by *A*0. We found out that the alternatives that adopt SD_{+600} can increase the peak flood level at Pavia; for example, with *A*34 in 81% of flood events the peak is greater by several centimeters (19 cm in the worst event, Section 14.1).

This undesired effect could probably be reduced, if not completely eliminated, by associating the structural action SD_{+600} with a rule that orders that the release from the lake be a function of the flow in the Po, when the lake level exceeds the upper extreme of the

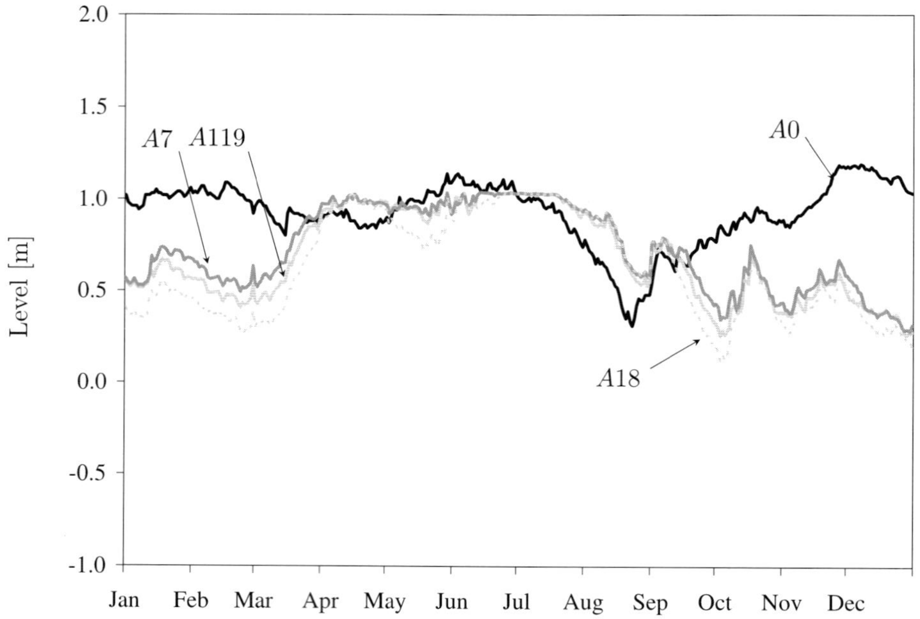

Figure 15.16: Trajectories of the median lake level at Sesto Calende with alternatives in the ($SD_{CUR}/RANGE_{CUR}$) class and with $A0$.

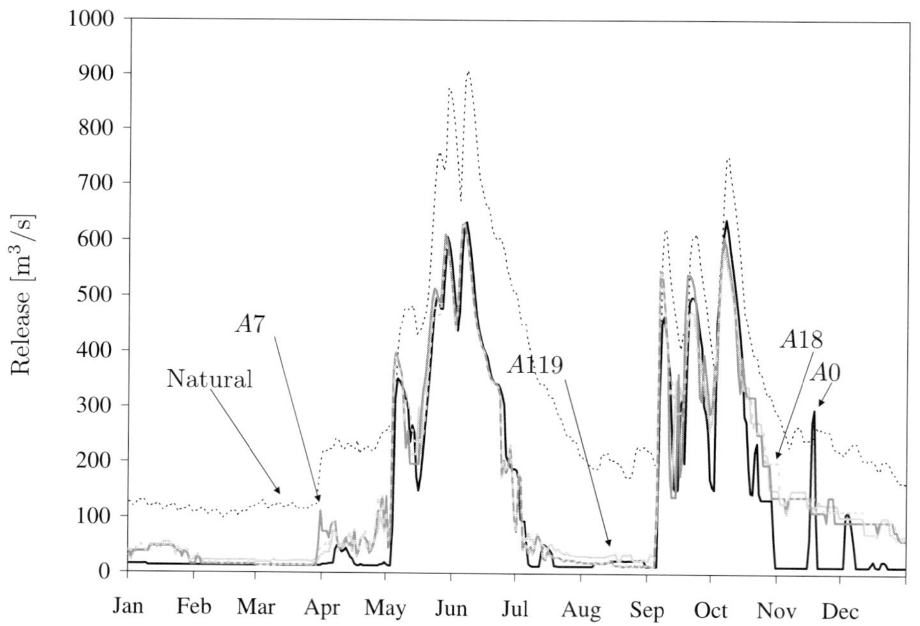

Figure 15.17: Trajectories of the release into the Ticino in 1984 produced by the alternatives in the ($SD_{CUR}/RANGE_{CUR}$) class and by $A0$, and natural outflow trajectory in the same year.

Table 15.7. The final set of reasonable alternatives

Alternative	SD	RANGE	MEF	Mitigation measure	Weights		
					λ_{fl}	λ_{irr}	λ_{env}
$A0$	ACT	ACT	13_{CON}	–	–	–	–
$A7$	ACT	ACT	13_{CON}	–	0.00	0.99	0.01
$A18$	ACT	ACT	13_{CON}	–	0.01	0.98	0.01
$A119$	ACT	ACT	25_{MOD}	–	0.01	0.98	0.01
$M54$	+600	ACT	13_{CON}	REDUCTION WITH PO BOLLE RESERVE	0.99	0.01	0.00
$M168$	+600	ACT	25_{MOD}	REDUCTION WITH PO BOLLE RESERVE	0.99	0.01	0.00
$M34$	+600	ENL	13_{CON}	REDUCTION WITH PO BOLLE RESERVE	0.99	0.01	0.00
$M81$	+600	ENL	13_{CON}	REDUCTION WITH PO BOLLE RESERVE	0.60	0.38	0.02
$M122$	+600	ENL	20_{MOD}	REDUCTION WITH PO BOLLE RESERVE	0.60	0.39	0.01
$M110$	+600	ENL	25_{MOD}	REDUCTION WITH PO BOLLE RESERVE	0.55	0.35	0.10

regulation range. However, the search for the optimal rule was considered to be a task beyond the limits of the Project. For this reason we only conducted a few trials (Section 14.2), with the aim of providing some useful indications that would guide any future research into the matter. We simulated the effects of a rule (which we called REDUCTION WITH PO) that envisages the complete opening of the Miorina dam, when the upper extreme of the regulation range is reached, only as long as the release is less than a given threshold $\bar{r}_t$. For higher values, the dam is regulated in such a way that the release is equal to the minimum value between $\bar{r}_t$ and the release that would occur with SD_{CUR}, at the same lake level. By doing so, when the lake level exceeds the level $\bar{h}_t$ at which the new stage–discharge relation would release $\bar{r}_t$, the lake would be driven by the current stage–discharge relation, i.e. the dynamics of the lake would remain the same as today. The trials that were performed showed that the threshold $\bar{r}_t$ should not be fixed: it should depend on the forecast $\hat{q}_{t+1}^{\text{Po}}$ of the flow in the River Po for the current day $[t, t+1)$. For example, the threshold could be chosen with the following rules:

(1) if $\hat{q}_{t+1}^{\text{Po}} < 5000$ m^3/s, then $\bar{r}_t = \infty$, which means that the release is determined by SD_{+600};

(2) if $\hat{q}_{t+1}^{\text{Po}} \in [5000, 6000)$, then $\bar{r}_t = 2300$ m^3/s;

(3) if $\hat{q}_{t+1}^{\text{Po}} \in [6000, 7000)$, then $\bar{r}_t = 1500$ m^3/s;

(4) if $\hat{q}_{t+1}^{\text{Po}} > 7000$, then $\bar{r}_t = 0$ m^3/s, which means that the release is determined by SD_{CUR}.

By combining this rule with alternative $A34$, alternative $M34$ is obtained, which, in all the events except for one, does not increase the peak at Pavia or even reduces it, while

Table 15.8. The positions of the Stakeholders with respect to the reasonable alternatives. The precise meaning of the terms is defined in Section 10.1

SD	RANGE	MEF	Alternative	Stakeholders			
				Support	Accept	Oppose	Tolerate
CUR	CUR	13_{CON}	*A0*	GRAIA Society ENEL Company	Province of Pavia	Municipalities of Verbania and Locarno Lake Navigation Company Bolle di Magadino Foundation East Sesia Consortium East Ticino Consortium Ticino Parks	
		13_{CON}	*A7*	Bolle di Magadino Foundation	GRAIA Society Ticino Parks Province of Pavia	Municipalities of Verbania and Locarno Lake Navigation Company ENEL Company	East Sesia Consortium East Ticino Consortium
		13_{CON}	*A18*	Bolle di Magadino Foundation	GRAIA Society Ticino Parks Province of Pavia	East Sesia Consortium East Ticino Consortium ENEL Company	Municipalities of Verbania and Locarno Lake Navigation Company
		25_{MOD}	*A119*	Bolle di Magadino Foundation Ticino Parks	GRAIA Society Province of Pavia	Municipalities of Verbania and Locarno Lake Navigation Company East Sesia Consortium East Ticino Consortium ENEL Company	

(*continued on next page*)

Table 15.8. *(Continued)*

SD	RANGE	MEF	Alternative	Stakeholders			
				Support	Accept	Oppose	Tolerate
+600	CUR	13_{CON}	$M54$	Municipalities of Verbania and Locarno Lake Navigation Company	GRAIA Society East Sesia Consortium East Ticino Consortium ENEL Company Province of Pavia	Ticino Parks Bolle di Magadino Foundation	
		25_{MOD}	$M168$	Municipalities of Verbania and Locarno Lake Navigation Company	GRAIA Society ENEL Company Province of Pavia	East Sesia Consortium East Ticino Consortium Bolle di Magadino Foundation	Ticino Parks
	ENL	13_{CON}	$M34$	Municipalities of Verbania and Locarno Lake Navigation Company East Sesia Consortium East Ticino Consortium	GRAIA Society ENEL Company Province of Pavia	Ticino Parks Bolle di Magadino Foundation	
		13_{CON}	$M81$	Municipalities of Verbania and Locarno Lake Navigation Company	GRAIA Society East Sesia Consortium East Ticino Consortium Province of Pavia	ENEL Company Bolle di Magadino Foundation	Ticino Parks
		20_{MOD}	$M122$	Municipalities of Verbania and Locarno Lake Navigation Company	GRAIA Society East Sesia Consortium East Ticino Consortium Province of Pavia	ENEL Company Bolle di Magadino Foundation	Ticino Parks
		25_{MOD}	$M110$	Municipalities of Verbania and Locarno Lake Navigation Company	GRAIA Society Province of Pavia	East Sesia Consortium East Ticino Consortium ENEL Company Bolle di Magadino Foundation	Ticino Parks

maintaining the same index value for *Upstream Flooding* as $A34$ (for more details see Section 14.2.2).

Unfortunately, the evaluation horizon contains only two flood events that are really meaningful for evaluating the effects of the alternatives at Pavia, because only on two occasions did the peak exceed the flood level (3.50 m at the Covered Bridge hydrometric station). In order to increase the number of events with which the mitigation measure REDUCTION WITH PO could be evaluated, we decided to examine the flood event of October 2000 (Section 14.2.2.3). This event is particularly interesting, both because it was not used in the design of the alternatives, and so simulating the policy in that event means validating it, and because the event was exceptional. It has a return time of about 70 years and it caused greater damage than any other event considered so far (in Switzerland alone the damage amounted to around M€ 110). The results obtained are very promising: as shown in Tables 14.11 and 14.12, the lake level peak would have been reduced by 69 cm with respect to the level that actually occurred ($A0$) and the Ticino level peak at Pavia would have been reduced by 10 cm (from 3.86 to 3.76 m).

These results should not be considered conclusive (for many reasons, including the fact that the effects that might be obtained with a real predictor of the flow in the Po were not studied); they are only a stimulus for further investigations, which should move in two directions: modifying the dam structure and changing the regulation policy. Structural changes to the dam should be thought of in order to allow for regulation even in flood conditions; new regulation policies should be designed by explicitly considering the objective of reducing downstream flooding and without constraining the release in correspondence to particular lake levels, as the regulation range does.

15.4.2 Mitigation for the Upstream Environment sector

With all the reasonable alternatives, the sector index of *Upstream Environment* can be increased by introducing a set of mitigation measures. The measures should be specifically designed for each of the areas along the perimeter of the lake that have high environmental value, but since this task exceeds the limits of the Project, we studied only the set of measures (BOLLE RESERVE) to be applied to the Bolle di Magadino reserve, because this area is by far the largest and of the greatest environmental value.

The interventions that were defined (described in Section 14.3) can be summarized as follows:

- the creation of shelves at higher levels than the current ones;
- acquisition of 40 to 60 hectares of land that is currently used for agriculture behind the marsh area, with the aim of allowing the marshland ecosystem to move inland, when the lake level regime rises because of the adoption of $\text{RANGE}_{\text{ENL}}$;
- reactivation of the delta dynamics, by removing a part of the Ticino banks at its mouth, so that the river would again influence the lands acquired.

The cost of these interventions was estimated at 10÷15 million Euro.

By applying these interventions to alternative $A34$, for example, the index of *Upstream Environment* increases from 0.27 ($A34$) to 0.85: a significant result. What was not studied, however, and this should be done, is how to manage the transition from the current condition to the new steady-state regime: a too fast transition could provoke serious damage, since the ecosystem would not have the time needed to adapt to the new situation (Section 2.2.1.6 of the DVD). If this study demonstrated that the transition could be managed safely, the Bolle di Magadino Foundation declared that it would support all the alternatives that were mitigated in this way. To conclude the discussion we note that analogous measures must be studied also for the other areas of environmental importance.

15.5 Monitoring

The choice of the best compromise alternative would have to be accompanied by the definition of the monitoring system that would check its actual effects over time. However, the monitoring of the effects induced by the management of the Verbano is precisely the task of the Italian–Swiss Commission, as stated in the Act of its foundation. Moreover, the data that should be collected by the monitoring system are the same data that are currently collected by the Ticino Consortium as part of its remit. Thus, no further interventions have to be defined for this purpose.

Chapter 16

The Final Decision

EB, FC and RSS

Once the reasonable alternatives have been identified, the Italian-Swiss International Commission, which oversees the use of Verbano's waters, will have to express an opinion about the suitability of the Project and the reasonable alternatives. If they are considered to be well founded, the choice will be in the hands of the final Decision Makers (DMs), the Italian and Swiss governments, who will negotiate to identify the *best compromise alternative*. In this chapter we analyse how these last negotiations might be carried out.

16.1 The decision

The *Final Decision* problem is a Multi-Decision-Maker Problem, since both Italy and Switzerland are autonomous Decision Makers, even if their decisions are conditioned in some cases by the other's agreement. In fact, Switzerland cannot decide to excavate the lake outlet without Italy's agreement, since the outlet is not in its territory. On the other hand, Italy could perhaps decide to excavate autonomously, but certainly could not consider modifying the regulation range without Swiss agreement, since Verbano is an international lake.

The alternatives from which Italy and Switzerland will choose are those presented in Table 14.15, with respect to which the Stakeholders position themselves as shown in Table 14.16. In order to identify the best compromise alternative we cannot use one of the traditional decision methods, like, for example, the MAVT (Chapter 9), because they are suited for the single Decision Maker case. Instead we can use the Pareto Race (Section 10.1), in the modified version that we used in the *Comparison* phase (Elementary Negotiation Procedure, ENP). Nevertheless the negotiations have not yet occurred and without the real Decision Makers the ENP cannot be employed. This is why the analysis of what might come out of negotiations between Italy and Switzerland must be carried out with other tools.

16.1.1 A two Decision Makers game: Italy and Switzerland

The analysis of the negotiation process must be dynamic, since the process is made up of a sequence of proposals and counterproposals between the two Parties until they get to an equilibrium point. Thus negotiations can be represented as a game with two players, who cooperate with each other, or better, who seem to cooperate, since each of them pursues the aim of maximizing his/her own satisfaction. Negotiations are concluded when they get to a situation in which neither of the players can take a decision that increases his/her own satisfaction without being vetoed by the other, who opposes it because his/her satisfaction would be diminished.

- **First move:** *Switzerland proposes excavating the lake outlet.*

  The first move has already been made by Switzerland, who put forward the request for excavation several years ago: it was that very request that led to this Project. From the results that have been obtained, we must now evaluate whether it is plausible that Switzerland will maintain the proposal.

  To do this, we will examine the conflict between the two groups of Swiss Stakeholders who are in favour of and contrary to the proposal. We already know that the conflict is very modest because the upstream Stakeholders agree that the excavation is a positive thing to do, since it increases the index of almost all the upstream sectors. Actually, there are a few alternatives that diminish the *Upstream Tourism* and *Upstream Environment* sector indices, with respect to $A0$. Nevertheless, as for *Upstream Tourism*, internal compensation holds: in fact, as shown in Figures 4.2 and 4.3, the Stakeholders interested in the *Upstream Tourism* sector are also interested in *Upstream Flooding*, so that the disadvantages in the first can be compensated by the big advantages for the second. As for the *Upstream Environment* sector, with the introduction of the measure BOLLE RESERVE, the internal conflict would almost disappear, because that measure makes the index higher than $A0$'s for all the alternatives with SD_{+600}. Except for the Bolle di Magadino, along the Swiss shores of the lake there are only a few, small areas with natural value, the interests of which could not realistically influence the decision. We can thus conclude that it is more than plausible that Switzerland will maintain the proposal for excavation, and ask Italy to agree to the stage–discharge relation that we have called SD_{+600}.

  By observing Figure 16.1 it can be noted that, once the excavation has been carried out and if the regulation range remains the current one ($\text{RANGE}_{\text{CUR}}$), Switzerland is almost indifferent as to which regulation policy is applied. In fact, the increase in the *Upstream Flooding* index is such as to make the differences caused by the different regulation policies negligible. This is still true even when all the other sectors that obtain an advantage from the intervention are considered. We can thus conclude that at first Switzerland will maintain its request for excavating the lake outlet without adding any normative condition and will leave the choice of the regulation policy to the Ticino Consortium as it is at present.

- **Second move:** *Italy responds by saying that excavating the outlet can only be considered if it is accompanied by the enlargement of the regulation range* (i.e. action $\text{RANGE}_{\text{ENL}}$).

  Unlike Switzerland, the internal conflict in Italy is very marked and particularly heated between the *Irrigation* and ENEL *Power* sectors on one side and the *Downstream Environment* sector on the other.

  By observing Figure 16.2, which shows the values of the *Irrigation* sector index, it can be noted that alternative $A0$ is not the most satisfactory for this sector. Moreover, in the classes of alternatives ($\text{SD}_{+600}/\text{RANGE}_{\text{CUR}}$) and ($\text{SD}_{+600}/\text{RANGE}_{\text{ENL}}$), the policies that produce the highest values for *Irrigation* are not among the reasonable

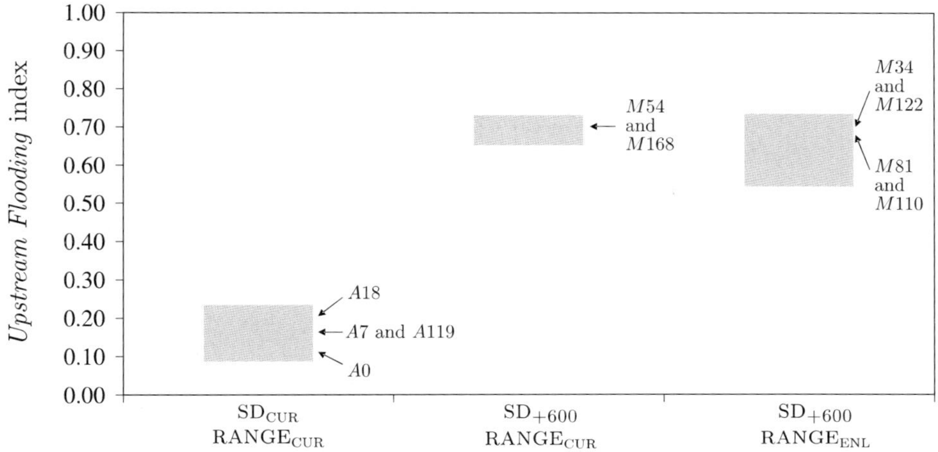

Figure 16.1: Values of the *Upstream Flooding* sector index for different classes of alternatives. For each class, the bar represents the range of index values produced by changing the policy. The values that correspond to the reasonable alternatives are highlighted.

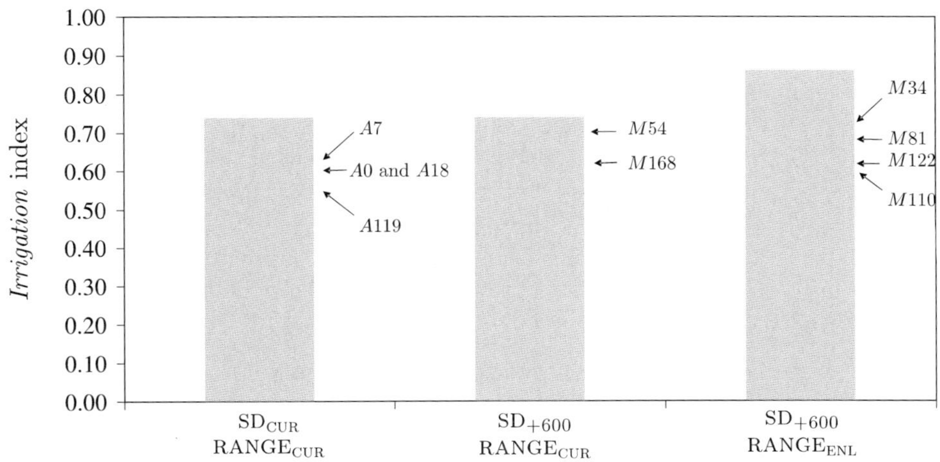

Figure 16.2: Values of the *Irrigation* sector index for different classes of alternatives. For each class, the bar represents the range of index values produced by changing the policy. The values that correspond to the reasonable alternatives are highlighted.

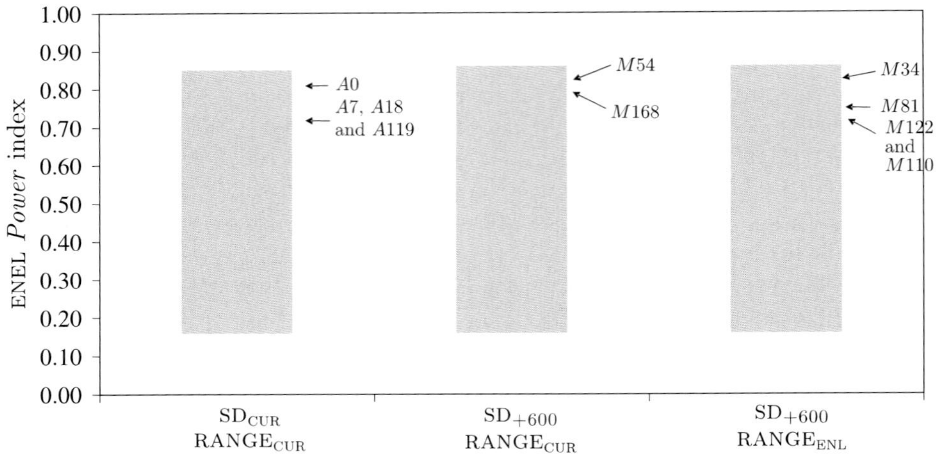

Figure 16.3: Values of the ENEL *Power* sector index for different classes of alternatives. For each class, the bar represents the range of index values produced by changing the policy. The values that correspond to the reasonable alternatives are highlighted.

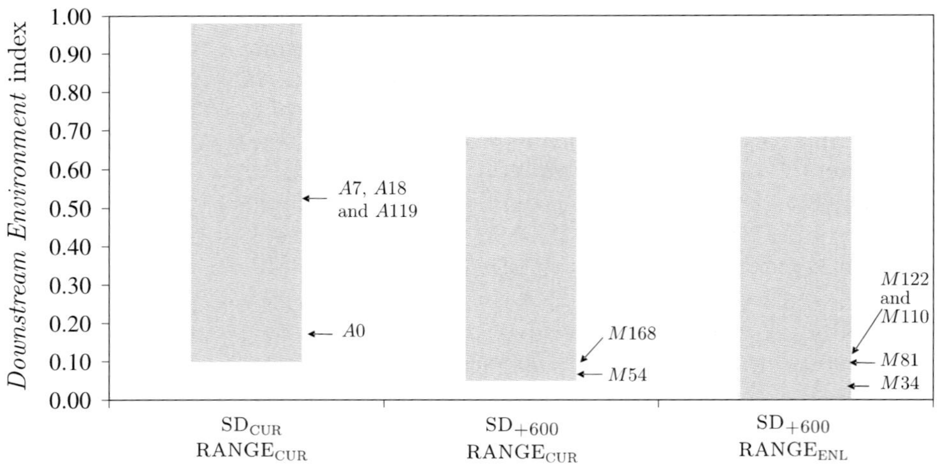

Figure 16.4: Values of the *Downstream Environment* sector index for different classes of alternatives. For each class, the bar represents the range of index values produced by changing the policy. The values that correspond to the reasonable alternatives are highlighted.

alternatives; however, alternatives $M54$, $M34$ and $M81$ increase the *Irrigation* index with respect to $A0$. The last two alternatives adopt $\text{RANGE}_{\text{ENL}}$, a choice that the Irrigation Consortia favour.

The position of ENEL is not as simple as the position of the Irrigation Consortia because ENEL is undecided between the classes ($\text{SD}_{+600}/\text{RANGE}_{\text{ENL}}$) and ($\text{SD}_{+600}/\text{RANGE}_{\text{CUR}}$), as shown in Figure 16.3.[1] The choice depends, in fact, on the alternative being considered: alternative $M34$ is preferred to $M54$ (even if it is not very noticeable in the figure, $M34$ is slightly superior to $M54$) but, among the alternatives with higher MEF values, ENEL prefers $M168$ to $M81$, $M110$ and $M122$.

The position of the Stakeholders interested in *Downstream Environment* is completely the opposite. In fact, they oppose the alternatives in which excavating is considered because the index associated to them is very low as shown in Figure 16.4.

The Stakeholders interested in *Downstream Flooding* are afraid that excavating might cause an increase in the flood peaks in the Ticino downstream from the lake and so they consider this action dangerous a priori. Only the implementation of the REDUCTION WITH PO measure induces them to accept the alternatives that contemplate the excavation.

Since it is such a serious conflict, the Italian decision depends on the real power relationship between the opposing sectors: thus it is plausible that, before deciding, Italy would like to know if the hypothesis of widening the regulation range is acceptable to Switzerland.

- **Third move:** *Switzerland proposes the implementation of alternative* $M54$ *or* $M81$.

  Among the reasonable alternatives, the one that guarantees the maximum satisfaction for the *Upstream Flooding* is $M54$, which has also the distinct advantage of producing a high value for *Upstream Tourism* index. Furthermore, it does not require that the upper extreme of the regulation range be raised, which is an action that the Swiss public does not approve. Therefore the Swiss government would be prone to choose this alternative, but at the same time fears that if Italy were in doubt between the classes ($\text{SD}_{+600}/\text{RANGE}_{\text{ENL}}$) and ($\text{SD}_{\text{CUR}}/\text{RANGE}_{\text{CUR}}$), the option on $M54$, i.e. on an alternative of the class ($\text{SD}_{+600}/\text{RANGE}_{\text{CUR}}$), might cause it to reject the excavation. Thus it is improbable that Switzerland would declare that it is not willing to accept an enlargement of the regulation range.

  Figure 16.1 shows, however, that among the alternatives from the class ($\text{SD}_{+600}/\text{RANGE}_{\text{ENL}}$) there are alternatives that produce much lower values for *Upstream Flooding* than the minimum value that can be obtained with alternatives from the class ($\text{SD}_{+600}/\text{RANGE}_{\text{CUR}}$). To avoid running into one of these alternatives, it is probable that Switzerland will decide to indicate a precise alternative also in this class: the choice might be $M81$, which seems to be a good compromise between all the interests at stake.

- **Fourth move:** *Italy must definitively decide between the three classes of alternatives.*

  It is likely that the class ($\text{SD}_{+600}/\text{RANGE}_{\text{CUR}}$) is rejected by Italy, because that class is principally interesting for Upstream Stakeholders, while the other two classes draw

[1] Remember that, because of the Galli Law, with all the alternatives the ENEL *Power* sector index is lower than with $A0$.

the most interest from the downstream Stakeholders. Therefore, Italy's choice will be between those.

  - If *Italy decided upon the class* ($\text{SD}_{+600}/\text{RANGE}_{\text{ENL}}$), *the choice would probably be of alternative* $M122$. In fact, the Ticino Parks are opposed to $M34$ and they are not in favour of $M81$ (in fact, it is included among their least-bad alternatives), not only because with these alternatives the value of their index drops below the value they obtain with $A0$, but also because $M34$ and $M81$ ignore their request to raise the MEF to 25 m^3/s. The Italian choice could be $M110$, which is similar to $M81$ and has a MEF equal to 25_{MOD}, but the Irrigation Consortia and ENEL are opposed to it. Therefore, the Italian choice will most probably be $M122$ (even if ENEL is opposed to it) because it is characterized by a MEF of 20_{MOD}, is accepted by the Irrigation Consortia and is a least-bad alternative for the Ticino Parks. Moreover it takes account of the needs expressed by the Stakeholders interested in the *Downstream Flooding* sector, given that it includes the REDUCTION WITH PO mitigation measure.
  - If *Italy decided for the class* ($\text{SD}_{\text{CUR}}/\text{RANGE}_{\text{CUR}}$), *it would probably opt for alternative* $A18$.

    If Italy refused to excavate, the Irrigation Consortia would no longer be willing to give the Ticino Parks a MEF of 25 m^3/s, because that would cause them a loss in value with respect to alternative $A0$ (compare the performances of $A119$ and $A0$ in Figure 16.2). It is thus plausible that the Italian choice would go to $A18$, which is characterized by MEF 13_{CON}.

- **Fifth move:** *Switzerland accepts the Italian choice.*
  - If Italy proposed $M122$, the value of the *Upstream Flooding* sector index would be comparable to the one associated to $M81$ (Figure 16.1), and so Switzerland would have no reason to refuse the Italian proposal.
  - If Italy proposed $A18$, the value of *Upstream Flooding* sector index would be higher than that associated to the current situation ($A0$): again, Switzerland would have no reason to refuse. Note that in this case the structural and normative conditions remain unchanged with respect to the current situation. However, Switzerland would be interested to establish which policy should be adopted: if it didn't, it would not have any guarantee of not falling back into $A0$, which has worse performances than $A18$.

This description is the most probable evolution of the negotiation process between the two nations. It is evident that the key decision is the Italian one, which will depend significantly upon the power relations that exist between the *Irrigation/Power* sectors on the one hand and the *Downstream Flooding/Downstream Environment* on the other.

16.2 Which future for the decision?

The considerations above show that there is an alternative that is preferable to $A0$ for both final Decision Makers. But will the two governments really make a decision? It is more rational to think that the two governments will postpone the final decision and request that the study be developed further for the critical points that were highlighted and not yet analysed. By doing so the Decision Makers will have more elements upon which to base their decision.

At this point (July 2006) the members of the Italian-Swiss International Commission have not yet been able to express their opinion about the reasonable alternatives: in 2001, on the termination of their mandate, the Italian delegates of the Commission were not re-elected because of a conflict which arose between Government and Regions about which had the power to nominate the delegates; at time of writing this conflict is still not resolved. However, in a conference[2] which was held in Ascona in November 2001, political comments were expressed:

(1) a representative from the Federal Water Office of the Swiss government declared that he was in favour of the alternatives in the class (SD_{600}/$RANGE_{ENL}$), mitigated with REDUCTION WITH PO and BOLLE RESERVE;

(2) a representative of the 'Magistrato per il Po', the Agency concerned with Po floods which has a great deal of influence on the Italian decision, affirmed that, in the absence of further studies on the repercussions of lake regulation on the entire Po valley, no change in the current situation could be taken into consideration.

If the members of the Commission take similar positions, the Commission will certainly request further studies. Therefore let us see what they might be and how they would have to be structured.

16.2.1 Which studies?

There are several different studies that are suited to increasing the information available for the *Final Decision*, but they would all be very costly if they had to be carried out for a wide spectrum of alternatives. This is why it would be advisable that the Commission indicated a restricted number of alternatives for which to proceed with these studies. We list the studies in a random order, since, in practice, no relative importance can be expressed among them without expressing a political preference.

1. **Study of the transient period between the current regulation and the regulation envisaged by a given alternative**
   In the Verbano Project only steady-state effects were considered. To overcome this limitation, for several alternatives,[3] which should be indicated by the Commission or, at most, for all the reasonable alternatives, a regulation policy for the transient period could be designed and, if the alternative includes the action SD_{+600}, the effects of the excavation phase estimated. The regulation policy for the transient period will not be periodic and it has to be designed by considering a finite time horizon.

2. **Study of the effects of the REDUCTION WITH PO measure**
   First, the possibility of using the Miorina dam for the regulation of the lake at levels higher than 1.50 m (above zero at Sesto Calende) must be analysed. If the dam proves to be unsuitable for this task, which is likely to be the case, changes to the dam structure must be studied and the costs of the modifications estimated. Then the regulation policies would have to be designed, by adding a fourth indicator for the *reduction of downstream flooding* criterion to the three design indicators that

[2] *Verbano: which future? An ecosystem's perils and means of defence*, 9th/10th November 2001, Ascona, Switzerland.

[3] The operation is very onerous in terms of computing time: it would thus be necessary to limit it to a small number of 'interesting' alternatives.

were considered in the present Project. As explained in Section 14.2.3, the design of these policies could be carried out with the *Q-learning planning* algorithm (see Section 13.4 of THEORY). The study could be limited to considering the effects on the level at Pavia or be extended, as requested in the Ascona conference in November 2001, to the entire Po valley. In this last case it would be necessary to identify a model to describe the propagation of the flood wave in the Po and introduce a new sector (*Po valley*).

3. **Study of the effects of the enlarged regulation range**
   To overcome any reservations about adopting the $\text{RANGE}_{\text{ENL}}$, for each alternative, the range of lake levels that it usually produces can be determined. As we have highlighted several times, it is the regulation policy that strongly influences the performances, so that, for the same structural and normative actions, the effects can be very different. Some regulation policies (for example those that favour the *Irrigation* and ENEL *Power* sectors) tend to keep the lake level close to the upper extreme of the regulation range, at least in some periods; others (for example those that favour *Upstream Flooding*) tend to keep the level of the lake constantly low and do not fully exploit the freedom allowed by the regulation range. Identifying the range of lake levels usually produced by a given alternative thus means identifying the effective regulation range that characterizes the alternative.

4. **Analysis of other interventions or scenarios**
   The Verbano Project considered only alternatives that were designed by combining the interventions proposed by the Stakeholders and they were analysed in the scenario that was supposed to be most probable. In light of the results that were obtained, the Decision Makers (i.e. the Italian-Swiss Commission) might think it was interesting to explore other interventions or scenarios. For example, if the Commission believed that alternative $M34$ was an interesting alternative because it achieved a wide agreement, it might suggest reviewing the distribution policy in favour of *Downstream Environment*, with the aim of transforming the opposition from the Ticino Parks into acceptance. Clearly this would lead to a worsening of the situation for other sectors and the Decision Makers should indicate which those would be. For example, they could suggest reducing the water supply to the users that can turn to other sources and specify how the substitution could take place. These users could be, for example, the ENEL hydropower plants along the Ticino: a reduction in their Abstraction Licence would lead to a reduction in power production, but the loss could be compensated by an increase in power production in other plants that are owned by ENEL. Or the Commission might consider reducing the licence to one or more of the Irrigation Consortia, provided that it managed to identify another source of water for them or it subsidized the transformation of the irrigation system, which could be changed for many crops, from flood to sprinkler irrigation.

Chapter 17

Implementation

FC and RSS

Once the final Decision Makers have chosen the best compromise alternative and both the regulation policy to adopt in the transient period and the way in which any excavation is to be carried out have been defined, the alternative A that results from this must be implemented. This means that the structural, normative and regulation actions that characterize it must be realized. At the same time, other minor interventions can be designed and realized, in order to reduce the damage from floods and the discomfort to the population. We will describe these interventions in the first section of this chapter.

It is easy to imagine how structural actions are implemented and to understand that the normative action takes form in a new Regulation Licence. It is less easy to imagine how the regulation action is implemented. Both the time-variant policy for the transient and the steady-state policy (i.e. the one associated to the best compromise alternative) must be applied day by day in the management of the lake. Every day, in fact, the Regulator must decide the release from the Miorina dam and the flow to be left to the Ticino at the Panperduto diversion dam for environmental purposes. Implementing these decisions requires making other decisions, typical of the *operational control level* (see Section 1.4), such as how to position the gates of the dam in order to release the water volume that was decided, over the following twenty-four hours. The operational control decisions can be made using intuition or, better, by imposing and solving an operative Control Problem, which is generally formulated by assuming that the inflow is deterministic. We will not explore these operational aspects, which are more related to Hydraulics; instead, we will focus on the decision-making problem at the management level in the second section of this chapter. The description will be maintained at a simple, introductory level; the reader will find a deeper treatment and references in Chapter 14 of THEORY, together with an example of application to the Verbano.

17.1 Designing other interventions

Section 3.6 of the DVD describes a set of interventions for reducing the damage produced by floods and the discomfort to the population:

- creating flood protection structures for roads and buildings;
- interventions to protect the drainage network and sewerage system;

- interventions for protecting the electricity network;
- setting norms for installing oil tanks;
- setting norms for designing water treatment plants;
- setting norms for waste disposal.

As we explain in the above mentioned section, these interventions were not considered in the Verbano Project because they cannot be designed until the best compromise alternative has been chosen. In fact, designing them requires the definition of the maximum elevation $\bar{h}$ up to which the interventions and norms must provide protection. Since floods are a random phenomenon, the elevation $\bar{h}$ is chosen depending on the level of risk that is considered acceptable. The latter is defined either by the probability p_{fail} that the protection might fail (for example because the flood overflows the barrier and inundates the protected buildings) or by the return time T_{ret} of that event ($T_{\text{ret}} = 1/p_{\text{fail}}$).

Once the level of acceptable risk p_{fail} has been set, the elevation $\bar{h}$ is obtained with the following procedure:

(1) by deterministic simulation, the lake level trajectory produced by alternative A is obtained. As inflow scenario, the time series of historical inflow to the lake or a *synthetic series* of inflow produced by a model is adopted;

(2) the probability distribution of the maximum annual lake level is identified;

(3) the lake level that is exceeded with probability p_{fail} is determined and from this the corresponding elevation $\bar{h}$.

Once the elevation $\bar{h}$ is obtained, the interventions can finally be designed.

An alternative to the procedure just described is that the elevation $\bar{h}$ is defined through the economic evaluation described in Section 3.6 of the DVD, in correspondence to each intervention, but, notice, also this evaluation requires us to know the probability distribution of the lake level.

17.2 Management and the POLFC scheme

The simplest and most immediate way to make management decisions is to feed the regulation policy of the adopted alternative A with the current information $\Im_t$ (current day t, lake storage s_t and, if required by the policy, forecast $\hat{q}_{t+1}^{\text{Po}}$ of the average flow in the Po in the next 24 hours, i.e. in the interval $[t, t+1)$, see 14.2.3) and to choose a control $\mathbf{u}_t$ in the set M_t of controls that the policy provides (see Section 5.2.1). Remember that the control $\mathbf{u}_t$ is a vector with two components: the release decision u_t^{MIO}, which controls the release from Miorina, and the environmental allotment decision u_t^{TIC}, which controls the flow to be left to the Ticino at the Panperduto diversion dam for environmental purposes.[1]

Actually this is a poor approach, since the regulation policy associated to alternative A was designed by assuming a set of information (*design scenario*, Section 7.5) that can be different from the one existing at time t, when the decision is implemented. For example, in

[1]For those alternatives that do not foresee the environmental allotment decision, such as $M168$ for example, the control is a scalar quantity.

the design scenario the water demand was given an a priori value, but in real time the Regulator might know that the season is further along and that most of the corn has already been harvested, or that the rice planting is late because the year is colder than usual, and therefore that actual water demand is different from the a priori one. Furthermore, in the design scenario we assumed that the hydropower plants are always operable, but it is possible that at time t some of them are inoperable. It would be foolish not to take these anomalies into account and to make a decision that would be optimal in 'normal' conditions but not in the real ones. Therefore, it is advisable to make the decision by means of an *adaptive control* scheme. In Control Theory terminology this means that the set M_t is computed *on-line* (i.e. at the same time as the decision) and in *real time* (i.e. the computation is fast enough to provide a decision that is not obsolete; in practice, if the decision step is a day, the computation must be completed in ten minutes at most).

The adaptive scheme that best lends itself to our case is the *Partial Open Loop Feedback Control* (POLFC) scheme (Bertsekas, 1976, see Section 14.3.3 of THEORY). It requires that, at each time t, an (Optimal) Control Problem be solved, in which the values of the weights λ_{fl}, λ_{irr} and λ_{env} associated to the design indicators are the same as in the Control Problem used to design alternative A, and whose scenario is derived from the design scenario of alternative A, taking into account the anomalies that occur at time t. Since the Problem must be solved in real time, its time horizon must necessarily be limited to only a few days, from t to $t+h$. It is not necessary, however, that the anomaly finish before $t+h$, because the Problem will be formulated and solved again on each of the following days and only the decision that corresponds to time t is actually used. The end of the horizon is thus never reached. To prevent excessively short horizons (i.e. low values of h) negatively influencing the decision at time t, it is opportune to introduce a *penalty* on the final state s_{t+h}, which expresses the expected cost-to-go after the end of the horizon, as a function of the finale state (see $g_h(s_h)$ in equation (7.5)). By definition, this penalty is the optimal cost-to-go $H^*_{t+h}(\cdot)$ defined by the Bellman function $H^*_{\cdot}(\cdot)$ associated to alternative A (Section 7.7).

More precisely,[2] let

$$i_{\text{fl}} = \sum_{\tau=t}^{t+h-1} S_{\text{fl}}^{\text{Tot}}(s_t) \tag{17.1a}$$

$$i_{\text{irr}} = \sum_{\tau=t}^{t+h-1} \left(d_{\tau+1}^{\text{VC}} + d_{\tau+1}^{\text{REC}} + d_{\tau+1}^{\text{IC}}\right)^{12} \tag{17.1b}$$

$$i_{\text{Enel}} = \sum_{\tau=t}^{t+h-1} \theta_\tau D_{\tau+1}^{\text{Enel}} \tag{17.1c}$$

$$i_{\text{env}} = \sum_{\tau=t}^{t+h-1} \left(q_{\tau+1}^{\text{TIC}} - r_\tau^{n,\text{median}}\right)^2 \tag{17.1d}$$

be the indicators that correspond to the four objectives defined over the horizon $[t, t+h)$, where the deficits $d_{\tau+1}^{\text{VC}}$, $d_{\tau+1}^{\text{REC}}$, $d_{\tau+1}^{\text{IC}}$ and $D_{\tau+1}^{\text{Enel}}$ are computed with respect to the water demand values that are actually expected.

[2] We describe the case in which the current information $\Im_t$ is constituted of the current day t and the lake storage s_t. The extension to the case in which it also includes the forecast $\hat{q}_{t+1}^{\text{Po}}$ of tomorrow's average Po flow rate is left to the reader.

The Control Problem is lexicographic (Section 7.6.3) and so it is defined by a Primary Problem and a Secondary Problem. The Primary Problem is the following

$$\min_{p}\left[\lambda_{\text{fl}} J_{\text{fl}}(p) + \lambda_{\text{irr}} J_{\text{irr}}(p) + \lambda_{\text{env}} J_{\text{env}}(p) + H^{*\text{prim}}_{t+h}(s_{t+h})\right] \tag{17.2a}$$

subject to

$$s_{\tau+1} = f_\tau(s_\tau, \mathbf{u}_\tau, \varepsilon_{\tau+1}) \qquad \tau = t, t+1, \ldots, t+h-1 \tag{17.2b}$$

$$\mathbf{u}_\tau \in M_{\tau/t}(s_\tau) \subseteq U_\tau(s_\tau) \qquad \tau = t, t+1, \ldots, t+h-1 \tag{17.2c}$$

$$\varepsilon_{\tau+1} \sim N(0,1) \qquad \tau = t, t+1, \ldots, t+h-1 \tag{17.2d}$$

$$p = \left\{M_{\tau/t}(\cdot);\ \tau = t, t+1, \ldots, t+h-1\right\} \tag{17.2e}$$

in which, as we have said, the weights λ_{fl}, λ_{irr} and λ_{env} assume the same values as in the Primary Control Problem (7.32), whose solution is the policy of alternative A; the objectives $J_{\text{fl}}(p)$, $J_{\text{irr}}(p)$ and $J_{\text{env}}(p)$ are the same as in that Problem; and $H^{*\text{prim}}(\cdot)$ is its Bellman function. The solution to Problem (17.2) provides an efficient policy p^*, which is composed of a sequence of set-valued control laws $M^*_{\tau/t}(\cdot)$ over the horizon $[t, t+h)$, where the second subscript t is a reminder that these laws are computed by solving the Problem formulated at time t. As we explained in Section 5.2.1, given a time t and a state s_t of the system, the law $M^*_{t/t}(\cdot)$ provides a set $M^*_{t/t}(s_t)$ of optimal controls, from which to choose the ones that minimize the lost revenue for hydropower plants. For this we define the following Secondary Control Problem, sequentially after the previous one

$$\min_{p}\left[J_{\text{Enel}}(p) + H^{*\text{sec}}_{t+h}(s_{t+h})\right] \tag{17.3a}$$

subject to the constraints

$$s_{\tau+1} = f_\tau(s_\tau, \mathbf{u}_\tau, \varepsilon_{\tau+1}) \qquad \tau = t, t+1, \ldots, t+h-1 \tag{17.3b}$$

$$\mathbf{u}_\tau \in M_{\tau/t}(s_\tau) \subseteq M^*_{\tau/t}(s_\tau) \qquad \tau = t, t+1, \ldots, t+h-1 \tag{17.3c}$$

$$\varepsilon_{\tau+1} \sim N(0,1) \qquad \tau = t, t+1, \ldots, t+h-1 \tag{17.3d}$$

$$p = \left\{M_{\tau/t}(\cdot);\ \tau = t, t+1, \ldots, t+h-1\right\} \tag{17.3e}$$

where $H^{*\text{sec}}(\cdot)$ is the Bellman function of the Secondary Problem (7.33) associated to alternative A. The solution to the Secondary Problem (17.3) provides a set $M^{**}_{t/t}(s_t)$ of optimal controls for time t and state s_t, from which the Regulator will choose the one that he believes is the most suitable at time t. At time $t+1$, the two Problems will be re-formulated on the basis of the new information available at that time and their solution will provide the set of controls $M^{**}_{t+1/t+1}(s_{t+1})$. From it, once again the Regulator will have to choose the control to adopt at time $t+1$.

The POLFC scheme also makes it possible to use hydro-meteorological information I_t: it is sufficient that the *a priori distribution* $N(0,1)$ of the disturbance be replaced by the *a posteriori distributions* $\phi_\tau(\cdot|I_t)$ (with $\tau = t, t+1, \ldots, t+h-1$), which are computed on the basis of the hydro-meteorological information I_t.

Clearly, the use of an adaptive control scheme such as POLFC is more time-consuming than simply using the regulation policy associated to alternative A. There are two main reasons for this. In the first place, it requires the availability of a computer dedicated to the on-line solution of the Lexicographic Control Problem (17.2)–(17.3); in the second place that computer must be fed each day with the information necessary to define the Problem itself (see Figures 5.3 and 5.8).

The system that provides this information is composed of two systems in cascade: the *information system* and the *forecast system*. Every day t, the first must provide, at least, the measure of the lake storage, the current hydropower water demand and news about the state of the crops, from which the irrigation demand can be deduced. It would even be preferable that it also supplied hydro-meteorological information such as precipitation, water level in the tributaries, snowpack depth, depth of the water table, etc. This data, along with the past records stored in its data base, should be passed to the forecast system, which forecasts the inflow to the lake in the days from t to $t+h$ (more precisely, it estimates the probability distribution of the inflow in each one of these days). The data collected in the information system and the forecasts elaborated by the forecast system feed the *management system*. This is constituted of the solver of the Lexicographic Control Problem (17.2)–(17.3), which provides the set $M^{**}_{t/t}(s_t)$ of efficient controls for the current day, and of a simulator, which makes it possible for the Regulator to estimate the effects of the different controls in $M^{**}_{t/t}(s_t)$. The three systems (information, forecast and management) constitute the Multi-Objective Decision Support System for the management level (MODSS/M), which should not be confused with the MODSS/P used at the planning level (see Section 1.5).

The forecast elaborated by the forecast system can also be sent to all the organizations that can make good use of it, such as the territorial authorities (Cities, Provinces, Regions, Cantons) and Civil Protection services.

References

The bibliography contains almost exclusively texts concerning with the Verbano Project. Methodological references can be found in THEORY.

Abbott, M.B., 1979, *Computational Hydraulics, Elements of the Theory of Free Surface Flows*, Pitman Publishing Limited, London, UK.

Anastasi, G., 1992, *Studio di fattibilità per un incremento del deflusso del fiume Ticino a Sesto Calende*, Studio d'ingegneria Anastasi, Bellinzona, CH.

Anastasi, G., 1994, *Studio di fattibilità per un incremento del deflusso del fiume Ticino a Sesto Calende*, Studio d'ingegneria Anastasi, Bellinzona, CH.

Anthony, R.N., 1965, Planning and control systems: a framework for analysis, PhD thesis, Graduate School of Business Administration, Harvard University, Cambridge, MA.

Aufiero, A., Soncini-Sessa, R., Weber, E., 2001, Set-valued control laws in minmax control problem, *in: Proceedings of IFAC Workshop Modelling and Control in Environmental Issues*, August 22–23, Elsevier, Yokohama, J.

Barto, A., Sutton, R., 1998, *Reinforcement Learning: An Introduction*, MIT Press, Boston, MA.

Beinat, E., 1997, *Value Functions for Environmental Management*, Kluwer Academic Publishers, Dordrecht, NL.

Bertola, P.F., 2001, Difese puntiformi: le barriere, *in: Il Verbano: quale futuro? Insidie di un ecosistema e mezzi di difesa*, SUPSI, Ascona, CH, 207-212.

Bertsekas, D.P., 1976, *Dynamic Programming and Stochastic Control*, Academic Press, New York, NY.

Bianchi, A., Laniado, E., 1983, Messa a punto di un metodo per la valutazione a priori dei danni delle piene lacuali, *in: Memorie e studi dell'Istituto di Idraulica e Costruzioni Idrauliche*, Politecnico di Milano, Milano, I.

Bishop, C., 1995, *Neural Networks for Pattern Recognition*, Oxford University Press, Oxford, UK.

Brown, L.R., 2001, How water scarcity will shape the new century, *Water Science and Technology*, **43** (4), 17-22.

Brutsaert, W., 2005, *Hydrology, an Introduction*, Cambridge University Press, New York, NY.

Budge, I., 1996, *The New Challenge of Direct Democracy*, Polity Press, Cambridge, UK.

Camani, M., 2001, Esondazioni e rischi per l'ambiente, *in: Il Verbano: quale futuro? Insidie di un ecosistema e mezzi di difesa*, SUPSI, Ascona, CH, 21-23.

Castelletti, A., Corani, G., Rizzoli, A.E., Soncini-Sessa, R., Weber, E., 2001, A reinforcement learning approach for the operational management of a water system, *in: Proceedings of IFAC Workshop Modelling and Control in Environmental Issues*, August 22–23, Elsevier, Yokohama, J.

Cattaneo, C., 1844, *Notizie naturali e civili su la Lombardia*, Tip. Bernardoni, Milano, Italia.

Cattaneo, M., Maione, U., Mignosa, P., Tomirotti, M., 2000, L'evento di piena dell'ottobre 2000 sul bacino del Ticino, *L'Acqua*, **6**, 7-16.

Chow, V.T., Maidment, D.R., Mays, L.V., 1998, *Applied Hydrology*, McGraw-Hill, New York, NY.

CIRF – Centro Italiano di Riqualificazione Fluviale, 2004, *La riqualificazione fluviale in Italia*, Mazzanti Editore, Venezia, I.

Citrini, D., 1973, Le piene del Lago Maggiore e del Ticino emissario nel primo trentennio di regolazione, *in: Pubblicazione n. 14*, Consorzio del Ticino, Milano, I.

Citrini, D., Cozzo, G., 1981, Influenza dei serbatoi alpini sulla regolazione del Lago Maggiore, *in: Pubblicazione n. 15*, Consorzio del Ticino, Milano, I.

Citrini, D., Cozzo, G., 1982a, Influenza dei serbatoi alpini sulla regolazione del Lago Maggiore, *in: Memorie e studi dell'Istituto di Idraulica e Costruzioni Idrauliche del Politecnico di Milano*, Politecnico di Milano, Milano, I.

Citrini, D., Cozzo, G., 1982b, Influenza delle modalità di regolazione del Lago Maggiore nei riguardi delle portate erogabili e dei livelli di piena, *in: Pubblicazione n. 16*, Consorzio del Ticino, Milano, I.

Commissione internazionale per la protezione delle acque italo-svizzere, 1979, Rapporti su studi e ricerche condotti nel bacino del Lago Maggiore, Campagna, 1979.

Connor, D.M., 1997, *Public Participation: A Manual*, Development Press, Victoria, BC, CND.

Consorzio del Ticino, 1968, La regolazione del Lago Maggiore nel primo venticinquennio di esercizio 1943–1967, *in: Pubblicazione n. 13*, Consorzio del Ticino, Milano, I.

Cybenco, G., 1989, Approximation by superpositions of a sigmoidal function, *Mathematics of Control Signals and Systems*, **2** (4), 303-314.

De Marchi, G., 1950a, Determinazione dei benefici ritraibili dall'elevamento del limite superiore della trattenuta attiva dall'altezza di +1.00 m a quella di +1.50 m sopra lo zero dell'idrometro di Sesto Calende, *in: Pubblicazione n. 3*, Consorzio del Ticino, Milano, I.

De Marchi, G., 1950b, Ripercussione della regolazione del Lago Maggiore sulle piene del lago e su quelle del Ticino a Sesto Calende, *in: Pubblicazione n. 4*, Consorzio del Ticino, Milano, I.

De Marchi, G., 1950c, Variazioni provocate dalla regolazione dei deflussi nelle tenute dei livelli del Lago Maggiore, *in: Pubblicazione n. 5*, Consorzio del Ticino, Milano, I.

Del Furia, L., Rizzoli, A.E., Arditi, R., 1995, Lakemaker: a general object-oriented software tool for modeling the eutrophication process in lakes, *Environmental Software*, **10** (1), 43-64.

Delli Priscoli, J., 2004, What is public participation in water resources management and why is it important? *Water International*, **29** (2), 221-227.

Egger, G., Muhar, S., Schmutz, S., Petutschnig, J., Aangermann, K., Trimmel, S., Preis, S., Ömer, B., Strigl, A., 2003, Guiding visions ('leitbilder') and measures for river basins as developed for the Möll River, *Österreichische Wasser und Abfallwirtschaft*.

Elarabawy, M., Attia, B., Tosswell, P., 1998, Water resources in Egypt: strategies for next century, *Journal of Water Resources Planning and Management – ASCE*, **124** (6), 310-319.

Ellis, H.M., 1970, The application of decision analysis to the problem of choosing an air pollution control program for New York City, PhD thesis, Graduate School of Business Administration, Harvard University, Cambridge, MA.

ENEL, 1999, *Contributo Enel alla definizione del minimo deflusso costante vitale a valle della traversa del Panperduto sul Ticino*, Divisione Produzione Ingegneria Civile ed Idraulica Unità di Idrologia, Castellanza, I.

Engelhardt, A., 2001, Acque reflue, *in: Il Verbano: quale futuro? Insidie di un ecosistema e mezzi di difesa*, SUPSI, Ascona, CH, 207-212.

European Commission, 2000, Directive 2000/60/EC of the European Parliament and of the Council establishing a framework for Community action in the field of water policy, *in: Official Journal*, European Commission, Brussels, B.

European Commission, 2001, Directive 2001/42/EC of the European Parliament and of the Council of 27 June 2001 on the assessment of the effects of certain plans and programmes on the environment, *in: Official Journal*, European Commission, Brussels, B.

European Commission, 2003a, Common Implementation Strategy for the Water Framework Directive, Guidance Document n. 11: Planning Process. Produced by Working Group 2.9.

European Commission, 2003b, Directive 2003/35/EC of the European Parliament and of the Council providing for public participation in respect of the drawing up of certain plans and programmes relating to the environment and amending with regard to public participation and access to justice Council Directives 85/337/ECC and 96/61/EC, *in: Official Journal*, European Commission, Brussels, B.

European Environmental Agency, EEA, 1999, Environmental indicators: Typology and overview, Technical Report No. 25, EEA.

Fantoli, G., 1897, *Sul regime idraulico dei laghi*, Hoepli. Milano, I.

FAO, 1986, *Yield Response to Water Deficit. Irrigation and Drainage Papers*, **33**, Food and Agriculture Organization, Rome, I.

FAO, 1994, *CLIMWAT: A Climatic Database for CROPWAT. Irrigation and Drainage Papers*, **49**, Food and Agriculture Organization, Rome, I.

FAO, 1995, *CROPWAT. A Computer Program for Irrigation Planning and Management. Irrigation and Drainage Papers*, **46**, Food and Agriculture Organization, Rome, I.

French, S., 1988, *Decision Theory: An Introduction to the Mathematics of Rationality*, Ellis Horwood, Chichester, UK.

Funahashi, K., 1989, On the approximate realization of continuous mappings by neural networks, *Neural Networks*, **2** (3), 183-192.

Gherardelli, L., 1964, Influenza dei serbatoi alpini sulla regolazione del Lago Maggiore, *in: Pubblicazione n. 9*, Consorzio del Ticino, Milano, I.

Graf, W.H., 1998, *Fluvial Hydraulics: Flow and Transport Processes in Channels of Simple Geometry*, John Wiley & Sons, New York, NY.

Guariso, G., 1987, Valutazione di impatto ambientale della normativa di regolazione del Lago Maggiore, *Terra*, **2**, 37-41.

Guariso, G., Laniado, E., 1983, Agricultural impact of Lake Maggiore management, *International Journal of Water Research and Development*, **1**, 299-310.

Guariso, G., Laniado, E., 1984, Lake Maggiore water management: a two objective analysis, *in: VIII European Session of the European Soc. of Agricultural Economists*, Milano, I.

Guariso, G., Laniado, E., Rinaldi, S., 1985a, Legal constraints and conflict resolution in reservoir control, *in: Systems Analysis Applied to Water and Related Land Resources, IFAC Conference, Lisbon*, P.

Guariso, G., Laniado, E., Rinaldi, S., 1985b, The management of Lake Maggiore: conflict analysis and the price of non cooperation, *Operations Research Spektrum*, **7**, 101-109.

GWP – Global Water Partnership, 2003, ToolBox Integrated Water Resources Management, http://gwpforum.netmasters05.netmasters.nl/en/index.html.

Hammond, J., Keeney, R., Raiffa, H., 1999, *Smart Choices*, Harvard Business School Press, Cambridge, MA.

Hare, M., Letcher, R.A., Jakeman, A.J., 2003, Participatory modelling in natural resource management: A comparison of four case studies, *Integrated Assessment*, **4** (2), 62-72.

Hornik, K., Stinchcombe, M., White, H., 1989, Multilayer feedforward networks are universal approximators, *Neural Networks*, **2**, 359-366.

Jungwirth, M., Muhar, S., Schmutz, S., 2000, *Assessing the Ecological Integrity of Running Waters*, Kluwer Academic Publishers, Dordrecht, NL.

Kaelbling, L.P., Littman, M.L., Moore, A.W., 1996, Reinforcement learning: a survey, *Journal of Artificial Intelligence Research*, **4**, 237-285.

Keeney, R.L., 1992, *Value Focused Thinking*, Harvard University Press, Boston, MA.

Keeney, R.L., Raiffa, H., 1976, *Decision with Multiple Objectives: Preferences and Value Trade-offs*, John Wiley & Sons, New York, NY.

Korhonen, P., Laakso, J., 1986, A visual interactive method for solving the multiple criteria problems, *European Journal of Operational Research*, **24** (2), 277-287.

Maas, A., Hufschmidt, M.M., Dorfam, R., Thomas, H.A., Marglin, S.A., Fair , G.M., 1962, *Design of Water Resource Systems*, Harvard University Press, Boston, MA.

Maidment, D.R., 1993, *Handbook of Hydrology*, McGraw-Hill, New York, NY.

Maione, U., Mignosa, P., 1995, Il funzionamento idraulico dell'incile del Lago Maggiore: modello matematico dal lago allo sbarramento di Porto della Torre e ricostruzione delle piene dell'autunno 1991 e 1993, *in: Pubblicazione n. 19*, Consorzio del Ticino, Milano, I.

Maione, U., Mignosa, P., 2001, Effetti sulle piene del Verbano e del Ticino provocati da un aumento della capacit di deflusso dell'incile, *in: Il Verbano: quale futuro? Insidie di un ecosistema e mezzi di difesa*, SUPSI, Ascona, CH, 93-109.

Marchetti, M., Raffa, U., 1965, Le portate dell'incile del Lago Maggiore in regime libero prima e dopo la sistemazione del suo alveo, *in: Memorie e Studi dell'Istituto di Idraulica e Costruzioni Idrauliche*, Pubblicazione n. 232, Politecnico di Milano, Milano, I.

Mariotta, P., 2001, Adeguamento rete elettrica, *in: Il Verbano: quale futuro? Insidie di un ecosistema e mezzi di difesa*, SUPSI, Ascona, CH, 207-212.

McCulloch, W.S., Pitts, W.H., 1943, A logical calculus of the ideas immanent in nervous activity, *Bulletin of Mathematical Biophysics*, **5**, 115-133.

Miller, G.A., 1956, The magic number seven, plus or minus two: some limits on our capacity for processing information, *Psychological Review*, **63**, 81-97.

Mostert, E., 2003, The challenge of public participation, *Water Policy*, **5** (2), 81-97.

Muhar, S., Schwarz, M., 2000, Assessing rivers with high and good habitat integrity in the Alpine region as foundation for conservation programmes: applications in Austria, *Verhandlungen Internationale Vereinigung für Theoretische und Angewandte Limnologie*, **27**, 669-673.

Nardini, A., 2005, *Decidere l'ambiente con l'approccio partecipato. Collezione CIRF*, Mazzanti Editore, Venezia, I.

Nardini, A., Fahmy, H., 2005, Integrated evaluation of Egypt's water resources plans: a framework to cope with sustainability, *Water International*, **30** (3), 314-328.

NAWQAM, 1999, Egypt's current water balance. Inception report, Technical Report 3, MPWWR-PS, Cairo, ET.

OECD – Organisation for Economic Co-operation and Development, 1994, Environmental indicators, Technical Report, OECD, Paris, F.

Peterka, V., 1986, Control of uncertain processes: applied theory and algorithms, *Kybernetika*, **22**, 1-102.

Piatti, A., Martini, O., 2001, Le dighe e la laminazione. influenza dei bacini idroelettrici sulle piene del Lago Maggiore, *in: Il Verbano: quale futuro? Insidie di un ecosistema e mezzi di difesa*, SUPSI, Ascona, CH, 75-92.

Raiffa, H., Richardson, J., Metcalfe, D., 2002, *Negotiation Analysis*, The Belknap Press of Harvard University Press, Cambridge, MA.

Renn, O., 1995, Public participation in impact assessment: a social learning perspective, *Environmental Impact Assesment Review*, **15** (3), 443-463.

Renn, O., Webler, T., Rakel, H., Dienel, P., Johnson, B., 1993, Public participation in decision-making: a three-step procedure, *Policy Science*, **26** (3), 189-214.

Renn, O., Webler, T., Wiedemann, P., Eds., 1995, *Fairness and Competence in Citizen Participation. Evaluating Models for Environmental Discourse*, Kluwer Academic Publishers, Dordrecht, NL.

Rinaldi, S., 1980, A method for the real time forecast of the outflow from a lake, *Applied Mathematical Modelling*, **4**, 322-324.

Ripley, B., 1996, *Pattern Recognition and Neural Networks*, Cambridge University Press, Cambridge, UK.

Rubin, J.Z., Pruitt, D.G., Kim, S.H., 1994, *Social Conflict: Escalation, Stalemate, and Settlement*, second ed., McGraw-Hill, New York, NY.

Salvadè, G., 2001, Monitoraggio e modello idrologico del bacino del Verbano, *in: Il Verbano: quale futuro? Insidie di un ecosistema e mezzi di difesa*, SUPSI, Ascona, CH, 121-146.

Shannon, C.E., 1949, Communication in the presence of noise, *Proceedings of the Institute of Radio Engineers*, **37**, 10-21.

Soncini-Sessa, R., Nardini, A., Gandolfi, C., Kraszewski, A., 1990, Computer aided water reservoir management: a prototype two level DSS, *in: Proceedings of the NATO ARW on Computer Aided Support Systems in Water Resources Research and Management*, September 23–28, Ericeira, P.

Soncini-Sessa, R., Rizzoli, A.E., Villa, L., Weber, E., 1999, TwoLe: a software tool for planning and management of water reservoir networks, *Hydrological Science Journal*, **44** (4), 619-631.

UNCSD – United Nations Commission on Sustainable Development, 1996, Indicators of sustainable development framework and methodologies, Technical Report, UNCSD, New York, NY.

UNECE, 1990, *Post-Project Analysis in Environmental Impact Assessment*, United Nations Economic Commission for Europe ECE/ENVWA/11, Geneva, CH.

White, D.J., 1963, Dynamic programming, Markov chains, and the method of successive approximations, *Journal of Mathematical Analysis and Applications*, **6**, 373-376.

Wiersma, B.G., Ed., 2004, *Environmental Monitoring*, CRC Press.

Index